Theory of Factorial Design

Single- and Multi-Stratum Experiments

MONOGRAPHS ON STATISTICS AND APPLIED PROBABILITY

General Editors

F. Bunea, V. Isham, N. Keiding, T. Louis, R. L. Smith, and H. Tong

Monographs on Statistics and Applied Probability 131

Theory of Factorial Design

Single- and Multi-Stratum Experiments

Ching-Shui Cheng

University of California, Berkeley, USA

and

Academia Sinica, Taiwan

CRC Press
Taylor & Francis Group
Boca Raton London New York

CRC Press is an imprint of the
Taylor & Francis Group, an **informa** business
A CHAPMAN & HALL BOOK

CRC Press
Taylor & Francis Group
6000 Broken Sound Parkway NW, Suite 300
Boca Raton, FL 33487-2742

First issued in paperback 2019

© 2014 by Taylor & Francis Group, LLC
CRC Press is an imprint of Taylor & Francis Group, an Informa business

No claim to original U.S. Government works

ISBN-13: 978-1-4665-0557-5 (hbk)
ISBN-13: 978-0-367-37898-1 (pbk)

Visit the Taylor & Francis Web site at
http://www.taylorandfrancis.com

and the CRC Press Web site at
http://www.crcpress.com

Preface

Factorial designs are widely used in many scientific and industrial investigations. The objective of this book is to provide a rigorous, systematic, and up-to-date treatment of the theoretical aspects of this subject. Despite its long history, research in factorial design has grown considerably in the past two decades. Several books covering these advances are available; nevertheless new discoveries continued to emerge. There are also useful old results that seem to have been overlooked in recent literature and, in my view, deserve to be better known.

Factorial experiments with multiple error terms (strata) that result from complicated structures of experimental units are common in agriculture. In recent years, the design of such experiments also received much attention in industrial applications. A theory of orthogonal block structures that goes back to John Nelder provides a unifying framework for the design and analysis of multi-stratum experiments. One feature of the present book is to present this elegant and general theory which, once understood, is simple to use, and can be applied to various structures of experimental units in a unified and systematic way. The mathematics required to understand this theory is perhaps what, in Rosemary Bailey's words, "obscured the essential simplicity" of the theory. In this book, I tried to minimize the mathematics needed, and did not present the theory in the most general form as developed by Bailey and her coauthors. To prepare readers for the general theory, a unified treatment of some simple designs such as completely randomized designs, block designs, and row-column designs is presented first. Therefore the book also covers these elementary non-factorial-design topics. It is suitable as a reference book for researchers and as a textbook for graduate students who have taken a first course in the design of experiments. Since the book is self-contained and includes many examples, it should also be accessible to readers with minimal previous exposure to experimental design as long as they have good mathematical and statistical backgrounds. Readers are required to be familiar with linear algebra. A review of linear model theory is given in Chapter 2, and a brief survey of some basic algebraic results on finite groups and fields can be found in the Appendix. Sections that can be skipped, at least on the first reading, without affecting the understanding of the material in later parts of the book are marked with stars.

In addition to a general theory of multi-stratum factorial design, the book covers many other topics and results that have not been reported in books. These include, among others, the useful method of design key for constructing multi-stratum factorial designs, the methods of partial foldover and doubling for constructing two-level resolution IV designs, some results on the structures of two-level resolution IV designs taken from the literature of projective geometry, the extension of minimum aberration to nonregular designs, and the minimum moment aberration criterion, which is equivalent to minimum aberration.

The book does not devote much space to the analysis of factorial designs due to its theoretical nature, and also because excellent treatment of strategies for data analysis can be found in several more applied books. Another subject that does not receive a full treatment is the so-called nonregular designs. It is touched upon in Chapter 8 when orthogonal arrays are introduced, and some selected topics are surveyed in Chapter 15. The research on nonregular designs is still very active and expands rapidly. It deserves another volume.

The writing of this book originated from a ten-lecture workshop on "Recent Developments in Factorial Design" I gave in June 2002 at the Institute of Statistical Science, Academia Sinica, in Taiwan. I thank Chen-Hsin Chen, Director of the institute at the time, for his invitation. The book was written over a long period of time while I taught at the University of California, Berkeley, and also during visits to the National Center for Theoretical Sciences in Hsinchu, Taiwan, and the Issac Newton Institute for Mathematical Sciences in Cambridge, United Kingdom. The support of these institutions and the US National Science Foundation is acknowledged. The book could not have been completed without the help of many people. It contains results from joint works with Rosemary Bailey, Dursun Bulutoglu, Hegang Chen, Lih-Yuan Deng, Mike Jacroux, Bobby Mee, Rahul Mukerjee, Nam-Ky Nguyen, David Steinberg, Don Sun, Boxin Tang, Pi-Wen Tsai, Hongquan Xu, and Oksoun Yee. I had the privilege of working with them. I also had the fortune to know Rosemary Bailey early in my career. Her work has had a great impact on me, and this book uses the framework she had developed. Boxin Tang read the entire book, and both Rosemary Bailey and Don Ylvisaker read more than half of it. They provided numerous detailed and very helpful comments as well as pointing out many errors. Hegang Chen, Chen-Tuo Liao, and Hongquan Xu helped check the accuracy of some parts of the book. As a LaTex novice, I am very grateful to Pi-Wen Tsai for her help whenever I ran into problems with LaTex. She also read and commented on earlier versions of several chapters. Yu-Ting Chen and Chiun-How Kao helped fix some figures. I would also like to acknowledge our daughter Adelaide for her endearing love and support as well as her upbeat reminder to always see the bright side. Last but not least, I am most grateful to my wife Suzanne Pan for her thankless support and care over the years and for patiently reading this "Tian Shu" from cover to cover.

Additional material for the book will be maintained at http://www.crcpress.com/product/isbn/9781466505575/ and http://www.stat.sinica.edu.tw/factorial-design/.

Dedicated to the memory of

Leonard Chung-Wei Cheng

Contents

Preface

Factorial designs are widely used in many scientific and industrial investigations. The objective of this book is to provide a rigorous, systematic, and up-to-date treatment of the theoretical aspects of this subject. Despite its long history, research in factorial design has grown considerably in the past two decades. Several books covering these advances are available; nevertheless new discoveries continued to emerge. There are also old useful results that seem to have been overlooked in recent literature and, in my view, deserve to be better known.

Factorial experiments with multiple error terms (strata) that result from complicated structures of experimental units are common in agriculture. In recent years, the design of such experiments also received much attention in industrial applications. A theory of orthogonal block structures that goes back to John Nelder provides a unifying framework for the design and analysis of multi-stratum experiments. One feature of the present book is to present this elegant and general theory which, once understood, is simple to use, and can be applied to various structures of experimental units in a unified and systematic way. The mathematics required to understand this theory is perhaps what, in Rosemary Bailey's words, "obscured the essential simplicity" of the theory. In this book, I tried to minimize the mathematics needed, and did not present the theory in the most general form as developed by Bailey and her coauthors. To prepare readers for the general theory, a unified treatment of some simple designs such as completely randomized designs, block designs, and row-column designs is presented first. Therefore the book also covers these elementary non-factorial-design topics. It is suitable as a reference book for researchers and as a textbook for graduate students who have taken a first course in the design of experiments. Since the book is self-contained and includes many examples, it should also be accessible to readers with minimal previous exposure to experimental design as long as they have good mathematical and statistical backgrounds. Readers are required to be familiar with linear algebra. A review of linear model theory is given in Chapter 2, and a brief survey of some basic algebraic results on finite groups and fields can be found in the Appendix. Sections that can be skipped, at least on the first reading, without affecting the understanding of the material in later parts of the book are marked with stars.

In addition to a general theory of multi-stratum factorial design, the book covers many other topics and results that have not been reported in books. These include, among others, the useful method of design key for constructing multi-stratum factorial designs, the methods of partial foldover and doubling for constructing two-level resolution IV designs, some results on the structures of two-level resolution IV designs taken from the literature of projective geometry, the extension of minimum aberration to nonregular designs, and the minimum moment aberration criterion, which is equivalent to minimum aberration.

The book does not devote much space to the analysis of factorial designs due to its theoretical nature, and also because excellent treatment of strategies for data analysis can be found in several more applied books. Another subject that does not receive a full treatment is the so-called nonregular designs. It is touched upon in Chapter 8 when orthogonal arrays are introduced, and some selected topics are surveyed in Chapter 15. The research on nonregular designs is still very active and expands rapidly. It deserves another volume.

The writing of this book originated from a ten-lecture workshop on "Recent Developments in Factorial Design" I gave in June 2002 at the Institute of Statistical Science, Academia Sinica, in Taiwan. I thank Chen-Hsin Chen, Director of the institute at the time, for his invitation. The book was written over a long period of time while I taught at the University of California, Berkeley, and also during visits to the National Center for Theoretical Sciences in Hsinchu, Taiwan, and the Issac Newton Institute for Mathematical Sciences in Cambridge, United Kingdom. The support of these institutions and the US National Science Foundation is acknowledged. The book could not have been completed without the help of many people. It contains results from joint works with Rosemary Bailey, Dursun Bulutoglu, Hegang Chen, Lih-Yuan Deng, Mike Jacroux, Bobby Mee, Rahull Mukerjee, Nam-Ky Nguyen, David Steinberg, Don Sun, Boxin Tang, Pi-Wen Tsai, Hongquan Xu, and Oksoun Yee. I had the privilege of working with them. I also had the fortune to know Rosemary Bailey early in my career. Her work has had a great impact on me, and this book uses the framework she had developed. Boxin Tang read the entire book, and both Rosemary Bailey and Don Ylvisaker read more than half of it. They provided numerous detailed and very helpful comments as well as pointing out many errors. Hegang Chen, Chen-Tuo Liao, and Hongquan Xu helped check the accuracy of some parts of the book. As a LaTex novice, I am very grateful to Pi-Wen Tsai for her help whenever I ran into problems with LaTex. She also read and commented on earlier versions of several chapters. Yu-Ting Chen and Chiun-How Kao helped fix some figures. I would also like to acknowledge our daughter Adelaide for her endearing love and support as well as her upbeat reminder to always see the bright side. Last but not least, I am most grateful to my wife Suzanne Pan for her thankless support and care over the years and for patiently reading this "Tian Shu" from cover to cover.

Additional material for the book will be maintained at http://www.crcpress.com/ product/isbn/9781466505575/ and http://www.stat.sinica.edu.tw/factorial-design/.

Chapter 1

Introduction

Many of the fundamental ideas and principles of experimental design were developed by Sir R. A. Fisher at the Rothamsted Experimental Station (Fisher, 1926). This agricultural background is reflected in some terminology of experimental design that is still being used today. Agricultural experiments are conducted, e.g., to compare different varieties of a certain crop or different fertilizers. In general, those that are under comparison in an experiment are called *treatments*. Manufacturing processes in industrial experiments and drugs in pharmaceutical studies are examples of treatments. In an agricultural experiment, the varieties or fertilizers are assigned to plots, and the yields are compared after harvesting. Each plot is called an *experimental unit* (or unit). In general, an experimental unit can be defined as the smallest division of the experimental material such that different units may receive different treatments (Cox, 1958, p. 2). At the design stage, a treatment is chosen for each experimental unit.

One fundamental difficulty in such *comparative experiments* is inherent variability of the experimental units. No two plots have exactly the same soil quality, and there are other variations beyond the experimenter's control such as weather conditions. Consequently, effects of the treatments may be biased by uncontrolled variations. A solution is to assign the treatments randomly to the units. In addition to guarding against potential systematic biases, *randomization* also provides a basis for appropriate statistical analysis.

The simplest kind of randomized experiment is one in which treatments arc assigned to units completely at random. In a *completely randomized experiment*, the precision of a treatment comparison depends on the overall variability of the experimental units. When the experimental units are highly variable, the treatment comparisons do not have good precision. In this case, the method of *blocking* is an effective way to reduce experimental error. The idea is to divide the experimental units into more homogeneous groups called blocks. When the treatments are compared on the units within each block, the precision is improved since it depends on the smaller within-block variability.

Suppose the experimental units are grouped into b blocks of size k. Even though efforts are made for the units in the same block to be as alike as possible, they are still not the same. Given an initial assignment of the treatments to the bk unit labels based on statistical, practical and/or other considerations, randomization is carried out by randomly permuting the unit labels within each block (done independently from block to block), and also randomly permuting the block labels. The additional step of randomly permuting block labels is to assure that an observation intended on a given treatment is equally likely to occur at any of the experimental units.

Under a completely randomized experiment, the experimental units are considered to be unstructured. The structure of the experimental units under a block design is an example of *nesting*. Suppose there are b blocks each consisting of k units; then each experimental unit can be labeled by a pair (i, j), $i = 1, \ldots, b$, $j = 1, \ldots, k$. This involves two factors with b and k levels, respectively. Here if $i \neq i'$, unit (i, j) bears no relation to unit (i', j); indeed, within-block randomization renders positions of the units in each block immaterial. We say that the k-level factor is nested in the b-level factor, and denote this structure by b/k or block/unit if the two factors involved are named "block" and "unit," respectively.

Another commonly encountered structure of the experimental units involves two blocking factors. For example, in agricultural experiments the plots may be laid out in rows and columns, and we try to eliminate from the treatment comparisons the spatial variations due to row-to-row and column-to-column differences. In experiments that are carried out on several different days and in several different time slots on each day, the observed responses might be affected by day-to-day and time-to-time variations. In this case each experimental run can be represented by a cell of a rectangular grid with those corresponding to experimental runs taking place on the same day (respectively, in the same time slot) falling in the same row (respectively, the same column). In general, suppose rc experimental units can be arranged in r rows and c columns such that any two units in the same row have a definite relation, and so do those in the same column. Then we have an example of *crossing*. This structure of experimental units is denoted by $r \times c$ or row $\times$ column if the two factors involved are named "row" and "column," respectively. In such a *row-column experiment*, given an initial assignment of the treatments to the rc unit labels, randomization is carried out by randomly permuting the row labels and, independently, randomly permuting the column labels. This assures that the structure of the experimental units is preserved: two treatments originally assigned to the same row (respectively, column) remain in the same row (respectively, column) after randomization.

For example, suppose there are four different manufacturing processes compared in four time slots on each of four days. With the days represented by rows and times represented by columns, a possible design is

1	2	3	4
2	1	4	3
3	4	1	2
4	3	2	1

where the four numbers 1, 2, 3, and 4 are labels of the treatments assigned to the units represented by the 16 row-column combinations. We see that each of the four numbers appears once in each row and once in each column. Under such a design, called a *Latin square*, all the treatments can be compared on each of the four days as well as in each of the four time slots. If the random permutation is such that the first, second, third, and fourth rows of the Latin square displayed above are mapped to the first, fourth, second, and third rows, respectively, and the first, second, third, and fourth columns are mapped to the fourth, third, first, and second columns, respectively, then it results in the following Latin square to be used in actual experimentation.

3	4	2	1
1	2	4	3
2	1	3	4
4	3	1	2

The structures of experimental units are called *block structures*. Block and row-column designs are based on the two simplest block structures involving nesting and crossing, respectively. Nelder (1965a) defined *simple block structures* to be those that can be obtained by iterations of nesting ($/$) and crossing ($\times$) operators. For example, $n_1/(n_2 \times n_3)$ represents the block structure under a *nested row-column design*, where $n_1 n_2 n_3$ experimental units are grouped into n_1 blocks of size $n_2 n_3$, and within each block the $n_2 n_3$ units are arranged in n_2 rows and n_3 columns. Randomization of such an experiment can be done by randomly permuting the block labels and carrying out the appropriate randomization for the block structure $n_2 \times n_3$ within each block, that is, randomly permuting the row labels and column labels separately. In an experiment with the block structure $n_1/(n_2/n_3)$, $n_1 n_2 n_3$ experimental units are grouped into n_1 blocks, and within each block the $n_2 n_3$ units are further grouped into n_2 smaller blocks (often called *whole-plots*) of n_3 units (often called *subplots*). To randomize such a *blocked split-plot experiment*, we randomly permute the block labels and carry out the appropriate randomization for the block structure n_2/n_3 within each block, that is, randomly permute the whole-plot labels within each block, and randomly permute the subplot labels within each whole-plot. Note that $n_1/(n_2/n_3)$ is the same as $(n_1/n_2)/n_3$.

In general, randomization of an experiment with a simple block structure is carried out according to the appropriate randomization for nesting or crossing at each stage of the block structure formula.

Like experimental units, the treatments may also have a structure. One can compare treatments by examining the pairwise differences of treatment effects. When the treatments do not have a structure (for example, when they are different varieties of a crop), one may be equally interested in all the pairwise comparisons of treatment effects. However, if they do have a certain structure, then some comparisons may be more important than others. For example, suppose one of the treatments is a control. Then one may be more interested in the comparisons between the control and new treatments.

In this book, treatments are to have a *factorial structure*: each treatment is a combination of multiple factors (variables) called *treatment factors*. Suppose there are n treatment factors and the ith factor has s_i values or settings to be studied. Each of these values or settings is called a *level*. The treatments, also called *treatment com-*

binations in this context, consist of all $s_1 \cdots s_n$ possible combinations of the factor levels. The experiment is called an $s_1 \times \cdots \times s_n$ factorial experiment, and is called an s^n experiment when $s_1 = \cdots = s_n = s$. For example, a fertilizer may be a combination of the levels of three factors N (nitrogen), P (phosphate), and K (potash), and a chemical process might involve temperature, pressure, concentration of a catalyst, etc. Fisher (1926) introduced factorial design to agricultural experiments, and Yates (1935, 1937) made significant contributions to its early development.

When the treatments have a factorial structure, typically we are interested in the effects of individual factors, as well as how the factors interact with one another. Special functions of the treatment effects, called *main effects* and *interactions*, can be defined to represent such effects of interest. We say that the treatment factors do not interact if, when the levels of a factor are changed while those of the other factors are kept constant, the changes in the treatment effects only depend on the levels of the varying factor. In this case, we can separate the effects of individual factors, and the effect of each treatment combination can be obtained by summing up these individual effects. Under such *additivity* of the treatment factors, for example, to determine the combination of N, P, and K with the highest average yield, one can simply find the best level of each of the three factors separately. Otherwise, the factors need to be considered simultaneously. Roughly speaking, the main effect of a treatment factor is its effects averaged over the levels of the other factors, and the interaction effects measure departures from additivity. Precise definitions of these effects, collectively called *factorial effects*, will be given in Chapter 6.

A factorial experiment with each treatment combination observed once is called a *complete* factorial experiment. We also refer to it as a single-replicate complete factorial experiment. The analysis of completely randomized experiments in which each treatment combination is observed the same number of times, to be presented in Chapter 6, is straightforward. It becomes more involved if the experimental units have a more complicated block structure and/or if not all the treatment combinations can be observed.

When a factorial experiment is blocked, with each block consisting of one replicate of all the treatment combinations, the analysis is still very simple. As will be discussed in Chapter 6, in this case all the treatment main effects and interactions can be estimated in the same way as if there were no blocking, except that the variances of these estimators depend on the within-block variability instead of the overall variability of the experimental units. Since the total number of treatment combinations increases rapidly as the number of factors becomes large, a design that accommodates all the treatment combinations in each block requires large blocks whose homogeneity is difficult to control. In order to achieve smaller within-block variability, we cannot accommodate all the treatment combinations in the same block and must use *incomplete blocks*. It may also be impractical to carry out experiments in large blocks. Then, since not all the treatment combinations appear in the same block, the estimates of some treatment factorial effects cannot be based on within-block comparisons alone. This may result in less precision for such estimates. For example, suppose an experiment on two two-level factors A_1 and A_2 is to be run on two different days with the two combinations $(0,0)$ and $(1,1)$ of the levels of A_1 and

A_2 observed on one day, and the other two combinations $(0,1)$ and $(1,0)$ observed on the other day, where 0 and 1 are the two factor levels. Then estimates of the main effect (comparison of the two levels) of factor A_1 and the main effect of A_2 are based on within-block comparisons, but as will be seen in Chapter 7, the interaction of the two factors would have to be estimated by comparing the observations on the first day with those on the second day, resulting in less precision. We say that this two-factor interaction is *confounded* with blocks.

When a factorial experiment must be run in incomplete blocks, we choose a design in such a way that only those factorial effects that are less important or are known to be negligible are confounded with blocks. Typically the main effects are deemed more important, and one would avoid confounding them with blocks. However, due to practical constraints, sometimes one must confound certain main effects with blocks. For instance, it may be difficult to change the levels of some factors. In the aforementioned example, if a factor must be kept at the same level on each day, then the main effect of that factor can only be estimated by a more variable between-day comparison.

Often the number of treatment combinations is so large that it is practically possible to observe only a small subset of the treatment combinations. This is called a *fractional factorial design*. Then, since not all the treatment combinations are observed, some factorial effects are mixed up and cannot be distinguished. We say that they are *aliased*. For example, when only 16 treatment combinations are to be observed in an experiment involving six two-level factors, there are 63 factorial effects (6 main effects, 15 two-factor interactions, 20 three-factor interactions, 15 four-factor interactions, 6 five-factor interactions, and 1 six-factor interaction), but only 15 degrees of freedom are available for estimating them. This is possible if many of the factorial effects are negligible. One design issue is which 16 of the 64 treatment combinations are to be selected.

An important property of a fractional factorial design, called *resolution*, pertains to the extent to which the lower-order effects are mixed up with higher-order effects. For example, under a design of resolution III, no main effect is aliased with other main effects, but some main effects are aliased with two-factor interactions; under a design of resolution IV, no main effect is aliased with other main effects or two-factor interactions, but some two-factor interactions are aliased with other two-factor interactions; under a design of resolution V, no main effects and two-factor interactions are aliased with one another. When the experimenter has little knowledge about the relative importance of the factorial effects, it is common to assume that the lower-order effects are more important than higher-order effects (the main effects are more important than interactions, and two-factor interactions are more important than three-factor interactions, etc.), and effects of the same order are equally important. Under such a *hierarchical assumption*, it is desirable to have a design with high resolution. A popular criterion of selecting fractional factorial designs and a refinement of maximum resolution, called *minimum aberration*, is based on the idea of minimizing the aliasing among the more important lower-order effects.

When the experimental units have a certain block structure, in addition to picking a fraction of the treatment combinations, we also have to decide how to assign

the selected treatment combinations to the units. In highly fractionated factorial experiments with complicated block structures, we have complex aliasing of treatment
factorial effects as well as multiple levels of precision for their estimates. The bulk of
this book is about the study of such designs, including their analysis, selection, and
construction. The term "multi-stratum" in the subtitle of the book refers to multiple
sources of errors that arise from complicated block structures, while "single-stratum"
is synonymous with "complete randomization" where there is one single error term.

Treatment and block structures are two important components of a randomized
experiment. Nelder (1965a,b) emphasized their distinction and developed a theory
for the analysis of randomized experiments with simple block structures. Simple
block structures cover most, albeit not all the block structures that are commonly
encountered in practice. Speed and Bailey (1982) and Tjur (1984) further developed
the theory to cover the more general *orthogonal block structures*. This theory, an
account of which can be found in Bailey (2008), provides the basis for the approach
adopted in this book.

We turn to five examples of factorial experiments to motivate some of the topics
to be discussed in the book. The first three examples involve simple block structures.
The block structures in Examples 1.4 and 1.5 are not simple block structures, but the
theory developed by Speed and Bailey (1982) and Tjur (1984) is applicable. We will
return to these examples from time to time in later chapters to illustrate applications
of the theory as it is developed.

Our first example is a replicated complete factorial experiment with a relatively
complicated block structure.

Example 1.1. Loughin (2005) studied the design of an experiment on weed control.
Herbicides can kill the weeds that reduce soybean yields, but they can also kill soybeans. On the other hand, soybean varieties can be bred or engineered to be resistant
to certain herbicides. An experiment is to be carried out to study what factors in-
fluence weed control and yield of genetically altered soybean varieties. Four factors
studied in the experiment are *soybean variety/herbicide combinations* in which the
herbicide is safe for the soybean variety, *dates* and *rates* of herbicide application,
and *weed species*. There are three variety/herbicide combinations, two dates (early
and late), three rates (1/4, 1/2, and 1), and seven weed species, giving a total of 126
treatments with a $3 \times 2 \times 3 \times 7$ factorial structure. Soybeans and weeds are planted
together and a herbicide safe for the soybean variety is sprayed at the designated
time and rate. Then weed properties (numbers, density, mass) and soybean yields are
measured. However, there are some practical constraints on how the experiment can
be run. Due to herbicide drift, different varieties cannot be planted too close together
and buffer zones between varieties are needed, but the field size is not large enough to
allow for 126 plots of adequate size with large buffers between each pair of adjacent
plots. Therefore, for efficient use of space, one needs to plant all of a given soybean
variety contiguously so that fewer buffers are needed. Additional drift concerns lead
to a design described as follows. First the field is divided into four blocks to accommodate four replications:

Each block is split into three plots with two buffer zones, and the variety/herbicide combinations are randomly assigned to the plots within blocks:

Var 1	Var 3	Var 2

Each plot is then split into two subplots, with application times randomly assigned to subplots within plots:

Late	Early	Early	Late	Late	Early

Furthermore each subplot is split into three sub-subplots, with application rates randomly assigned to sub-subplots within subplots:

$\frac{1}{2}$	$\frac{1}{4}$	1	1	$\frac{1}{4}$	$\frac{1}{2}$	$\frac{1}{2}$	1	$\frac{1}{4}$	1	$\frac{1}{2}$	$\frac{1}{4}$	1	$\frac{1}{2}$	$\frac{1}{4}$	$\frac{1}{4}$	1	$\frac{1}{2}$

Each block is divided into seven horizontal strips, with the weed species randomly assigned to the strips within blocks:

We end up with 504 combinations of sub-subplots and strips:

Each of the 126 treatment combinations appears once in each of the four blocks. To summarize, we have four replicates of a complete $3 \times 2 \times 3 \times 7$ factorial experiment with the block structure $4/[(3/2/3) \times 7]$. Both subplots in the same plot are assigned the same variety/herbicide combination, all the sub-subplots in the same subplot are assigned the same herbicide application time, and all the sub-subplot and strip intersections in the same strip are assigned the same weed species. Various aspects of the analysis of this design will be discussed in Sections 12.1, 12.9, 12.10, and 13.10.

In Example 1.1, there are 18 sub-subplots in each block. If different soybean varieties were to be assigned to neighboring sub-subplots, then 17 buffer zones would be needed in each block. With only two buffer zones per block under the proposed design, comparisons of soybean varieties are based on between-plot comparisons, which are expected to be more variable than those between subplots and sub-subplots. The precision of the estimates of such effects is sacrificed in order to satisfy the practical constraints.

Example 1.2. McLeod and Brewster (2004) discussed an experiment for identifying key factors that would affect the quality of a chrome-plating process. Suppose six two-level treatment factors are to be considered in the experiment: A, chrome concentration; B, chrome to sulfate ratio; C, bath temperature; S, etching current density; T, plating current density; and U, part geometry. The response variables include, e.g., the numbers of pits and cracks. The chrome plating is done in a bath (tank), which contains several rectifiers, but only two will be used. On any given day the levels of A, B, and C cannot be changed since they represent characteristics of the bath. On the other hand, the levels of factors S, T, and U can be changed at the rectifier level. The experiment is to be run on 16 days, with four days in each of four weeks. Therefore there are a total of 32 runs with the block structure (4 weeks)/(4 days)/(2 runs), and one has to choose 32 out of the $2^6 = 64$ treatment combinations. Weeks, days, and runs can be considered as blocks, whole-plots, and subplots, respectively. The three factors A, B, and C must have constant levels on the two experimental runs on the same day, and are called whole-plot treatment factors. The other three factors S, T, and U are not subject to this constraint and are called subplot treatment factors. We will return to this example in Sections 12.9, 13.2, 13.4, 13.5, 13.7, 14.4, and 14.13.

Example 1.3. Miller (1997) described a laundry experiment for investigating methods of reducing the wrinkling of clothes. Suppose the experiment is to be run in two blocks, with four washers and four dryers to be used. After four cloth samples have been washed in each washer, the 16 samples are divided into four groups with each group containing one sample from each washer. Each of these groups is then assigned to one dryer. The extent of wrinkling on each sample is evaluated at the end of the experiment. This results in 32 experimental runs that can be thought to have the $2/(4 \times 4)$ block structure shown in Figure 1.1, where each cell represents a cloth sample, rows represent sets of samples that are washed together, and columns represent sets of samples that are dried together. There are ten two-level treatment factors,

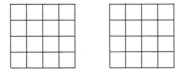

Figure 1.1 *A $2/(4 \times 4)$ block structure*

six of which (A, B, C, D, E, F) are configurations of washers and four (S, T, U, V) are configurations of dryers. One has to choose 32 out of the $2^{10} = 1024$ treatment combinations. Furthermore, since the experimental runs on the cloth samples in the same row are conducted in the same washing cycle, each of A, B, C, D, E, F must have a constant level in each row. Likewise, each of S, T, U, V must have a constant level in each column. Thus in each block, four combinations of the levels of A, B, C, D, E, F are chosen, one for each row, and four combinations of the levels of S, T, U, V are chosen, one for each column. The four combinations of washer settings are then coupled with the four combinations of dryer settings to form 16 treatment combinations of the ten treatment factors in the same block. An experiment run in this way requires only four washer loads and four dryer loads in each block. If one were to do complete randomization in each block, then four washer loads and four dryer loads could produce only four observations. The trade-off is that the main effect of each treatment factor is confounded with either rows or columns. Construction and analysis of designs for such *blocked strip-plot experiments* will be discussed in Sections 12.2, 12.9, 12.10, 13.3, 13.4, 13.5, 13.6, 13.7, 14.5, 14.6, and 14.14.

Federer and King (2006) gave a comprehensive treatment of split-plot and strip-plot designs and their many variations. In this book, we present a unifying theory that can be systematically applied to a very general class of multi-stratum experiments.

Example 1.3 is an experiment with two processing stages: washing and drying. Many industrial experiments involve a sequence of processing stages, with the levels of various treatment factors assigned and processed at different stages. At each stage the experimental units are partitioned into disjoint classes. Those in the same class, which will be processed together, are assigned the same level of each of the treatment factors that are to be processed at that stage. We call the treatment factors processed at the ith stage the ith-stage treatment factors. In Example 1.3, levels of the six washer factors are set at the first stage and those of the four dryer factors are set at the second stage. So the washer configurations are first-stage treatment factors and the dryer configurations are second-stage treatment factors. Such an experiment with two processing stages can be thought to have experimental units with a row-column structure.

In Examples 1.1–1.3, the experimental units can be represented by *all* the level combinations of some *unit factors*. In the next two examples, we present experiments in which the experimental units are a *fraction* of unit-factor level combinations.

Example 1.4. Mee and Bates (1998) discussed designs of experiments with multiple processing stages in the fabrication of integrated circuits. Suppose that at the first stage 16 batches of material are divided into four groups of equal size, with the same level of each first-stage treatment factor assigned to all the batches in the same group. At the second stage they are rearranged into another four groups of equal size, again with the same level of each second-stage treatment factor assigned to all the batches in the same group. As in Example 1.3, the groupings at the two stages can be represented by rows and columns. Then each of the first-stage groups and each of the second-stage groups have exactly one batch in common. This is a desirable property whose advantage will be explained in Section 12.13. Now suppose there is a third stage. Then we need a third grouping of the batches. One possibility is to group according to the numbers in the Latin square shown earlier:

1	2	3	4
2	1	4	3
3	4	1	2
4	3	2	1

One can assign the same level of each third-stage treatment factor to all the units (batches) corresponding to the same number in the Latin square. One advantage is that each of the third-stage groups has exactly one unit in common with any group at the first two stages. If a fourth stage is needed, then one may group according to the following Latin square:

1	2	3	4
4	3	2	1
2	1	4	3
3	4	1	2

This and the previous Latin square have the property that when one is superimposed on the other, each of the 16 pairs of numbers (i, j), $1 \leq i, j \leq 4$, appears in exactly one cell. We say that these two Latin squares are *orthogonal* to each other. If the fourth-stage grouping is done according to the numbers in the second Latin square, then each of the fourth-stage groups also has exactly one unit in common with each group at any of the first three stages. This kind of block structure cannot be obtained by iterations of nesting and crossing operators. To be a simple block structure, with four groups at each of three or four stages, one would need $4^3 = 64$ or $4^4 = 256$ units, respectively. Thus the 16 units can be regarded as a quarter or one-sixteenth fraction of the combinations of three or four 4-level factors, respectively. The following is a complete 2^4 design which can be used for experiments in which the levels of the four treatment factors are set at four stages, one factor per stage: the first factor has a constant level in each row, the second factor has a constant level in each column, the third factor has a constant level in each cell occupied by the same number in the first Latin square, and the fourth factor has a constant level in each cell occupied by the same number in the second Latin square.

0000	0011	0101	0110
0010	0001	0111	0100
1001	1010	1100	1111
1011	1000	1110	1101

We will return to this example in Sections 12.5, 12.10, 12.13, and 13.11.

Example 1.5. Bingham, Sitter, Kelly, Moore, and Olivas (2008) discussed experiments with multiple processing stages where more groups are needed at each stage, which makes it impossible for all the groups at different stages to share common units. For example, in an experiment with two processing stages, suppose 32 experimental units are to be partitioned into 8 groups of size 4 at each of the two stages. One possibility is to partition the 32 units as in Figure 1.1. The eight rows of size 4, four of which from each of the two blocks, together constitute the eight first-stage groups, and the eight columns in the two blocks together constitute the eight second-stage groups. As shown in the following figure, the 32 starred experimental units are a *fraction* of the 64 units in a completely crossed 8×8 square.

*	*	*	*				
*	*	*	*				
*	*	*	*				
*	*	*	*				
				*	*	*	*
				*	*	*	*
				*	*	*	*
				*	*	*	*

As in Example 1.4, the 32 units do not have a simple block structure. An important difference, however, is that in the current setting not all the first-stage groups can meet with every second-stage group, causing some complications in the design and analysis (to be discussed in Section 13.12). The figure shows that the 32 units are divided into two groups of size 16, which we call *pseudo* blocks since they are not part of the originally intended block structure. We will revisit this example in Sections 12.13 and 13.12, and show that it can be treated as an experiment with the block structure $2/(4 \times 4)$. A similar problem was studied in Vivacqua and Bisgaard (2009), to be discussed in Section 14.7.

An overview

Some introductory material is presented in Chapters 2–5. Chapter 2 is a review of some results on linear models, with emphasis on one-way and two-way layout models and a geometric characterization of the condition of proportional frequencies between two factors. Under the assumption of treatment-unit additivity, randomization models are developed in Chapter 3 for some designs with simple block structures, including block designs and row-column designs. In Chapter 4, the condition of proportional frequencies is extended to a notion of orthogonal factors that plays an important role in the block structures studied in this book. Some mathematical results on factors that are needed throughout the book are also gathered there. A condition that entails simple analysis of a randomized design (Theorem 5.1) is established in Chapter 5. This result is used to present a unified treatment of the analyses of three classes of orthogonal designs (completely randomized designs, complete block designs, and Latin square designs) under the randomization models derived in Chapter 3. It is also a key result for developing, in later chapters, a general theory of orthogonal designs for experiments with more complicated block structures.

The treatment factorial structure is introduced in Chapter 6. It is shown how certain special functions of the treatment effects can be defined to represent main effects and interactions of the treatment factors. Unless all the treatment factors have two levels, the choices of such functions are not unique. Several methods of constructing them based on orthogonal polynomials, finite Euclidean geometry, and Abelian groups are presented. The discussion of complete factorial designs is continued in Chapter 7, for experiments that are conducted in incomplete blocks or row-column layouts, including split-plot and strip-plot designs. In this case, there is more than one error term and some factorial effects are confounded with blocks, rows, or columns. A construction method based on design keys is presented in addition to a commonly used method, and is shown to enjoy several advantages.

Fractional factorial designs under complete randomization are treated in Chapters 8–11. In Chapter 8, the important combinatorial structure of orthogonal arrays is introduced. Some basic properties of orthogonal arrays as fractional factorial designs, including upper bounds on the number of factors that can be accommodated by orthogonal arrays of a given run size, are derived. We also present several methods of constructing orthogonal arrays, in particular the foldover method and the construction via difference matrices. The chapter is concluded with a brief discussion of applications of orthogonal arrays to computer experiments and three variants of orthogonal arrays recently introduced for this purpose. The emphasis of this book is mainly on the so-called regular fractional factorial designs, which are easy to construct and analyze, and have nice structures and a rich theory. In Chapter 9 we provide a treatment of their basics, including design construction, aliasing and estimability of factorial effects, resolution, a search algorithm for finding designs under which some required effects can be estimated, and the connection with the linear codes of coding theory. The criterion of minimum aberration and some related criteria for selecting regular fractional factorial designs are discussed in Chapter 10. The statistical meaning of minimum aberration is clarified via its implications on the aliasing pattern of factorial effects. It is shown that this criterion produces designs with good properties under model uncertainty and good lower-dimensional projections. The connection to coding theory provides two powerful tools for constructing minimum aberration designs: the MacWilliams identities can be used to establish a complementary design theory that is useful for determining minimum aberration designs when there are many factors; the Pless power moment identities lead to the criterion of minimum moment aberration, which is equivalent to minimum aberration. Besides the theoretical interest, this equivalence is useful for analytical characterization and algorithmic construction of minimum aberration designs. A Bayesian approach to the design and analysis of factorial experiments, also applicable to nonregular designs, is presented at the end of Chapter 10. Regular designs are also closely related to finite projective geometries. The connection is made in two optional sections in Chapter 9, and is used to characterize and construct minimum aberration designs in Chapter 10. The geometric connection culminates in an elegant theory of the construction and structures of resolution IV designs in Chapter 11. While foldover is a well-known method of constructing resolution IV designs, many resolution IV designs cannot be constructed by this method. We translate the geometric results into design language, and among other topics, present the methods of doubling and partial foldover for constructing them.

In Chapters 12–14, we turn to factorial designs with more complicated block structures called multi-stratum designs. Some basic results on Nelder's simple block structures and the more general orthogonal block structures are derived in Chapter 12. A general theory for the design and analysis of orthogonal multi-stratum complete factorial designs is developed in Chapter 13. This theory is applied to several settings, including blocked split-plot designs, blocked strip-plot designs, and design of experiments with multiple processing stages. Chapter 14 is devoted to the construction of multi-stratum fractional factorial designs and criteria for their selection under model uncertainty in the spirit of minimum aberration. The five motivating examples presented above are revisited.

We survey a few nonregular design topics in Chapter 15. Under nonregular designs, the factorial effects are aliased in a complicated way, but their run sizes are more flexible than regular designs. At the initial stage of experimentation, often only a small number of the potential factors are important. Due to their run-size economy, nonregular designs are suitable for conducting factor screening experiments under the *factor sparsity* principle. In this context, it is useful to study the property of the design when it is projected onto small subsets of factors. We also discuss the relevant topics of search designs and supersaturated designs. The objective of a search design is to identify and discriminate nonnegligible effects under the assumption that the number of nonnegligible effects is small. Supersaturated designs have more unknown parameters than the degrees of freedom available for estimating them and are useful for screening active factors. In addition to these and other miscellaneous topics, we show how some of the results presented in earlier chapters can be extended to nonregular designs. For example, coding theory again proves useful for providing a way to extend minimum aberration to nonregular designs.

Throughout this book, the starred sections can be skipped, at least on the first reading. Relevant exercises are also marked with stars.

Chapter 2

Linear Model Basics

In this chapter we review some basic results from linear model theory, including least squares estimation, the Gauss–Markov Theorem, and tests of linear hypotheses, with applications to the analysis of fixed-effect one-way and additive two-way layout models. We show how the analysis of an additive two-way layout model is simplified when certain vector spaces associated with the two factors are orthogonal. This geometric condition is shown to be equivalent to that the two factors satisfy the condition of proportional frequencies.

2.1 Least squares

Consider the linear model

$$y_i = \sum_{j=1}^{p} x_{ij}\theta_j + \varepsilon_i, \ i = 1, \ldots, N, \tag{2.1}$$

where $x_{ij}, 1 \leq i \leq N, 1 \leq j \leq p$, are known constants, $\theta_1, \ldots, \theta_p$ are unknown parameters, and $\varepsilon_1, \ldots, \varepsilon_N$ are uncorrelated random variables with zero mean and common variance σ^2. Let $\mathbf{y} = (y_1, \ldots, y_N)^T$, $\boldsymbol{\theta} = (\theta_1, \ldots, \theta_p)^T$, $\boldsymbol{\varepsilon} = (\varepsilon_1, \ldots, \varepsilon_N)^T$, and $\mathbf{X}$ be the $N \times p$ matrix with the (i, j)th entry equal to x_{ij}, where T stands for "transpose." Then (2.1) can be written as

$$\mathbf{y} - \mathbf{X}\boldsymbol{\theta} + \boldsymbol{\varepsilon}, \tag{2.2}$$

with

$$E(\boldsymbol{\varepsilon}) = \mathbf{0}, \text{ and } \text{cov}(\boldsymbol{\varepsilon}) = \sigma^2 \mathbf{I}_N, \tag{2.3}$$

where $E(\boldsymbol{\varepsilon}) = (E(\varepsilon_1), \ldots, E(\varepsilon_N))^T$, $\text{cov}(\boldsymbol{\varepsilon})$ is the covariance matrix of $\boldsymbol{\varepsilon}$, $\mathbf{0}$ is the vector of zeros, and $\mathbf{I}_N$ is the identity matrix of order N. We call $\mathbf{X}$ the *model matrix*.

Least squares estimators $\widehat{\theta}_1, \ldots, \widehat{\theta}_p$ of $\theta_1, \ldots, \theta_p$ are obtained by minimizing $\sum_{i=1}^{N}(y_i - \sum_{j=1}^{p} x_{ij}\theta_j)^2 = \|\mathbf{y} - \mathbf{X}\boldsymbol{\theta}\|^2$.

Let $E(\mathbf{y}) = (E(y_1), \ldots, E(y_N))^T$. Then under (2.2) and (2.3), $E(\mathbf{y}) = \mathbf{X}\boldsymbol{\theta}$. This implies that $E(\mathbf{y})$ is a linear combination of the column vectors of $\mathbf{X}$. Let $R(\mathbf{X})$, called the column space of $\mathbf{X}$, be the linear space generated by the column vectors of $\mathbf{X}$. Then $E(\mathbf{y}) \in R(\mathbf{X})$. The least squares method is to find a vector $\widehat{\mathbf{y}} = \mathbf{X}\widehat{\boldsymbol{\theta}}$ in $R(\mathbf{X})$ such that $\|\mathbf{y} - \widehat{\mathbf{y}}\|$ is minimized. This is achieved if $\widehat{\mathbf{y}}$ is the orthogonal projection of $\mathbf{y}$ onto

$R(\mathbf{X})$. Denoting the orthogonal projection matrix onto a space V by $\mathbf{P}_V$, we have

$$\widehat{\mathbf{y}} = \mathbf{X}\widehat{\boldsymbol{\theta}} = \mathbf{P}_{R(\mathbf{X})}\mathbf{y}.$$

Here $\mathbf{y} - \mathbf{X}\widehat{\boldsymbol{\theta}}$, called the *residual*, is the orthogonal projection of $\mathbf{y}$ onto $R(\mathbf{X})^{\perp}$, where $R(\mathbf{X})^{\perp} = \{\mathbf{x} \in \mathbb{R}^N : \mathbf{x}$ is orthogonal to all the vectors in $R(\mathbf{X})\}$ is the orthogonal complement of $R(\mathbf{X})$ in $\mathbb{R}^N$. Therefore $\mathbf{y} - \mathbf{X}\widehat{\boldsymbol{\theta}}$ is orthogonal to all the column vectors of $\mathbf{X}$, and it follows that

$$\mathbf{X}^T\left(\mathbf{y} - \mathbf{X}\widehat{\boldsymbol{\theta}}\right) = \mathbf{0}$$

or

$$\mathbf{X}^T\mathbf{X}\widehat{\boldsymbol{\theta}} = \mathbf{X}^T\mathbf{y}. \tag{2.4}$$

Equation (2.4) is called the *normal equation*. If $\mathbf{X}^T\mathbf{X}$ is invertible, then $\widehat{\boldsymbol{\theta}} = (\mathbf{X}^T\mathbf{X})^{-1}\mathbf{X}^T\mathbf{y}$, with

$$E(\widehat{\boldsymbol{\theta}}) = \boldsymbol{\theta} \text{ and } \text{cov}(\widehat{\boldsymbol{\theta}}) = \sigma^2(\mathbf{X}^T\mathbf{X})^{-1}. \tag{2.5}$$

Let the rank of $\mathbf{X}$ be r. Then $\mathbf{X}^T\mathbf{X}$ is invertible if and only if $r = p$. Unless $r = p$, not all the parameters in $\boldsymbol{\theta}$ are identifiable, and solutions to (2.4) are not unique. In this case, (2.4) can be solved by using generalized inverses.

A *generalized inverse* of a matrix $\mathbf{A}$ is defined as any matrix $\mathbf{A}^-$ such that $\mathbf{A}\mathbf{A}^-\mathbf{A} = \mathbf{A}$. For any generalized inverse $(\mathbf{X}^T\mathbf{X})^-$ of $\mathbf{X}^T\mathbf{X}$, $\widehat{\boldsymbol{\theta}} = (\mathbf{X}^T\mathbf{X})^-\mathbf{X}^T\mathbf{y}$ is a solution to the normal equation. Even though $\widehat{\boldsymbol{\theta}}$ may not be unique, since $\mathbf{y}$ has a unique orthogonal projection onto $R(\mathbf{X})$, $\mathbf{X}\widehat{\boldsymbol{\theta}} = \mathbf{X}(\mathbf{X}^T\mathbf{X})^-\mathbf{X}^T\mathbf{y}$ is unique and does not depend on the choice of $(\mathbf{X}^T\mathbf{X})^-$. A byproduct of this is an explicit expression for the orthogonal projection matrix onto the column space of any matrix.

Theorem 2.1. *The orthogonal projection matrix onto $R(\mathbf{X})$ is $\mathbf{X}(\mathbf{X}^T\mathbf{X})^-\mathbf{X}^T$, where $(\mathbf{X}^T\mathbf{X})^-$ is any generalized inverse of $\mathbf{X}^T\mathbf{X}$.*

A linear function $\mathbf{c}^T\boldsymbol{\theta} = \sum_{i=1}^p c_i\theta_i$ of the unknown parameters is said to be *estimable* if there exists an $N \times 1$ vector $\mathbf{a}$ such that $E(\mathbf{a}^T\mathbf{y}) = \mathbf{c}^T\boldsymbol{\theta}$ for all $\boldsymbol{\theta}$. Such an estimator of $\mathbf{c}^T\boldsymbol{\theta}$ is called a a linear unbiased estimator. Since $E(\mathbf{a}^T\mathbf{y}) = \mathbf{a}^T\mathbf{X}\boldsymbol{\theta}$, it is equal to $\mathbf{c}^T\boldsymbol{\theta}$ for all $\boldsymbol{\theta}$ if and only if $\mathbf{c} = \mathbf{X}^T\mathbf{a}$. This shows that $\mathbf{c}^T\boldsymbol{\theta}$ is estimable if and only if $\mathbf{c} \in R(\mathbf{X}^T)$. From matrix algebra, it is known that $R(\mathbf{X}^T) = R(\mathbf{X}^T\mathbf{X})$. Therefore,

$$\mathbf{c}^T\boldsymbol{\theta} \text{ is estimable if and only if } \mathbf{c} \in R(\mathbf{X}^T\mathbf{X}). \tag{2.6}$$

If $\mathbf{c}^T\boldsymbol{\theta}$ is estimable, then $\mathbf{c}^T\widehat{\boldsymbol{\theta}}$ does not depend on the solution to the normal equation.

Theorem 2.2. *(Gauss–Markov Theorem) Under (2.2)–(2.3), if $\mathbf{c}^T\boldsymbol{\theta}$ is estimable, then for any solution $\widehat{\boldsymbol{\theta}}$ to (2.4), $\mathbf{c}^T\widehat{\boldsymbol{\theta}}$ has the smallest variance among all the linear unbiased estimators of $\mathbf{c}^T\boldsymbol{\theta}$.*

It is common to refer to $\mathbf{c}^T\widehat{\boldsymbol{\theta}}$ as the best linear unbiased estimator, or BLUE, of $\mathbf{c}^T\boldsymbol{\theta}$.

For any estimable function $\mathbf{c}^T\boldsymbol{\theta}$, we have

$$\text{var}\left(\mathbf{c}^T\widehat{\boldsymbol{\theta}}\right) = \sigma^2\mathbf{c}^T(\mathbf{X}^T\mathbf{X})^-\mathbf{c},$$

where $(\mathbf{X}^T\mathbf{X})^-$ is any generalized inverse of $\mathbf{X}^T\mathbf{X}$. We call $\mathbf{X}^T\mathbf{X}$ the *information matrix* for $\boldsymbol{\theta}$.

Suppose the model matrix $\mathbf{X}$ is design dependent. We say that a design is D-, A-, or E-optimal if it minimizes, respectively, the determinant, trace, or the largest eigenvalue of the covariance matrix of $\widehat{\boldsymbol{\theta}}$ among all the competing designs. By (2.5), these optimality criteria, which are referred to as the D-, A-, and E-criterion, respectively, are equivalent to maximizing $\det(\mathbf{X}^T\mathbf{X})$, minimizing $\text{tr}(\mathbf{X}^T\mathbf{X})^{-1}$, and maximizing the smallest eigenvalue of $\mathbf{X}^T\mathbf{X}$, respectively.

2.2   Estimation of σ^2

We have

$$\mathbf{y} = \mathbf{X}\widehat{\boldsymbol{\theta}} + (\mathbf{y} - \mathbf{X}\widehat{\boldsymbol{\theta}}), \tag{2.7}$$

where $\mathbf{X}\widehat{\boldsymbol{\theta}}$ and $\mathbf{y} - \mathbf{X}\widehat{\boldsymbol{\theta}}$ are orthogonal to each other. By the Pythagorean Theorem,

$$\|\mathbf{y}\|^2 = \left\|\mathbf{X}\widehat{\boldsymbol{\theta}}\right\|^2 + \left\|\mathbf{y} - \mathbf{X}\widehat{\boldsymbol{\theta}}\right\|^2.$$

The last term in this identity, $\left\|\mathbf{y} - \mathbf{X}\widehat{\boldsymbol{\theta}}\right\|^2$, is called the *residual sum of squares*. Since $\mathbf{y} - \mathbf{X}\widehat{\boldsymbol{\theta}} = \mathbf{P}_{R(\mathbf{X})^\perp}\mathbf{y}$,

$$
\begin{aligned}
\text{E}\left[\left\|\mathbf{y} - \mathbf{X}\widehat{\boldsymbol{\theta}}\right\|^2\right] &= \text{E}\left[(\mathbf{P}_{R(\mathbf{X})^\perp}\mathbf{y})^T(\mathbf{P}_{R(\mathbf{X})^\perp}\mathbf{y})\right] \\
&= \text{E}\left(\mathbf{y}^T\mathbf{P}_{R(\mathbf{X})^\perp}\mathbf{y}\right) \\
&= [\text{E}(\mathbf{y})]^T\mathbf{P}_{R(\mathbf{X})^\perp}[\text{E}(\mathbf{y})] + \sigma^2\text{tr}\left(\mathbf{P}_{R(\mathbf{X})^\perp}\right) \\
&= \sigma^2\text{tr}\left(\mathbf{P}_{R(\mathbf{X})^\perp}\right) \\
&= \sigma^2\dim\left(R(\mathbf{X})^\perp\right) \\
&= \sigma^2(N - r),
\end{aligned}
$$

where the second equality follows from (A.2) and (A.3) in the Appendix, the third equality follows from (A.5), the fourth equality holds since $\text{E}(\mathbf{y}) \in R(\mathbf{X})$ and $R(\mathbf{X}) \perp R(\mathbf{X})^\perp$, and the fifth equality follows from (A.4).

Let $s^2 = \left\|\mathbf{y} - \mathbf{X}\widehat{\boldsymbol{\theta}}\right\|^2 /(N - r)$. Then s^2, called the residual *mean square*, is an unbiased estimator of σ^2. The dimension of $R(\mathbf{X})^\perp$, $N - r$, is called the *degrees of freedom* associated with the residual sum of squares.

Under the assumption that $\mathbf{y}$ has a normal distribution, $\frac{1}{\sigma^2}\left\|\mathbf{y} - \mathbf{X}\widehat{\boldsymbol{\theta}}\right\|^2$ has a χ^2-distribution with $N - r$ degrees of freedom. Thus, for any estimable function $\mathbf{c}^T\boldsymbol{\theta}$,

$$\frac{\mathbf{c}^T\widehat{\boldsymbol{\theta}}}{s\sqrt{\mathbf{c}^T(\mathbf{X}^T\mathbf{X})^-\mathbf{c}}} \text{ has a } t\text{-distribution with } N - r \text{ degrees of freedom.}$$

Therefore a $100(1 - \alpha)\%$ confidence interval for $\mathbf{c}^T\boldsymbol{\theta}$ is $\mathbf{c}^T\widehat{\boldsymbol{\theta}} \pm t_{N-r;1-\alpha/2}s \cdot \sqrt{\mathbf{c}^T(\mathbf{X}^T\mathbf{X})^-\mathbf{c}}$, where $t_{N-r;1-\alpha/2}$ is the $(1 - \alpha/2)$th quantile of the t-distribution with $N - r$ degrees of freedom.

In the rest of the book, we often abbreviate degrees of freedom, sum of squares, and mean square as d.f., SS, and MS, respectively.

2.3 F-test

We have seen that under linear model (2.2)–(2.3), $E(\mathbf{y}) \in R(\mathbf{X})$. Suppose V is a subspace of $R(\mathbf{X})$ with $\dim(V) = q$. Let $R(\mathbf{X}) \ominus V = \{\mathbf{y} \in R(\mathbf{X}) : \mathbf{y} \text{ is orthogonal to } V\}$ be the orthogonal complement of V relative to $R(\mathbf{X})$. Then $R(\mathbf{X}) \ominus V$ has dimension $r - q$, and $\mathbf{X}\widehat{\boldsymbol{\theta}} = \mathbf{P}_{R(\mathbf{X})}\mathbf{y}$ can be decomposed as

$$\mathbf{P}_{R(\mathbf{X})}\mathbf{y} = \mathbf{P}_V\mathbf{y} + \mathbf{P}_{R(\mathbf{X})\ominus V}\mathbf{y}. \tag{2.8}$$

By combining (2.7) and (2.8), we have

$$\mathbf{y} - \mathbf{P}_V\mathbf{y} = \mathbf{P}_{R(\mathbf{X})\ominus V}\mathbf{y} + \mathbf{P}_{R(\mathbf{X})^\perp}\mathbf{y}, \tag{2.9}$$

where the two components on the right side are orthogonal. Thus

$$\left\|\mathbf{y} - \mathbf{P}_V\mathbf{y}\right\|^2 = \left\|\mathbf{P}_{R(\mathbf{X})\ominus V}\mathbf{y}\right\|^2 + \left\|\mathbf{P}_{R(\mathbf{X})^\perp}\mathbf{y}\right\|^2. \tag{2.10}$$

Under the assumption that $\mathbf{y}$ has a normal distribution, a test of the null hypothesis $H_0\colon E(\mathbf{y}) \in V$ against the alternative hypothesis that $E(\mathbf{y}) \notin V$ is based on the ratio

$$F = \frac{\left\|\mathbf{P}_{R(\mathbf{X})\ominus V}\mathbf{y}\right\|^2/(r - q)}{\left\|\mathbf{P}_{R(\mathbf{X})^\perp}\mathbf{y}\right\|^2/(N - r)}. \tag{2.11}$$

It can be shown that the likelihood ratio test is to reject H_0 for large values of F. Under H_0, F has an F-distribution with $r - q$ and $N - r$ degrees of freedom. Therefore the null hypothesis is rejected at level α if $F > F_{r-q,N-r;1-\alpha}$, where $F_{r-q,N-r;1-\alpha}$ is the $(1 - \alpha)$th quantile of the F-distribution with $r - q$ and $N - r$ degrees of freedom.

The left side of (2.10) is the residual sum of squares when H_0 is true, and the second term on the right side is the residual sum of squares under the full model (2.2). Thus the sum of squares $\left\|\mathbf{P}_{R(\mathbf{X})\ominus V}\mathbf{y}\right\|^2$ that appears in the numerator of the test statistic in (2.11) is the difference of the residual sums of squares under the full model and the reduced model specified by H_0.

2.4 One-way layout

Let $\mathbf{y} = (y_{11}, \ldots, y_{1r_1}, \ldots, y_{t1}, \ldots, y_{tr_t})^T$, where $y_{11}, \ldots, y_{1r_1}, \ldots, y_{t1}, \ldots, y_{tr_t}$ are uncorrelated random variables with constant variance σ^2, and $E(y_{lh}) = \alpha_l$, for all $1 \le h \le r_l$, $1 \le l \le t$. Then we can express $\mathbf{y}$ as in (2.2)–(2.3), with $\boldsymbol{\theta} = (\alpha_1, \ldots, \alpha_t)^T$, and

$$
\mathbf{X} = \begin{bmatrix}
\mathbf{1}_{r_1} & \mathbf{0} & \cdots & \mathbf{0} \\
\mathbf{0} & \mathbf{1}_{r_2} & \cdots & \mathbf{0} \\
\vdots & \vdots & \ddots & \vdots \\
\mathbf{0} & \mathbf{0} & \mathbf{0} & \mathbf{1}_{r_t}
\end{bmatrix},
$$

where $\mathbf{1}_{r_l}$ is the $r_l \times 1$ vector of ones. This model arises, for example, when $y_{11}, \ldots, y_{1r_l}$ are a random sample from a population with mean α_l and variance σ^2. It is also commonly used as a model for analyzing a completely randomized experiment, where the lth treatment, $1 \le l \le t$, is replicated r_l times, $y_{11}, \ldots, y_{1r_l}$ are the observations on the lth treatment, and $\alpha_1, \ldots, \alpha_t$ are effects of the treatments.

Clearly $\mathbf{X}^T\mathbf{X} = \mathrm{diag}(r_1, \ldots, r_t)$, the $t \times t$ diagonal matrix with $r_1, \ldots, r_t$ as the diagonal entries; therefore (2.4) has a unique solution: $\widehat{\alpha}_l = y_{l\cdot}$, where $y_{l\cdot} = \frac{1}{r_l}\sum_{h=1}^{r_l} y_{lh}$. For any function $\sum_{l=1}^t c_l \alpha_l$, we have $\sum_{l=1}^t c_l \widehat{\alpha}_l = \sum_{l=1}^t c_l y_{l\cdot}$, with

$$
\mathrm{var}\left(\sum_{l=1}^t c_l \widehat{\alpha}_l\right) = \sigma^2 \sum_{l=1}^t \frac{c_l^2}{r_l}.
$$

The projection $\mathbf{P}_{R(\mathbf{X})}\mathbf{y} = \mathbf{X}\widehat{\boldsymbol{\theta}}$ has its first r_1 components equal to $y_{1\cdot}$, the next r_2 components equal to $y_{2\cdot}, \ldots$, etc. Therefore the residual sum of squares can be expressed as

$$
\left\|\mathbf{y} - \mathbf{X}\widehat{\boldsymbol{\theta}}\right\|^2 = \sum_{l=1}^t \sum_{h=1}^{r_l} (y_{lh} - y_{l\cdot})^2.
$$

Call this the *within-group* sum of squares and denote it by W. We have $N = \sum_{l=1}^t r_l$, and $\mathrm{rank}(\mathbf{X}) = t$. Therefore the residual sum of squares has $N - t$ degrees of freedom, and so if $s^2 = W/(N-t)$, then $E(s^2) = \sigma^2$.

Now we impose the normality assumption and consider the test of $H_0: \alpha_1 = \cdots = \alpha_t$. Under H_0, $E(\mathbf{y}) \in V$, where V is the one-dimensional space consisting of all the vectors with constant entries. Then $\mathbf{P}_V\mathbf{y}$ is the vector with all the entries equal to the overall average $y_{\cdot\cdot} = \frac{1}{N}\sum_{l=1}^t \sum_{h=1}^{r_l} y_{lh}$. Componentwise, (2.9) can be expressed as

$$
y_{lh} - y_{\cdot\cdot} = (y_{l\cdot} - y_{\cdot\cdot}) + (y_{lh} - y_{l\cdot})
$$

and, in the present context, (2.10) reduces to

$$
\sum_{l=1}^t \sum_{h=1}^{r_l} (y_{lh} - y_{\cdot\cdot})^2 = \sum_{l=1}^t r_l(y_{l\cdot} - y_{\cdot\cdot})^2 + \sum_{l=1}^t \sum_{h=1}^{r_l} (y_{lh} - y_{l\cdot})^2.
$$

Thus $\|\mathbf{P}_{R(\mathbf{X}) \ominus V}\mathbf{y}\|^2 = \sum_{l=1}^t r_l(y_{l\cdot} - y_{\cdot\cdot})^2$, which has $t - 1$ degrees of freedom and is called the *between-group* sum of squares. Denote it by B. Then the F-test statistic is $\frac{B/(t-1)}{W/(N-t)}$. In the application to completely randomized experiments, the between-group sum of squares is also called the treatment sum of squares.

These results can be summarized in Table 2.1, called an ANOVA (Analysis of Variance) table.

Table 2.1 *ANOVA table for one-way layout*

source	sum of squares	d.f.	mean square
Between groups	$\sum_{l=1}^{t} r_l(y_{l.} - y_{..})^2$	$t-1$	$\frac{1}{t-1}\sum_{l=1}^{t} r_l(y_{l.} - y_{..})^2$
Within groups	$\sum_{l=1}^{t}\sum_{h=1}^{r_l}(y_{lh} - y_{l.})^2$	$N-t$	$\frac{1}{N-t}\sum_{l=1}^{t}\sum_{h=1}^{r_l}(y_{lh} - y_{l.})^2$
Total	$\sum_{l=1}^{t}\sum_{h=1}^{r_l}(y_{lh} - y_{..})^2$	$N-1$	

Remark 2.1. If we write the model as $E(y_{lh}) = \mu + \alpha_l$, then the model matrix $\mathbf{X}$ has an extra column of 1's and $p = t + 1$. Since rank$(\mathbf{X}) = t < p$, the parameters themselves are not identifiable. By using (2.6), one can verify that $\sum_{l=1}^{t} c_l\alpha_l$ is estimable if and only if $\sum_{l=1}^{t} c_l = 0$. Such functions are called *contrasts*. The pairwise differences $\alpha_l - \alpha_{l'}$, $1 \leq l \neq l' \leq t$, are examples of contrasts and are referred to as elementary contrasts. Because of the constraint $\sum_{l=1}^{t} c_l = 0$, the treatment contrasts form a $(t-1)$-dimensional vector space that is generated by the elementary contrasts. Functions such as $\alpha_1 - \frac{1}{2}(\alpha_2 + \alpha_3)$ and $\frac{1}{2}(\alpha_1 + \alpha_2) - \frac{1}{3}(\alpha_3 + \alpha_4 + \alpha_5)$ are also contrasts. If $\sum_{l=1}^{t} c_l\alpha_l$ is a contrast, then $\sum_{l=1}^{t} c_l\alpha_l = \sum_{l=1}^{t} c_l(\mu + \alpha_l)$, and it can be shown that, as before, $\sum_{l=1}^{t} c_l\widehat{\alpha}_l = \sum_{l=1}^{t} c_l y_{l.}$. Therefore, if the interest is in estimating the contrasts, then it does not matter whether $E(y_{lh})$ is written as α_l or $\mu + \alpha_l$. A contrast $\sum_{l=1}^{t} c_l\alpha_l$ is said to be normalized if $\sum_{l=1}^{t} c_l^2 = 1$.

2.5 Estimation of a subset of parameters

Suppose $\boldsymbol{\theta}$ is partitioned as $\boldsymbol{\theta} = (\boldsymbol{\theta}_1^T\ \boldsymbol{\theta}_2^T)^T$, where $\boldsymbol{\theta}_1$ is $q \times 1$ and $\boldsymbol{\theta}_2$ is $(p-q) \times 1$; e.g., the parameters in $\boldsymbol{\theta}_2$ are nuisance parameters, or the components of $\boldsymbol{\theta}_1$ and $\boldsymbol{\theta}_2$ are effects of the levels of two different factors. Partition $\mathbf{X}$ as $[\mathbf{X}_1\ \mathbf{X}_2]$ accordingly. Then (2.2) can be written as

$$\mathbf{y} = \mathbf{X}_1\boldsymbol{\theta}_1 + \mathbf{X}_2\boldsymbol{\theta}_2 + \boldsymbol{\varepsilon}, \tag{2.12}$$

and (2.4) is the same as

$$\mathbf{X}_1^T\mathbf{X}_1\widehat{\boldsymbol{\theta}}_1 + \mathbf{X}_1^T\mathbf{X}_2\widehat{\boldsymbol{\theta}}_2 = \mathbf{X}_1^T\mathbf{y}, \tag{2.13}$$

and

$$\mathbf{X}_2^T\mathbf{X}_1\widehat{\boldsymbol{\theta}}_1 + \mathbf{X}_2^T\mathbf{X}_2\widehat{\boldsymbol{\theta}}_2 = \mathbf{X}_2^T\mathbf{y}. \tag{2.14}$$

If $R(\mathbf{X}_1)$ and $R(\mathbf{X}_2)$ are orthogonal, $(\mathbf{X}_1^T\mathbf{X}_2 = \mathbf{0})$, then $\widehat{\boldsymbol{\theta}}_1$ and $\widehat{\boldsymbol{\theta}}_2$ can be computed by solving $\mathbf{X}_1^T\mathbf{X}_1\widehat{\boldsymbol{\theta}}_1 = \mathbf{X}_1^T\mathbf{y}$ and $\mathbf{X}_2^T\mathbf{X}_2\widehat{\boldsymbol{\theta}}_2 = \mathbf{X}_2^T\mathbf{y}$ separately. In this case, least squares estimators of estimable functions of $\boldsymbol{\theta}_1$ are the same regardless of whether $\boldsymbol{\theta}_2$ is in the model. Likewise, dropping $\boldsymbol{\theta}_1$ from (2.12) does not change the least squares estimators of estimable functions of $\boldsymbol{\theta}_2$.

If $R(\mathbf{X}_1)$ and $R(\mathbf{X}_2)$ are not orthogonal, then for estimating $\boldsymbol{\theta}_1$ one needs to adjust for $\boldsymbol{\theta}_2$, and vice versa. By (2.14), $\widehat{\boldsymbol{\theta}}_2$ can be written as

$$\widehat{\boldsymbol{\theta}}_2 = \left(\mathbf{X}_2^T\mathbf{X}_2\right)^- \left[\mathbf{X}_2^T\mathbf{y} - \mathbf{X}_2^T\mathbf{X}_1\widehat{\boldsymbol{\theta}}_1\right]. \tag{2.15}$$

We eliminate $\widehat{\boldsymbol{\theta}}_2$ by substituting the right side of (2.15) for the $\widehat{\boldsymbol{\theta}}_2$ in (2.13). Then $\widehat{\boldsymbol{\theta}}_1$ can be obtained by solving

$$\left[\mathbf{X}_1^T\mathbf{X}_1 - \mathbf{X}_1^T\mathbf{X}_2\left(\mathbf{X}_2^T\mathbf{X}_2\right)^-\mathbf{X}_2^T\mathbf{X}_1\right]\widehat{\boldsymbol{\theta}}_1 = \mathbf{X}_1^T\mathbf{y} - \mathbf{X}_1^T\mathbf{X}_2\left(\mathbf{X}_2^T\mathbf{X}_2\right)^-\mathbf{X}_2^T\mathbf{y}$$

or

$$\left\{\mathbf{X}_1^T\left[\mathbf{I} - \mathbf{X}_2\left(\mathbf{X}_2^T\mathbf{X}_2\right)^-\mathbf{X}_2^T\right]\mathbf{X}_1\right\}\widehat{\boldsymbol{\theta}}_1 = \mathbf{X}_1^T\left[\mathbf{I} - \mathbf{X}_2\left(\mathbf{X}_2^T\mathbf{X}_2\right)^-\mathbf{X}_2^T\right]\mathbf{y}. \tag{2.16}$$

This is called the *reduced normal equation* for $\boldsymbol{\theta}_1$.

We write the reduced normal equation as

$$\mathbf{C}_1\widehat{\boldsymbol{\theta}}_1 = \mathbf{Q}_1, \tag{2.17}$$

where

$$\mathbf{C}_1 = \mathbf{X}_1^T\left[\mathbf{I} - \mathbf{X}_2\left(\mathbf{X}_2^T\mathbf{X}_2\right)^-\mathbf{X}_2^T\right]\mathbf{X}_1, \tag{2.18}$$

and

$$\mathbf{Q}_1 = \mathbf{X}_1^T\left[\mathbf{I} - \mathbf{X}_2\left(\mathbf{X}_2^T\mathbf{X}_2\right)^-\mathbf{X}_2^T\right]\mathbf{y}. \tag{2.19}$$

Since $\mathbf{X}_2(\mathbf{X}_2^T\mathbf{X}_2)^-\mathbf{X}_2^T$ is the orthogonal projection matrix onto $R(\mathbf{X}_2)$, $\mathbf{I} - \mathbf{X}_2(\mathbf{X}_2^T\mathbf{X}_2)^-\mathbf{X}_2^T$ is the orthogonal projection matrix onto $R(\mathbf{X}_2)^\perp$, and

$$\mathbf{C}_1 = \mathbf{X}_1^T\mathbf{P}_{R(\mathbf{X}_2)^\perp}\mathbf{X}_1.$$

If we put

$$\widetilde{\mathbf{X}}_1 - \mathbf{P}_{R(\mathbf{X}_2)^\perp}\mathbf{X}_1,$$

then we can express $\mathbf{C}_1$ as

$$\mathbf{C}_1 = \widetilde{\mathbf{X}}_1^T\widetilde{\mathbf{X}}_1,$$

and the reduced normal equation (2.16) can be written as

$$\widetilde{\mathbf{X}}_1^T\widetilde{\mathbf{X}}_1\widehat{\boldsymbol{\theta}}_1 = \widetilde{\mathbf{X}}_1^T\widetilde{\mathbf{y}},$$

where $\widetilde{\mathbf{y}} = \mathbf{P}_{R(\mathbf{X}_2)^\perp}\mathbf{y}$. Therefore the least squares estimators of estimable functions of $\boldsymbol{\theta}_1$ are functions of $\widetilde{\mathbf{y}} = \mathbf{P}_{R(\mathbf{X}_2)^\perp}\mathbf{y}$: to eliminate $\boldsymbol{\theta}_2$, (2.12) is projected onto $R(\mathbf{X}_2)^\perp$ to become

$$\widetilde{\mathbf{y}} = \widetilde{\mathbf{X}}_1\boldsymbol{\theta}_1 + \widetilde{\boldsymbol{\varepsilon}},$$

where $\widetilde{\boldsymbol{\varepsilon}} = \mathbf{P}_{R(\mathbf{X}_2)^\perp}\boldsymbol{\varepsilon}$.

Theorem 2.3. *Under (2.12) and (2.3), a linear function $\mathbf{c}_1^T \boldsymbol{\theta}_1$ of $\boldsymbol{\theta}_1$ is estimable if and only if $\mathbf{c}_1$ is a linear combination of the column vectors of $\mathbf{C}_1$. If $\mathbf{c}_1^T \boldsymbol{\theta}_1$ is estimable, then $\mathbf{c}_1^T \widehat{\boldsymbol{\theta}}_1$ is its best linear unbiased estimator, and*

$$var\left(\mathbf{c}_1^T \widehat{\boldsymbol{\theta}}_1\right) = \sigma^2 \mathbf{c}_1^T \mathbf{C}_1^- \mathbf{c}_1,$$

where $\widehat{\boldsymbol{\theta}}_1$ is any solution to (2.17), and $\mathbf{C}_1^-$ is any generalized inverse of $\mathbf{C}_1$.

In particular, if $\mathbf{C}_1$ is invertible, then all the parameters in $\boldsymbol{\theta}_1$ are estimable, with

$$\mathrm{cov}\left(\widehat{\boldsymbol{\theta}}_1\right) = \sigma^2 \mathbf{C}_1^{-1}. \tag{2.20}$$

The matrix $\mathbf{C}_1$ is called the information matrix for $\boldsymbol{\theta}_1$.

Suppose the parameters in $\boldsymbol{\theta}_2$ are nuisance parameters, and we are only interested in estimating $\boldsymbol{\theta}_1$. Then a design is said to be D_s-, A_s-, or E_s-optimal if it minimizes, respectively, the determinant, trace, or the largest eigenvalue of the covariance matrix of $\widehat{\boldsymbol{\theta}}_1$ among all the competing designs. By (2.20), these optimality criteria are equivalent to maximizing $\det(\mathbf{C}_1)$, minimizing $\mathrm{tr}(\mathbf{C}_1^{-1})$, and maximizing the smallest eigenvalue of $\mathbf{C}_1$, respectively. Here s refers to a *subset* of parameters.

2.6 Hypothesis testing for a subset of parameters

Under (2.12) and (2.3), suppose we further assume that $\boldsymbol{\varepsilon}$ has a normal distribution. Consider testing the null hypothesis

$$H_0: \mathrm{E}(\mathbf{y}) = \mathbf{X}_2 \boldsymbol{\theta}_2.$$

Under (2.12), $\mathrm{E}(\mathbf{y}) \in R(\mathbf{X}_1) + R(\mathbf{X}_2)$, and under H_0, $\mathrm{E}(\mathbf{y}) \in R(\mathbf{X}_2)$. By (2.11), the F-test statistic in this case is

$$F = \frac{\left\|\mathbf{P}_{[R(\mathbf{X}_1)+R(\mathbf{X}_2)]\ominus R(\mathbf{X}_2)}\mathbf{y}\right\|^2 / \dim([R(\mathbf{X}_1)+R(\mathbf{X}_2)]\ominus R(\mathbf{X}_2))}{\left\|\mathbf{P}_{[R(\mathbf{X}_1)+R(\mathbf{X}_2)]^\perp}\mathbf{y}\right\|^2 / (N - \dim[R(\mathbf{X}_1)+R(\mathbf{X}_2)])}.$$

This is based on the decomposition

$$\mathbb{R}^N = R(\mathbf{X}_2) \oplus ([R(\mathbf{X}_1)+R(\mathbf{X}_2)] \ominus R(\mathbf{X}_2)) \oplus [R(\mathbf{X}_1)+R(\mathbf{X}_2)]^\perp. \tag{2.21}$$

It can be shown that

$$\left\|\mathbf{P}_{[R(\mathbf{X}_1)+R(\mathbf{X}_2)]\ominus R(\mathbf{X}_2)}\mathbf{y}\right\|^2 = \widehat{\boldsymbol{\theta}}_1^T \mathbf{Q}_1, \tag{2.22}$$

and

$$\dim([R(\mathbf{X}_1)+R(\mathbf{X}_2)] \ominus R(\mathbf{X}_2)) = \mathrm{rank}(\mathbf{C}_1). \tag{2.23}$$

We leave the proofs of these as an exercise.

A test of the null hypothesis $H_0: \mathrm{E}(\mathbf{y}) = \mathbf{X}_1 \boldsymbol{\theta}_1$ is based on the decomposition

$$\mathbb{R}^N = R(\mathbf{X}_1) \oplus ([R(\mathbf{X}_1)+R(\mathbf{X}_2)] \ominus R(\mathbf{X}_1)) \oplus [R(\mathbf{X}_1)+R(\mathbf{X}_2)]^\perp. \tag{2.24}$$

In this case,

$$\left\|\mathbf{P}_{[R(\mathbf{X}_1)+R(\mathbf{X}_2)]\ominus R(\mathbf{X}_1)}\mathbf{y}\right\|^2 = \widehat{\boldsymbol{\theta}}_2^T\mathbf{Q}_2,$$

where

$$\mathbf{Q}_2 = \mathbf{X}_2^T\left[\mathbf{I} - \mathbf{X}_1\left(\mathbf{X}_1^T\mathbf{X}_1\right)^{-}\mathbf{X}_1^T\right]\mathbf{y}.$$

When $\mathbf{X}_1^T\mathbf{X}_2 = \mathbf{0}$, both (2.21) and (2.24) reduce to

$$\mathbb{R}^N = R(\mathbf{X}_1) \oplus R(\mathbf{X}_2) \oplus [R(\mathbf{X}_1) \oplus R(\mathbf{X}_2)]^{\perp}.$$

2.7   Adjusted orthogonality

Consider the model

$$\mathbf{y} = \mathbf{X}_1\boldsymbol{\theta}_1 + \mathbf{X}_2\boldsymbol{\theta}_2 + \mathbf{X}_3\boldsymbol{\theta}_3 + \boldsymbol{\varepsilon}, \tag{2.25}$$

with

$$E(\boldsymbol{\varepsilon}) = \mathbf{0}, \text{ and } \mathrm{cov}(\boldsymbol{\varepsilon}) = \sigma^2\mathbf{I}_N.$$

Suppose the parameters in $\boldsymbol{\theta}_2$ are nuisance parameters. To estimate $\boldsymbol{\theta}_1$ and $\boldsymbol{\theta}_3$, we eliminate $\boldsymbol{\theta}_2$ by projecting $\mathbf{y}$ onto $R(\mathbf{X}_2)^{\perp}$. This results in a reduced normal equation for $\widehat{\boldsymbol{\theta}}_1$ and $\widehat{\boldsymbol{\theta}}_3$:

$$\begin{bmatrix} \tilde{\mathbf{X}}_1^T\tilde{\mathbf{X}}_1 & \tilde{\mathbf{X}}_1^T\tilde{\mathbf{X}}_3 \\ \tilde{\mathbf{X}}_3^T\tilde{\mathbf{X}}_1 & \tilde{\mathbf{X}}_3^T\tilde{\mathbf{X}}_3 \end{bmatrix}\begin{bmatrix} \widehat{\boldsymbol{\theta}}_1 \\ \widehat{\boldsymbol{\theta}}_3 \end{bmatrix} = \begin{bmatrix} \tilde{\mathbf{X}}_1^T\tilde{\mathbf{y}} \\ \tilde{\mathbf{X}}_3^T\tilde{\mathbf{y}} \end{bmatrix},$$

where

$$\tilde{\mathbf{X}}_i = \mathbf{P}_{R(\mathbf{X}_2)^{\perp}}\mathbf{X}_i, i = 1, 3, \text{ and } \tilde{\mathbf{y}} = \mathbf{P}_{R(\mathbf{X}_2)^{\perp}}\mathbf{y}.$$

If $\tilde{\mathbf{X}}_1^T\tilde{\mathbf{X}}_3 = \mathbf{0}$, then $\widehat{\boldsymbol{\theta}}_1$ and $\widehat{\boldsymbol{\theta}}_3$ can be computed by solving the equations $\tilde{\mathbf{X}}_1^T\tilde{\mathbf{X}}_1\widehat{\boldsymbol{\theta}}_1 = \tilde{\mathbf{X}}_1^T\tilde{\mathbf{y}}$ and $\tilde{\mathbf{X}}_3^T\tilde{\mathbf{X}}_3\widehat{\boldsymbol{\theta}}_3 = \tilde{\mathbf{X}}_3^T\tilde{\mathbf{y}}$ separately. In this case, least squares estimators of estimable functions of $\boldsymbol{\theta}_1$ under (2.25) are the same regardless of whether $\boldsymbol{\theta}_3$ is in the model. Likewise, dropping $\boldsymbol{\theta}_1$ from (2.25) does not change the least squares estimators of estimable functions of $\boldsymbol{\theta}_3$. Loosely we say that $\boldsymbol{\theta}_1$ and $\boldsymbol{\theta}_3$ are orthogonal adjusted for $\boldsymbol{\theta}_2$.

Note that $\tilde{\mathbf{X}}_1^T\tilde{\mathbf{X}}_3 = \mathbf{0}$ is equivalent to

$$P_{R(\mathbf{X}_2)^{\perp}}[R(\mathbf{X}_1)] \text{ is orthogonal to } P_{R(\mathbf{X}_2)^{\perp}}[R(\mathbf{X}_3)]. \tag{2.26}$$

Under normality, the F-test statistic for the null hypothesis that $E(\mathbf{y}) = \mathbf{X}_2\boldsymbol{\theta}_2 + \mathbf{X}_3\boldsymbol{\theta}_3$ is

$$\frac{\left\|\mathbf{P}_{[R(\mathbf{X}_1)+R(\mathbf{X}_2)+R(\mathbf{X}_3)]\ominus[R(\mathbf{X}_2)+R(\mathbf{X}_3)]}\mathbf{y}\right\|^2/\mathrm{dim}([R(\mathbf{X}_1)+R(\mathbf{X}_2)+R(\mathbf{X}_3)]\ominus[R(\mathbf{X}_2)+R(\mathbf{X}_3)])}{\left\|\mathbf{P}_{[R(\mathbf{X}_1)+R(\mathbf{X}_2)+R(\mathbf{X}_3)]^{\perp}}\mathbf{y}\right\|^2/(N-\mathrm{dim}[R(\mathbf{X}_1)+R(\mathbf{X}_2)+R(\mathbf{X}_3)])}.$$

If (2.26) holds, then we have

$$R(\mathbf{X}_1) + R(\mathbf{X}_2) + R(\mathbf{X}_3) = R(\mathbf{X}_2) \oplus P_{R(\mathbf{X}_2)^{\perp}}[R(\mathbf{X}_1)] \oplus P_{R(\mathbf{X}_2)^{\perp}}[R(\mathbf{X}_3)].$$

In this case,

$$[R(\mathbf{X}_1)+R(\mathbf{X}_2)+R(\mathbf{X}_3)] \ominus [R(\mathbf{X}_2)+R(\mathbf{X}_3)] = P_{R(\mathbf{X}_2)^\perp}[R(\mathbf{X}_1)]$$
$$= [R(\mathbf{X}_1)+R(\mathbf{X}_2)] \ominus R(\mathbf{X}_2).$$

So the sum of squares in the numerator of the F-test statistic is equal to the quantity $\widehat{\boldsymbol{\theta}}_1^T \mathbf{Q}_1$ given in (2.22), with its degrees of freedom equal to rank($\mathbf{C}_1$), where $\mathbf{C}_1$ and $\mathbf{Q}_1$ are as in (2.18) and (2.19), respectively. A similar conclusion can be drawn for testing the hypothesis that $E(\mathbf{y}) = \mathbf{X}_1\boldsymbol{\theta}_1 + \mathbf{X}_2\boldsymbol{\theta}_2$.

2.8   Additive two-way layout

In a two-way layout, the observations are classified according to the levels of two factors. Suppose the two factors have t and b levels, respectively. At each level combination (i, j), $1 \le i \le t$, $1 \le j \le b$, there are n_{ij} observations y_{ijh}, $0 \le h \le n_{ij}$, such that

$$y_{ijh} = \alpha_i + \beta_j + \varepsilon_{ijh}, \tag{2.27}$$

where the ε_{ijh}'s are uncorrelated random variables with zero mean and constant variance σ^2. We require $\sum_{j=1}^{b} n_{ij} > 0$ for all i, and $\sum_{i=1}^{t} n_{ij} > 0$ for all j, so that there is at least one observation on each level of the two factors, but some n_{ij}'s may be zero. This is called an additive two-way layout model, which is commonly used for analyzing block designs, where each y_{ijh} is an observation on the ith treatment in the jth block. With this in mind, we call the two factors treatment and block factors, and denote them by $\mathcal{T}$ and $\mathcal{B}$, respectively; then $\alpha_1, \ldots, \alpha_t$ are the treatment effects and $\beta_1, \ldots, \beta_b$ are the block effects. Let $N = \sum_{i=1}^{t} \sum_{j=1}^{b} n_{ij}$, and think of the observations as taken on N units that are grouped into b blocks. Define an $N \times t$ matrix $\mathbf{X}_{\mathcal{T}}$ with 0 and 1 entries such that the (v, i)th entry of $\mathbf{X}_{\mathcal{T}}$, $1 \le v \le N$, $1 \le i \le t$, is 1 if and only if the ith treatment is assigned to the vth unit. Similarly, let $\mathbf{X}_{\mathcal{B}}$ be the $N \times b$ matrix with 0 and 1 entries such that the (v, j)th entry of $\mathbf{X}_{\mathcal{B}}$, $1 \le v \le N$, $1 \le j \le b$, is 1 if and only if the vth unit is in the jth block. The two matrices $\mathbf{X}_{\mathcal{T}}$ and $\mathbf{X}_{\mathcal{B}}$ are called unit-treatment and unit-block incidence matrices, respectively. Then we can write (2.27) as

$$\mathbf{y} = \mathbf{X}_{\mathcal{T}}\boldsymbol{\alpha} + \mathbf{X}_{\mathcal{B}}\boldsymbol{\beta} + \boldsymbol{\varepsilon},$$

where $\boldsymbol{\alpha} = (\alpha_1, \ldots, \alpha_t)^T$ and $\boldsymbol{\beta} = (\beta_1, \ldots, \beta_b)^T$.

We use the results in Sections 2.5 and 2.6 to derive the analysis for such models. Then we apply the results in Section 2.7 to show in Section 2.9 that the analysis can be much simplified when certain conditions are satisfied.

Let $\mathbf{N}$ be the $t \times b$ matrix whose (i, j)th entry is n_{ij}, $n_{i+} = \sum_{j=1}^{b} n_{ij}$, and $n_{+j} = \sum_{i=1}^{t} n_{ij}$. For block designs, n_{i+} is the number of observations on the ith treatment, and n_{+j} is the size of the jth block. Also, let $y_{i++} = \sum_{j=1}^{b} \sum_{h} y_{ijh}$ and $y_{+j+} = \sum_{i=1}^{t} \sum_{h} y_{ijh}$, the ith treatment total and jth block total, respectively. Then $\mathbf{X}_{\mathcal{T}}^T \mathbf{X}_{\mathcal{T}}$ is the diagonal matrix with diagonal entries $n_{1+}, \ldots, n_{t+}$, $\mathbf{X}_{\mathcal{B}}^T \mathbf{X}_{\mathcal{B}}$ is the diagonal matrix with diagonal entries $n_{+1}, \ldots, n_{+b}$, $\mathbf{X}_{\mathcal{T}}^T \mathbf{y} = (y_{1++}, \ldots, y_{t++})^T$, $\mathbf{X}_{\mathcal{B}}^T \mathbf{y} = (y_{+1+}, \ldots, y_{+b+})^T$, and

$$\mathbf{X}_{\mathcal{T}}^T \mathbf{X}_{\mathcal{B}} = \mathbf{N}. \tag{2.28}$$

By (2.17), (2.18), and (2.19), the reduced normal equations for $\boldsymbol{\alpha}$ and $\boldsymbol{\beta}$ are, respectively,

$$\mathbf{C}_{\mathcal{T}}\widehat{\boldsymbol{\alpha}} = \mathbf{Q}_{\mathcal{T}} \tag{2.29}$$

and

$$\mathbf{C}_{\mathcal{B}}\widehat{\boldsymbol{\beta}} = \mathbf{Q}_{\mathcal{B}},$$

where

$$\mathbf{C}_{\mathcal{T}} = \begin{bmatrix} n_{1+} & \cdots & 0 \\ \vdots & \ddots & \vdots \\ 0 & \cdots & n_{t+} \end{bmatrix} - \mathbf{N} \begin{bmatrix} \frac{1}{n_{+1}} & \cdots & 0 \\ \vdots & \ddots & \vdots \\ 0 & \cdots & \frac{1}{n_{+b}} \end{bmatrix} \mathbf{N}^T,$$

$$\mathbf{Q}_{\mathcal{T}} = \begin{bmatrix} y_{1++} \\ \vdots \\ y_{t++} \end{bmatrix} - \mathbf{N} \begin{bmatrix} \frac{1}{n_{+1}} & \cdots & 0 \\ \vdots & \ddots & \vdots \\ 0 & \cdots & \frac{1}{n_{+b}} \end{bmatrix} \begin{bmatrix} y_{+1+} \\ \vdots \\ y_{+b+} \end{bmatrix},$$

$$\mathbf{C}_{\mathcal{B}} = \begin{bmatrix} n_{+1} & \cdots & 0 \\ \vdots & \ddots & \vdots \\ 0 & \cdots & n_{+b} \end{bmatrix} - \mathbf{N}^T \begin{bmatrix} \frac{1}{n_{1+}} & \cdots & 0 \\ \vdots & \ddots & \vdots \\ 0 & \cdots & \frac{1}{n_{t+}} \end{bmatrix} \mathbf{N},$$

and

$$\mathbf{Q}_{\mathcal{B}} = \begin{bmatrix} y_{+1+} \\ \vdots \\ y_{+b+} \end{bmatrix} - \mathbf{N}^T \begin{bmatrix} \frac{1}{n_{1+}} & \cdots & 0 \\ \vdots & \ddots & \vdots \\ 0 & \cdots & \frac{1}{n_{t+}} \end{bmatrix} \begin{bmatrix} y_{1++} \\ \vdots \\ y_{t++} \end{bmatrix}.$$

It can be verified that both $\mathbf{C}_{\mathcal{T}}$ and $\mathbf{C}_{\mathcal{B}}$ have zero column sums. Therefore $\text{rank}(\mathbf{C}_{\mathcal{T}}) \leq t - 1$, $\text{rank}(\mathbf{C}_{\mathcal{B}}) \leq b - 1$, and, by Theorem 2.3, if $\sum_{i=1}^{t} c_i \alpha_i$ is estimable, then $\sum_{i=1}^{t} c_i = 0$. All such contrasts are estimable if and only if $\text{rank}(\mathbf{C}_{\mathcal{T}}) = t - 1$. Similarly, if $\sum_{j=1}^{b} d_j \beta_j$ is estimable, then $\sum_{j=1}^{b} d_j = 0$, and all such contrasts are estimable if and only if $\text{rank}(\mathbf{C}_{\mathcal{B}}) = b - 1$.

Theorem 2.4. *All the contrasts of* $\alpha_1, \ldots, \alpha_t$ *are estimable if and only if for any* $1 \leq i \neq i' \leq t$, *there is a sequence*

$$i_1, j_1, i_2, j_2, \ldots, i_k, j_k, i_{k+1}$$

such that $i_1 = i$, $i_{k+1} = i'$, *and for all* $1 \leq s \leq k$, $n_{i_s j_s} > 0$ *and* $n_{i_{s+1}, j_s} > 0$.

Proof. Suppose the condition in the theorem holds. We need to show that all the contrasts of $\alpha_1, \ldots, \alpha_t$ are estimable. Since the space of all the contrasts is generated by the pairwise differences, it suffices to show that all the pairwise differences $\alpha_i - \alpha_{i'}$ are estimable.

Let y_{ij} be any of the observations at the level combination (i, j). Then

$$
\begin{aligned}
& E(y_{i_1 j_1} - y_{i_2 j_1} + y_{i_2 j_2} - \cdots + y_{i_k j_k} - y_{i_{k+1} j_k}) \\
= \ & (\alpha_{i_1} + \beta_{j_1}) - (\alpha_{i_2} + \beta_{j_1}) + (\alpha_{i_2} + \beta_{j_2}) - \cdots + (\alpha_{i_k} + \beta_{j_k}) - (\alpha_{i_{k+1}} + \beta_{j_k}) \\
= \ & \alpha_{i_1} - \alpha_{i_{k+1}} \\
= \ & \alpha_i - \alpha_{i'}.
\end{aligned}
$$

This shows that $\alpha_i - \alpha_{i'}$ is estimable.

Conversely, if the condition in the theorem does not hold, then the treatments can be partitioned into two disjoint sets such that any treatment from one set never appears in the same block with any treatment from the other set. Then it can be seen that $\mathbf{C}_T$ is of the form

$$
\begin{bmatrix} \mathbf{C}_1 & \mathbf{0} \\ \mathbf{0} & \mathbf{C}_2 \end{bmatrix}.
$$

Since $\mathbf{C}_T$ has zero column sums, both $\mathbf{C}_1$ and $\mathbf{C}_2$ also have zero column sums. It follows that $\mathrm{rank}(\mathbf{C}_T) = \mathrm{rank}(\mathbf{C}_1) + \mathrm{rank}(\mathbf{C}_2) \leq t - 2$; therefore not all the contrasts of $\alpha_1, \ldots, \alpha_t$ are estimable. $\square$

If any two treatments can be connected by a chain of alternating treatments and blocks as in Theorem 2.4, then any two blocks can also be connected by such a sequence. Therefore the condition in Theorem 2.4 is also a necessary and sufficient condition for all the contrasts of $\beta_1, \ldots, \beta_b$ to be estimable. In particular, all the contrasts of $\alpha_1, \ldots, \alpha_t$ are estimable if and only if all the contrasts of $\beta_1, \ldots, \beta_b$ are estimable.

Throughout the rest of this section, we assume that the condition in Theorem 2.4 holds; therefore $\mathrm{rank}(C_T) = t - 1$ and $\mathrm{rank}(C_B) = b - 1$. This is the case, for example, when there is at least one observation at each level combination of the two factors. In view of Theorem 2.4, designs with $\mathrm{rank}(C_T) = t - 1$ are called *connected* designs.

Suppose we would like to test the hypothesis that $\alpha_1 = \cdots = \alpha_t$. Under the null hypothesis, let the common value of the α_i's be α; then $E(y_{ijh}) = \alpha + \beta_j$. This reduces (2.27) to a one-way layout model. Absorb α into β_j (see Remark 2.1); then it is the same as to test that $E(\mathbf{y}) = \mathbf{X}_B \boldsymbol{\beta}$. Therefore the results in Section 2.6 can be applied. In particular, the sum of squares that appears in the numerator of the F-test statistic is equal to $\widehat{\boldsymbol{\alpha}}^T \mathbf{Q}_T = \widehat{\boldsymbol{\alpha}}^T \mathbf{C}_T \widehat{\boldsymbol{\alpha}}$, with $t - 1$ degrees of freedom. The residual sum of squares can be computed by subtracting $\widehat{\boldsymbol{\alpha}}^T \mathbf{Q}_T$ from the residual (within-group) sum of squares under the one-way layout model containing block effects only. Let $y_{i\cdot\cdot} = \frac{1}{n_{i+}} \sum_{j=1}^{b} \sum_h y_{ijh}$, $y_{\cdot j\cdot} = \frac{1}{n_{+j}} \sum_{i=1}^{t} \sum_h y_{ijh}$, and $y_{\cdots} = \frac{1}{N} \sum_{i=1}^{t} \sum_{j=1}^{b} \sum_h y_{ijh}$ be the ith treatment mean, jth block mean, and overall mean, respectively. Then we have the ANOVA in Table 2.2.

A test of the hypothesis that $E(\mathbf{y}) = \mathbf{X}_T \boldsymbol{\alpha}$ can be based on an ANOVA similar to that in Table 2.2 with the roles of treatments and blocks reversed.

Table 2.2 *ANOVA table for an additive two-way layout*

source	sum of squares	d.f.	mean square
Blocks (ignoring treatments)	$\sum_j n_{+j}(y_{\cdot j\cdot} - y_{\cdots})^2$	$b-1$	$\frac{1}{b-1}\sum_j n_{+j}(y_{\cdot j\cdot} - y_{\cdots})^2$
Treatments (adjusted for blocks)	$\widehat{\alpha}^T \mathbf{Q}_{\mathcal{T}}$	$t-1$	$\frac{1}{t-1}\widehat{\alpha}^T \mathbf{Q}_{\mathcal{T}}$
Residual	By subtraction	$N-b-t+1$	$\frac{1}{N-b-t+1}$(Residual SS)
Total	$\sum_i \sum_j \sum_h (y_{ijh} - y_{\cdots})^2$	$N-1$	

2.9 The case of proportional frequencies

As explained in Remark 2.1, model (2.27) can be written as

$$y_{ijh} = \mu + \alpha_i + \beta_j + \varepsilon_{ijh} \tag{2.30}$$

without changing the least squares estimators of estimable functions of $(\alpha_1, \ldots, \alpha_t)^T$, least squares estimators of estimable functions of $(\beta_1, \ldots, \beta_b)^T$, and the reduced normal equations. Write (2.30) in the form of (2.25) with $\boldsymbol{\theta}_1 = (\alpha_1, \ldots, \alpha_t)^T$, $\boldsymbol{\theta}_3 = (\beta_1, \ldots, \beta_b)^T$, and $\boldsymbol{\theta}_2 = \mu$. Then $\mathbf{X}_1 = \mathbf{X}_{\mathcal{T}}$, $\mathbf{X}_3 = \mathbf{X}_{\mathcal{B}}$, and $\mathbf{X}_2 = \mathbf{1}_N$. In this case, the adjusted orthogonality condition (2.26) is that

$$P_{R(\mathbf{1}_N)^\perp}[R(\mathbf{X}_{\mathcal{T}})] \text{ is orthogonal to } P_{R(\mathbf{1}_N)^\perp}[R(\mathbf{X}_{\mathcal{B}})]. \tag{2.31}$$

Since only one treatment can be assigned to each unit, each row of $\mathbf{X}_{\mathcal{T}}$ has exactly one entry equal to 1, and all the other entries are zero. It follows that the sum of all the columns of $\mathbf{X}_{\mathcal{T}}$ is equal to $\mathbf{1}_N$. Therefore $R(\mathbf{1}_N) \subseteq R(\mathbf{X}_{\mathcal{T}})$, and hence $P_{R(\mathbf{1}_N)^\perp}[R(\mathbf{X}_{\mathcal{T}})] = R(\mathbf{X}_{\mathcal{T}}) \ominus R(\mathbf{1}_N)$. Likewise, $P_{R(\mathbf{1}_N)^\perp}[R(\mathbf{X}_{\mathcal{B}})] = R(\mathbf{X}_{\mathcal{B}}) \ominus R(\mathbf{1}_N)$.

We say that $\mathcal{T}$ and $\mathcal{B}$ satisfy the condition of *proportional frequencies* if

$$n_{ij} = \frac{n_{i+}n_{+j}}{N} \text{ for all } i, j; \tag{2.32}$$

here n_{ij}/n_{i+} does not depend on i, and n_{ij}/n_{+j} does not depend on j. In particular, if n_{ij} is a constant for all i and j, then (2.32) holds; in the context of block designs, this means that all the treatments appear the same number of times in each block.

Theorem 2.5. *Factors $\mathcal{T}$ and $\mathcal{B}$ satisfy the condition of proportional frequencies if and only if $R(\mathbf{X}_{\mathcal{T}}) \ominus R(\mathbf{1}_N)$ is orthogonal to $R(\mathbf{X}_{\mathcal{B}}) \ominus R(\mathbf{1}_N)$.*

Proof. We first note that (2.31) is equivalent to

$$\left(\mathbf{P}_{R(\mathbf{1}_N)^\perp}\mathbf{X}_{\mathcal{T}}\right)^T \mathbf{P}_{R(\mathbf{1}_N)^\perp}\mathbf{X}_{\mathcal{B}} = \mathbf{0}. \tag{2.33}$$

Since $\mathbf{P}_{R(\mathbf{1}_N)} = \mathbf{1}_N(\mathbf{1}_N^T\mathbf{1}_N)^{-1}\mathbf{1}_N^T = \frac{1}{N}\mathbf{J}_N$, where $\mathbf{J}_N$ is the $N \times N$ matrix of 1's, we have $\mathbf{P}_{R(\mathbf{1}_N)^\perp} = \mathbf{I} - \frac{1}{N}\mathbf{J}_N$. Then (2.33) is the same as

$$\mathbf{X}_T^T \left(\mathbf{I} - \frac{1}{N}\mathbf{J}_N\right)\mathbf{X}_\mathcal{B} = \mathbf{0}$$

or

$$\mathbf{X}_T^T\mathbf{X}_\mathcal{B} = \frac{1}{N}\mathbf{X}_T^T\mathbf{J}_N\mathbf{X}_\mathcal{B}. \tag{2.34}$$

By (2.28), the left-hand side of (2.34) is $\mathbf{N}$. The (i,j)th entry of the right-hand side is $n_{i+}n_{+j}/N$. Therefore (2.33) holds if and only if the two factors satisfy the condition of proportional frequencies. $\qquad\square$

By the results in Section 2.7 and Theorem 2.5, if the treatment and block factors satisfy the condition of proportional frequencies, then the least squares estimator of any contrast $\sum_{i=1}^t c_i\alpha_i$ is the same as that under the one-way layout

$$y_{ijh} = \mu + \alpha_i + \varepsilon_{ijh}.$$

So $\sum_{i=1}^t c_i\widehat{\alpha}_i = \sum_{i=1}^t c_iy_{i..}$, with variance $\sigma^2(\sum_{i=1}^t c_i^2/n_{i+})$. Likewise, the least squares estimator of any contrast $\sum_{j=1}^b d_j\beta_j$ is $\sum_{j=1}^b d_jy_{.j.}$, with variance $\sigma^2(\sum_{j=1}^b d_j^2/n_{+j})$.

In this case, we have the decomposition

$$R^N = R(\mathbf{1}_N) \oplus [R(\mathbf{X}_T) \ominus R(\mathbf{1}_N)] \oplus [R(\mathbf{X}_\mathcal{B}) \ominus R(\mathbf{1}_N)] \oplus [R(\mathbf{X}_T) + R(\mathbf{X}_\mathcal{B})]^\perp.$$

Thus

$$\mathbf{y} - \mathbf{P}_{R(\mathbf{1}_N)}\mathbf{y} = \mathbf{P}_{R(\mathbf{X}_T)\ominus R(\mathbf{1}_N)}\mathbf{y} + \mathbf{P}_{R(\mathbf{X}_\mathcal{B})\ominus R(\mathbf{1}_N)}\mathbf{y} + \mathbf{P}_{[R(\mathbf{X}_T)+R(\mathbf{X}_\mathcal{B})]^\perp}\mathbf{y}.$$

Componentwise, we have

$$y_{ijh} - y_{...} = (y_{i..} - y_{...}) + (y_{.j.} - y_{...}) + (y_{ijh} - y_{i..} - y_{.j.} + y_{...}).$$

The identity

$$\left\|\mathbf{y} - \mathbf{P}_{R(\mathbf{1}_N)}\mathbf{y}\right\|^2 = \left\|\mathbf{P}_{R(\mathbf{X}_T)\ominus R(\mathbf{1}_N)}\mathbf{y}\right\|^2 + \left\|\mathbf{P}_{R(\mathbf{X}_\mathcal{B})\ominus R(\mathbf{1}_N)}\mathbf{y}\right\|^2 + \left\|\mathbf{P}_{[R(\mathbf{X}_T)+R(\mathbf{X}_\mathcal{B})]^\perp}\mathbf{y}\right\|^2$$

gives

$$\sum_i\sum_j\sum_h (y_{ijh} - y_{...})^2 = \sum_i n_{i+}(y_{i..} - y_{...})^2 + \sum_j n_{+j}(y_{.j.} - y_{...})^2$$
$$+ \sum_i\sum_j\sum_h (y_{ijh} - y_{i..} - y_{.j.} + y_{...})^2.$$

This leads to a single ANOVA table.

source	sum of squares	d.f.	mean square
Blocks	$\sum_j n_{+j}(y_{\cdot j\cdot} - y_{\cdots})^2$	$b-1$	$\frac{1}{b-1}\sum_j n_{+j}(y_{\cdot j\cdot} - y_{\cdots})^2$
Treatments	$\sum_i n_{i+}(y_{i\cdot\cdot} - y_{\cdots})^2$	$t-1$	$\frac{1}{t-1}\sum_i n_{i+}(y_{i\cdot\cdot} - y_{\cdots})^2$
Residual	By subtraction	$N-b-t+1$	$\frac{1}{N-b-t+1}$(Residual SS)
Total	$\sum_i\sum_j\sum_h(y_{ijh} - y_{\cdots})^2$	$N-1$	

Exercises

2.1 Prove (2.22) and (2.23).

2.2 Suppose $\widehat{\boldsymbol{\theta}}_1$ and $\widehat{\boldsymbol{\theta}}_2$ are the least squares estimators of $\boldsymbol{\theta}_1$ and $\boldsymbol{\theta}_2$ under the model $\mathbf{y} = \mathbf{X}_1\boldsymbol{\theta}_1 + \mathbf{X}_2\boldsymbol{\theta}_2 + \boldsymbol{\varepsilon}$, and $\boldsymbol{\theta}_1^*$ is the least squares estimator of $\boldsymbol{\theta}_1$ under $\mathbf{y} = \mathbf{X}_1\boldsymbol{\theta}_1 + \boldsymbol{\varepsilon}$. Show that $\mathbf{X}_1\boldsymbol{\theta}_1^*$, the prediction of $\mathbf{y}$ based on the model $\mathbf{y} = \mathbf{X}_1\boldsymbol{\theta}_1 + \boldsymbol{\varepsilon}$, is equal to

$$\mathbf{X}_1\widehat{\boldsymbol{\theta}}_1 + \mathbf{X}_1\left(\mathbf{X}_1^T\mathbf{X}_1\right)^{-}\mathbf{X}_1^T\mathbf{X}_2\widehat{\boldsymbol{\theta}}_2.$$

Comment on this result.

2.3 Show that the two matrices $\mathbf{C}_T$ and $\mathbf{C}_B$ defined in Section 2.8 have zero column sums. Therefore, if $\sum_{i=1}^t c_i\alpha_i$ is estimable, then $\sum_{i=1}^t c_i = 0$, and if $\sum_{j=1}^b d_j\beta_j$ is estimable, then $\sum_{j=1}^b d_j = 0$.

2.4 (Optimality of complete block designs) In the setting of Section 2.8, suppose $n_{+j} = k$ for all $j = 1,\ldots,b$. Such a two-way layout can be considered as a block design with b blocks of constant size k. Suppose $\text{rank}(\mathbf{C}_T) = t - 1$.

(a) Let $\mathbf{C}_T = \sum_{i=1}^{t-1} \mu_i\boldsymbol{\xi}_i\boldsymbol{\xi}_i^T$, where $\mu_1,\ldots,\mu_{t-1}$ are the nonzero eigenvalues of $\mathbf{C}_T$, and $\boldsymbol{\xi}_1,\ldots,\boldsymbol{\xi}_{t-1}$ are the associated orthonormal eigenvectors. Show that $\sum_{i=1}^{t-1} \mu_i^{-1}\boldsymbol{\xi}_i\boldsymbol{\xi}_i^T$ is a generalized inverse of $\mathbf{C}_T$.

(b) Use the generalized inverse given in (a) to show that

$$\sum_{1\le i<i'\le t} \text{var}\left(\widehat{\alpha}_i - \widehat{\alpha}_{i'}\right) = \sigma^2 t\sum_{i=1}^{t-1} \mu_i^{-1}.$$

(c) Consider the case $k = t$. In this case, a design with $n_{ij} = 1$ for all i,j is called a complete block design. Show that for a complete block design, $\mu_1 = \cdots = \mu_{t-1}$.

(d) Show that a complete block design maximizes $\text{tr}(\mathbf{C}_T) = \sum_{i=1}^{t-1} \mu_i$ among all the block designs with $n_{+j} = k (= t)$ for all $1 \le j \le b$.

(e) Use (b), (c), (d), and the fact that $f(x) = x^{-1}$ is a convex function to show that a complete block design minimizes $\sum_{1\le i<i'\le t}\text{var}(\widehat{\alpha}_i - \widehat{\alpha}_{i'})$ among all the block designs with $n_{+j} = k (= t)$ for all $1 \le j \le b$.

2.5 (Balanced incomplete block designs and their optimality) Continuing Exercise 2.4, suppose $k < t$.

(a) Show that if $n_{ij} = 0$ or 1 for all i, j, then the ith diagonal entry of $\mathbf{C}_T$ is equal to $(k-1)q_i/k$, and the (i,i')th off-diagonal entry is equal to $-\lambda_{ii'}/k$, where q_i is the number of times the ith treatment appears, and $\lambda_{ii'}$ is the number of blocks in which both the ith and i'th treatments appear.

(b) In addition to $n_{ij} = 0$ or 1 for all i, j, suppose $q_i = q$ for all i, and $\lambda_{ii'} = \lambda$ for all $1 \le i \ne i' \le t$. Such designs are called balanced incomplete block designs; see Section A.8. Use Exercise 2.3 to show that $(t-1)\lambda = (k-1)q$.

(c) Show that under a balanced incomplete block design, $\frac{k}{t\lambda}\mathbf{I}_t$ is a generalized inverse of $\mathbf{C}_T$. Use this to provide an expression for the least squares estimator of a pairwise comparison $\alpha_i - \alpha_{i'}$ of treatment effects, and show that $\mathrm{var}(\widehat{\alpha}_i - \widehat{\alpha}_{i'}) = \frac{2k}{t\lambda}\sigma^2$ for all $1 \le i \ne i' \le t$.

(d) Show that the properties in (c) and (d) of Exercise 2.4 also hold for balanced incomplete block designs, and hence for given b, t, and k, if there exists a balanced incomplete block design, then it minimizes $\sum_{1 \le i < i' \le t} \mathrm{var}(\widehat{\alpha}_i - \widehat{\alpha}_{i'})$ among all the block designs with $n_{+j} = k$ for all $1 \le j \le b$. [Kiefer (1958, 1975)]

2.6   (Löwner ordering of matrices and comparison of experiments) Given two symmetric matrices $\mathbf{A}_1$ and $\mathbf{A}_2$, we say that $\mathbf{A}_1 \ge \mathbf{A}_2$ if $\mathbf{A}_1 - \mathbf{A}_2$ is nonnegative definite. Consider two linear models

$$\text{model 1:}\quad \mathrm{E}(\mathbf{y}) = \mathbf{X}_1\boldsymbol{\theta},\ \mathrm{cov}(\mathbf{y}) = \sigma^2\mathbf{I}_N,$$
$$\text{model 2:}\quad \mathrm{E}(\mathbf{y}) = \mathbf{X}_2\boldsymbol{\theta},\ \mathrm{cov}(\mathbf{y}) = \sigma^2\mathbf{I}_N.$$

Show that $\mathbf{X}_1^T\mathbf{X}_1 \ge \mathbf{X}_2^T\mathbf{X}_2$ if and only if

(a) all linear functions $\mathbf{c}^T\boldsymbol{\theta}$ of $\boldsymbol{\theta}$ that are estimable under model 2 are also estimable under model 1, and

(b) for any $\mathbf{c}^T\boldsymbol{\theta}$ that is estimable under model 2, the variance of its least squares estimator under model 2 is at least as large as that under model 1. [Ehrenfeld (1956)]

2.7   Consider the linear models

$$\text{model 1:}\quad \mathrm{E}(\mathbf{y}) = \mathbf{X}_1\boldsymbol{\theta}_1 + \mathbf{X}_2\boldsymbol{\theta}_2,\ \mathrm{cov}(\mathbf{y}) = \sigma^2\mathbf{I}_N,$$
$$\text{model 2:}\quad \mathrm{E}(\mathbf{y}) = \mathbf{X}_1\boldsymbol{\theta}_1 + \mathbf{X}_2\boldsymbol{\theta}_2 + \mathbf{X}_3\boldsymbol{\theta}_3,\ \mathrm{cov}(\mathbf{y}) = \sigma^2\mathbf{I}_N.$$

Let $\mathbf{C}_1$ and $\mathbf{C}_1^*$ be the information matrices for $\boldsymbol{\theta}_1$ under models 1 and 2, respectively. Show that $\mathbf{C}_1 \ge \mathbf{C}_1^*$, and $\mathbf{C}_1 = \mathbf{C}_1^*$ if (2.26) holds.

2.8   Consider the linear model

$$\mathrm{E}(\mathbf{y}) = \mathbf{X}\boldsymbol{\theta},\ \mathrm{cov}(\mathbf{y}) = \sigma^2\mathbf{I}_N,$$

where $\mathbf{X}$ is $N \times p$ and $\boldsymbol{\theta}$ is $p \times 1$. Suppose $\|\mathbf{x}\| \le a$ for all columns $\mathbf{x}$ of $\mathbf{X}$, where $\|\mathbf{x}\|^2 = \mathbf{x}^T\mathbf{x}$. Show that for any unknown parameter θ_i, $\mathrm{var}(\widehat{\theta}_i) \ge \frac{1}{a^2}\sigma^2$, and that the equality is attained if $\mathbf{X}^T\mathbf{X} = a^2\mathbf{I}_p$. In particular, if $|x_{ij}| \le 1$ for all the entries x_{ij} of $\mathbf{X}$, then $\mathrm{var}(\widehat{\theta}_i) \ge \frac{1}{N}\sigma^2$, and the equality is attained if $\mathbf{X}^T\mathbf{X} = N\mathbf{I}_p$. Such a matrix must have all its entries equal to 1 or -1. (An $N \times N$ matrix $\mathbf{X}$ with 1 and -1 entries such that $\mathbf{X}^T\mathbf{X} = N\mathbf{I}_N$ is called a *Hadamard matrix*.)

Chapter 3

Randomization and Blocking

We begin with a discussion of randomization and blocking, and present statistical models for completely randomized designs, randomized block designs, row-column designs, nested row-column designs, and blocked split-plot designs. Based on an assumption of additivity between treatments and experimental units, these models can be justified by appropriate randomizations that preserve the block structures (Grundy and Healy, 1950; Nelder, 1965a; Bailey, 1981, 1991). The same approach can be applied to experiments with more general block structures, including all the simple block structures, to be discussed in Chapter 12.

3.1 Randomization

Throughout this book, we denote the set of experimental units by Ω and the set of treatments by $\mathfrak{T}$. We also denote the number of treatments and the number of experimental units by t and N, respectively. The units can be labeled by integers $1, \ldots, N$, and the treatments are labeled by $1, \ldots, t$. For any finite set A, we denote the number of elements in A by $|A|$.

Under complete randomization, the allocation of treatments to the units is completely random. Suppose the lth treatment is to be assigned to r_l units, where $l = 1, \ldots, t$, with $\sum_{l=1}^{t} r_l = N$. One can choose r_1 of the N units randomly and assign to them the first treatment, then choose r_2 of the remaining $N - r_1$ units randomly and assign to them the second treatment, etc. A mathematically equivalent method that can be generalized conveniently to more complicated block structures is to start with an arbitrary initial design in which the lth treatment is assigned to r_l unit labels, $l = 1, \ldots, t$. One permutation of the N unit labels is then drawn randomly from the $N!$ possible permutations. The chosen random permutation is used to determine which r_l units will actually receive the lth treatment.

In general, given an initial assignment of the treatment labels to unit labels, a random permutation of the unit labels is applied to obtain a randomized experimental plan for use in actual experimentation. The experimental units are considered to be unstructured under complete randomization. When they have a certain structure, we restrict to permutations of unit labels that preserve the structure. Such permutations are called *allowable permutations*. Randomization is carried out by drawing a permutation randomly from the set of allowable permutations only.

Each initial design can be described by an $N \times t$ unit-treatment incidence matrix

$\mathbf{X}_T$ whose rows correspond to unit labels and columns correspond to treatment labels, with the (w, l)th entry equal to 1 if the lth treatment label is assigned to the wth unit label, and 0 otherwise. For example, under the completely randomized design described above, $\mathbf{X}_T$ is an arbitrary $N \times t$ matrix with 0 and 1 entries such that there is one 1 in each row and r_l 1's in the lth column.

3.2   Assumption of additivity and models for completely randomized designs

We make the assumption of additivity between treatments and units: an observation on unit w when treatment l is applied there, $1 \leq w \leq N$, $1 \leq l \leq t$, is assumed to be

$$\alpha_l + \delta_w, \tag{3.1}$$

where α_l is an unknown constant representing the effect of the lth treatment, and δ_w is the effect of the wth unit. The treatment effects $\alpha_1, \ldots, \alpha_t$ are parameters of interest. We allow δ_w to be a random variable, incorporating measurement errors, uncontrolled variations, etc. It is also assumed that δ_w has a finite variance and $\text{cov}(\delta_w, \delta_{w'})$ only depends on the units.

Let $\phi(w), 1 \leq \phi(w) \leq t$, be the treatment applied to the wth unit, and y_w be the observation on that unit. Then y_w can be expressed as

$$y_w = \alpha_{\phi(w)} + \delta_w, w = 1, \ldots, N. \tag{3.2}$$

Some simplifying assumptions on the joint distribution of the δ_w's are needed to carry out statistical inference for the treatment effects. A common assumption for completely randomized designs is that $\delta_1, \ldots, \delta_N$ are uncorrelated and identically distributed. Under this assumption, with $\mu = \text{E}(\delta_w)$, $\sigma^2 = \text{var}(\delta_w)$, and $\varepsilon_w = \delta_w - \mu$, we have

$$y_w = \mu + \alpha_{\phi(w)} + \varepsilon_w, w = 1, \ldots, N, \tag{3.3}$$

where

$$\text{E}(\varepsilon_w) = 0 \text{ and } \text{var}(\varepsilon_w) = \sigma^2, \ \text{cov}(\varepsilon_w, \varepsilon_{w'}) = 0 \text{ for all } w \neq w'. \tag{3.4}$$

This is the usual one-way layout model discussed in Section 2.4.

A more general model replaces the assumption that the ε_w's are uncorrelated with

$$\text{cov}(\varepsilon_w, \varepsilon_{w'}) = \rho\sigma^2 \text{ for all } w \neq w'.$$

In matrix notation, we have

$$\mathbf{y} = \mu\mathbf{1}_N + \mathbf{X}_T\boldsymbol{\alpha} + \boldsymbol{\varepsilon}, \tag{3.5}$$

and

$$\text{E}(\boldsymbol{\varepsilon}) = \mathbf{0}, \mathbf{V} = \text{cov}(\boldsymbol{\varepsilon}) = \sigma^2(1 - \rho)\mathbf{I}_N + \rho\sigma^2\mathbf{J}_N. \tag{3.6}$$

In the future, we will omit the subscripts in $\mathbf{I}_N$ and $\mathbf{J}_N$ when the dimension is obvious from the context. Note that model (3.3)–(3.4) is a special case with $\rho = 0$.

Model (3.5)–(3.6) can be justified by complete randomization; see Section 3.6.

3.3   Randomized block designs

Suppose the experimental units are grouped into b blocks of size k. Given an initial assignment of the treatments to the bk unit labels, randomization is carried out by randomly permuting the unit labels within each block (done independently from block to block), and also randomly permuting the block labels. This is equivalent to drawing a permutation randomly, not from all the $N!$ permutations of the $N = bk$ unit labels as in the case of complete randomization, but from the $(b!)(k!)^b$ allowable permutations that preserve the structure of the units: two unit labels are in the same block if and only if they remain in the same block after permutation.

A commonly used linear model for block designs assumes that the δ_w's in (3.2) are uncorrelated random variables with a constant variance and that $E(\delta_w)$ only depends on the block to which the unit belongs. Let y_{ij}, $i = 1, \ldots, b$, $j = 1, \ldots, k$, be the observation on the jth unit in the ith block. Then such a model can be expressed as

$$y_{ij} = \mu + \alpha_{\phi(i,j)} + \beta_i + \varepsilon_{ij}, \tag{3.7}$$

where $\phi(i, j)$ is the treatment applied to the jth unit in the ith block, $\beta_1, \cdots, \beta_b$ are unknown constants, and $\{\varepsilon_{ij}\}$ are mutually uncorrelated random variables with

$$E(\varepsilon_{ij}) = 0 \text{ and } \mathrm{var}(\varepsilon_{ij}) = \sigma^2. \tag{3.8}$$

This is the usual *fixed-effect* additive two-way layout model discussed in Section 2.8.

Another model assumes that $E(\delta_w) = 0$ and $\mathrm{var}(\delta_w) = \sigma^2$, but $\mathrm{cov}(\delta_w, \delta_{w'}) = \rho_1 \sigma^2$ for any two units w and w' in the same block, and $\mathrm{cov}(\delta_w, \delta_{w'}) = \rho_2 \sigma^2$ if units w and w' are in different blocks. Such a model can be written as

$$y_{ij} = \mu + \alpha_{\phi(i,j)} + \delta_{ij}, \tag{3.9}$$

where

$$E(\delta_{ij}) = 0, \; \mathrm{cov}(\delta_{ij}, \delta_{i'j'}) = \begin{cases} \sigma^2, & \text{if } i = i', \, j = j'; \\ \rho_1 \sigma^2, & \text{if } i = i', \, j \neq j'; \\ \rho_2 \sigma^2, & \text{if } i \neq i'. \end{cases} \tag{3.10}$$

Typically the units in the same block are more alike than those in different blocks; then one would have

$$\rho_1 > \rho_2. \tag{3.11}$$

The patterns-of-covariance form of (3.10) arises, e.g., when $\{\beta_i\}$ and $\{\varepsilon_{ij}\}$ in (3.7) are mutually uncorrelated random variables with

$$E(\beta_i) = E(\varepsilon_{ij}) = 0, \text{ and } \mathrm{var}(\beta_i) = \sigma_B^2, \; \mathrm{var}(\varepsilon_{ij}) = \sigma_{\mathcal{E}}^2. \tag{3.12}$$

Let $\delta_{ij} = \beta_i + \varepsilon_{ij}$. Then (3.9) and (3.10) hold with

$$\sigma^2 = \sigma_B^2 + \sigma_{\mathcal{E}}^2, \; \rho_1 = \frac{\sigma_B^2}{\sigma_B^2 + \sigma_{\mathcal{E}}^2}, \text{ and } \rho_2 = 0. \tag{3.13}$$

In this case, we do have (3.11).

We show in Section 3.6 that model (3.9)–(3.10) can be justified by randomization.

Remark 3.1. Both the model induced by randomization and model (3.7) with random effects satisfying (3.12) can be expressed in the same mathematical form as in (3.9) and (3.10). However, they are philosophically different. For example, under the latter, we have $\rho_1 > \rho_2$. Such a constraint, however, may not hold for a general randomization model.

Remark 3.2. The between-block randomization is possible only if all the blocks are of the same size. For unequal block sizes, (3.10) cannot be justified by randomization and has to be assumed.

3.4 Randomized row-column designs

Suppose the experimental units are arranged in r rows and c columns. Given an initial assignment of the treatment labels to the rc unit labels, randomization can be carried out by randomly permuting the row labels and, independently, randomly permuting the column labels. This is equivalent to drawing a permutation of the unit labels randomly from the $r!c!$ allowable permutations: two unit labels are in the same row (column) if and only if they remain in the same row (column, respectively) after permutation.

As in the case of block designs, we present two commonly used models for row-column designs, one with fixed and the other with random row and column effects. Let y_{ij}, $i = 1, \ldots, r$, $j = 1, \ldots, c$, be the observation on the unit at the intersection of the ith row and jth column. Then a model with fixed row and column effects assumes that

$$y_{ij} = \mu + \alpha_{\phi(i,j)} + \beta_i + \gamma_j + \varepsilon_{ij}, \tag{3.14}$$

where $\phi(i, j)$ is the treatment assigned to the unit at the intersection of the ith row and the jth column, β_i and γ_j are unknown constants, and $\{\varepsilon_{ij}\}$ are mutually uncorrelated random variables with

$$E(\varepsilon_{ij}) = 0 \text{ and } \text{var}(\varepsilon_{ij}) = \sigma^2. \tag{3.15}$$

Another model assumes that

$$y_{ij} = \mu + \alpha_{\phi(i,j)} + \delta_{ij}, \tag{3.16}$$

where

$$E(\delta_{ij}) = 0, \ \text{cov}(\delta_{ij}, \delta_{i'j'}) = \begin{cases} \sigma^2, & \text{if } i = i', \ j = j'; \\ \rho_1\sigma^2, & \text{if } i = i', \ j \neq j'; \\ \rho_2\sigma^2, & \text{if } i \neq i', \ j = j'; \\ \rho_3\sigma^2, & \text{if } i \neq i', \ j \neq j'. \end{cases} \tag{3.17}$$

That is, all the observations have the same variance, and there are three correlations ρ_1, ρ_2, and ρ_3 depending on whether the two observations involved are in the same row and different columns, same column and different rows, or different rows and different columns. Typically, we expect to have

$$\rho_1 > \rho_3 \text{ and } \rho_2 > \rho_3. \tag{3.18}$$

For example, if $\{\beta_i\}$, $\{\gamma_j\}$, and $\{\varepsilon_{ij}\}$ in (3.14) are mutually uncorrelated random variables with

$$E(\beta_i) = E(\gamma_j) = E(\varepsilon_{ij}) = 0, \ \text{var}(\beta_i) = \sigma_{\mathcal{R}}^2, \ \text{var}(\gamma_j) = \sigma_{\mathcal{C}}^2, \ \text{var}(\varepsilon_{ij}) = \sigma_{\mathcal{E}}^2,$$

and $\delta_{ij} = \beta_i + \gamma_j + \varepsilon_{ij}$, then (3.16) and (3.17) hold with

$$\sigma^2 = \sigma_{\mathcal{R}}^2 + \sigma_{\mathcal{C}}^2 + \sigma_{\mathcal{E}}^2, \ \rho_1 = \frac{\sigma_{\mathcal{R}}^2}{\sigma_{\mathcal{R}}^2 + \sigma_{\mathcal{C}}^2 + \sigma_{\mathcal{E}}^2}, \ \rho_2 = \frac{\sigma_{\mathcal{C}}^2}{\sigma_{\mathcal{R}}^2 + \sigma_{\mathcal{C}}^2 + \sigma_{\mathcal{E}}^2}, \ \text{and} \ \rho_3 = 0. \quad (3.19)$$

In this case, we do have (3.18).

We show in Section 3.6 that (3.16)–(3.17) can be justified by randomization.

3.5   Nested row-column designs and blocked split-plot designs

Consider a randomized experiment with the block structure $n_1/(n_2 \times n_3)$. Let y_{ijk} be the observation on the unit at the intersection of the jth row and kth column in the ith block. We assume the following model for such a randomized experiment.

$$y_{ijk} = \mu + \alpha_{\phi(i,j,k)} + \varepsilon_{ijk}, 1 \leq i \leq n_1, 1 \leq j \leq n_2, 1 \leq k \leq n_3, \quad (3.20)$$

where $\phi(i,j,k)$ is the treatment applied to unit (i,j,k), and

$$E(\varepsilon_{ijk}) = 0, \ \text{cov}(\varepsilon_{ijk}, \varepsilon_{i'j'k'}) = \begin{cases} \sigma^2, & \text{if } i = i', j = j', k = k'; \\ \rho_1 \sigma^2, & \text{if } i = i', j = j', k \neq k'; \\ \rho_2 \sigma^2, & \text{if } i = i', j \neq j', k = k'; \\ \rho_3 \sigma^2, & \text{if } i = i', j \neq j', k \neq k'; \\ \rho_4 \sigma^2, & \text{if } i \neq i'. \end{cases} \quad (3.21)$$

Thus there are four correlations depending on whether the two observations involved are in the same row and different columns of the same block, the same column and different rows of the same block, different rows and different columns of the same block, or different blocks. These reflect four different relations between any pair of experimental units.

On the other hand, in an experiment with the block structure $(n_1/n_2)/n_3$, let y_{ijk} be the observation on the kth subplot of the jth whole-plot in the ith block. Then a commonly used model assumes that

$$y_{ijk} = \mu + \alpha_{\phi(i,j,k)} + \varepsilon_{ijk}, 1 \leq i \leq n_1, 1 \leq j \leq n_2, 1 \leq k \leq n_3, \quad (3.22)$$

where $\phi(i,j,k)$ is the treatment applied to unit (i,j,k), and

$$E(\varepsilon_{ijk}) = 0, \ \text{cov}(\varepsilon_{ijk}, \varepsilon_{i'j'k'}) = \begin{cases} \sigma^2, & \text{if } i = i', j = j', k = k'; \\ \rho_1 \sigma^2, & \text{if } i = i', j = j', k \neq k'; \\ \rho_2 \sigma^2, & \text{if } i = i', j \neq j'; \\ \rho_3 \sigma^2, & \text{if } i \neq i'. \end{cases} \quad (3.23)$$

In this case there are three correlations depending on whether the two observations involved are in the same whole-plot, different whole-plots of the same block, or different blocks.

We justify (3.20)–(3.21) and (3.22)–(3.23) by randomization in Section 3.6.

3.6   Randomization model*

Under the additivity assumption (3.1), we show that randomization models for completely randomized designs, randomized block designs, row-column designs, nested row-column designs, and blocked split-plot designs can be described by (3.5)–(3.6), (3.9)–(3.10), (3.16)–(3.17), (3.20)–(3.21), and (3.22)–(3.23), respectively.

Let $\mathfrak{P}$ be a set of permutations of the unit labels. Suppose the randomization is carried out by permuting the unit labels according to a permutation drawn randomly from $\mathfrak{P}$. Denote such a random permutation by Π, so $P(\Pi = \pi) = 1/|\mathfrak{P}|$ for all $\pi \in \mathfrak{P}$. For each $w = 1,\dots,N$, let y_w be the observation that takes place at unit $\Pi(w)$. Then under the additivity assumption (3.1), the randomization model can be written as

$$y_w = \alpha_{\phi(w)} + \sum_{v=1}^{N} \delta_v d_{wv}, w = 1,\dots,N,$$

where $d_{wv} = 1$ if $\Pi(w) = v$, and is 0 otherwise. Note that for each w, $\sum_{v=1}^{N} d_{wv} = 1$.
   Let

$$\eta_w = \sum_{v=1}^{N} \delta_v d_{wv}.$$

Then

$$y_w = \alpha_{\phi(w)} + \eta_w, w = 1,\dots,N. \tag{3.24}$$

We say that $\mathfrak{P}$ is *transitive* if for any two units v and w, there exists at least one permutation $\pi \in \mathfrak{P}$ such that $\pi(w) = v$. For any $\pi_1, \pi_2 \in \mathfrak{P}$, let $\pi_1 * \pi_2$ be the permutation such that $\pi_1 * \pi_2(w) = \pi_1(\pi_2(w))$ for all w. Throughout this section, we assume that $\mathfrak{P}$ is a group under this operation, with the permutation $\pi(w) = w$ for all w as the identity element. For the definition and properties of groups, see Section A.1.
   We first prove a lemma.

Lemma 3.1. *Suppose $\mathfrak{P}$ is a transitive group under $*$. Then η_w, $w = 1,\dots,N$, have identical distributions.*

Proof. For any $1 \leq w \leq N$, let $\mathfrak{P}_w = \{\pi \in \mathfrak{P}: \pi(w) = w\}$. Then it is easy to see that $\mathfrak{P}_w$ is a subgroup of $\mathfrak{P}$. We first show that for any $\pi_1, \pi_2 \in \mathfrak{P}$, $\pi_1(w) = \pi_2(w)$ if and only if $\pi_1 * \mathfrak{P}_w = \pi_2 * \mathfrak{P}_w$, so either both $\pi_1, \pi_2 \in \mathfrak{P}_w$ (in which case both $\pi_1(w)$ and $\pi_2(w)$ are equal to w), or they are in the same coset of $\mathfrak{P}_w$. Suppose $\pi_1 * \mathfrak{P}_w = \pi_2 * \mathfrak{P}_w$. Then $\pi_1 = \pi_2 * \pi$ for some $\pi \in \mathfrak{P}_w$. This implies that $\pi_1(w) = \pi_2(\pi(w)) = \pi_2(w)$. On the other hand, if $\pi_1(w) = \pi_2(w)$, then $(\pi_1^{-1} * \pi_2)(w) = \pi_1^{-1}[\pi_2(w)] = \pi_1^{-1}[\pi_1(w)] = w$. Then $\pi_1^{-1} * \pi_2 \in \mathfrak{P}_w$, so $\pi_1 * \mathfrak{P}_w = \pi_2 * \mathfrak{P}_w$.
   Since all the permutations in $\mathfrak{P}_w$ map w to w, all those in the same coset of $\mathfrak{P}_w$

map w to the same v (which is different from w), and those in different cosets map w to different v's, it follows from the transitivity of $\mathfrak{P}$ that the N unit labels are in one-to-one correspondence with $\mathfrak{P}_w$ and its cosets. Furthermore, since $\mathfrak{P}_w$ and all the cosets are of the same size (see Section A.1), we have $|\mathfrak{P}| = |\mathfrak{P}_w| N$ and that for any $1 \leq v \leq N$, there are $|\mathfrak{P}_w| = \frac{|\mathfrak{P}|}{N}$ permutations $\pi \in \mathfrak{P}$ such that $\pi(w) = v$. Then for any $1 \leq v, w \leq N$, $P(\eta_w = \delta_v) = P(\Pi(w) = v) = \frac{1}{|\mathfrak{P}|} \frac{|\mathfrak{P}|}{N} = \frac{1}{N}$. This implies that all the η_w, $w = 1, \ldots, N$, have identical distributions. $\square$

Next we study the covariance structure of $\eta_1, \ldots, \eta_N$. Given any two pairs (w_1, w_2) and (v_1, v_2) of units, $1 \leq v_1, v_2, w_1, w_2 \leq N$, we say that they are in the same orbit if there is a permutation $\pi \in \mathfrak{P}$ such that $\pi(w_1) = v_1$ and $\pi(w_2) = v_2$. As in the proof of Lemma 3.1, for any $1 \leq w_1, w_2 \leq N$, the set of permutations $\{\pi \in \mathfrak{P} : \pi(w_1) = w_1 \text{ and } \pi(w_2) = w_2\}$ is a subgroup of $\mathfrak{P}$. For any (w_1, w_2), let O be the orbit that contains (w_1, w_2). Then the same kind of argument employed in the proof of Lemma 3.1 can be used to show that the number of permutations in $\mathfrak{P}$ that map (w_1, w_2) to any pair of units (v_1, v_2) in O is equal to $\frac{|\mathfrak{P}|}{|O|}$. Then $P[(\eta_{w_1}, \eta_{w_2}) = (\delta_{v_1}, \delta_{v_2})] = \frac{1}{|\mathfrak{P}|} \frac{|\mathfrak{P}|}{|O|} = \frac{1}{|O|}$ for all $(v_1, v_2) \in O$. This shows that if (w_1, w_2) and (w_1', w_2') are in the same orbit, then (η_{w_1}, η_{w_2}) and $(\eta_{w_1'}, \eta_{w_2'})$ have the same joint distribution. In particular, $\text{cov}(\eta_{w_1}, \eta_{w_2}) = \text{cov}(\eta_{w_1'}, \eta_{w_2'})$.

Since $\eta_1, \ldots, \eta_N$ have identical probability distributions, they have the same mean. Let $\mu = E(\eta_w)$ and $\varepsilon_w = \eta_w - \mu$. Then under the randomization model (3.24), we have

$$y_w = \mu + \alpha_{\phi(w)} + \varepsilon_w, \quad w = 1, \ldots, N, \tag{3.25}$$

where

$$E(\varepsilon_w) = 0, \tag{3.26}$$

and

$$\text{cov}(\varepsilon_{w_1}, \varepsilon_{w_2}) = \text{cov}(\varepsilon_{w_1'}, \varepsilon_{w_2'}) \text{ if } (w_1, w_2) \text{ and } (w_1', w_2') \text{ are in the same orbit.} \tag{3.27}$$

The transitivity of $\mathfrak{P}$ implies that all the (w_1, w_2) with $w_1 = w_2$ form a single orbit. In this case (3.27) reduces to that $\eta_1, \ldots, \eta_N$ have the same variance. That $\eta_1, \ldots, \eta_N$ have the same variance also follows from Lemma 3.1.

We summarize the discussions in the above as a theorem.

Theorem 3.2. *Under the additivity assumption (3.1), if randomization is carried out by drawing a permutation randomly from a set of permutations of the unit labels that is a transitive group, then the randomization model can be expressed as (3.25), with the mean and covariance structure of the error term given by (3.26) and (3.27), respectively.*

Typically $\mathfrak{P}$ consists of the allowable permutations that preserve a block structure. In Chapter 12, the notion of allowable permutations will be precisely defined

and the allowable permutations for simple block structures will be explicitly determined. Under a b/k or $r \times c$ block structure, it is easy to see that the allowable permutations form a transitive group. For b/k, we can map a unit to any other unit in
the same block by a within-block permutation, and to any unit in a different block
by a between-block permutation followed by a within-block permutation. Likewise,
for $r \times c$, we can map a unit to any unit in the same row and a different column
by a column permutation, to any unit in the same column and a different row by a
row permutation, and to any unit in a different row and different column by a column permutation followed by a row permutation. For each experiment with a simple
block structure, since the randomization is carried out by performing the randomization procedure for nesting or crossing at each stage of the block structure formula,
it can be shown by mathematical induction that the allowable permutations form a
transitive group. Therefore Theorem 3.2 applies to all the simple block structures.

Under complete randomization, there are two orbits $\{(v, w) : v = w\}$ and $\{(v, w) :
v \neq w\}$. Under b/k, there are three orbits determined by whether the two units v
and w are equal, different but in the same block, or in different blocks. Under $r \times c$,
there are four orbits determined by whether v and w are equal, in the same row
and different columns, different rows and the same column, or different rows and
different columns. Under $n_1/(n_2 \times n_3)$, there are five orbits determined by whether v
and w are equal, in the same row and different columns of the same block, different
rows and the same column of the same block, different rows and different columns of
the same block, or different blocks. Finally, under $(n_1/n_2)/n_3$, there are four orbits
determined by whether v and w are equal, different but in the same whole-plot, in
different whole-plots of the same block, or in different blocks. These observations
lead to the covariance structures specified in (3.6), (3.10), (3.17), (3.21), and (3.23).

In Sections 12.3 and 12.14, for an arbitrary simple block structure, we will
characterize the orbits of pairs of experimental units under allowable permutations,
thereby determining the associated randomization model explicitly.

Exercises

3.1   Verify the equivalences stated in the first paragraphs of Sections 3.3 and 3.4.

3.2   Verify (3.19).

3.3   Describe a model with random effects that will lead to (3.20) and (3.21). What
      constraints does such a model impose on ρ_1, ρ_2, ρ_3, and ρ_4? Repeat for model
      (3.22)–(3.23).

3.4   Describe the covariance structure of an appropriate randomization model for
      the experiment discussed in Example 1.1.

3.5   Show that there are four orbits of pairs of experimental units for the block
      structure $n_1/(n_2/n_3)$ as described in Section 3.6. Repeat for the block structure
      $n_1/(n_2 \times n_3)$.

Chapter 4

Factors

Each block or treatment structure can be described by a set of factors defined on the experimental units or treatments, respectively. In this chapter we review the concept of factors and gather some useful mathematical results that are needed throughout the book. For two factors defined on the same set, we have given in Section 2.9 a characterization of the condition of proportional frequencies in terms of the orthogonality of some vector spaces associated with the factors. This is further extended to a notion of orthogonal factors. Hasse diagrams are introduced as a graphical tool for representing block (or treatment) structures. The infimum and supremum of two factors are defined in Section 4.6. These concepts and the notion of orthogonal factors are not needed until Chapter 12 (except the optional Section 6.8). The other material in Sections 4.6 and 4.7 is also not needed until Chapter 12. Readers may postpone reading these sections. We mostly follow the terminology and approach in Bailey (1996, 2008).

4.1 Factors as partitions

An $n_{\mathcal{F}}$-level factor $\mathcal{F}$ on Ω can be defined in two equivalent ways. We can think of $\mathcal{F}$ as a function $\mathcal{F} : \Omega \to \{1, \ldots, n_{\mathcal{F}}\}$ that assigns a *level* of $\mathcal{F}$ to each element of Ω, with each of the $n_{\mathcal{F}}$ levels assigned to at least one element of Ω. Equivalently, $\mathcal{F}$ can also be thought of as a partition of Ω into $n_{\mathcal{F}}$ disjoint nonempty subsets, with each of the subsets consisting of the units that have the same level of $\mathcal{F}$. These $n_{\mathcal{F}}$ subsets of Ω are called $\mathcal{F}$-classes. If all the $\mathcal{F}$-classes are of the same size, which must be equal to $N/n_{\mathcal{F}}$, then $\mathcal{F}$ is called a *uniform* factor. For example, under the block structure b/k, each unit belongs to one of b blocks. This defines a b-level uniform factor. The b blocks are a partition of the units into b disjoint classes of size k.

Two factors $\mathcal{F}_1$ and $\mathcal{F}_2$ are said to be *equivalent*, denoted by $\mathcal{F}_1 \equiv \mathcal{F}_2$, if they induce the same partition of the experimental units. If $\mathcal{F}_1$ and $\mathcal{F}_2$ are not equivalent and each $\mathcal{F}_1$-class is contained in some $\mathcal{F}_2$-class, which implies that any two units taking the same level of $\mathcal{F}_1$ also have the same level of $\mathcal{F}_2$, then we say that $\mathcal{F}_1$ is *finer* than $\mathcal{F}_2$ ($\mathcal{F}_2$ is coarser than $\mathcal{F}_1$), or $\mathcal{F}_1$ is *nested* in $\mathcal{F}_2$. We write $\mathcal{F}_1 \prec \mathcal{F}_2$ if $\mathcal{F}_1$ is finer than $\mathcal{F}_2$. We also write $\mathcal{F}_1 \preceq \mathcal{F}_2$ if $\mathcal{F}_1 \prec \mathcal{F}_2$ or $\mathcal{F}_1 \equiv \mathcal{F}_2$. For instance, in Example 1.2, the four weeks define a four-level factor that partitions the 32 units into four disjoint classes of size eight, and the 16 days define a 16-level factor that partitions the units into 16 disjoint classes of size two. Clearly the day factor is nested in the week factor: the two runs on the same day are also in the same week.

The coarsest among all the factors has all the experimental units in one single class. On the other hand, for the finest factor, each class consists of one single unit. We call the former the *universal* factor and the latter the *equality* factor, and denote them by $\mathcal{U}$ and $\mathcal{E}$, respectively.

In this chapter, we state the results mostly in terms of factors defined on the experimental units though they are also applicable to factors defined on any set, including the set of treatments. To distinguish factors defined on the units from those defined on the treatments, we use letters with different fonts to denote them. For example, the universal and equality factors on the set of treatments are denoted by U and E, respectively, whereas we have used $\mathcal{U}$ and $\mathcal{E}$ to denote their counterparts for experimental units. Similarly, A, B, C, etc., are reserved for factors defined on the treatments, and $\mathcal{A}$, $\mathcal{B}$, $\mathcal{C}$, etc., denote factors defined on the experimental units.

4.2 Block structures and Hasse diagrams

The block structure b/k is determined by a bk-level equality factor $\mathcal{E}$ that specifies a total of bk units and a b-level factor $\mathcal{B}$ that partitions the bk units into b classes of size k. We also include the trivial factor $\mathcal{U}$, and say that the structure b/k is defined by $\mathfrak{F} = \{\mathcal{U}, \mathcal{B}, \mathcal{E}\}$. The relation $\prec$ defines the order $\mathcal{E} \prec \mathcal{B} \prec \mathcal{U}$ among the three factors.

In general, a block structure is defined as a set $\mathfrak{F}$ of factors on the experimental units, and we always include $\mathcal{E}$ and $\mathcal{U}$ in every block structure. Thus a set of unstructured units can be regarded as having the block structure $\mathfrak{F} = \{\mathcal{U}, \mathcal{E}\}$.

The structure $r \times c$ is defined by $\{\mathcal{U}, \mathcal{R}, \mathcal{C}, \mathcal{E}\}$, where $\mathcal{R}$ is an r-level factor that partitions the rc units into r rows of size c, and $\mathcal{C}$ is a c-level factor that partitions the units into c columns of size r. We have $\mathcal{E} \prec \mathcal{R} \prec \mathcal{U}$ and $\mathcal{E} \prec \mathcal{C} \prec \mathcal{U}$. The two factors $\mathcal{R}$ and $\mathcal{C}$ are not nested in each other; instead, the intersection of each $\mathcal{R}$-class and each $\mathcal{C}$-class contains exactly one unit.

An $n_1/(n_2/n_3)$ block structure such as the one in Example 1.2 is defined by a set of four factors $\{\mathcal{U}, \mathcal{B}, \mathcal{P}, \mathcal{E}\}$ with $\mathcal{E} \prec \mathcal{P} \prec \mathcal{B} \prec \mathcal{U}$, where $\mathcal{B}$ is an n_1-level factor that partitions the $n_1 n_2 n_3$ units into n_1 blocks of size $n_2 n_3$ and $\mathcal{P}$ is an $n_1 n_2$-level factor that divides the $n_1 n_2 n_3$ units into $n_1 n_2$ whole-plots of size n_3, with n_2 whole-plots in each of the n_1 blocks.

An $n_1/(n_2 \times n_3)$ block structure such as the one in Example 1.3 is defined by $\{\mathcal{U}, \mathcal{B}, \mathcal{R}', \mathcal{C}', \mathcal{E}\}$ with $\mathcal{E} \prec \mathcal{R}' \prec \mathcal{B} \prec \mathcal{U}$ and $\mathcal{E} \prec \mathcal{C}' \prec \mathcal{B} \prec \mathcal{U}$, where $\mathcal{B}$ is an n_1-level factor that partitions the $n_1 n_2 n_3$ units into n_1 blocks of size $n_2 n_3$, $\mathcal{R}'$ is an $n_1 n_2$-level factor that partitions the $n_1 n_2 n_3$ units into $n_1 n_2$ rows of size n_3 with n_2 rows in each block, and $\mathcal{C}'$ is an $n_1 n_3$-level factor that partitions the $n_1 n_2 n_3$ units into $n_1 n_3$ columns of size n_2 with n_3 columns in each block.

A formal definition of Nelder's simple block structures as collections of partitions of the experimental units will be given in Section 12.2.

The relation $\preceq$ defines a *partial order* on each block structure $\mathfrak{F}$ in the sense that

(i) (reflexivity) $\mathcal{F} \preceq \mathcal{F}$ for each $\mathcal{F} \in \mathfrak{F}$;

(ii) (antisymmetry) if $\mathcal{F} \preceq \mathcal{G}$ and $\mathcal{G} \preceq \mathcal{F}$, then $\mathcal{F} \equiv \mathcal{G}$;

(iii) (transitivity) if $\mathcal{F} \preceq \mathcal{G}$ and $\mathcal{G} \preceq \mathcal{H}$, then $\mathcal{F} \preceq \mathcal{H}$.

It is common to represent a partially ordered set by a *Hasse diagram*. In the present context, there is a node for each factor, and if $\mathcal{F} \prec \mathcal{G}$, then the node for $\mathcal{F}$ is below that for $\mathcal{G}$, with a line segment or a sequence of line segments going upwards from $\mathcal{F}$ to $\mathcal{G}$. The following is the Hasse diagram for the block structure b/k:

Figure 4.1 *Hasse diagram for b/k*

There is a node for each of the three factors, with the name of the factor and its number of levels attached to the node. Since $\mathcal{E} \prec \mathcal{B}$, $\mathcal{B}$ sits above $\mathcal{E}$ with a line segment connecting the two nodes; similarly for $\mathcal{B}$ and $\mathcal{U}$.

The block structure $r \times c$ can be represented by the following Hasse diagram:

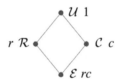

Figure 4.2 *Hasse diagram for r × c*

In this Hasse diagram there is no line segment connecting $\mathcal{R}$ and $\mathcal{C}$ since neither is nested in the other.

The Hasse diagrams for $(n_1/n_2)/n_3$ and $n_1/(n_2 \times n_3)$ are shown in Figure 4.3.

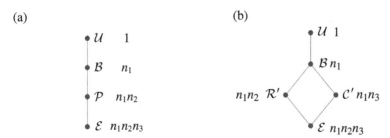

Figure 4.3 *Hasse diagrams for $n_1/n_2/n_3$ and $n_1/(n_2 \times n_3)$*

4.3 Some matrices and spaces associated with factors

The relation between the units and the levels (classes) of a factor $\mathcal{F}$ can be described by an $N \times n_{\mathcal{F}}$ *incidence matrix* $\mathbf{X}_{\mathcal{F}} = [x_{wj}]$, where

$$x_{wj} = \begin{cases} 1, & \text{if the } j\text{th level of } \mathcal{F} \text{ is assigned to unit } w; \\ 0, & \text{otherwise.} \end{cases}$$

The relation between the units with respect to $\mathcal{F}$ can be represented by an $N \times N$ *relation matrix* $\mathbf{R}_{\mathcal{F}} = [r_{vw}]$, where

$$r_{vw} = \begin{cases} 1, & \text{if the same level of } \mathcal{F} \text{ is assigned to units } v \text{ and } w; \\ 0, & \text{otherwise.} \end{cases}$$

The factor $\mathcal{F}$ can be a treatment factor $(\mathcal{T})$, block factor $(\mathcal{B})$, row factor $(\mathcal{R})$, or column factor $(\mathcal{C})$, etc. Then $\mathbf{X}_{\mathcal{F}}$ and $\mathbf{R}_{\mathcal{F}}$ are denoted by $\mathbf{X}_{\mathcal{T}}$, $\mathbf{X}_{\mathcal{B}}$, $\mathbf{X}_{\mathcal{R}}$, $\mathbf{X}_{\mathcal{C}}$ and $\mathbf{R}_{\mathcal{T}}$, $\mathbf{R}_{\mathcal{B}}$, $\mathbf{R}_{\mathcal{R}}$, $\mathbf{R}_{\mathcal{C}}$, and the numbers of levels are denoted by $n_{\mathcal{T}}, n_{\mathcal{B}}, n_{\mathcal{R}}, n_{\mathcal{C}}$, respectively. We also denote $n_{\mathcal{T}}, n_{\mathcal{B}}, n_{\mathcal{R}}$, and $n_{\mathcal{C}}$ by t, b, r, and c, respectively. Note that $\mathbf{X}_{\mathcal{T}}$ and $\mathbf{X}_{\mathcal{B}}$ have been defined in Section 2.8.

Example 4.1. Consider the following assignment of four treatments to four blocks each of size three:

$$
\begin{array}{cc}
\begin{array}{c} 1 \\ 2 \\ 3 \end{array} & \begin{array}{|c|} \hline 3 \\ 2 \\ 1 \\ \hline \end{array}
\end{array}
\quad
\begin{array}{cc}
\begin{array}{c} 4 \\ 5 \\ 6 \end{array} & \begin{array}{|c|} \hline 2 \\ 4 \\ 1 \\ \hline \end{array}
\end{array}
\quad
\begin{array}{cc}
\begin{array}{c} 7 \\ 8 \\ 9 \end{array} & \begin{array}{|c|} \hline 4 \\ 2 \\ 1 \\ \hline \end{array}
\end{array}
\quad
\begin{array}{cc}
\begin{array}{c} 10 \\ 11 \\ 12 \end{array} & \begin{array}{|c|} \hline 1 \\ 4 \\ 3 \\ \hline \end{array}
\end{array},
$$

where the numbers inside each block are treatment labels, and the numbers to the left of each block are unit labels. Then

$$
\mathbf{R}_{\mathcal{B}} = \begin{bmatrix} \mathbf{J}_3 & \mathbf{0} & \mathbf{0} & \mathbf{0} \\ \mathbf{0} & \mathbf{J}_3 & \mathbf{0} & \mathbf{0} \\ \mathbf{0} & \mathbf{0} & \mathbf{J}_3 & \mathbf{0} \\ \mathbf{0} & \mathbf{0} & \mathbf{0} & \mathbf{J}_3 \end{bmatrix} \quad \text{and} \quad \mathbf{X}_{\mathcal{B}} = \begin{bmatrix} \mathbf{1}_3 & \mathbf{0} & \mathbf{0} & \mathbf{0} \\ \mathbf{0} & \mathbf{1}_3 & \mathbf{0} & \mathbf{0} \\ \mathbf{0} & \mathbf{0} & \mathbf{1}_3 & \mathbf{0} \\ \mathbf{0} & \mathbf{0} & \mathbf{0} & \mathbf{1}_3 \end{bmatrix}.
$$

Each row of $\mathbf{X}_{\mathcal{F}}$ has exactly one entry equal to 1 since each unit is in exactly one $\mathcal{F}$-class. Furthermore,

$$\mathbf{R}_{\mathcal{F}} = \mathbf{X}_{\mathcal{F}} \mathbf{X}_{\mathcal{F}}^{T}. \tag{4.1}$$

Therefore $R(\mathbf{X}_{\mathcal{F}}) = R(\mathbf{R}_{\mathcal{F}})$, where $R(\mathbf{X})$ is the column space of $\mathbf{X}$. This follows from (4.1), or can be seen from the fact that $\mathbf{X}_{\mathcal{F}}$ and $\mathbf{R}_{\mathcal{F}}$ consist of the same columns when repetitions are ignored. Let $R(\mathbf{X}_{\mathcal{F}})$ be denoted by $V_{\mathcal{F}}$ ($V_{\mathcal{T}}, V_{\mathcal{B}}, V_{\mathcal{R}}$, or $V_{\mathcal{C}}$ when $\mathcal{F}$ is the treatment, block, row, or column factor, respectively). Then

$$V_{\mathcal{F}} = \{\mathbf{y}: y_v = y_w \text{ if units } v \text{ and } w \text{ are in the same } \mathcal{F}\text{-class}\}.$$

That is, $V_{\mathcal{F}}$ consists of all the vectors whose entries only depend on the levels of $\mathcal{F}$. Clearly $V_{\mathcal{F}}$ is an $n_{\mathcal{F}}$-dimensional subspace of $\mathbb{R}^N$. In particular, for the universal factor $\mathcal{U}$, $\mathbf{X}_{\mathcal{U}} = \mathbf{1}_N$ and $\mathbf{R}_{\mathcal{U}} = \mathbf{J}_N$. This is because all the units are in the same $\mathcal{U}$-class. Therefore $V_{\mathcal{U}}$ is the one-dimensional subspace of $\mathbb{R}^N$ consisting of all the vectors with constant entries. On the other hand, for the equality factor, $\mathbf{R}_{\mathcal{E}} = \mathbf{I}_N$, $V_{\mathcal{E}} = \mathbb{R}^N$, and, for suitably ordered columns, $\mathbf{X}_{\mathcal{E}} = \mathbf{I}_N$.

Example 4.2. (Example 4.1 continued) The space $V_{\mathcal{B}}$ is a four-dimensional subspace of $\mathbb{R}^{12}$ consisting of all the 12×1 vectors $\mathbf{y} = (y_1, \ldots, y_{12})^T$ such that $y_1 = y_2 = y_3$, $y_4 = y_5 = y_6$, $y_7 = y_8 = y_9$, and $y_{10} = y_{11} = y_{12}$. On the other hand, the subspace $V_{\mathcal{T}}$ consists of all the 12×1 vectors $\mathbf{y} = (y_1, \ldots, y_{12})^T$ such that $y_1 = y_{12}$, $y_2 = y_4 = y_8$, $y_3 = y_6 = y_9 = y_{10}$, and $y_5 = y_7 = y_{11}$.

Remark 4.1. Under a completely randomized experiment, the covariance matrix in (3.6) can be expressed as

$$
\begin{aligned}
\mathbf{V} &= \sigma^2 \mathbf{I}_N + \rho \sigma^2 (\mathbf{J}_N - \mathbf{I}_N) \\
&= \sigma^2 \mathbf{R}_{\mathcal{E}} + \rho \sigma^2 (\mathbf{R}_{\mathcal{U}} - \mathbf{R}_{\mathcal{E}}) \\
&= \sigma^2 (1 - \rho) \mathbf{R}_{\mathcal{E}} + \rho \sigma^2 \mathbf{R}_{\mathcal{U}}.
\end{aligned}
$$

In other words, it is a linear combination of the relation matrices of $\mathcal{E}$ and $\mathcal{U}$. Similarly, for the block structure b/k, the covariance matrix in (3.10) can be written as

$$
\mathbf{V} = \sigma^2 \mathbf{I}_N + \rho_1 \sigma^2 (\mathbf{R}_{\mathcal{B}} - \mathbf{I}_N) + \rho_2 \sigma^2 (\mathbf{J}_N - \mathbf{R}_{\mathcal{B}}).
$$

This is because $\mathbf{R}_{\mathcal{B}} - \mathbf{I}_N$ and $\mathbf{J}_N - \mathbf{R}_{\mathcal{B}}$ are $(0, 1)$ matrices with the (v, w)th entry of $\mathbf{R}_{\mathcal{B}} - \mathbf{I}_N$ equal to 1 if and only if $v \neq w$ and units v and w belong to the same block, and the (v, w)th entry of $\mathbf{J}_N - \mathbf{R}_{\mathcal{B}}$ equal to 1 if and only if units v and w belong to different blocks. Since $\mathbf{J}_N = \mathbf{R}_{\mathcal{U}}$ and $\mathbf{I}_N = \mathbf{R}_{\mathcal{E}}$, we see that

$$
\begin{aligned}
\mathbf{V} &= \sigma^2 \mathbf{R}_{\mathcal{E}} + \rho_1 \sigma^2 (\mathbf{R}_{\mathcal{B}} - \mathbf{R}_{\mathcal{E}}) + \rho_2 \sigma^2 (\mathbf{R}_{\mathcal{U}} - \mathbf{R}_{\mathcal{B}}) \\
&= \sigma^2 (1 - \rho_1) \mathbf{R}_{\mathcal{E}} + \sigma^2 (\rho_1 - \rho_2) \mathbf{R}_{\mathcal{B}} + \rho_2 \sigma^2 \mathbf{R}_{\mathcal{U}}.
\end{aligned}
$$

Therefore it can be expressed as a linear combination of $\mathbf{R}_{\mathcal{E}}$, $\mathbf{R}_{\mathcal{B}}$, and $\mathbf{R}_{\mathcal{U}}$, the relation matrices of the three factors that define the block structure. We will show in Section 12.11 that this holds for all the simple block structures and is crucial for proving a key result (Theorem 12.7); see (12.48).

It follows immediately from the definition of $V_{\mathcal{F}}$ that

$$
\mathcal{F}_1 \preceq \mathcal{F}_2 \Rightarrow V_{\mathcal{F}_2} \subseteq V_{\mathcal{F}_1}. \tag{4.2}
$$

Since $\mathcal{U}$ is the coarsest factor, by (4.2), we have

$$
V_{\mathcal{U}} \subseteq V_{\mathcal{F}} \text{ for any factor } \mathcal{F}. \tag{4.3}
$$

Let $V_{\mathcal{F}} \ominus V_{\mathcal{U}}$ be the orthogonal complement of $V_{\mathcal{U}}$ relative to $V_{\mathcal{F}}$. Then $V_{\mathcal{F}} \ominus V_{\mathcal{U}}$, consisting of all the vectors $\mathbf{y}$ in $V_{\mathcal{F}}$ such that $\sum_{w=1}^{N} y_w = 0$, has dimension $n_{\mathcal{F}} - 1$. We denote $V_{\mathcal{F}} \ominus V_{\mathcal{U}}$ by $C_{\mathcal{F}}$, where C stands for *contrast*. This is because each vector in $C_{\mathcal{F}}$ defines a *unit contrast* whose coefficients only depend on the levels of $\mathcal{F}$. Then the two linear spaces $R(\mathbf{X}_T) \ominus R(\mathbf{1}_N)$ and $R(\mathbf{X}_B) \ominus R(\mathbf{1}_N)$ in Theorem 2.5 are, respectively, C_T and C_B.

4.4 Orthogonal projections, averages, and sums of squares

For any factor $\mathcal{F}$ and any $1 \le v \le N$, let $\Omega_v = \{1 \le w \le N : \text{unit } w \text{ is in the same } \mathcal{F}\text{-class as unit } v\}$. Then for each $\mathbf{y}$ in $\mathbb{R}^N$,

$$\text{the } v\text{th entry of } \mathbf{P}_{V_{\mathcal{F}}}\mathbf{y} = \frac{1}{|\Omega_v|} \sum_{w:w \in \Omega_v} y_w. \tag{4.4}$$

So the orthogonal projection $\mathbf{P}_{V_{\mathcal{F}}}\mathbf{y}$ replaces each y_v with the average of the y_w's over all the units w that are in the same $\mathcal{F}$-class as unit v. One simple way to show this is to use the fact that $\mathbf{P}_{V_{\mathcal{F}}} = \mathbf{X}_{\mathcal{F}}(\mathbf{X}_{\mathcal{F}}^T\mathbf{X}_{\mathcal{F}})^{-1}\mathbf{X}_{\mathcal{F}}^T$, and observe that $\mathbf{X}_{\mathcal{F}}^T\mathbf{X}_{\mathcal{F}}$ is the diagonal matrix with the ith diagonal entry equal to the size of the ith $\mathcal{F}$-class.

Example 4.3. (Examples 4.1 and 4.2 continued) Since the four blocks consist of units 1, 2, 3; 4, 5, 6; 7, 8, 9; and 10, 11, 12, respectively, the orthogonal projection of any $\mathbf{y} = (y_1,\ldots,y_{12})^T$ onto V_B is $(u_1, u_1, u_1, u_2, u_2, u_2, u_3, u_3, u_3, u_4, u_4, u_4)^T$, where $u_1 = (y_1 + y_2 + y_3)/3$, $u_2 = (y_4 + y_5 + y_6)/3$, $u_3 = (y_7 + y_8 + y_9)/3$, and $u_4 = (y_{10} + y_{11} + y_{12})/3$ are the *block means*. Similarly, the orthogonal projection of $\mathbf{y}$ onto V_T is $(v_3, v_2, v_1, v_2, v_4, v_1, v_4, v_2, v_1, v_1, v_4, v_3)^T$, where $v_1 = \frac{1}{4}(y_3 + y_6 + y_9 + y_{10})$, $v_2 = \frac{1}{3}(y_2 + y_4 + y_8)$, $v_3 = \frac{1}{2}(y_1 + y_{12})$, and $v_4 = \frac{1}{3}(y_5 + y_7 + y_{11})$ are the *treatment means*. Making the orthogonal projections $\mathbf{P}_{V_B}\mathbf{y}$ and $\mathbf{P}_{V_T}\mathbf{y}$ amounts to computing block and treatment means, respectively. It is easy to see that the entries of $\mathbf{R}_B\mathbf{y}$ are the *block totals*. Since each block contains three units, we have $\mathbf{P}_{V_B} = \frac{1}{3}\mathbf{R}_B$. However, this kind of relation does not hold between $\mathbf{P}_{V_T}$ and $\mathbf{R}_T$ since $\mathcal{T}$ is not a uniform factor.

In general, if $\mathcal{F}$ is a uniform factor, then

$$\mathbf{P}_{V_{\mathcal{F}}} = \frac{n_{\mathcal{F}}}{N}\mathbf{R}_{\mathcal{F}}. \tag{4.5}$$

This is because

$$\mathbf{P}_{V_{\mathcal{F}}} = \mathbf{X}_{\mathcal{F}}\left(\mathbf{X}_{\mathcal{F}}^T\mathbf{X}_{\mathcal{F}}\right)^{-1}\mathbf{X}_{\mathcal{F}}^T = \frac{n_{\mathcal{F}}}{N}\mathbf{X}_{\mathcal{F}}\mathbf{X}_{\mathcal{F}}^T = \frac{n_{\mathcal{F}}}{N}\mathbf{R}_{\mathcal{F}},$$

where the last equality follows from (4.1).

Since $\mathbf{R}_{\mathcal{U}} = \mathbf{J}_N$, by (4.5), the orthogonal projection matrix onto $V_{\mathcal{U}}$ is

$$\mathbf{P}_{V_{\mathcal{U}}} = \frac{1}{N}\mathbf{J}_N. \tag{4.6}$$

Thus, all the entries of $\mathbf{P}_{V_{\mathcal{U}}}\mathbf{y}$ are equal to $y. = \frac{1}{N}\sum_{w=1}^{N} y_w$, the *overall mean*.

The squared length of $\mathbf{P}_{V_{\mathcal{F}}}\mathbf{y}$, called a *sum of squares*, is equal to

$$\left\|\mathbf{P}_{V_{\mathcal{F}}}\mathbf{y}\right\|^2 = \sum_{i=1}^{n_{\mathcal{F}}}(\text{size of the } i\text{th } \mathcal{F}\text{-class})(\text{average of the } y_w\text{'s over the } i\text{th } \mathcal{F}\text{-class})^2.$$

(4.7)

From (4.7), we have

$$\left\|\mathbf{P}_{V_{\mathcal{U}}}\mathbf{y}\right\|^2 = N(y.)^2 = \frac{1}{N}\left(\sum_{w=1}^{N} y_w\right)^2.$$

(4.8)

The orthogonal projection matrix onto $C_{\mathcal{F}} = V_{\mathcal{F}} \ominus V_{\mathcal{U}}$ can be computed as

$$\mathbf{P}_{V_{\mathcal{F}}\ominus V_{\mathcal{U}}} = \mathbf{P}_{V_{\mathcal{F}}} - \mathbf{P}_{V_{\mathcal{U}}}.$$

(4.9)

By the Pythagorean Theorem, we obtain

$$\left\|\mathbf{P}_{C_{\mathcal{F}}}\mathbf{y}\right\|^2 = \left\|\mathbf{P}_{V_{\mathcal{F}}\ominus V_{\mathcal{U}}}\mathbf{y}\right\|^2 = \left\|\mathbf{P}_{V_{\mathcal{F}}}\mathbf{y}\right\|^2 - \left\|\mathbf{P}_{V_{\mathcal{U}}}\mathbf{y}\right\|^2,$$

(4.10)

where $\left\|\mathbf{P}_{V_{\mathcal{F}}}\mathbf{y}\right\|^2$ and $\left\|\mathbf{P}_{V_{\mathcal{U}}}\mathbf{y}\right\|^2$ are given in (4.7) and (4.8), respectively. This is a sum of squares adjusted for the mean.

For instance, by (4.9), a typical entry of $\mathbf{P}_{V_{\mathcal{B}}\ominus V_{\mathcal{U}}}\mathbf{y}$ is the deviation of a block mean from the overall mean. By (4.7), (4.8), and (4.10),

$$\left\|\mathbf{P}_{V_{\mathcal{B}}\ominus V_{\mathcal{U}}}\mathbf{y}\right\|^2 = \sum_{j=1}^{b}\frac{1}{k_j}B_j^2 - \frac{1}{N}\left(\sum_{w=1}^{N} y_w\right)^2,$$

where B_j and k_j are the total and size of the jth block, respectively.

We conclude this section by noting that

$$\mathbf{P}_{V_{\mathcal{F}}^{\perp}} = \mathbf{I} - \mathbf{P}_{V_{\mathcal{F}}}.$$

(4.11)

4.5 Condition of proportional frequencies

Let $\mathcal{F}_1$ and $\mathcal{F}_2$ be two factors on the same set of units. Then the relation between $\mathcal{F}_1$ and $\mathcal{F}_2$ can be described by an $n_{\mathcal{F}_1} \times n_{\mathcal{F}_2}$ incidence matrix

$$\mathbf{N} = \left[n_{ij}\right],$$

where n_{ij} is the number of units in the intersection of the ith $\mathcal{F}_1$-class and the jth $\mathcal{F}_2$-class. It is easy to see that

$$\mathbf{N} = \mathbf{X}_{\mathcal{F}_1}^T \mathbf{X}_{\mathcal{F}_2}.$$

(4.12)

For example, if $\mathcal{F}_1 = \mathcal{T}$ and $\mathcal{F}_2 = \mathcal{B}$, then n_{ij} is the number of times the ith treatment appears in the jth block, as in Section 2.8 where (4.12) has appeared as (2.28). The notion of proportional frequencies can be defined for any two factors. We say that $\mathcal{F}_1$ and $\mathcal{F}_2$ satisfy the condition of proportional frequencies if

$$n_{ij} = \frac{n_{i+}n_{+j}}{N} \text{ for all } i, j,$$

where $n_{i+} = \sum_{j=1}^{n_{\mathcal{F}_2}} n_{ij}$ and $n_{+j} = \sum_{i=1}^{n_{\mathcal{F}_1}} n_{ij}$. Then by Theorem 2.5, $\mathcal{F}_1$ and $\mathcal{F}_2$ satisfy the condition of proportional frequencies if and only if

$$V_{\mathcal{F}_1} \ominus V_{\mathcal{U}} \text{ and } V_{\mathcal{F}_2} \ominus V_{\mathcal{U}} \text{ are orthogonal.} \tag{4.13}$$

4.6   Supremums and infimums of factors

Given two factors $\mathcal{F}_1$ and $\mathcal{F}_2$ on the same set of units, we denote by $\mathcal{F}_1 \vee \mathcal{F}_2$ the factor (partition) such that (i) $\mathcal{F}_1 \preceq \mathcal{F}_1 \vee \mathcal{F}_2$ and $\mathcal{F}_2 \preceq \mathcal{F}_1 \vee \mathcal{F}_2$, and (ii) if $\mathcal{F}_1 \preceq \mathcal{G}$ and $\mathcal{F}_2 \preceq \mathcal{G}$, then $\mathcal{F}_1 \vee \mathcal{F}_2 \preceq \mathcal{G}$. In other words, $\mathcal{F}_1 \vee \mathcal{F}_2$ is the finest factor $\mathcal{G}$ satisfying $\mathcal{F}_1 \preceq \mathcal{G}$ and $\mathcal{F}_2 \preceq \mathcal{G}$. We call $\mathcal{F}_1 \vee \mathcal{F}_2$ the supremum of $\mathcal{F}_1$ and $\mathcal{F}_2$.

The classes of $\mathcal{F}_1 \vee \mathcal{F}_2$ can be determined as follows. Two units u and v are in the same $(\mathcal{F}_1 \vee \mathcal{F}_2)$-class if and only if there is a finite sequence of units

$$u = u_1, u_2, \ldots, u_n = v \tag{4.14}$$

such that for all i, u_i and u_{i+1} are in the same $\mathcal{F}_1$-class or the same $\mathcal{F}_2$-class. We leave the proof of this as an exercise. Alternatively, construct a graph with all the experimental units as its vertices. Connect two vertices by an edge if the corresponding units are in the same $\mathcal{F}_1$-class or in the same $\mathcal{F}_2$-class. Then all the vertices (units) that are in the same connected component of the graph are in the same $(\mathcal{F}_1 \vee \mathcal{F}_2)$-class. Using this, one can easily establish the following result.

Proposition 4.1. *For any two factors $\mathcal{F}_1$ and $\mathcal{F}_2$ on the same set of units, if each $\mathcal{F}_1$-class has a nonempty intersection with every $\mathcal{F}_2$-class, then $\mathcal{F}_1 \vee \mathcal{F}_2 = \mathcal{U}$.*

In particular, $\mathcal{R} \vee C = \mathcal{U}$ under the block structure $r \times c$. Similarly, $\mathcal{R}' \vee C' = \mathcal{B}$ under the $n_1/(n_2 \times n_3)$ block structure; this is because within each block, each row and each column intersect, but all the rows and columns from different blocks share no units. These two properties imply that from any unit in any row of a block, we can move according to (4.14) to any unit in any column of the same block, but we can never move out of the block.

Remark 4.2. Theorem 2.4 can be rephrased as follows: under a fixed-effect additive two-way layout model, all the contrasts of $\alpha_1, \ldots, \alpha_t$ are estimable if and only if $\mathcal{B} \vee \mathcal{T} = \mathcal{U}$.

The infimum of $\mathcal{F}_1$ and $\mathcal{F}_2$, denoted by $\mathcal{F}_1 \wedge \mathcal{F}_2$, is defined as the factor such that (i) $\mathcal{F}_1 \wedge \mathcal{F}_2 \preceq \mathcal{F}_1$ and $\mathcal{F}_1 \wedge \mathcal{F}_2 \preceq \mathcal{F}_2$, and (ii) if $\mathcal{G} \preceq \mathcal{F}_1$ and $\mathcal{G} \preceq \mathcal{F}_2$, then $\mathcal{G} \preceq \mathcal{F}_1 \wedge \mathcal{F}_2$. In other words, $\mathcal{F}_1 \wedge \mathcal{F}_2$ is the coarsest factor $\mathcal{G}$ satisfying $\mathcal{G} \preceq \mathcal{F}_1$ and $\mathcal{G} \preceq \mathcal{F}_2$. It is easy to see from this definition that every $(\mathcal{F}_1 \wedge \mathcal{F}_2)$-class is an intersection of an $\mathcal{F}_1$-class and an $\mathcal{F}_2$-class: two units are in the same class of $\mathcal{F}_1 \wedge \mathcal{F}_2$ if and only if they are in the same class of $\mathcal{F}_1$, and are also in the same class of $\mathcal{F}_2$. Thus, $\mathcal{R} \wedge \mathcal{C} = \mathcal{E}$ under the block structure $r \times c$, and we also have $\mathcal{R}' \wedge \mathcal{C}' = \mathcal{E}$ under $n_1/(n_2 \times n_3)$. As another example, suppose $\Omega = \{(x_1, x_2, x_3) : x_1 = 1, \cdots, n_1, x_2 = 1, \cdots, n_2, x_3 = 1, \cdots, n_3\}$, and $\mathcal{F}_1$ and $\mathcal{F}_2$ are partitions of Ω according to the values of x_1 and x_2, respectively. Then the infimum $\mathcal{F}_1 \wedge \mathcal{F}_2$ of the n_1-level factor $\mathcal{F}_1$ and the n_2-level factor $\mathcal{F}_2$ is the $(n_1 n_2)$-level factor whose levels are all the level combinations of $\mathcal{F}_1$ and $\mathcal{F}_2$.

We note that

$$\text{if } \mathcal{F}_1 \preceq \mathcal{F}_2, \text{ then } \mathcal{F}_1 \vee \mathcal{F}_2 = \mathcal{F}_2 \text{ and } \mathcal{F}_1 \wedge \mathcal{F}_2 = \mathcal{F}_1. \tag{4.15}$$

The supremum of two factors can be characterized geometrically.

Theorem 4.2. *For any two factors $\mathcal{F}_1$ and $\mathcal{F}_2$ on the same set of units, we have $V_{\mathcal{F}_1 \vee \mathcal{F}_2} = V_{\mathcal{F}_1} \cap V_{\mathcal{F}_2}$.*

Proof. Since $\mathcal{F}_1 \preceq \mathcal{F}_1 \vee \mathcal{F}_2$ and $\mathcal{F}_2 \preceq \mathcal{F}_1 \vee \mathcal{F}_2$, by (4.2), $V_{\mathcal{F}_1 \vee \mathcal{F}_2} \subseteq V_{\mathcal{F}_1}$ and $V_{\mathcal{F}_1 \vee \mathcal{F}_2} \subseteq V_{\mathcal{F}_2}$. Therefore $V_{\mathcal{F}_1 \vee \mathcal{F}_2} \subseteq V_{\mathcal{F}_1} \cap V_{\mathcal{F}_2}$.

To show that $V_{\mathcal{F}_1} \cap V_{\mathcal{F}_2} \subseteq V_{\mathcal{F}_1 \vee \mathcal{F}_2}$, let $\mathbf{x}$ be any vector in $V_{\mathcal{F}_1} \cap V_{\mathcal{F}_2}$; we need to show that $\mathbf{x} \in V_{\mathcal{F}_1 \vee \mathcal{F}_2}$. The vector $\mathbf{x}$, say $\mathbf{x} = (x_1, \ldots, x_N)^T$, can be used to define a partition $\mathcal{V}$ of the units: units v and w are in the same $\mathcal{V}$-class if and only if $x_v = x_w$. Then

$$\mathbf{x} \in V_\mathcal{V}. \tag{4.16}$$

Since $\mathbf{x} \in V_{\mathcal{F}_1} \cap V_{\mathcal{F}_2}$, $\mathbf{x}$ has constant entries over all the units in each $\mathcal{F}_1$-class as well as all the units in each $\mathcal{F}_2$-class. This implies that $\mathcal{F}_1 \preceq \mathcal{V}$ and $\mathcal{F}_2 \preceq \mathcal{V}$. By the definition of $\mathcal{F}_1 \vee \mathcal{F}_2$, $\mathcal{F}_1 \vee \mathcal{F}_2 \preceq \mathcal{V}$. Then by (4.2), $V_\mathcal{V} \subseteq V_{\mathcal{F}_1 \vee \mathcal{F}_2}$. It follows from this and (4.16) that $\mathbf{x} \in V_{\mathcal{F}_1 \vee \mathcal{F}_2}$. $\square$

Since $V_\mathcal{U}$ is contained in each of $V_{\mathcal{F}_1}$, $V_{\mathcal{F}_2}$ and $V_{\mathcal{F}_1 \vee \mathcal{F}_2}$, Theorem 4.2 can be rephrased as follows.

Corollary 4.3. *For any two factors $\mathcal{F}_1$ and $\mathcal{F}_2$ on the same set of units, $C_{\mathcal{F}_1 \vee \mathcal{F}_2} = C_{\mathcal{F}_1} \cap C_{\mathcal{F}_2}$.*

4.7 Orthogonality of factors

Theorem 2.5 shows that if $\mathcal{F}_1$ and $\mathcal{F}_2$ satisfy the condition of proportional frequencies, then $V_{\mathcal{F}_1} \ominus V_\mathcal{U}$ and $V_{\mathcal{F}_2} \ominus V_\mathcal{U}$ are orthogonal. This can be extended to a general definition of orthogonal factors.

Definition 4.1. Two factors $\mathcal{F}_1$ and $\mathcal{F}_2$ on the same set of units are said to be orthogonal if

$$V_{\mathcal{F}_1} \ominus (V_{\mathcal{F}_1} \cap V_{\mathcal{F}_2}) \text{ and } V_{\mathcal{F}_2} \ominus (V_{\mathcal{F}_1} \cap V_{\mathcal{F}_2}) \text{ are orthogonal.} \tag{4.17}$$

Since $V_{\mathcal{U}} \subseteq V_{\mathcal{F}_1} \cap V_{\mathcal{F}_2}$, clearly (4.13) implies (4.17).

Proposition 4.4. *If $\mathcal{F}_1$ and $\mathcal{F}_2$ have proportional frequencies, then they are orthogonal.*

In the next theorem we show that (4.17) is equivalent to that in each $(\mathcal{F}_1 \vee \mathcal{F}_2)$-class, $\mathcal{F}_1$ and $\mathcal{F}_2$ have proportional frequencies. This means that for each $(\mathcal{F}_1 \vee \mathcal{F}_2)$-class Γ, if both the ith $\mathcal{F}_1$-class and the jth $\mathcal{F}_2$-class are contained in Γ, then

$$n_{ij} = \frac{n_{i+}n_{+j}}{|\Gamma|}, \tag{4.18}$$

where n_{ij}, n_{i+}, and n_{+j} are as in Section 4.5. For example, the two factors $\mathcal{R}'$ and $\mathcal{C}'$ in the block structure $b/(r \times c)$ do not have proportional frequencies since not every $\mathcal{R}'$-class meets every $\mathcal{C}'$-class, but in each of the b blocks, every row meets every column at exactly one unit. Therefore, $\mathcal{R}'$ and $\mathcal{C}'$ satisfy the condition of proportional frequencies in each class of $\mathcal{B} = \mathcal{R}' \vee \mathcal{C}'$. By the following theorem, they are orthogonal.

Theorem 4.5. *Two factors $\mathcal{F}_1$ and $\mathcal{F}_2$ on the same set of units are orthogonal if and only if the condition of proportional frequencies holds in each $(\mathcal{F}_1 \vee \mathcal{F}_2)$-class.*

Proof. Similar to the proof of Theorem 2.5, we have that $V_{\mathcal{F}_1} \ominus (V_{\mathcal{F}_1} \cap V_{\mathcal{F}_2})$ and $V_{\mathcal{F}_2} \ominus (V_{\mathcal{F}_1} \cap V_{\mathcal{F}_2})$ are orthogonal if and only if

$$\left[\mathbf{P}_{(V_{\mathcal{F}_1} \cap V_{\mathcal{F}_2})^{\perp}} \mathbf{X}_{\mathcal{F}_1} \right]^{T} \left[\mathbf{P}_{(V_{\mathcal{F}_1} \cap V_{\mathcal{F}_2})^{\perp}} \mathbf{X}_{\mathcal{F}_2} \right] = \mathbf{0}. \tag{4.19}$$

Since $V_{\mathcal{F}_1} \cap V_{\mathcal{F}_2} = V_{\mathcal{F}_1 \vee \mathcal{F}_2}$, (4.19) is equivalent to

$$\mathbf{X}_{\mathcal{F}_1}^{T} \left[\mathbf{I} - \mathbf{P}_{V_{\mathcal{F}_1 \vee \mathcal{F}_2}} \right] \mathbf{X}_{\mathcal{F}_2} = \mathbf{0}$$

or

$$\mathbf{X}_{\mathcal{F}_1}^{T} \mathbf{X}_{\mathcal{F}_2} = \mathbf{X}_{\mathcal{F}_1}^{T} \mathbf{P}_{V_{\mathcal{F}_1 \vee \mathcal{F}_2}} \mathbf{X}_{\mathcal{F}_2}. \tag{4.20}$$

We show that (4.20) holds if and only if the condition of proportional frequencies is satisfied in each $(\mathcal{F}_1 \vee \mathcal{F}_2)$-class.

As in the proof of Theorem 2.5, the (i, j)th entry of the left side of (4.20) is n_{ij}. The (i, j)th entry of the right side is $\mathbf{y}_i^{T} \mathbf{P}_{V_{\mathcal{F}_1 \vee \mathcal{F}_2}} \mathbf{z}_j$, where $\mathbf{y}_i$ is the ith column of $\mathbf{X}_{\mathcal{F}_1}$ and $\mathbf{z}_j$ is the jth column of $\mathbf{X}_{\mathcal{F}_2}$. For any $1 \le w \le N$, the wth entry of $\mathbf{y}_i$ (respectively, $\mathbf{z}_j$) is equal to 1 if unit w is in the ith $\mathcal{F}_1$-class (respectively, jth $\mathcal{F}_2$-class); otherwise it is zero. By (4.4), if Γ is the $(\mathcal{F}_1 \vee \mathcal{F}_2)$-class that contains the jth $\mathcal{F}_2$-class, then the wth entry of $\mathbf{P}_{V_{\mathcal{F}_1 \vee \mathcal{F}_2}} \mathbf{z}_j$ is equal to $\frac{n_{+j}}{|\Gamma|}$ if $w \in \Gamma$; otherwise it is zero. A simple calculation shows that if the ith $\mathcal{F}_1$-class is also contained in Γ, then $\mathbf{y}_i^{T} \mathbf{P}_{V_{\mathcal{F}_1 \vee \mathcal{F}_2}} \mathbf{z}_j = \frac{n_{i+}n_{+j}}{|\Gamma|}$. Thus if (4.20) holds, and the ith $\mathcal{F}_1$-class and the jth $\mathcal{F}_2$-class are contained in the same $(\mathcal{F}_1 \vee \mathcal{F}_2)$-class Γ, then (4.18) holds.

Conversely, suppose (4.18) holds for all the pairs (i, j) such that the ith $\mathcal{F}_1$-class and the jth $\mathcal{F}_2$-class are contained in the same $(\mathcal{F}_1 \vee \mathcal{F}_2)$-class. Then for any such pair, the (i, j)th entry of both sides of (4.20) are equal. On the other hand, if the ith $\mathcal{F}_1$-class and the jth $\mathcal{F}_2$-class are in different $(\mathcal{F}_1 \vee \mathcal{F}_2)$-classes, then it is easy to see that the (i, j)th entries of both sides of (4.20) are zero. □

Remark 4.3. Theorem 4.5 can be stated as follows (Bailey, 1996). For each $\omega \in \Omega$, let $\mathcal{F}(\omega)$ be the $\mathcal{F}$-class containing ω. Then two factors $\mathcal{F}_1$ and $\mathcal{F}_2$ defined on Ω are orthogonal if and only if $|(\mathcal{F}_1 \wedge \mathcal{F}_2)(\omega)|\,|(\mathcal{F}_1 \vee \mathcal{F}_2)(\omega)| = |\mathcal{F}_1(\omega)|\,|\mathcal{F}_2(\omega)|$ for all $\omega \in \Omega$. Thus if $\mathcal{F}_1$ and $\mathcal{F}_2$ are orthogonal and uniform, then $\mathcal{F}_1 \vee \mathcal{F}_2$ is uniform if and only if $\mathcal{F}_1 \wedge \mathcal{F}_2$ is uniform.

We have another characterization of the orthogonality of factors.

Theorem 4.6. *Two factors $\mathcal{F}_1$ and $\mathcal{F}_2$ on the same set of units are orthogonal if and only if* $\mathbf{P}_{V_{\mathcal{F}_1}} \mathbf{P}_{V_{\mathcal{F}_2}} = \mathbf{P}_{V_{\mathcal{F}_2}} \mathbf{P}_{V_{\mathcal{F}_1}}$. *In this case,* $\mathbf{P}_{V_{\mathcal{F}_1}} \mathbf{P}_{V_{\mathcal{F}_2}} = \mathbf{P}_{V_{\mathcal{F}_2}} \mathbf{P}_{V_{\mathcal{F}_1}} = \mathbf{P}_{V_{\mathcal{F}_1 \vee \mathcal{F}_2}}$.

Proof. For any $\mathbf{y}$, $\mathbf{P}_{V_{\mathcal{F}_2}} \mathbf{y} = \mathbf{P}_{V_{\mathcal{F}_1} \cap V_{\mathcal{F}_2}} \mathbf{y} + \mathbf{P}_{V_{\mathcal{F}_2} \ominus (V_{\mathcal{F}_1} \cap V_{\mathcal{F}_2})} \mathbf{y}$. Therefore

$$\mathbf{P}_{V_{\mathcal{F}_1}} \mathbf{P}_{V_{\mathcal{F}_2}} \mathbf{y} = \mathbf{P}_{V_{\mathcal{F}_1}} \mathbf{P}_{V_{\mathcal{F}_1} \cap V_{\mathcal{F}_2}} \mathbf{y} + \mathbf{P}_{V_{\mathcal{F}_1}} \mathbf{P}_{V_{\mathcal{F}_2} \ominus (V_{\mathcal{F}_1} \cap V_{\mathcal{F}_2})} \mathbf{y}. \quad (4.21)$$

Since $V_{\mathcal{F}_1} \cap V_{\mathcal{F}_2} \subseteq V_{\mathcal{F}_1}$, the first term on the right side of (4.21) is equal to $\mathbf{P}_{V_{\mathcal{F}_1} \cap V_{\mathcal{F}_2}} \mathbf{y}$. On the other hand, by the orthogonality of $\mathcal{F}_1$ and $\mathcal{F}_2$, (4.17) holds. It follows that $V_{\mathcal{F}_1}$ is orthogonal to $V_{\mathcal{F}_2} \ominus (V_{\mathcal{F}_1} \cap V_{\mathcal{F}_2})$. Then the second term on the right side of (4.21) is $\mathbf{0}$. Therefore $\mathbf{P}_{V_{\mathcal{F}_1}} \mathbf{P}_{V_{\mathcal{F}_2}} \mathbf{y} = \mathbf{P}_{V_{\mathcal{F}_1} \cap V_{\mathcal{F}_2}} \mathbf{y}$ for all $\mathbf{y}$. This proves that $\mathbf{P}_{V_{\mathcal{F}_1}} \mathbf{P}_{V_{\mathcal{F}_2}} = \mathbf{P}_{V_{\mathcal{F}_1} \cap V_{\mathcal{F}_2}} = \mathbf{P}_{V_{\mathcal{F}_1 \vee \mathcal{F}_2}}$. Similarly, we have $\mathbf{P}_{V_{\mathcal{F}_2}} \mathbf{P}_{V_{\mathcal{F}_1}} = \mathbf{P}_{V_{\mathcal{F}_1 \vee \mathcal{F}_2}}$.

To prove the converse, suppose $\mathbf{P}_{V_{\mathcal{F}_1}} \mathbf{P}_{V_{\mathcal{F}_2}} = \mathbf{P}_{V_{\mathcal{F}_2}} \mathbf{P}_{V_{\mathcal{F}_1}}$. Then for any $\mathbf{y}$, we have $\mathbf{P}_{V_{\mathcal{F}_1}} \mathbf{P}_{V_{\mathcal{F}_2}} \mathbf{y} \in V_{\mathcal{F}_1} \cap V_{\mathcal{F}_2}$. If $\mathbf{y} \in V_{\mathcal{F}_1} \ominus (V_{\mathcal{F}_1} \cap V_{\mathcal{F}_2})$, then it follows from $\mathbf{P}_{V_{\mathcal{F}_1}} \mathbf{y} = \mathbf{y}$ that $\mathbf{P}_{V_{\mathcal{F}_2}} \mathbf{y} = \mathbf{P}_{V_{\mathcal{F}_2}} \mathbf{P}_{V_{\mathcal{F}_1}} \mathbf{y} \in V_{\mathcal{F}_1} \cap V_{\mathcal{F}_2}$. Thus for all $\mathbf{y} \in V_{\mathcal{F}_1} \ominus (V_{\mathcal{F}_1} \cap V_{\mathcal{F}_2})$, $\mathbf{P}_{V_{\mathcal{F}_2}} \mathbf{y}$ is orthogonal to $\mathbf{y}$. This is possible only if $\mathbf{y}$ is orthogonal to $V_{\mathcal{F}_2}$. So we have shown that $V_{\mathcal{F}_1} \ominus (V_{\mathcal{F}_1} \cap V_{\mathcal{F}_2})$ is orthogonal to $V_{\mathcal{F}_2}$. It follows that $V_{\mathcal{F}_1} \ominus (V_{\mathcal{F}_1} \cap V_{\mathcal{F}_2})$ and $V_{\mathcal{F}_2} \ominus (V_{\mathcal{F}_1} \cap V_{\mathcal{F}_2})$ are orthogonal. □

The following is a consequence of either Definition 4.1 or Theorem 4.6.

Corollary 4.7. *If $\mathcal{F}_1 \preceq \mathcal{F}_2$, then $\mathcal{F}_1$ and $\mathcal{F}_2$ are orthogonal.*

Corollary 4.8. *All the factors on Ω are orthogonal to $\mathcal{E}$ and $\mathcal{U}$.*

Corollary 4.8 holds since $\mathcal{E}$ and $\mathcal{U}$ are, respectively, the finest and coarsest factors.

The orthogonality of factors in the sense of Definition 4.1 is an important condition that will be imposed on the block structures studied in this book; see Chapter 12.

Example 4.4. All the four block structures we have seen so far, n_1/n_2, $n_1 \times n_2$, $(n_1/n_2)/n_3$, and $n_1/(n_2 \times n_3)$, have mutually orthogonal factors. For n_1/n_2 and $(n_1/n_2)/n_3$, this follows from Corollary 4.7, since $\mathcal{E} \prec \mathcal{B} \prec \mathcal{U}$ under n_1/n_2 and $\mathcal{E} \prec \mathcal{P} \prec \mathcal{B} \prec \mathcal{U}$ under $(n_1/n_2)/n_3$. For $r \times c$, since the intersection of each $\mathcal{R}$-class and each $\mathcal{C}$-class contains exactly one unit, $\mathcal{R}$ and $\mathcal{C}$ have proportional frequencies. This implies that $\mathcal{R}$ and $\mathcal{C}$ are orthogonal. The orthogonality of the other pairs of factors follows from Corollary 4.7. Finally, for $n_1/(n_2 \times n_3)$, the discussion in the paragraph preceding Theorem 4.5 shows that $\mathcal{R}'$ and $\mathcal{C}'$ are orthogonal. The orthogonality of the other pairs of factors also follows from Corollary 4.7.

We need the following result in Section 12.13.

Theorem 4.9. *Let* $\mathcal{F}$*,* $\mathcal{G}$*, and* $\mathcal{H}$ *be three pairwise orthogonal factors on the same set. Then* $\mathcal{F} \vee \mathcal{G}$ *and* $\mathcal{H}$ *are orthogonal.*

Proof. Since $\mathcal{F}$, $\mathcal{G}$, and $\mathcal{H}$ are pairwise orthogonal, by Theorem 4.6, $\mathbf{P}_{V_{\mathcal{F}\vee\mathcal{G}}}\mathbf{P}_{V_{\mathcal{H}}} = \mathbf{P}_{V_{\mathcal{F}}}\mathbf{P}_{V_{\mathcal{G}}}\mathbf{P}_{V_{\mathcal{H}}} = \mathbf{P}_{V_{\mathcal{F}}}\mathbf{P}_{V_{\mathcal{H}}}\mathbf{P}_{V_{\mathcal{G}}} = \mathbf{P}_{V_{\mathcal{H}}}\mathbf{P}_{V_{\mathcal{F}}}\mathbf{P}_{V_{\mathcal{G}}} = \mathbf{P}_{V_{\mathcal{H}}}\mathbf{P}_{V_{\mathcal{F}\vee\mathcal{G}}}$. Once again by Theorem 4.6, $\mathcal{F} \vee \mathcal{G}$ and $\mathcal{H}$ are orthogonal. $\qquad\square$

Exercises

4.1   Prove (4.4).

4.2   Provide an example to show that when both $\mathcal{F}_1$ and $\mathcal{F}_2$ are uniform, $\mathcal{F}_1 \vee \mathcal{F}_2$ and $\mathcal{F}_1 \wedge \mathcal{F}_2$ are not necessarily uniform.

4.3   Provide an example to show that for three pairwise orthogonal factors $\mathcal{F}$, $\mathcal{G}$, and $\mathcal{H}$ on the same set, $\mathcal{F} \wedge \mathcal{G}$ and $\mathcal{H}$ may not be orthogonal.

4.4   Verify that for each of the block structures $r \times c$, $n_1/(n_2 \times n_3)$, and $(n_1/n_2)/n_3$, the covariance matrix in (3.17), (3.21), and (3.23) can be expressed as a linear combination of the relation matrices of the factors that define the block structure.

4.5   Verify the characterizations of $(\mathcal{F}_1 \vee \mathcal{F}_2)$-classes stated in the paragraph preceding Proposition 4.1. Also, show that in (4.14), we can require that $n = 2h$, u_{2j-1} and u_{2j} are in the same $\mathcal{F}_1$-class for all $1 \le j \le h$, and u_{2j} and u_{2j+1} are in the same $\mathcal{F}_2$ class for all $1 \le j \le h-1$. This is the version given on p.173 of Bailey (2008).

4.6   Verify the claim in Remark 4.2.

4.7   Suppose $\Omega = \{\mathbf{x} = (x_1,\dots,x_m)^T : x_i = 1,\dots,n_i, 1 \le i \le m\}$. For each subset S of $\{1,\dots,m\}$, let $\mathcal{F}_S$ be the factor on Ω such that $\mathbf{x}$ and $\mathbf{y}$ are in the same $\mathcal{F}_S$-class if and only if $x_i = y_i$ for all $i \in S$. Show that for any two subsets S_1 and S_2 of $\{1,\dots,m\}$, $\mathcal{F}_{S_1}$ and $\mathcal{F}_{S_2}$ are orthogonal and $\mathcal{F}_{S_1} \vee \mathcal{F}_{S_2} = \mathcal{F}_{S_1 \cap S_2}$.

Chapter 5

Analysis of Some Simple Orthogonal Designs

In this chapter we derive the analysis of three classes of designs under the models presented in Chapter 3: completely randomized designs, randomized complete block designs, and randomized Latin square designs. All of these are examples of orthogonal designs that can be analyzed in a straightforward manner. Under a completely randomized design, the precision of the estimates of treatment contrasts depends on the overall variability of the experimental units. Under a randomized block design, the overall variability is decomposed into between-block variability and within-block variability; we say that there are two *strata* (two error terms or two sources of error). If each treatment appears the same number of times in each block, then the precision of the estimates of treatment contrasts depends only on the within-block variability. Similarly, under a randomized row-column design there are three strata corresponding to between-row variability, between-column variability, and between-unit variability adjusted for the between-row variability and between-column variability. If each treatment appears the same number of times in each row and the same number of times in each column, then the precision of the estimates of treatment contrasts depends only on the adjusted between-unit variability.

5.1 A general result

In Section 2.4, we have shown that under the usual one-way layout model

$$\mathbf{y} = \mu\mathbf{1}_N + \mathbf{X}_T\boldsymbol{\alpha} + \boldsymbol{\varepsilon} \text{ with } \mathrm{E}(\boldsymbol{\varepsilon}) = \mathbf{0} \text{ and } \mathrm{cov}(\boldsymbol{\varepsilon}) = \sigma^2\mathbf{I},$$

the best linear unbiased estimator of a treatment contrast $\sum_{l=1}^{t} c_l\alpha_l$ is $\sum_{l=1}^{t} c_l\overline{T}_l$, where $\overline{T}_l$ is the lth treatment mean, and $\mathrm{var}\left(\sum_{l=1}^{t} c_l\overline{T}_l\right) = \sigma^2 \sum_{l=1}^{t} c_l^2/r_l$, where r_l is the number of replications of the lth treatment. In this case, estimates of treatment contrasts are simply the same contrasts of treatment means, and their standard errors are easy to compute. We will prove that this holds for a more general covariance structure of the random error $\boldsymbol{\varepsilon}$. This generalization provides a sufficient condition that entails simple analyses. It can be applied to model (3.5)–(3.6) for completely randomized designs, model (3.9)–(3.10) for randomized complete block designs, and model (3.16)–(3.17) for randomized Latin square designs. It can also be applied to more general orthogonal designs to be defined and studied in Chapter 13.

51

Throughout this section, we assume that

$$\mathbf{y} = \mu\mathbf{1}_N + \mathbf{X}_T\boldsymbol{\alpha} + \boldsymbol{\varepsilon}, \text{ with } E(\boldsymbol{\varepsilon}) = \mathbf{0} \text{ and } \text{cov}(\boldsymbol{\varepsilon}) = \mathbf{V}. \tag{5.1}$$

Under (5.1), $E(\mathbf{y}) = \mu\mathbf{1}_N + \mathbf{X}_T\boldsymbol{\alpha}$. Since $\mathbf{1}_N \in V_T$ (by (4.3)) and $\mathbf{X}_T\boldsymbol{\alpha} \in V_T$, we have $E(\mathbf{y}) \in V_T$.

For any treatment contrast $\mathbf{c}^T\boldsymbol{\alpha} = \sum_{l=1}^t c_l\alpha_l$, define an $N \times 1$ vector $\mathbf{c}^* = (c_1^*, \ldots, c_N^*)^T$ such that

$$c_w^* = c_{\phi(w)}/r_{\phi(w)}, \ 1 \leq w \leq N, \tag{5.2}$$

where $\phi(w)$ is the treatment label assigned to unit w.

Definition 5.1. Two treatment contrasts $\mathbf{c}^T\boldsymbol{\alpha}$ and $\mathbf{d}^T\boldsymbol{\alpha}$ are said to be orthogonal if $\sum_{l=1}^t c_ld_l/r_l = 0$. In particular, if $r_1 = \cdots = r_t$, then $\mathbf{c}^T\boldsymbol{\alpha}$ and $\mathbf{d}^T\boldsymbol{\alpha}$ are orthogonal if and only if $\sum_{l=1}^t c_ld_l = 0$.

It follows from Definition 5.1 and (5.2) that

$$\mathbf{c}^T\boldsymbol{\alpha} \text{ and } \mathbf{d}^T\boldsymbol{\alpha} \text{ are orthogonal if and only if } (\mathbf{c}^*)^T\mathbf{d}^* = 0. \tag{5.3}$$

We first prove a general result that will be used repeatedly in the book.

Theorem 5.1. *Under (5.1), suppose $\mathbf{c}^T\boldsymbol{\alpha}$ is a treatment contrast such that $\mathbf{c}^*$ is an eigenvector of $\mathbf{V}$, where $\mathbf{c}^*$ is as defined in (5.2); in other words, there is a real number ξ such that $\mathbf{V}\mathbf{c}^* = \xi\mathbf{c}^*$. Then the best linear unbiased estimator of $\mathbf{c}^T\boldsymbol{\alpha}$ is $(\mathbf{c}^*)^T\mathbf{y} = \sum_{l=1}^t c_l\overline{T}_l$. Furthermore, $\text{var}\left(\sum_{l=1}^t c_l\overline{T}_l\right) = \xi\sum_{l=1}^t c_l^2/r_l$.*

Since $\mathbf{V}$ is nonnegative definite, we have $\xi \geq 0$. We already know that the conclusion of Theorem 5.1 holds if $\mathbf{V} = \xi\mathbf{I}$; this is a consequence of the Gauss–Markov Theorem applied to the usual uncorrelated one-way layout model. When $\mathbf{V}\mathbf{c}^* = \xi\mathbf{c}^*$, we have $\mathbf{V}\mathbf{c}^* = (\xi\mathbf{I})\mathbf{c}^*$, so $\mathbf{V}$ behaves like $\xi\mathbf{I}$ in the direction defined by $\mathbf{c}^*$. This is essentially the reason why the conclusion of Theorem 5.1 carries over from the case $\mathbf{V} = \xi\mathbf{I}$ to where $\mathbf{c}^*$ is an eigenvector of $\mathbf{V}$. The following proof is adapted from a proof of the Gauss–Markov Theorem.

Proof. We first note that by the definition of $\mathbf{c}^*$,

$$(\mathbf{c}^*)^T\mathbf{1}_N = \sum_{l=1}^t c_l = 0. \tag{5.4}$$

Let $\mathbf{a}^T\mathbf{y}$ be a linear unbiased estimator of $\mathbf{c}^T\boldsymbol{\alpha}$. Then $E(\mathbf{a}^T\mathbf{y}) = \mathbf{a}^T(\mu\mathbf{1} + \mathbf{X}_T\boldsymbol{\alpha}) = \mu\mathbf{a}^T\mathbf{1} + \mathbf{a}^T\mathbf{X}_T\boldsymbol{\alpha}$, which, by the definition of unbiased estimators, is equal to $\mathbf{c}^T\boldsymbol{\alpha}$ for all μ and $\boldsymbol{\alpha}$. Let $\boldsymbol{\alpha} = 0$; then $\mu\mathbf{a}^T\mathbf{1} = 0$ for all μ. This implies that $\mathbf{a}^T\mathbf{1} = 0$. Then $E(\mathbf{a}^T\mathbf{y}) = \mathbf{a}^T\mathbf{X}_T\boldsymbol{\alpha} = \mathbf{c}^T\boldsymbol{\alpha}$ for all $\boldsymbol{\alpha}$. It follows that $\mathbf{c} = \mathbf{X}_T^T\mathbf{a}$, or,

$$c_l = \sum_{w:\ \phi(w)=l} a_w \text{ for all } l = 1,\ldots,t. \tag{5.5}$$

By (5.5), (4.4), and the definition of $\mathbf{c}^*$,

$$\mathbf{c}^* = \mathbf{P}_{V_T}\mathbf{a}. \tag{5.6}$$

We have $(\mathbf{c}^*)^T\mathbf{y} = \sum_{l=1}^{t} c_l \overline{T}_l$, which is clearly an unbiased estimator of $\mathbf{c}^T\boldsymbol{\alpha}$.

By (5.6) and the assumption that $\mathbf{c}^*$ is an eigenvector of $\mathbf{V}$,

$$\mathbf{V}\mathbf{P}_{V_T}\mathbf{a} = \xi\mathbf{P}_{V_T}\mathbf{a}. \tag{5.7}$$

Now,

$$\begin{aligned}
\text{var}\left(\mathbf{a}^T\mathbf{y}\right) &= \text{var}\left[\left(\mathbf{P}_{V_T}\mathbf{a}\right)^T\mathbf{y}\right] + \text{var}\left[\left(\mathbf{a} - \mathbf{P}_{V_T}\mathbf{a}\right)^T\mathbf{y}\right] \\
&\quad + 2\text{cov}\left[\left(\mathbf{P}_{V_T}\mathbf{a}\right)^T\mathbf{y}, \left(\mathbf{a} - \mathbf{P}_{V_T}\mathbf{a}\right)^T\mathbf{y}\right].
\end{aligned} \tag{5.8}$$

We have

$$\begin{aligned}
\text{cov}\left[\left(\mathbf{P}_{V_T}\mathbf{a}\right)^T\mathbf{y}, \left(\mathbf{a} - \mathbf{P}_{V_T}\mathbf{a}\right)^T\mathbf{y}\right] &= \left(\mathbf{P}_{V_T}\mathbf{a}\right)^T\mathbf{V}\left(\mathbf{a} - \mathbf{P}_{V_T}\mathbf{a}\right) \\
&= \xi\left(\mathbf{P}_{V_T}\mathbf{a}\right)^T\left(\mathbf{a} - \mathbf{P}_{V_T}\mathbf{a}\right) \quad \text{[By (5.7)]} \\
&= 0.
\end{aligned}$$

By this and (5.8), $\text{var}\left(\mathbf{a}^T\mathbf{y}\right) \geq \text{var}\left[(\mathbf{P}_{V_T}\mathbf{a})^T\mathbf{y}\right] = \text{var}\left[(\mathbf{c}^*)^T\mathbf{y}\right]$. Therefore $(\mathbf{c}^*)^T\mathbf{y} = \sum_{l=1}^{t} c_l\overline{T}_l$ is the best linear unbiased estimator of $\mathbf{c}^T\boldsymbol{\alpha}$, with

$$\text{var}\left[(\mathbf{c}^*)^T\mathbf{y}\right] = (\mathbf{c}^*)^T\mathbf{V}\mathbf{c}^* = \xi(\mathbf{c}^*)^T(\mathbf{c}^*) = \xi\sum_{l=1}^{t} c_l^2/r_l.$$

$\square$

In addition to (5.1), now we assume that

$$\mathbf{V}\mathbf{x} = \xi\mathbf{x} \text{ for all } \mathbf{x} \in V_T \ominus V_\mathcal{U}. \tag{5.9}$$

That is, the $(t-1)$-dimensional space $V_T \ominus V_\mathcal{U}$, which corresponds to the space of all the treatment contrasts, is contained in an eigenspace of $\mathbf{V}$ with eigenvalue ξ.

Theorem 5.2. *Under (5.1) and (5.9), the best linear unbiased estimator of any treatment contrast $\sum_{l=1}^{t} c_l\alpha_l$ is $\sum_{l=1}^{t} c_l\overline{T}_l$, and $\text{var}\left(\sum_{l=1}^{t} c_l\overline{T}_l\right) = \xi\sum_{l=1}^{t} c_l^2/r_l$. Furthermore, if $\mathbf{c}^T\boldsymbol{\alpha}$ and $\mathbf{d}^T\boldsymbol{\alpha}$ are orthogonal contrasts, then $\text{cov}\left(\mathbf{c}^T\widehat{\boldsymbol{\alpha}}, \mathbf{d}^T\widehat{\boldsymbol{\alpha}}\right) = 0$, where $\widehat{\boldsymbol{\alpha}} = (\widehat{\alpha}_1, \ldots, \widehat{\alpha}_t)^T$.*

Proof. For any treatment contrast $\mathbf{c}^T\boldsymbol{\alpha}$, by (5.4), $(\mathbf{c}^*)^T\mathbf{1}_N = 0$. Also, by the definition of $\mathbf{c}^*$, $\mathbf{c}^* \in V_T$. Therefore

$$\mathbf{c}^* \in V_T \ominus V_\mathcal{U}. \tag{5.10}$$

By (5.9) and (5.10), the condition in Theorem 5.1 is satisfied by $\mathbf{c}$. It follows that the best linear unbiased estimator of $\mathbf{c}^T\boldsymbol{\alpha}$ and its variance are as given. If $\mathbf{c}^T\boldsymbol{\alpha}$ and $\mathbf{d}^T\boldsymbol{\alpha}$ are orthogonal, then by (5.3), $(\mathbf{c}^*)^T\mathbf{d}^* = 0$. Then $\text{cov}\left(\mathbf{c}^T\widehat{\boldsymbol{\alpha}}, \mathbf{d}^T\widehat{\boldsymbol{\alpha}}\right) = \text{cov}\left[(\mathbf{c}^*)^T\mathbf{y}, (\mathbf{d}^*)^T\mathbf{y}\right] = (\mathbf{c}^*)^T\mathbf{V}\mathbf{d}^* = \xi(\mathbf{c}^*)^T\mathbf{d}^* = 0$. $\square$

By (5.9), there is a subspace W of $V_{\mathcal{U}}^{\perp}$ such that

$$V_T \ominus V_{\mathcal{U}} \subset W, \text{ and } \mathbf{V}\mathbf{x} = \xi\mathbf{x} \text{ for all } \mathbf{x} \in W. \tag{5.11}$$

Such a space W can be decomposed as $W = (V_T \ominus V_{\mathcal{U}}) \oplus [W \ominus (V_T \ominus V_{\mathcal{U}})]$. For any $\mathbf{y} \in \mathbb{R}^N$, we have $\mathbf{P}_W\mathbf{y} = \mathbf{P}_{V_T \ominus V_{\mathcal{U}}}\mathbf{y} + \mathbf{P}_{W \ominus (V_T \ominus V_{\mathcal{U}})}\mathbf{y}$. Then

$$\left\|\mathbf{P}_W\mathbf{y}\right\|^2 = \left\|\mathbf{P}_{V_T \ominus V_{\mathcal{U}}}\mathbf{y}\right\|^2 + \left\|\mathbf{P}_{W \ominus (V_T \ominus V_{\mathcal{U}})}\mathbf{y}\right\|^2. \tag{5.12}$$

This gives an analysis of variance (ANOVA) identity. We call $\left\|\mathbf{P}_{V_T \ominus V_{\mathcal{U}}}\mathbf{y}\right\|^2$ the *treatment sum of squares* and $\left\|\mathbf{P}_{W \ominus (V_T \ominus V_{\mathcal{U}})}\mathbf{y}\right\|^2$ the *residual sum of squares*. By (4.7) and (4.10),

$$\left\|\mathbf{P}_{V_T \ominus V_{\mathcal{U}}}\mathbf{y}\right\|^2 = \sum_{l=1}^{t} r_l \left(\overline{T}_l - y.\right)^2 = \sum_{l=1}^{t} r_l \left(\overline{T}_l\right)^2 - N(y.)^2, \tag{5.13}$$

where $y.$ is the overall mean.

We have the following standard result on the expectations of the two sums of squares on the right side of (5.12).

Proposition 5.3. *Under (5.1) and (5.9),*

(i) $E\left(\left\|\mathbf{P}_{W \ominus (V_T \ominus V_{\mathcal{U}})}\mathbf{y}\right\|^2\right) = [dim(W) - (t-1)]\xi,$

(ii) $E\left(\left\|\mathbf{P}_{V_T \ominus V_{\mathcal{U}}}\mathbf{y}\right\|^2\right) = \sum_{l=1}^{t} r_l(\alpha_l - \overline{\alpha})^2 + (t-1)\xi,$ where $\overline{\alpha} = \frac{1}{N}\sum_{l=1}^{t} r_l\alpha_l.$

Proof. Since $E(\mathbf{y}) \in V_T$ and $W \subset V_{\mathcal{U}}^{\perp}$,

$$\mathbf{P}_{W \ominus (V_T \ominus V_{\mathcal{U}})}E(\mathbf{y}) = \mathbf{0}. \tag{5.14}$$

Consequently,

$$
\begin{aligned}
E\left(\left\|\mathbf{P}_{W \ominus (V_T \ominus V_{\mathcal{U}})}\mathbf{y}\right\|^2\right) &= E\left(\mathbf{y}^T\mathbf{P}_{W \ominus (V_T \ominus V_{\mathcal{U}})}\mathbf{y}\right) \\
&= [E(\mathbf{y})]^T\mathbf{P}_{W \ominus (V_T \ominus V_{\mathcal{U}})}E(\mathbf{y}) + \text{tr}\left(\mathbf{P}_{W \ominus (V_T \ominus V_{\mathcal{U}})}\mathbf{V}\right) \\
&= \text{tr}\left(\mathbf{P}_{W \ominus (V_T \ominus V_{\mathcal{U}})}\mathbf{V}\right) \qquad \text{[By (5.14)]} \\
&= \text{tr}\left[\xi\mathbf{P}_{W \ominus (V_T \ominus V_{\mathcal{U}})}\right] \\
&= dim[W \ominus (V_T \ominus V_{\mathcal{U}})]\xi \\
&= [dim(W) - (t-1)]\xi;
\end{aligned}
$$

$$
\begin{aligned}
E\left(\left\|\mathbf{P}_{V_T \ominus V_{\mathcal{U}}}\mathbf{y}\right\|^2\right) &= E\left(\mathbf{y}^T\mathbf{P}_{V_T \ominus V_{\mathcal{U}}}\mathbf{y}\right) \\
&= [E(\mathbf{y})]^T\mathbf{P}_{V_T \ominus V_{\mathcal{U}}}E(\mathbf{y}) + \text{tr}\left[\xi\mathbf{P}_{V_T \ominus V_{\mathcal{U}}}\right] \\
&= \sum_{l=1}^{t} r_l\left[E\left(\overline{T}_l\right) - E(y.)\right]^2 + (t-1)\xi \\
&= \sum_{l=1}^{t} r_l(\alpha_l - \overline{\alpha})^2 + (t-1)\xi.
\end{aligned}
$$

$\square$

It follows from Proposition 5.3 that if $\dim(W) > t - 1$, then the *residual mean square* $\left\| \mathbf{P}_{W \ominus (V_T \ominus V_\mathcal{U})} \mathbf{y} \right\|^2 / [\dim(W) - (t - 1)]$ is an unbiased estimator of ξ.

5.2 Completely randomized designs

Consider model (3.5)–(3.6) for a completely randomized design. We can write $\mathbf{V}$ as

$$
\begin{aligned}
\mathbf{V} &= \left[\sigma^2 (1 - \rho) \right] \left(\mathbf{I}_N - \frac{1}{N} \mathbf{J}_N \right) + \frac{1}{N} \left[\sigma^2 + (N - 1)\rho\sigma^2 \right] \mathbf{J}_N \\
&= \left[\sigma^2 (1 - \rho) \right] \mathbf{P}_{V_\mathcal{U}^\perp} + \left[\sigma^2 + (N - 1)\rho\sigma^2 \right] \mathbf{P}_{V_\mathcal{U}}, \tag{5.15}
\end{aligned}
$$

where (5.15) follows from (4.6) and (4.11). Since $V_\mathcal{U}$ and $V_\mathcal{U}^\perp$ are orthogonal and $\mathbb{R}^N = V_\mathcal{U} \oplus V_\mathcal{U}^\perp$, the expression in (5.15) gives the spectral decomposition of $\mathbf{V}$, with eigenvalues $\sigma^2 (1 - \rho)$ and $\sigma^2 + (N - 1)\rho\sigma^2$, and the associated eigenspaces $V_\mathcal{U}^\perp$ and $V_\mathcal{U}$, respectively. Denote these eigenspaces and eigenvalues as

$$
W_\mathcal{U} = V_\mathcal{U}, \quad W_\mathcal{E} = V_\mathcal{U}^\perp, \quad \xi_\mathcal{U} = \sigma^2 + (N - 1)\rho\sigma^2, \quad \xi_\mathcal{E} = \sigma^2 (1 - \rho).
$$

Here $\mathcal{U}$ and $\mathcal{E}$ refer to universal and equality factors, respectively. Note that the block structure for a completely randomized experiment is $\mathfrak{F} = \{\mathcal{U}, \mathcal{E}\}$. We see that there is an eigenspace corresponding to each of the two factors in $\mathfrak{F}$. The two eigenspaces are called *strata*, and the dimensions, 1 and $N - 1$, respectively, are called their degrees of freedom.

Remark 5.1. Strictly speaking, $W_\mathcal{U}$ and $W_\mathcal{E}$ are eigenspaces only if $\xi_\mathcal{U} \neq \xi_\mathcal{E}$. For convenience, we abuse the language by calling both $W_\mathcal{U}$ and $W_\mathcal{E}$ eigenspaces and count them as two strata. The same remark applies to all the block structures discussed throughout the book.

Since $V_T \ominus V_\mathcal{U} \subset V_\mathcal{U}^\perp = W_\mathcal{E}$, the condition in Theorem 5.2 is satisfied. Therefore the best linear unbiased estimator of a treatment contrast $\sum_{l=1}^t c_l \alpha_l$ is $\sum_{l=1}^t c_l \overline{T}_l$, with

$$
\mathrm{var}\left(\sum_{l=1}^t c_l \overline{T}_l \right) = \xi_\mathcal{E} \sum_{l=1}^t c_l^2 / r_l. \tag{5.16}
$$

The eigenvalue $\xi_\mathcal{E}$, which plays the same role as σ^2 in the usual one-way layout model with homoscedastic and uncorrelated errors, can be interpreted as follows. We have

$$
\left\| \mathbf{P}_{V_\mathcal{U}^\perp} \mathbf{y} \right\|^2 = \sum_{w=1}^N (y_w - y_\cdot)^2.
$$

If $\alpha_1 = \cdots = \alpha_t$, then $\sum_{w=1}^N (y_w - y_\cdot)^2$, usually called the *total sum of squares*, does not contain the treatment effects and therefore should only reflect the variability among the experimental units. In this case, by the same argument used to prove part (i) of Proposition 5.3, we have

$$E\left(\left\|\mathbf{P}_{V_{\mathcal{U}}^{\perp}}\mathbf{y}\right\|^2\right) = \mathrm{tr}\left(\mathbf{P}_{V_{\mathcal{U}}^{\perp}}\mathbf{V}\right) = \mathrm{tr}\left(\xi_{\mathcal{E}}\mathbf{P}_{V_{\mathcal{U}}^{\perp}}\right) = (N-1)\xi_{\mathcal{E}},$$

where the first equality holds since $E(\mathbf{y}) \in V_{\mathcal{U}}$ when $\alpha_1 = \cdots = \alpha_t$, and the second equality is a consequence of (5.15). Therefore

$$\text{if } \alpha_1 = \cdots = \alpha_t, \text{ then } \xi_{\mathcal{E}} = \frac{1}{N-1}E\left[\sum_{w=1}^{N}(y_w - y.)^2\right]. \tag{5.17}$$

This shows that $\xi_{\mathcal{E}}$ can be thought of as the overall variance among the experimental units. By (5.16), under a completely randomized design, the precision of the estimate of a treatment contrast depends on the overall variability of the experimental units.

Taking the space W in (5.11) to be $V_{\mathcal{U}}^{\perp}$, we have

$$\mathbf{P}_{V_{\mathcal{U}}^{\perp}}\mathbf{y} = \mathbf{P}_{V_{\mathcal{T}} \ominus V_{\mathcal{U}}}\mathbf{y} + \mathbf{P}_{V_{\mathcal{T}}^{\perp}}\mathbf{y},$$

with the breakdown of degrees of freedom

$$N - 1 = (t-1) + (N-t).$$

The ANOVA identity in (5.12) yields the following decomposition of the total sum of squares:

$$\left\|\mathbf{P}_{V_{\mathcal{U}}^{\perp}}\mathbf{y}\right\|^2 = \left\|\mathbf{P}_{V_{\mathcal{T}} \ominus V_{\mathcal{U}}}\mathbf{y}\right\|^2 + \left\|\mathbf{P}_{V_{\mathcal{T}}^{\perp}}\mathbf{y}\right\|^2. \tag{5.18}$$

The first term on the right side, the treatment sum of squares, can be computed as in (5.13). The second term, the residual sum of squares, can be computed as $\sum_{w=1}^{N}\left(y_w - \overline{T}_{\phi(w)}\right)^2$ or by subtraction.

Without the assumption that $\alpha_1 = \cdots = \alpha_t$, by Proposition 5.3,

$$E\left[\sum_{w=1}^{N}\left(y_w - \overline{T}_{\phi(w)}\right)^2\right] = (N-t)\xi_{\mathcal{E}}.$$

Therefore $s^2 = \sum_{w=1}^{N}\left(y_w - \overline{T}_{\phi(w)}\right)^2/(N-t)$ is an unbiased estimator of $\xi_{\mathcal{E}}$, and the standard error of $\sum_{l=1}^{t}c_l\widehat{\alpha}_l$ can be calculated as $s\sqrt{\sum_{l=1}^{t}(c_l^2/r_l)}$.

Similarly, by Proposition 5.3,

$$E\left(\frac{1}{t-1}\left\|\mathbf{P}_{V_{\mathcal{T}} \ominus V_{\mathcal{U}}}\mathbf{y}\right\|^2\right) = \xi_{\mathcal{E}} + \frac{1}{t-1}\left[\sum_{l=1}^{t}r_l(\alpha_l - \overline{\alpha})^2\right].$$

The identity in (5.18) yields the ANOVA for a completely randomized design in Table 5.1, where E(MS) stands for expected mean square. Except for differences in the notations, it is the same as the ANOVA table for one-way layout given in Section 2.4.

Hooper (1989) discussed experimental randomization and the validity of normal-theory inference, and provided an asymptotic result showing that randomization supports the validity of tests for general linear hypotheses.

Table 5.1 *ANOVA table for completely randomized designs*

source	sum of squares	d.f.	MS	E(MS)
Treatments	$\sum_{l=1}^{t} r_l \left(\overline{T}_l - y_.\right)^2$	$t-1$	$\cdots$	$\xi_{\mathcal{E}} + \frac{1}{t-1}\left[\sum_{l=1}^{t} r_l(\alpha_l - \overline{\alpha})^2\right]$
Residual	By subtraction	$N-t$	$\cdots$	$\xi_{\mathcal{E}}$
Total	$\sum_{w=1}^{N}(y_w - y_.)^2$	$N-1$		

5.3   Null ANOVA for block designs

Model (3.9)–(3.10) for a randomized block design can be expressed as

$$\mathbf{y} = \mu\mathbf{1}_N + \mathbf{X}_{\mathcal{T}}\boldsymbol{\alpha} + \boldsymbol{\varepsilon}, \tag{5.19}$$

with

$$E(\boldsymbol{\varepsilon}) = \mathbf{0}, \ \mathbf{V} = \sigma^2\mathbf{I} + \rho_1\sigma^2(\mathbf{R}_{\mathcal{B}} - \mathbf{I}) + \rho_2\sigma^2(\mathbf{J} - \mathbf{R}_{\mathcal{B}}); \tag{5.20}$$

for (5.20), see Remark 4.1. Substituting $\mathbf{R}_{\mathcal{B}}$ and $\mathbf{J}$ by $k\mathbf{P}_{V_{\mathcal{B}}}$ and $N\mathbf{P}_{V_{\mathcal{U}}}$, respectively (see (4.5)), we can write $\mathbf{V}$ as

$$
\begin{aligned}
\mathbf{V} \ = \ & \sigma^2(1-\rho_1)\mathbf{P}_{V_{\mathcal{B}}^{\perp}} + \sigma^2\left[1 + \rho_1(k-1) - \rho_2 k\right]\mathbf{P}_{V_{\mathcal{B}} \ominus V_{\mathcal{U}}} \\
& + \sigma^2\left[1 + \rho_1(k-1) + \rho_2(N-k)\right]\mathbf{P}_{V_{\mathcal{U}}}.
\end{aligned}
$$

Let

$$W_{\mathcal{E}} = V_{\mathcal{B}}^{\perp}, \ W_{\mathcal{B}} = V_{\mathcal{B}} \ominus V_{\mathcal{U}}, \ W_{\mathcal{U}} = V_{\mathcal{U}}, \ \xi_{\mathcal{E}} = \sigma^2(1 - \rho_1),$$

$$\xi_{\mathcal{B}} = \sigma^2\left[1 + \rho_1(k-1) - \rho_2 k\right], \ \xi_{\mathcal{U}} = \sigma^2\left[1 + \rho_1(k-1) + \rho_2(N-k)\right].$$

Then

$$\mathbf{V} = \xi_{\mathcal{E}}\mathbf{P}_{W_{\mathcal{E}}} + \xi_{\mathcal{B}}\mathbf{P}_{W_{\mathcal{B}}} + \xi_{\mathcal{U}}\mathbf{P}_{W_{\mathcal{U}}}. \tag{5.21}$$

Similar to (5.15), (5.21) is the spectral decomposition of $\mathbf{V}$. In this case, there are three eigenspaces $W_{\mathcal{E}}$, $W_{\mathcal{B}}$, and $W_{\mathcal{U}}$ with dimensions $b(k-1)$, $b-1$, and 1, respectively. (See Remark 5.1.) These eigenspaces are called strata. There is one stratum corresponding to each of the three factors $\mathcal{U}$, $\mathcal{B}$, and $\mathcal{E}$ that define the block structure.

Since $\mathbf{y} = \mathbf{P}_{W_{\mathcal{E}}}\mathbf{y} + \mathbf{P}_{W_{\mathcal{B}}}\mathbf{y} + \mathbf{P}_{W_{\mathcal{U}}}\mathbf{y}$,

$$\mathbf{y} - \mathbf{P}_{V_{\mathcal{U}}}\mathbf{y} = \mathbf{y} - \mathbf{P}_{W_{\mathcal{U}}}\mathbf{y} = \mathbf{P}_{W_{\mathcal{E}}}\mathbf{y} + \mathbf{P}_{W_{\mathcal{B}}}\mathbf{y}.$$

This gives a decomposition of the total sum of squares:

$$\left\|\mathbf{P}_{V_{\mathcal{U}}^{\perp}}\mathbf{y}\right\|^2 = \left\|\mathbf{P}_{W_{\mathcal{E}}}\mathbf{y}\right\|^2 + \left\|\mathbf{P}_{W_{\mathcal{B}}}\mathbf{y}\right\|^2. \tag{5.22}$$

The two sums of squares $\left\|\mathbf{P}_{W_{\mathcal{B}}}\mathbf{y}\right\|^2 = \left\|\mathbf{P}_{V_{\mathcal{B}} \ominus V_{\mathcal{U}}}\mathbf{y}\right\|^2 = \left\|\mathbf{P}_{V_{\mathcal{B}}}\mathbf{y}\right\|^2 - \left\|\mathbf{P}_{V_{\mathcal{U}}}\mathbf{y}\right\|^2$ and

$\left\|\mathbf{P}_{W_{\mathcal{E}}}\mathbf{y}\right\|^2 = \left\|\mathbf{P}_{V_{\mathcal{B}}^{\perp}}\mathbf{y}\right\|^2 = \left\|(\mathbf{I} - \mathbf{P}_{V_{\mathcal{B}}})\mathbf{y}\right\|^2$ can be computed by using (4.7). For convenience, index the jth unit in the ith block by double subscripts (i, j). Then

$$\left\|\mathbf{P}_{W_{\mathcal{B}}}\mathbf{y}\right\|^2 = \sum_{i=1}^{b} k(y_{i\cdot} - y_{\cdot\cdot})^2 = \sum_{i=1}^{b} k(y_{i\cdot})^2 - bk(y_{\cdot\cdot})^2,$$

$$\left\|\mathbf{P}_{W_{\mathcal{E}}}\mathbf{y}\right\|^2 = \sum_{i=1}^{b}\sum_{j=1}^{k} (y_{ij} - y_{i\cdot})^2 = \sum_{i=1}^{b}\sum_{j=1}^{k} y_{ij}^2 - k\sum_{i=1}^{b} y_{i\cdot}^2.$$

By the same argument as in the derivation of (5.17), when $\alpha_1 = \cdots = \alpha_t$,

$$\xi_{\mathcal{B}} = \mathrm{E}\left[\frac{1}{b-1}\sum_{i=1}^{b} k(y_{i\cdot} - y_{\cdot\cdot})^2\right], \tag{5.23}$$

and

$$\xi_{\mathcal{E}} = \mathrm{E}\left[\frac{1}{b(k-1)}\sum_{i=1}^{b}\sum_{j=1}^{k} (y_{ij} - y_{i\cdot})^2\right]. \tag{5.24}$$

In view of (5.23) and (5.24), the two eigenvalues $\xi_{\mathcal{B}}$ and $\xi_{\mathcal{E}}$ are expected mean squares measuring between-block variability and within-block variability and are called *interblock* and *intrablock* variance, respectively. The two eigenspaces $W_{\mathcal{B}} = V_{\mathcal{B}} \ominus V_{\mathcal{U}}$ and $W_{\mathcal{E}} = V_{\mathcal{B}}^{\perp}$ are called the interblock and intrablock stratum, respectively. The eigenspace $W_{\mathcal{U}} = V_{\mathcal{U}}$ corresponds to the mean and is called the mean stratum. We often refer to $W_{\mathcal{E}}$ as the *bottom* stratum since $\mathcal{E}$ sits at the bottom of the Hasse diagram.

The identity in (5.22) leads to the *null* ANOVA for a block design in Table 5.2. The null ANOVA, analysis of variance without the treatments, provides an analysis of the variability among the experimental units.

Table 5.2 *Null ANOVA for block designs*

source	sum of squares	d.f.	mean square	E(MS)
Interblock	$\sum_{i=1}^{b} k(y_{i\cdot} - y_{\cdot\cdot})^2$	$b-1$	$\frac{1}{b-1}\sum_{i=1}^{b} k(y_{i\cdot} - y_{\cdot\cdot})^2$	$\xi_{\mathcal{B}}$
Intrablock	$\sum_{i=1}^{b}\sum_{j=1}^{k} (y_{ij} - y_{i\cdot})^2$	$b(k-1)$	$\frac{1}{b(k-1)}\sum_{i=1}^{b}\sum_{j=1}^{k} (y_{ij} - y_{i\cdot})^2$	$\xi_{\mathcal{E}}$
Total	$\sum_{i=1}^{b}\sum_{j=1}^{k} (y_{ij} - y_{\cdot\cdot})^2$	$bk-1$		

If $\mathbf{a}^T\mathbf{y}$ is a contrast of the observations ($\mathbf{a}^T\mathbf{1}_N = 0$), then

$$\mathrm{var}\left(\mathbf{a}^T\mathbf{y}\right) = \mathbf{a}^T\mathbf{V}\mathbf{a} = \mathbf{a}^T\left(\xi_{\mathcal{E}}\mathbf{P}_{W_{\mathcal{E}}} + \xi_{\mathcal{B}}\mathbf{P}_{W_{\mathcal{B}}} + \xi_{\mathcal{U}}\mathbf{P}_{W_{\mathcal{U}}}\right)\mathbf{a} = \xi_{\mathcal{B}}\mathbf{a}^T\mathbf{P}_{W_{\mathcal{B}}}\mathbf{a} + \xi_{\mathcal{E}}\mathbf{a}^T\mathbf{P}_{W_{\mathcal{E}}}\mathbf{a}.$$

This implies that

$$\text{if } \mathbf{a} \in W_{\mathcal{B}}, \text{ then } \mathrm{var}\left(\mathbf{a}^T\mathbf{y}\right) = \xi_{\mathcal{B}}\left\|\mathbf{a}\right\|^2,$$

and

$$\text{if } \mathbf{a} \in W_{\mathcal{E}}, \text{ then var } \left(\mathbf{a}^T \mathbf{y}\right) = \xi_{\mathcal{E}} \left\|\mathbf{a}\right\|^2.$$

When $\mathbf{a} \in W_{\mathcal{B}}$, $\mathbf{a}^T \mathbf{y} = \mathbf{a}^T \mathbf{P}_{W_{\mathcal{B}}} \mathbf{y} = \mathbf{a}^T \left(\mathbf{P}_{V_{\mathcal{B}}} \mathbf{y} - \mathbf{P}_{V_{\mathcal{U}}} \mathbf{y}\right) = \mathbf{a}^T \mathbf{P}_{V_{\mathcal{B}}} \mathbf{y}$ is a contrast of the block means. In this case, we say that $\mathbf{a}^T \mathbf{y}$ is confounded with blocks. On the other hand, if $\mathbf{a} \in W_{\mathcal{E}}$, then we say that $\mathbf{a}^T \mathbf{y}$ is orthogonal to blocks. In successful blocking, $\xi_{\mathcal{E}}$ is less than $\xi_{\mathcal{B}}$; then the normalized contrasts of $\mathbf{y}$ that are orthogonal to blocks have smaller variances than those that are confounded with blocks. We also note that if $\mathbf{a}_1 \in W_{\mathcal{B}}$ and $\mathbf{a}_2 \in W_{\mathcal{E}}$, then $\text{cov}(\mathbf{a}_1^T \mathbf{y}, \mathbf{a}_2^T \mathbf{y}) = 0$. This follows from the orthogonality of $W_{\mathcal{B}}$ and $W_{\mathcal{E}}$.

Under model (3.7) and (3.12) with random block effects, it can be seen, by applying (3.13) and the computation that leads to (5.21), that

$$\xi_{\mathcal{B}} = k\sigma_{\mathcal{B}}^2 + \sigma_{\mathcal{E}}^2, \tag{5.25}$$

and

$$\xi_{\mathcal{E}} = \sigma_{\mathcal{E}}^2. \tag{5.26}$$

We leave this as an exercise. In this case indeed $\xi_{\mathcal{B}} > \xi_{\mathcal{E}}$. For a more general result, see Section 12.12

5.4 Randomized complete block designs

Under a complete block design, $k = t$ and every treatment appears once in every block. When $k = t$ and all the treatment comparisons are of equal importance, intuitively this is the best design. It can be shown that a complete block design is optimal with respect to some statistical criteria. (See Exercise 2.4.) Another advantage of making each treatment appear once in every block is that it results in a simple analysis. This is because the condition of proportional frequencies is satisfied by the treatment and block factors. By Theorem 2.5, $V_T \ominus V_{\mathcal{U}} \perp V_{\mathcal{B}} \ominus V_{\mathcal{U}}$. It follows that $V_T \ominus V_{\mathcal{U}} \subset V_{\mathcal{B}}^{\perp} = W_{\mathcal{E}}$. By Theorem 5.2, the best linear unbiased estimator of a treatment contrast $\sum_{l=1}^{t} c_l \alpha_l$ is $\sum_{l=1}^{t} c_l \overline{T}_l$, with $\text{var}(\sum_{l=1}^{t} c_l \overline{T}_l) = \xi_{\mathcal{E}} \sum_{l=1}^{t} c_l^2 / b$. Thus the precision of the estimator of a treatment contrast depends on the intrablock variability. In this case the information for treatment contrasts is contained in the intrablock stratum, which is expected to have a smaller variance than the interblock stratum.

Since $V_T \ominus V_{\mathcal{U}} \subset W_{\mathcal{E}}$, with the space W in (5.11) equal to $W_{\mathcal{E}}$, the intrablock sum of squares $\left\|\mathbf{P}_{W_{\mathcal{E}}} \mathbf{y}\right\|^2 = \left\|\mathbf{P}_{V_{\mathcal{B}}^{\perp}} \mathbf{y}\right\|^2$ can be decomposed as

$$\left\|\mathbf{P}_{W_{\mathcal{E}}} \mathbf{y}\right\|^2 = \left\|\mathbf{P}_{V_T \ominus V_{\mathcal{U}}} \mathbf{y}\right\|^2 + \left\|\mathbf{P}_{V_{\mathcal{B}}^{\perp} \ominus (V_T \ominus V_{\mathcal{U}})} \mathbf{y}\right\|^2, \tag{5.27}$$

with the breakdown of the degrees of freedom

$$b(t-1) = (t-1) + (b-1)(t-1).$$

Again the treatment sum of squares $\left\|\mathbf{P}_{V_T \ominus V_{\mathcal{U}}} \mathbf{y}\right\|^2$ can be computed as

$\sum_{l=1}^{t} b(\overline{T}_l - y_{..})^2$, where $y_{..} = \frac{1}{bt} \sum_{i=1}^{b} \sum_{j=1}^{t} y_{ij}$, and the residual sum of squares $\left\| \mathbf{P}_{V_{\mathcal{B}}^{\perp} \ominus (V_{\mathcal{T}} \ominus V_{\mathcal{U}})} \mathbf{y} \right\|^2$ can be obtained by subtraction. By Proposition 5.3,

$$\mathrm{E} \left(\left\| \mathbf{P}_{V_{\mathcal{B}}^{\perp} \ominus (V_{\mathcal{T}} \ominus V_{\mathcal{U}})} \mathbf{y} \right\|^2 \right) = (b-1)(t-1)\xi_{\mathcal{E}}.$$

Thus the residual mean square $s^2 = \left\| \mathbf{P}_{V_{\mathcal{B}}^{\perp} \ominus (V_{\mathcal{T}} \ominus V_{\mathcal{U}})} \mathbf{y} \right\|^2 / (b-1)(t-1)$ is an unbiased estimator of $\xi_{\mathcal{E}}$. Similarly,

$$\mathrm{E} \left(\frac{1}{t-1} \left\| \mathbf{P}_{V_{\mathcal{T}} \ominus V_{\mathcal{U}}} \mathbf{y} \right\|^2 \right) = \xi_{\mathcal{E}} + \frac{1}{t-1} \left[\sum_{l=1}^{t} b(\alpha_l - \alpha_.)^2 \right]$$

where $\alpha_. = \frac{1}{t} \sum_{l=1}^{t} \alpha_l$.

Replacing the intrablock sum of squares in the null ANOVA table with the two sums of squares on the right side of (5.27), we have the full ANOVA table for a randomized complete block design in Table 5.3. It is the same as the ANOVA table for the fixed-effect additive two-way layout model given in Section 2.9, specialized to the case where there is a single observation at each combination of the treatment and block factors.

Table 5.3 *ANOVA for randomized complete block designs*

source	sum of squares	d.f.	MS	E(MS)
Blocks	$\sum_{i=1}^{b} t(y_{i.} - y_{..})^2$	$b-1$	$\cdots$	$\xi_{\mathcal{B}}$
Treatments	$\sum_{l=1}^{t} b(\overline{T}_l - y_{..})^2$	$t-1$	$\cdots$	$\xi_{\mathcal{E}} + \frac{1}{t-1} \left[\sum_{l=1}^{t} b(\alpha_l - \alpha_.)^2 \right]$
Residual	By subtraction	$(b-1)(t-1)$	$\cdots$	$\xi_{\mathcal{E}}$
Total	$\sum_{i=1}^{b} \sum_{j=1}^{t} (y_{ij} - y_{..})^2$	$bt-1$		

5.5 Randomized Latin square designs

Under model (3.16)–(3.17) for a randomized row-column design with r rows and c columns, an argument similar to what was used to derive (5.15) and (5.21) shows that the covariance matrix $\mathbf{V}$ has spectral decomposition

$$\mathbf{V} = \xi_{\mathcal{U}} \mathbf{P}_{W_{\mathcal{U}}} + \xi_{\mathcal{R}} \mathbf{P}_{W_{\mathcal{R}}} + \xi_{\mathcal{C}} \mathbf{P}_{W_{\mathcal{C}}} + \xi_{\mathcal{E}} \mathbf{P}_{W_{\mathcal{E}}}, \tag{5.28}$$

where

$$W_{\mathcal{U}} = V_{\mathcal{U}}, \ W_{\mathcal{R}} = V_{\mathcal{R}} \ominus V_{\mathcal{U}}, \ W_{\mathcal{C}} = V_{\mathcal{C}} \ominus V_{\mathcal{U}}, \text{ and } W_{\mathcal{E}} = (V_{\mathcal{R}} + V_{\mathcal{C}})^{\perp}. \tag{5.29}$$

We leave this as an exercise.

Therefore there are three eigenspaces other than $V_{\mathcal{U}}$: $W_{\mathcal{R}}$, $W_{\mathcal{C}}$, and $W_{\mathcal{E}}$, with dimensions $r-1$, $c-1$, and $(r-1)(c-1)$, respectively. Note that $\mathbf{P}_{W_{\mathcal{E}}} = \mathbf{I} - \mathbf{P}_{V_{\mathcal{R}}} - \mathbf{P}_{V_{\mathcal{C}}} + \mathbf{P}_{V_{\mathcal{U}}}$; this is because $\mathbf{P}_{W_{\mathcal{U}}} + \mathbf{P}_{W_{\mathcal{R}}} + \mathbf{P}_{W_{\mathcal{C}}} + \mathbf{P}_{W_{\mathcal{E}}} = \mathbf{I}$, $\mathbf{P}_{W_{\mathcal{U}}} = \mathbf{P}_{V_{\mathcal{U}}}$, $\mathbf{P}_{W_{\mathcal{R}}} = \mathbf{P}_{V_{\mathcal{R}}} - \mathbf{P}_{V_{\mathcal{U}}}$, and $\mathbf{P}_{W_{\mathcal{C}}} = \mathbf{P}_{V_{\mathcal{C}}} - \mathbf{P}_{V_{\mathcal{U}}}$. Again there is one stratum corresponding to each of the four factors $\mathcal{U}$, $\mathcal{R}$, $\mathcal{C}$, and $\mathcal{E}$ that define the block structure of a row-column design.

The total sum of squares can be decomposed as

$$\left\|\mathbf{P}_{V_{\mathcal{U}}^{\perp}}\mathbf{y}\right\|^2 = \left\|\mathbf{P}_{W_{\mathcal{R}}}\mathbf{y}\right\|^2 + \left\|\mathbf{P}_{W_{\mathcal{C}}}\mathbf{y}\right\|^2 + \left\|\mathbf{P}_{W_{\mathcal{E}}}\mathbf{y}\right\|^2 .$$

For convenience, index the unit at the intersection of the ith row and jth column by double subscripts (i, j). Then

$$\left\|\mathbf{P}_{W_{\mathcal{R}}}\mathbf{y}\right\|^2 = \left\|\mathbf{P}_{V_{\mathcal{R}}\ominus V_{\mathcal{U}}}\mathbf{y}\right\|^2 = \sum_{i=1}^{r} c(y_{i\cdot} - y_{..})^2,$$

$$\left\|\mathbf{P}_{W_{\mathcal{C}}}\mathbf{y}\right\|^2 = \left\|\mathbf{P}_{V_{\mathcal{C}}\ominus V_{\mathcal{U}}}\mathbf{y}\right\|^2 = \sum_{j=1}^{c} r(y_{\cdot j} - y_{..})^2,$$

and

$$\left\|\mathbf{P}_{W_{\mathcal{E}}}\mathbf{y}\right\|^2 = \left\|(\mathbf{I} - \mathbf{P}_{V_{\mathcal{R}}} - \mathbf{P}_{V_{\mathcal{C}}} + \mathbf{P}_{V_{\mathcal{U}}})\mathbf{y}\right\|^2 = \sum_{i=1}^{r}\sum_{j=1}^{c}(y_{ij} - y_{i\cdot} - y_{\cdot j} + y_{..})^2.$$

Similar to (5.17), when $\alpha_1 = \cdots = \alpha_t$,

$$\xi_{\mathcal{R}} = \mathrm{E}\left[\frac{1}{r-1}\sum_{i=1}^{r} c(y_{i\cdot} - y_{..})^2\right],$$

$$\xi_{\mathcal{C}} = \mathrm{E}\left[\frac{1}{c-1}\sum_{j=1}^{c} r(y_{\cdot j} - y_{..})^2\right],$$

$$\xi_{\mathcal{E}} = \mathrm{E}\left[\frac{1}{(r-1)(c-1)}\sum_{i=1}^{r}\sum_{j=1}^{c}(y_{ij} - y_{i\cdot} - y_{\cdot j} + y_{..})^2\right].$$

Thus $\xi_{\mathcal{R}}$ is the *between-row* variance, $\xi_{\mathcal{C}}$ is the *between-column* variance, and $\xi_{\mathcal{E}}$ is the *between-unit* variance adjusted for the between-row variability and between-column variability. The three eigenspaces $W_{\mathcal{R}}$, $W_{\mathcal{C}}$, and $W_{\mathcal{E}}$ are called row, column, and unit stratum, respectively. Again, $W_{\mathcal{U}} = V_{\mathcal{U}}$ is called the mean stratum, and we also call $W_{\mathcal{E}}$ the bottom stratum. Table 5.4 shows the *null* ANOVA for a row-column design.

If both r and c are multiples of t and all the treatment comparisons are equally important, then an optimal design is to have all the treatments appear the same number of times in each row and the same number of times in each column. (A result similar to the optimality of complete block designs in Exercise 2.4 can be established; see Exercise 5.7.) In this case the condition of proportional frequencies is

Table 5.4 *Null ANOVA for row-column designs*

source	sum of squares	d.f.	mean square	E(MS)
Rows	$\sum_{i=1}^{r} c(y_{i\cdot} - y_{\cdot\cdot})^2$	$r-1$	$\cdots$	$\xi_{\mathcal{R}}$
Columns	$\sum_{j=1}^{c} r(y_{\cdot j} - y_{\cdot\cdot})^2$	$c-1$	$\cdots$	$\xi_{\mathcal{C}}$
Units	$\sum_{i=1}^{r} \sum_{j=1}^{c} (y_{ij} - y_{i\cdot} - y_{\cdot j} + y_{\cdot\cdot})^2$	$(r-1)(c-1)$	$\cdots$	$\xi_{\mathcal{E}}$
Total	$\sum_{i=1}^{r} \sum_{j=1}^{c} (y_{ij} - y_{\cdot\cdot})^2$	$rc-1$		

satisfied by the treatment and row factors, as well as the treatment and column factors. By Theorem 2.5, $V_T \ominus V_{\mathcal{U}} \perp V_{\mathcal{R}} \ominus V_{\mathcal{U}}$ and $V_T \ominus V_{\mathcal{U}} \perp V_{\mathcal{C}} \ominus V_{\mathcal{U}}$. It follows that $(V_T \ominus V_{\mathcal{U}}) \perp [V_{\mathcal{U}} \oplus (V_{\mathcal{R}} \ominus V_{\mathcal{U}}) \oplus (V_{\mathcal{C}} \ominus V_{\mathcal{U}})]$, so

$$V_T \ominus V_{\mathcal{U}} \subset (V_{\mathcal{R}} + V_{\mathcal{C}})^{\perp} = W_{\mathcal{E}}.$$

By Theorem 5.2, the best linear unbiased estimator of a treatment contrast $\sum_{l=1}^{t} c_l \alpha_l$ is again $\sum_{l=1}^{t} c_l \overline{T}_l$, and $\mathrm{var}\left(\sum_{l=1}^{t} c_l \overline{T}_l\right) = \xi_{\mathcal{E}} \sum_{l=1}^{t} c_l^2 / q$, where $q = rc/t$.

Since $V_T \ominus V_{\mathcal{U}} \subset W_{\mathcal{E}}$, with the space W in (5.11) equal to $W_{\mathcal{E}}$, the stratum sum of squares $\left\|\mathbf{P}_{W_{\mathcal{E}}} \mathbf{y}\right\|^2 = \left\|\mathbf{P}_{(V_{\mathcal{R}}+V_{\mathcal{C}})^{\perp}} \mathbf{y}\right\|^2$ can be decomposed as

$$\left\|\mathbf{P}_{W_{\mathcal{E}}} \mathbf{y}\right\|^2 = \left\|\mathbf{P}_{V_T \ominus V_{\mathcal{U}}} \mathbf{y}\right\|^2 + \left\|\mathbf{P}_{(V_{\mathcal{R}}+V_{\mathcal{C}})^{\perp} \ominus (V_T \ominus V_{\mathcal{U}})} \mathbf{y}\right\|^2,$$

where the treatment sum of squares $\left\|\mathbf{P}_{V_T \ominus V_{\mathcal{U}}} \mathbf{y}\right\|^2$ is equal to $\sum_{l=1}^{t} \frac{rc}{t}(\overline{T}_l - y_{\cdot\cdot})^2$, and the residual sum of squares $\left\|\mathbf{P}_{(V_{\mathcal{R}}+V_{\mathcal{C}})^{\perp} \ominus (V_T \ominus V_{\mathcal{U}})} \mathbf{y}\right\|^2$ can be obtained by subtraction. By Proposition 5.3,

$$\mathrm{E}\left(\left\|\mathbf{P}_{(V_{\mathcal{R}}+V_{\mathcal{C}})^{\perp} \ominus (V_T \ominus V_{\mathcal{U}})} \mathbf{y}\right\|^2\right) = [(r-1)(c-1) - (t-1)]\xi_{\mathcal{E}}.$$

Thus $s^2 = \frac{1}{[(r-1)(c-1)-(t-1)]} \left\|\mathbf{P}_{(V_{\mathcal{R}}+V_{\mathcal{C}})^{\perp} \ominus (V_T \ominus V_{\mathcal{U}})} \mathbf{y}\right\|^2$ is an unbiased estimator of $\xi_{\mathcal{E}}$.

When $r = c = t$, a row-column design such that each treatment appears once in every row and once in every column is called a Latin square. We have the ANOVA for a randomized Latin square design in Table 5.5.

5.6   Decomposition of the treatment sum of squares

When the treatments are structured, some contrasts may be more important than the others. For example, in a factorial experiment, the experimenters are typically interested in treatment contrasts that represent main effects and interactions of various factors. We show in this section how the treatment sum of squares can be further decomposed into single-degree-of-freedom components corresponding to the contrasts of interest.

Table 5.5 *ANOVA for Latin square designs*

source	sum of squares	d.f.	MS	E(MS)
Rows	$\sum_{i=1}^{t} t(y_{i\cdot} - y_{\cdot\cdot})^2$	$t-1$	$\cdots$	$\xi_{\mathcal{R}}$
Columns	$\sum_{j=1}^{t} t(y_{\cdot j} - y_{\cdot\cdot})^2$	$t-1$	$\cdots$	$\xi_{\mathcal{C}}$
Treatments	$\sum_{l=1}^{t} t(\overline{T}_l - y_{\cdot\cdot})^2$	$t-1$	$\cdots$	$\xi_{\mathcal{E}} + \frac{1}{t-1}\left[\sum_{l=1}^{t} t(\alpha_l - \alpha_{\cdot})^2\right]$
Residual	By subtraction	$(t-1)(t-2)$	$\cdots$	$\xi_{\mathcal{E}}$
Total	$\sum_{i=1}^{t}\sum_{j=1}^{t}(y_{ij} - y_{\cdot\cdot})^2$	$t^2 - 1$		

Throughout this section, we assume that (5.1) and (5.9) hold. Then by Theorem 5.2, the best linear unbiased estimator of a treatment contrast $l = \mathbf{c}^T\boldsymbol{\alpha}$ is $\widehat{l} = \sum_{l=1}^{t} c_l \overline{T}_l$, and $\mathrm{var}\left(\widehat{l}\right) = \xi \sum_{l=1}^{t} c_l^2/r_l$. The corresponding t-statistic is $\widehat{l}/\left(s^2 \sum_{l=1}^{t} c_l^2/r_l\right)^{1/2}$, where $s^2 = \left\|\mathbf{P}_{W\ominus(V_{\mathcal{T}}\ominus V_{\mathcal{U}})}\mathbf{y}\right\|^2/[\dim(W) - (t-1)]$, and W is as in (5.11).

We define the sum of squares associated with l as

$$SS(l) = \frac{\widehat{l}^{\,2}}{\left(\sum_{l=1}^{t} c_l^2/r_l\right)}. \qquad (5.30)$$

Let $\mathbf{c}^*$ be as defined in (5.2). Then $\widehat{l} = (\mathbf{c}^*)^T\mathbf{y}$, and by (5.30),

$$SS(l) = \frac{1}{\|\mathbf{c}^*\|^2}\left[(\mathbf{c}^*)^T\mathbf{y}\right]^2. \qquad (5.31)$$

Since $SS(l) = SS(al)$ for any nonzero constant a, we may normalize l so that $\|\mathbf{c}^*\| = 1$.

Theorem 5.4. *Suppose* $l_1 = \mathbf{c}_1^T\boldsymbol{\alpha}, \ldots, l_{t-1} = \mathbf{c}_{t-1}^T\boldsymbol{\alpha}$ *are* $t-1$ *mutually orthogonal contrasts. Then the treatment sum of squares* $\left\|\mathbf{P}_{V_{\mathcal{T}}\ominus V_{\mathcal{U}}}\mathbf{y}\right\|^2 - \sum_{g=1}^{t-1} SS(l_g)$.

Proof. We have $\mathbf{c}_g^* \in V_{\mathcal{T}}\ominus V_{\mathcal{U}}$ for all $g = 1, \ldots, t-1$. Furthermore, by (5.3), $(\mathbf{c}_g^*)^T\mathbf{c}_{g'}^* = 0$ for all $1 \le g \ne g' \le t-1$. Hence $\mathbf{c}_1^*, \ldots, \mathbf{c}_{t-1}^*$ are an orthogonal basis of $V_{\mathcal{T}}\ominus V_{\mathcal{U}}$. Therefore

$$\left\|\mathbf{P}_{V_{\mathcal{T}}\ominus V_{\mathcal{U}}}\mathbf{y}\right\|^2 = \sum_{g=1}^{t-1}\frac{1}{\|\mathbf{c}_g^*\|^2}\left[(\mathbf{c}_g^*)^T\mathbf{y}\right]^2 = \sum_{g=1}^{t-1} SS(l_g),$$

where the last identity follows from (5.31). $\qquad\qquad\square$

5.7 Orthogonal polynomials

For $t > 2$, there are infinitely many sets of $t-1$ mutually orthogonal treatment contrasts. When the treatments are different values of a quantitative factor, say a set of t different temperature levels, it is useful to choose a set of mutually orthogonal treatment contrasts based on orthogonal polynomials. This amounts to expressing the treatment effects as a polynomial of the quantitative levels. In this section we assume that the quantitative treatment levels are equireplicate ($r_1 = \cdots = r_t$).

Definition 5.2. A system of orthogonal polynomials on t points (quantitative levels) $w_1, \ldots, w_t$ is a set $\{P_0, P_1, \ldots, P_{t-1}\}$ of polynomials such that $\sum_{i=1}^{t} P_k(w_i)P_{k'}(w_i) = 0$, for all $k \neq k'$, where $P_0 \equiv 1$ and P_k is a polynomial of degree k.

Orthogonal polynomials can be determined recursively from the lowest degree. For example, suppose $t = 3$ and w_1, w_2, w_3 are three equally spaced points, say $w_1 = -1$, $w_2 = 0$, $w_3 = 1$. Let $P_1(x)$ be $a + bx$. Then since $P_0(-1) = P_0(0) = P_0(1) = 1$, it follows from the orthogonality of P_0 and P_1 that $(a - b, a, a + b)^T$ is orthogonal to $(1, 1, 1)^T$. This implies that $a = 0$. Then $P_1(x) = bx$. Without loss of generality, take $b = 1$. Then $P_1(w_1) = -1$, $P_1(w_2) = 0$, and $P_1(w_3) = 1$. Suppose the effects of the three levels (treatments) are α_1, α_2, and α_3. Define a linear function of the treatment effects with the values of $P_1(w_1)$, $P_1(w_2)$, and $P_1(w_3)$ as its coefficients. Then we obtain the contrast $\alpha_3 - \alpha_1$, called the *linear contrast*.

Now suppose $P_2(x) = a + bx + cx^2$. Then $(a - b + c, a, a + b + c)^T$ is orthogonal to both $(1, 1, 1)^T$ and $(-1, 0, 1)^T$. The resulting equations lead to $b = 0$ and $a = -2c/3$. Let $c = 3$. Then $a = -2$, and so $P_2(w_1) = 1$, $P_2(w_2) = -2$, and $P_2(w_3) = 1$. This induces the contrast $\alpha_1 - 2\alpha_2 + \alpha_3$, called the *quadratic contrast*. The two contrasts $\alpha_3 - \alpha_1$ and $\alpha_1 - 2\alpha_2 + \alpha_3$ are orthogonal.

For three equally spaced points that are not necessarily -1, 0, and 1, say $w_1 = m - \Delta$, $w_2 = m$, and $w_3 = m + \Delta$, the aforementioned construction yields two orthogonal polynomials $P_1(x) = (x - m)/\Delta$ and $P_2(x) = 3\left[(\frac{x-m}{\Delta})^2 - \frac{2}{3}\right]$ with degrees 1 and 2, respectively.

In general, for equally spaced points, the orthogonal polynomials are unique up to constant multiples and can be scaled so that all the values $P_k(w_i)$ are integers. For three equally spaced levels, $\alpha_3 - 2\alpha_2 + \alpha_1 = \alpha_3 - \alpha_2 - (\alpha_2 - \alpha_1)$ measures the change of slopes, and it is equal to zero if and only if the effects α_1, α_2, and α_3 are a linear function of w_1, w_2, and w_3. If the quadratic contrast is zero, then the linear contrast $\alpha_3 - \alpha_1$ is equal to zero if and only if $\alpha_1 = \alpha_2 = \alpha_3$.

For $t = 4$ equally spaced quantitative levels w_1, w_2, w_3, and w_4, similar calculations lead to orthogonal polynomials with $P_0(w_1) = P_0(w_2) = P_0(w_3) = P_0(w_4) = 1$, $P_1(w_1) = -3$, $P_1(w_2) = -1$, $P_1(w_3) = 1$, $P_1(w_4) = 3$, $P_2(w_1) = 1$, $P_2(w_2) = -1$, $P_2(w_3) = -1$, $P_2(w_4) = 1$, $P_3(w_1) = -1$, $P_3(w_2) = 3$, $P_3(w_3) = -3$, and $P_3(w_4) = 1$. For treatment effects α_1, α_2, α_3, and α_4, these polynomials define three mutually orthogonal contrasts: $l_1 = -3\alpha_1 - \alpha_2 + \alpha_3 + 3\alpha_4$ (linear contrast), $l_2 = \alpha_1 - \alpha_2 - \alpha_3 + \alpha_4$ (quadratic contrast), and $l_3 = -\alpha_1 + 3\alpha_2 - 3\alpha_3 + \alpha_4$ (cubic contrast). Since $(1, 1, 1, 1)^T$, $(-3, -1, 1, 3)^T$, $(1, -1, -1, 1)^T$, and $(-1, 3, -3, 1)^T$ are mutually orthogonal, we can write $(\alpha_1, \alpha_2, \alpha_3, \alpha_4)^T$ as

$$(\alpha_1, \alpha_2, \alpha_3, \alpha_4)^T = h_0 \cdot (1, 1, 1, 1)^T + h_1 \cdot (-3, -1, 1, 3)^T$$
$$+ h_2 \cdot (1, -1, -1, 1)^T + h_3 \cdot (-1, 3, -3, 1)^T,$$

where $h_0 = l_0/4$, $h_1 = l_1/20$, $h_2 = l_2/4$, and $h_3 = l_3/20$. So $l_3 = 0$ if and only if the effects α_1, α_2, α_3, α_4 are a quadratic function of the levels. When $l_3 = 0$, we have $l_2 = 0$ if and only if the effects are a linear function of the levels, and when both l_3 and l_2 are zero, $l_1 = 0$ if and only if $\alpha_1 = \alpha_2 = \alpha_3 = \alpha_4$. This observation holds generally for higher-degree orthogonal polynomials.

Given a system of $t - 1$ orthogonal polynomials, one can construct a set of $t - 1$ mutually orthogonal treatment contrasts and apply the results in Section 5.6 to decompose the treatment sum of squares into single-degree-of-freedom components. We will discuss in Chapter 6 how orthogonal polynomials can be used to define contrasts representing factorial effects when the treatments have a factorial structure.

5.8 Orthogonal and nonorthogonal designs

The three classes of designs discussed in Sections 5.2–5.5 are *orthogonal designs* in the sense that $V_T \ominus V_U$ is contained in a certain eigenspace of the covariance matrix. This has the important consequence that the information for treatment contrasts is contained in the projection of the observation vector onto that eigenspace (stratum), resulting in simple analyses. The calculation of the estimates of treatment contrasts and their standard errors as well as the construction of ANOVA tables can be done in a straightforward manner. When $V_T \ominus V_U$ is not entirely in one stratum, it has nontrivial projections in more than one stratum; then there is information for the treatment contrasts in different strata, producing estimators with different precisions.

For example, suppose there are limitations on the block size so that incomplete blocks must be used. Then, since not all the treatments can appear in the same block, the condition of proportional frequencies no longer holds for the treatment and block factors. It follows that $V_T \ominus V_U$ is not orthogonal to $W_B = V_B \ominus V_U$; hence $V_T \ominus V_U$ is not a subspace of $W_\mathcal{E}$. On the other hand, $V_T \ominus V_U \subseteq W_B$ if and only if one has the "silly" design in which the same treatment is assigned to all the units in the same block (we say that the treatments are confounded with blocks). If this is to be avoided, then $V_T \ominus V_U$ has nontrivial projections in both the inter- and intrablock strata. In this case both projections $\mathbf{P}_{W_\mathcal{E}}\mathbf{y}$ and $\mathbf{P}_{W_B}\mathbf{y}$ contain treatment information. We say that a treatment contrast is estimable in the intrablock (respectively, interblock) stratum if it has an unbiased estimator of the form $\mathbf{a}^T\mathbf{P}_{W_\mathcal{E}}\mathbf{y}$ (respectively, $\mathbf{a}^T\mathbf{P}_{W_B}\mathbf{y}$). The best linear unbiased estimator of a treatment contrast based on $\mathbf{P}_{W_\mathcal{E}}\mathbf{y}$ (respectively, $\mathbf{P}_{W_B}\mathbf{y}$) is called the *intrablock* (respectively, *interblock*) estimator, whose variance depends on the intrablock (respectively, interblock) variance.

Projecting model (5.19)–(5.20) (which is the same as model (3.9)–(3.10)) for randomized block designs onto the intrablock stratum, we obtain

$$\mathbf{P}_{W_\mathcal{E}}\mathbf{y} = \mathbf{P}_{W_\mathcal{E}}\mathbf{X}_T\boldsymbol{\alpha} + \mathbf{P}_{W_\mathcal{E}}\boldsymbol{\varepsilon}, \tag{5.32}$$

with

$$E(\mathbf{P}_{W_\mathcal{E}}\boldsymbol{\varepsilon}) = \mathbf{0}, \ \text{cov}(\mathbf{P}_{W_\mathcal{E}}\boldsymbol{\varepsilon}) = \xi_\varepsilon \mathbf{P}_{W_\mathcal{E}}. \tag{5.33}$$

Intrablock estimators of the treatment effects can be obtained by solving the normal equation

$$(\mathbf{P}_{W_\mathcal{E}}\mathbf{X}_T)^T(\mathbf{P}_{W_\mathcal{E}}\mathbf{X}_T)\widehat{\boldsymbol{\alpha}}^\varepsilon = (\mathbf{P}_{W_\mathcal{E}}\mathbf{X}_T)^T(\mathbf{P}_{W_\mathcal{E}}\mathbf{y}),$$

which reduces to

$$\mathbf{X}_T^T \mathbf{P}_{W_\mathcal{E}} \mathbf{X}_T \widehat{\boldsymbol{\alpha}}^\mathcal{E} = \mathbf{X}_T^T \mathbf{P}_{W_\mathcal{E}} \mathbf{y}. \tag{5.34}$$

As in Theorem 2.3, a treatment contrast $\mathbf{c}^T \boldsymbol{\alpha}$ is estimable in the intrablock stratum if and only if $\mathbf{c}$ is a linear combination of the column vectors of $\mathbf{X}_T^T \mathbf{P}_{W_\mathcal{E}} \mathbf{X}_T$. If $\mathbf{c}^T \boldsymbol{\alpha}$ is estimable in the intrablock stratum, then $\mathbf{c}^T \widehat{\boldsymbol{\alpha}}^\mathcal{E}$ is its intrablock estimator, and

$$\mathrm{var} \left(\mathbf{c}^T \widehat{\boldsymbol{\alpha}}^\mathcal{E} \right) = \xi_\mathcal{E} \mathbf{c}^T \left(\mathbf{X}_T^T \mathbf{P}_{W_\mathcal{E}} \mathbf{X}_T \right)^- \mathbf{c},$$

where $\widehat{\boldsymbol{\alpha}}^\mathcal{E}$ is any solution to (5.34), and $(\mathbf{X}_T^T \mathbf{P}_{W_\mathcal{E}} \mathbf{X}_T)^-$ is any generalized inverse of $\mathbf{X}_T^T \mathbf{P}_{W_\mathcal{E}} \mathbf{X}_T$. On the other hand, interblock estimators of the treatment effects can be obtained by solving

$$\mathbf{X}_T^T \mathbf{P}_{W_\mathcal{B}} \mathbf{X}_T \widehat{\boldsymbol{\alpha}}^\mathcal{B} = \mathbf{X}_T^T \mathbf{P}_{W_\mathcal{B}} \mathbf{y}, \tag{5.35}$$

and for any solution $\widehat{\boldsymbol{\alpha}}^\mathcal{B}$ to (5.35) and any treatment contrast $\mathbf{c}^T \boldsymbol{\alpha}$ that is estimable in the interlock stratum,

$$\mathrm{var} \left(\mathbf{c}^T \widehat{\boldsymbol{\alpha}}^\mathcal{B} \right) = \xi_\mathcal{B} \mathbf{c}^T \left(\mathbf{X}_T^T \mathbf{P}_{W_\mathcal{B}} \mathbf{X}_T \right)^- \mathbf{c}.$$

Analysis of such "nonorthogonal" designs is more complicated. One needs to solve normal equations (5.34) and (5.35) to compute the inter- and intrablock estimates and can no longer just plug the treatment means into the contrast formula. Since the strata are mutually orthogonal, $\mathrm{cov}(\mathbf{a}^T \mathbf{P}_{W_\mathcal{B}} \mathbf{y}, \mathbf{b}^T \mathbf{P}_{W_\mathcal{E}} \mathbf{y}) = \mathbf{a}^T \mathbf{P}_{W_\mathcal{B}} \mathbf{V} \mathbf{P}_{W_\mathcal{E}} \mathbf{b} = 0$. In general, estimators in different strata are uncorrelated. If a treatment contrast can be estimated in more than one stratum, one may combine these uncorrelated estimators. The best linear combination has the weights inversely proportional to the variances of the estimators from different strata. One needs to estimate the stratum variances when they are unknown. In the literature of block designs, this is called *recovery of interblock information* (Yates, 1940).

However, one may be able to design an experiment so that the information for each treatment contrast of interest is contained in one single stratum. Then each of these contrasts is estimated in only one stratum, although different contrasts may be estimated in different strata, resulting in different precisions. For example, suppose $V_T \ominus V_\mathcal{U}$ can be decomposed as

$$V_T \ominus V_\mathcal{U} = Z_1 \oplus Z_2 \tag{5.36}$$

such that $Z_1 \subseteq W_\mathcal{B}$, $Z_2 \subseteq W_\mathcal{E}$, and for each treatment contrast $\mathbf{c}^T \boldsymbol{\alpha}$ of interest, either $\mathbf{c}^* \in Z_1$ or $\mathbf{c}^* \in Z_2$; then Theorem 5.1 applies, and each treatment contrast of interest can be estimated by the same contrast of the treatment means. In this case, even though $V_T \ominus V_\mathcal{U}$ is not entirely in one stratum, we still call the design orthogonal. We have seen in Chapter 1 a simple example of factorial experiments in which the main effects are estimated in the intrablock stratum and the interaction is estimated in the interblock stratum.

Under (5.36), the treatment sum of squares can be split into

$$\left\| \mathbf{P}_{V_T \ominus V_\mathcal{U}} \mathbf{y} \right\|^2 = \left\| \mathbf{P}_{Z_1} \mathbf{y} \right\|^2 + \left\| \mathbf{P}_{Z_2} \mathbf{y} \right\|^2.$$

The first term $\left\|\mathbf{P}_{Z_1}\mathbf{y}\right\|^2$ gives the treatment sum of squares in the interblock stratum, and $\left\|\mathbf{P}_{Z_2}\mathbf{y}\right\|^2$ is the treatment sum of squares in the intrablock stratum. The interblock and intrablock sums of squares in the null ANOVA (5.22) can be decomposed as

$$\left\|\mathbf{P}_{W_{\mathcal{B}}}\mathbf{y}\right\|^2 = \left\|\mathbf{P}_{Z_1}\mathbf{y}\right\|^2 + \left\|\mathbf{P}_{(V_{\mathcal{B}}\ominus V_{\mathcal{U}})\ominus Z_1}\mathbf{y}\right\|^2, \tag{5.37}$$

$$\left\|\mathbf{P}_{W_{\mathcal{E}}}\mathbf{y}\right\|^2 = \left\|\mathbf{P}_{Z_2}\mathbf{y}\right\|^2 + \left\|\mathbf{P}_{V_{\mathcal{B}}^{\perp}\ominus Z_2}\mathbf{y}\right\|^2, \tag{5.38}$$

where $\left\|\mathbf{P}_{(V_{\mathcal{B}}\ominus V_{\mathcal{U}})\ominus Z_1}\mathbf{y}\right\|^2$ and $\left\|\mathbf{P}_{V_{\mathcal{B}}^{\perp}\ominus Z_2}\mathbf{y}\right\|^2$ are the residual sums of squares in the inter- and intrablock strata, respectively. As in Proposition 5.3, one can show that $\left\|\mathbf{P}_{(V_{\mathcal{B}}\ominus V_{\mathcal{U}})\ominus Z_1}\mathbf{y}\right\|^2/[b-1-\dim(Z_1)]$ and $\left\|\mathbf{P}_{V_{\mathcal{B}}^{\perp}\ominus Z_2}\mathbf{y}\right\|^2/[b(k-1)-\dim(Z_2)]$ are unbiased estimators of $\xi_{\mathcal{B}}$ and $\xi_{\mathcal{E}}$, respectively, provided that $b-1 > \dim(Z_1)$ and $b(k-1) > \dim(Z_2)$. A full ANOVA table can be obtained from the null ANOVA in Section 5.3 by replacing the interblock and intrablock sums of squares with the two sums of squares on the right-hand sides of (5.37) and (5.38), respectively.

Suppose $l_1 = \mathbf{c}_1^T\boldsymbol{\alpha}, \ldots, l_{t-1} = \mathbf{c}_{t-1}^T\boldsymbol{\alpha}$ are a set of mutually orthogonal treatment contrasts such that each $\mathbf{c}_j^*$ belongs to either Z_1 or Z_2. Then

$$\left\|\mathbf{P}_{Z_h}\mathbf{y}\right\|^2 = \sum_{j:\mathbf{c}_j^*\in Z_h} \mathrm{SS}(l_j), \text{ for } h = 1, 2.$$

All of these discussions apply to row-column designs as well.

We have seen that a block design has two nontrivial strata (other than the mean stratum) corresponding to inter- and intrablock errors, and a row-column design has three nontrivial strata. A general theory for the analysis of experiments with multiple strata will be developed in Chapters 12 and 13. We will define and discuss orthogonal designs for general block structures in Chapter 13. Under an orthogonal design, either $V_{\mathcal{T}}\ominus V_{\mathcal{U}}$ is entirely in one stratum, or it can be broken into orthogonal components so that each component falls in one stratum and all the components together cover the contrasts of interest. This has important applications in factorial experiments where, as we will see in Chapter 6, $V_{\mathcal{T}}\ominus V_{\mathcal{U}}$ can be decomposed into orthogonal components corresponding to various main effects and interactions.

5.9   Models with fixed block effects

Consider the fixed-effect model (3.7)–(3.8) where the block effects β_i are unknown constants. In matrix form,

$$\mathbf{y} = \mu\mathbf{1} + \mathbf{X}_{\mathcal{T}}\boldsymbol{\alpha} + \mathbf{X}_{\mathcal{B}}\boldsymbol{\beta} + \boldsymbol{\varepsilon},$$

where

$$\mathrm{E}(\boldsymbol{\varepsilon}) = \mathbf{0}, \ \mathrm{cov}(\boldsymbol{\varepsilon}) = \sigma^2\mathbf{I}.$$

Under this model, in order to estimate the treatment effects, we need to project $\mathbf{y}$ onto $V_{\mathcal{B}}^{\perp} = W_{\mathcal{E}}$ to eliminate the nuisance parameters $\boldsymbol{\beta}$; see Section 2.5. This results in the projected model

$$\mathbf{P}_{W_{\mathcal{E}}}\mathbf{y} = \mathbf{P}_{W_{\mathcal{E}}}\mathbf{X}_T\boldsymbol{\alpha} + \mathbf{P}_{W_{\mathcal{E}}}\boldsymbol{\varepsilon}, \tag{5.39}$$

with

$$E(\mathbf{P}_{W_{\mathcal{E}}}\boldsymbol{\varepsilon}) = \mathbf{0}, \ \text{cov}(\mathbf{P}_{W_{\mathcal{E}}}\boldsymbol{\varepsilon}) = \sigma^2 \mathbf{P}_{W_{\mathcal{E}}}. \tag{5.40}$$

By comparing the two projected models (5.32)–(5.33) and (5.39)–(5.40), it is clear that the best linear unbiased estimator of a treatment contrast under the fixed-effect model (3.7)–(3.8) is the same as its intrablock estimator under the randomization (or mixed-effect) model (3.9)–(3.10), with $\xi_{\mathcal{E}}$ playing the same role as σ^2 in the former case. Such an estimator can be obtained by solving the normal equation (5.34), which is the same as (2.29).

Under a randomized complete block design, the fact that all the information for estimating treatment contrasts is contained in the intrablock stratum implies that, as we have seen, the best linear unbiased estimators of treatment contrasts under the fixed- and mixed-effect (randomization) models are the same. We have also seen that their standard errors can be computed in exactly the same way, and that we have the same ANOVA under both models.

Likewise the best linear unbiased estimators of treatment contrasts under the fixed-effect model (3.14)–(3.15) for row-column designs are the same as their estimators based on the projection of $\mathbf{y}$ onto $W_{\mathcal{E}} = (V_{\mathcal{R}} + V_C)^{\perp}$ under model (3.16)–(3.17). Such estimators can be obtained by solving (5.34) with $\mathbf{P}_{W_{\mathcal{E}}} = \mathbf{I} - \mathbf{P}_{V_{\mathcal{R}}} - \mathbf{P}_{V_C} + \mathbf{P}_{V_{\mathcal{U}}}$. If each treatment appears the same number of times in each row and the same number of times in each column (e.g., under a Latin square design), then the best linear unbiased estimators of treatment contrasts and their standard errors as well as the ANOVA tables are the same under the fixed- and mixed-effect (randomization) models.

Example 5.1. Wu and Hamada (2009, p. 91) discussed an experiment to compare four methods for predicting the shear strength for steel plate girders. The experiment used nine girders, and all four methods were applied to each girder, resulting in a complete block design with each girder as a block of size four. In using R to analyze the data, one has two options for the model formula in the aov function: aov(y~block+method) or aov(y~method+Error(block)), where y is the response variable. For the data in Wu and Hamada (2009), the former yields the following ANOVA table:

	Df	Sum Sq	Mean Sq	F value	Pr(>F)
girder	8	0.08949	0.01119	1.6189	0.1717
method	3	1.51381	0.50460	73.0267	3.296e-12
Residuals	24	0.16584	0.00691		

The latter produces the following:

```
Error: block

              Df     Sum Sq    Mean Sq    F value       Pr(>F)
Residuals     8     0.08949    0.01119


Error: within

              Df     Sum Sq    Mean Sq    F value       Pr(>F)
method        3     1.51381    0.50460    73.0267     3.296e-12
Residuals    24     0.16584    0.00691
```

The layouts of the two ANOVA tables are different, but the entries are the same. The first table is based on the fixed-effect model, while the latter shows the two strata. Under a complete block design, there is no treatment information in the interblock stratum, so the residual sum of squares in the interblock stratum (0.08949) is the same as the interblock sum of squares $\left\| \mathbf{P}_{W_B} \mathbf{y} \right\|^2$. The corresponding mean square (0.01119) is an estimate of the interblock variance ξ_B. On the other hand, the residual mean square in the intrablock stratum (0.00691) is an estimate of the intrablock variance ξ_ε, which, as shown in the first ANOVA table, is equal to the estimate of σ^2 under the fixed-effect model (3.7)–(3.8). Estimates of treatment effects and their standard errors are also the same under the two models.

For nonorthogonal designs (such as incomplete block designs), the second option mentioned in Example 5.1 will also provide interblock estimates of the treatment effects and ANOVA in the interblock stratum.

In GenStat, one needs to supply block structure and treatment structure formulas. One can use "block/unit" (or simply "block") for the block structure formula and "method" for the treatment structure formula. This also produces the second ANOVA table shown above.

Exercises

5.1 Verify (5.25) and (5.26).

5.2 Verify (5.28) and (5.29).

5.3 For each of the two block structures $n_1/(n_2/n_3)$ and $n_1/(n_2 \times n_3)$, derive the strata, write down the null ANOVA table, and interpret the stratum variances.

5.4 Show that (5.35) can be expressed as $\mathbf{C}\widehat{\boldsymbol{\alpha}} = \mathbf{Q}$, with

$$\mathbf{C} = \frac{1}{k}\mathbf{N}\mathbf{N}^T - \frac{1}{bk}\mathbf{q}\mathbf{q}^T, \quad \mathbf{Q} = \frac{1}{k}\mathbf{N}S_B - y_{..}\mathbf{q},$$

where $\mathbf{N}$ is the $t \times b$ treatment-block incidence matrix, $\mathbf{q} = (q_1, \dots, q_t)^T$, q_i is the number of replications of the ith treatment, $\mathbf{S}_B$ is the $b \times 1$ vector of block totals, and $y_{..}$ is the overall mean.

5.5 It was shown in Exercise 2.5 that under a balanced incomplete block design, the intrablock estimator of any pairwise treatment comparison has variance $\frac{2k}{t\lambda}\xi_{\mathcal{E}}$. Show that the variance of the interblock estimator of any pairwise treatment comparison is equal to $\frac{2k}{q-\lambda}\xi_B$.

5.6 Show that the least squares estimators of treatment effects under the fixed effect model (3.14)–(3.15) for row-column designs can be obtained by solving the equation $\mathbf{C}\widehat{\boldsymbol{\alpha}} = \mathbf{Q}$, with

$$\mathbf{C} = \mathrm{diag}(q_1,\ldots,q_t) - \frac{1}{c}\mathbf{N}\mathbf{N}^T - \frac{1}{r}\mathbf{M}\mathbf{M}^T + \frac{1}{rc}\mathbf{q}\mathbf{q}^T,$$

and

$$\mathbf{Q} = \mathbf{S}_T - \frac{1}{c}\mathbf{N}\mathbf{S}_{\mathcal{R}} - \frac{1}{r}\mathbf{M}\mathbf{S}_C + y_{..}\mathbf{q},$$

where $\mathbf{N}$ is the treatment-row incidence matrix, $\mathbf{M}$ is the treatment-column incidence matrix, $\mathbf{S}_{\mathcal{R}}$ is the $r \times 1$ vector of row totals, $\mathbf{S}_C$ is the $c \times 1$ vector of column totals, $\mathbf{S}_T$ is the $t \times 1$ vector of treatment totals, and $q_1,\ldots,q_t$, $\mathbf{q}$, and $y_{..}$ are as in Exercise 5.4. Also show that if each treatment appears the same number of times in each row, then the equation reduces to (2.29) with the columns considered as blocks.

5.7 In Exercise 5.6, suppose both r and c are multiples of t. Show that a design such that each treatment appears c/t times in each row and r/t times in each column minimizes $\sum_{1 \leq i < i' \leq t}\mathrm{var}(\widehat{\alpha}_i - \widehat{\alpha}_{i'})$ among all possible row-column designs.

5.8 In Exercise 5.6, suppose $r < t$ and c is a multiple of t. Show that if a row-column design is such that all the treatments appear the same number of times in each row, and it is a balanced incomplete block design when the columns are considered as blocks, then it minimizes $\sum_{1 \leq i < i' \leq t}\mathrm{var}(\widehat{\alpha}_i - \widehat{\alpha}_{i'})$ among all possible row-column designs.

Chapter 6

Factorial Treatment Structure and Complete Factorial Designs

Suppose each treatment is a combination of the levels of n factors $A_1, \ldots, A_n$ with $s_1, \ldots, s_n$ levels, respectively. Then the $s_1 \cdots s_n$ treatment combinations define $s_1 \cdots s_n - 1$ degrees of freedom. In this chapter, we show how to choose a set of $s_1 \cdots s_n - 1$ mutually orthogonal treatment contrasts, representing *main effects* and *interactions*, that are suitable for factorial experiments. These contrasts measure the effects of individual factors on the response as well as how they interact with one another. The analysis of complete factorial experiments under simple orthogonal designs such as completely randomized designs, randomized complete block designs, and randomized Latin square designs is presented. When $s_1 = \cdots = s_n = s$, where s is a prime number or power of a prime number, finite geometries can be used to choose an orthogonal basis of main-effect and interaction contrasts that is useful for design construction and other purposes. By using Abelian groups, this can be extended to the asymmetrical case where different treatment factors may have different numbers of levels that are not necessarily prime numbers or powers of prime numbers.

6.1 Factorial effects for two and three two-level factors

In an $s_1 \times \cdots \times s_n$ experiment, each treatment combination can be represented by a vector $\mathbf{x} = (x_1, \ldots, x_n)^T$, where $x_i \in S_i \equiv \{0, 1, \ldots, s_i - 1\}$ is the level of the ith factor. The effect of a treatment combination $\mathbf{x} = (x_1, \ldots, x_n)^T$ is denoted by $\alpha_{x_1 \ldots x_n}$, and the resulting $(s_1 \cdots s_n) \times 1$ vector with $\alpha_{x_1 \ldots x_n}$ as its $(x_1, \ldots, x_n)$th component is denoted by $\boldsymbol{\alpha}$. Sometimes $\alpha_{x_1 \ldots x_n}$ is also written as $\alpha(x_1, \ldots, x_n)$. In this notation, α is considered as a response function on the set $S = S_1 \times \cdots \times S_n = \{(x_1, \ldots, x_n)^T : x_i \in S_i\}$ of treatment combinations. This is particularly useful when the treatment factors are quantitative. The response function α and the vector $\boldsymbol{\alpha}$ represent the same thing, and we use both notations interchangeably. For $s_i = 2$, often we also denote the two levels by 1 and -1. From the context there should be no danger of confusion.

Suppose there are two two-level treatment factors A_1 and A_2. Then $t = 4$ and the four treatment combinations can be represented by $(0, 0)^T$, $(0, 1)^T$, $(1, 0)^T$, and $(1, 1)^T$. The corresponding treatment effects are α_{00}, α_{01}, α_{10}, and α_{11}, respectively.

We call 0 and 1 low and high levels, respectively. Then $\alpha_{10} - \alpha_{00}$ measures the effect of changing the level of A_1 from low to high while keeping A_2 at the low level. We call $\alpha_{10} - \alpha_{00}$ the main effect of A_1 when A_2 is at the low level and denote it by $A_1|A_2-$. Similarly, $A_1|A_2+ = \alpha_{11} - \alpha_{01}$. The average $\frac{1}{2}[(A_1|A_2-)+(A_1|A_2+)]$ measures the overall effect of changing the level of A_1 from low to high and is called the *main effect* of A_1. Abusing the notation, we also denote the main effect of factor A_1 by A_1. Then

$$A_1 = \frac{1}{2}[(\alpha_{11} - \alpha_{01})+(\alpha_{10} - \alpha_{00})] = \frac{1}{2}[(\alpha_{11} + \alpha_{10}) - (\alpha_{01} + \alpha_{00})]. \qquad (6.1)$$

Similarly, the main effect of A_2 is

$$A_2 = \frac{1}{2}[(\alpha_{11} + \alpha_{01}) - (\alpha_{10} + \alpha_{00})]. \qquad (6.2)$$

The deviation $(A_1|A_2+) - A_1$ is called the *interaction* of factors A_1 and A_2 and is denoted by A_1A_2. Then

$$A_1A_2 = \alpha_{11} - \alpha_{01} - A_1 = \frac{1}{2}[(\alpha_{11} + \alpha_{00}) - (\alpha_{01} + \alpha_{10})]. \qquad (6.3)$$

We also have

$$A_1A_2 = (A_2|A_1+) - A_2. \qquad (6.4)$$

From (6.3) and (6.4), if $A_1A_2 = 0$, then $A_1|A_2+ = A_1|A_2- = A_1$ and $A_2|A_1+ = A_2|A_1- = A_2$, so the main effect of A_1 does not depend on the levels of A_2 and the main effect of A_2 also does not depend on the levels of A_1. In this case, we say that the two factors *do not interact*.

Each of A_1, A_2, and A_1A_2 is a treatment contrast. Furthermore, they are mutually orthogonal in the sense of Definition 5.1 under equal replication of the treatments and thus form an orthogonal basis of the three-dimensional space of treatment contrasts.

Suppose there is a third two-level factor A_3. Following (6.1), we can define the main effect of A_1 as

$$A_1 = \frac{1}{4}[(\alpha_{100} - \alpha_{000})+(\alpha_{101} - \alpha_{001})+(\alpha_{110} - \alpha_{010})+(\alpha_{111} - \alpha_{011})],$$

which is the average of the comparisons of the two levels of A_1 over the four combinations of the other two factors. Then

$$A_1 = \frac{1}{4}[(\alpha_{100} + \alpha_{101} + \alpha_{110} + \alpha_{111}) - (\alpha_{000} + \alpha_{001} + \alpha_{010} + \alpha_{011})], \qquad (6.5)$$

where the four terms with positive coefficients correspond to the treatment combinations with A_1 at the high level. Similarly, the main effects of A_2 and A_3 are defined as

$$A_2 = \frac{1}{4}[(\alpha_{010} + \alpha_{011} + \alpha_{110} + \alpha_{111}) - (\alpha_{000} + \alpha_{001} + \alpha_{100} + \alpha_{101})], \qquad (6.6)$$

and

$$A_3 = \frac{1}{4}[(\alpha_{001} + \alpha_{011} + \alpha_{101} + \alpha_{111}) - (\alpha_{000} + \alpha_{010} + \alpha_{100} + \alpha_{110})]. \qquad (6.7)$$

By (6.3), the interaction of A_1 and A_2 when A_3 is at the high level is

$$(A_1A_2|A_3+) = \frac{1}{2}[(\alpha_{001} + \alpha_{111}) - (\alpha_{011} + \alpha_{101})].$$

Likewise,

$$(A_1A_2|A_3-) = \frac{1}{2}[(\alpha_{000} + \alpha_{110}) - (\alpha_{010} + \alpha_{100})].$$

The interaction of A_1 and A_2 can be defined as the average of $A_1A_2|A_3+$ and $A_1A_2|A_3-$, which gives

$$A_1A_2 = \frac{1}{4}[(\alpha_{111} + \alpha_{110} + \alpha_{001} + \alpha_{000}) - (\alpha_{011} + \alpha_{010} + \alpha_{101} + \alpha_{100})]. \tag{6.8}$$

Similarly,

$$A_1A_3 = \frac{1}{4}[(\alpha_{111} + \alpha_{101} + \alpha_{010} + \alpha_{000}) - (\alpha_{011} + \alpha_{001} + \alpha_{110} + \alpha_{100})], \tag{6.9}$$

and

$$A_2A_3 = \frac{1}{4}[(\alpha_{111} + \alpha_{011} + \alpha_{100} + \alpha_{000}) - (\alpha_{101} + \alpha_{001} + \alpha_{110} + \alpha_{010})]. \tag{6.10}$$

The three-factor interaction $A_1A_2A_3$ can then be defined as $(A_1A_2|A_3+) - A_1A_2$, or $(A_1A_3|A_2+) - A_1A_3$, or $(A_2A_3|A_1+) - A_2A_3$. All give

$$A_1A_2A_3 = \frac{1}{4}[(\alpha_{111} + \alpha_{100} + \alpha_{010} + \alpha_{001}) - (\alpha_{110} + \alpha_{101} + \alpha_{011} + \alpha_{000})]. \tag{6.11}$$

If $A_1A_2A_3 = 0$, then $(A_1A_2|A_3+) = (A_1A_2|A_3-) = A_1A_2$, so the interaction of factors A_1 and A_2 does not depend on the levels of A_3. In this case, we also have $(A_1A_3|A_2+) = (A_1A_3|A_2-) = A_1A_3$ and $(A_2A_3|A_1+) = (A_2A_3|A_1-) = A_2A_3$.

The seven linear functions of treatment effects defined in (6.5)–(6.11) are mutually orthogonal contrasts and thus form an orthogonal basis of the seven-dimensional space of treatment contrasts. Let μ be the average of the eight treatment effects and write $\frac{1}{2}A_1$, $\frac{1}{2}A_2$, $\frac{1}{2}A_3$, $\frac{1}{2}A_1A_2$, $\frac{1}{2}A_1A_3$, $\frac{1}{2}A_2A_3$, and $\frac{1}{2}A_1A_2A_3$ as β_1, β_2, β_3, β_{12}, β_{13}, β_{23}, and β_{123}, respectively. Then we have

$$
\begin{bmatrix} \mu \\ \beta_1 \\ \beta_2 \\ \beta_{12} \\ \beta_3 \\ \beta_{13} \\ \beta_{23} \\ \beta_{123} \end{bmatrix} = \frac{1}{8}
\begin{bmatrix}
1 & 1 & 1 & 1 & 1 & 1 & 1 & 1 \\
-1 & 1 & -1 & 1 & -1 & 1 & -1 & 1 \\
-1 & -1 & 1 & 1 & -1 & -1 & 1 & 1 \\
1 & -1 & -1 & 1 & 1 & -1 & -1 & 1 \\
-1 & -1 & -1 & -1 & 1 & 1 & 1 & 1 \\
1 & -1 & 1 & -1 & -1 & 1 & -1 & 1 \\
1 & 1 & -1 & -1 & -1 & -1 & 1 & 1 \\
-1 & 1 & 1 & -1 & 1 & -1 & -1 & 1
\end{bmatrix}
\begin{bmatrix} \alpha_{000} \\ \alpha_{100} \\ \alpha_{010} \\ \alpha_{110} \\ \alpha_{001} \\ \alpha_{101} \\ \alpha_{011} \\ \alpha_{111} \end{bmatrix}. \tag{6.12}
$$

The 8×8 matrix in (6.12), denoted by $\mathbf{H}$, can be obtained as follows. The first row corresponds to μ and has all the entries equal to 1. One then writes down the rows corresponding to the main effects, one for each factor: an entry is 1 (or -1) if the factor is at the high (or low) level in the corresponding treatment combination. Once the rows corresponding to the main effects have been completed, each of those corresponding to an interaction is the componentwise product (also called *Hadamard product*) of the rows corresponding to the main effects of the factors involved in the interaction. For example, the last row (corresponding to the three-factor interaction $A_1 A_2 A_3$) is the componentwise product of the second, third, and fifth rows, which correspond to the main effects of factors A_1, A_2, and A_3, respectively. From this construction, we see that the coefficients of a main-effect or interaction contrast depend only on the levels of the factors that are involved.

Throughout this book, we denote the Hadamard product $(x_1 y_1, \ldots, x_n y_n)$ of two vectors $\mathbf{x} = (x_1, \ldots, x_n)$ and $\mathbf{y} = (y_1, \ldots, y_n)$ by $\mathbf{x} \odot \mathbf{y}$.

It can be verified that $\mathbf{H}$ is a *Hadamard matrix* defined as follows.

Definition 6.1. A square matrix with all the entries equal to 1 or -1 is called a Hadamard matrix if its rows are mutually orthogonal.

We have $\mathbf{H}\mathbf{H}^T = 8\mathbf{I}_N$. Multiplying both sides of (6.12) by $\mathbf{H}^T = 8\mathbf{H}^{-1}$, we can express the treatment effects in terms of the main effects and interactions:

$$
\begin{bmatrix} \alpha_{000} \\ \alpha_{100} \\ \alpha_{010} \\ \alpha_{110} \\ \alpha_{001} \\ \alpha_{101} \\ \alpha_{011} \\ \alpha_{111} \end{bmatrix} =
\begin{bmatrix}
1 & -1 & -1 & 1 & -1 & 1 & 1 & -1 \\
1 & 1 & -1 & -1 & -1 & -1 & 1 & 1 \\
1 & -1 & 1 & -1 & -1 & 1 & -1 & 1 \\
1 & 1 & 1 & 1 & -1 & -1 & -1 & -1 \\
1 & -1 & -1 & 1 & 1 & -1 & -1 & 1 \\
1 & 1 & -1 & -1 & 1 & 1 & -1 & -1 \\
1 & -1 & 1 & -1 & 1 & -1 & 1 & -1 \\
1 & 1 & 1 & 1 & 1 & 1 & 1 & 1
\end{bmatrix}
\begin{bmatrix} \mu \\ \beta_1 \\ \beta_2 \\ \beta_{12} \\ \beta_3 \\ \beta_{13} \\ \beta_{23} \\ \beta_{123} \end{bmatrix}.
\tag{6.13}
$$

Since the 8×8 matrix in (6.13) is $\mathbf{H}^T$, by the discussion following (6.12), if we denote the low level by -1, instead of 0, and write the eight treatment effects as $\alpha(x_1, x_2, x_3)$, where x_1, x_2, $x_3 = 1$ or -1, then columns 2–8 of the matrix in (6.13) represent the values of $x_1, x_2, x_1 x_2, x_3, x_1 x_3, x_2 x_3$, and $x_1 x_2 x_3$, respectively. Therefore (6.13) is equivalent to that $\alpha(x_1, x_2, x_3)$ is the polynomial function

$$
\begin{aligned}
\alpha(x_1, x_2, x_3) = {} & \mu + \beta_1 x_1 + \beta_2 x_2 + \beta_3 x_3 + \beta_{12} x_1 x_2 \\
& + \beta_{13} x_1 x_3 + \beta_{23} x_2 x_3 + \beta_{123} x_1 x_2 x_3.
\end{aligned}
\tag{6.14}
$$

The main effects and interactions appear as coefficients of this polynomial function, with

$$
\beta_i = \frac{1}{8} \left\{ \sum_{(x_1, x_2, x_3):\, x_i = 1} \alpha(x_1, x_2, x_3) - \sum_{(x_1, x_2, x_3):\, x_i = -1} \alpha(x_1, x_2, x_3) \right\},
$$

$$\beta_{ij} = \frac{1}{8} \left\{ \sum_{(x_1,x_2,x_3): \, x_i x_j = 1} \alpha(x_1,x_2,x_3) - \sum_{(x_1,x_2,x_3): \, x_i x_j = -1} \alpha(x_1,x_2,x_3) \right\},$$

$$\beta_{123} = \frac{1}{8} \left\{ \sum_{(x_1,x_2,x_3): \, x_1 x_2 x_3 = 1} \alpha(x_1,x_2,x_3) - \sum_{(x_1,x_2,x_3): \, x_1 x_2 x_3 = -1} \alpha(x_1,x_2,x_3) \right\}.$$

Each of these β's is a treatment contrast with the corresponding row vector of $\frac{1}{8}\mathbf{H}$ as its coefficient vector.

From (6.14), if the three-factor interaction is equal to zero, then α is at most a second-degree polynomial. If, in addition, the two-factor interactions are also zero, then α is a linear function; in this case, there is no interaction and we have a simple *additive* treatment model.

The effects on both sides of (6.12) and (6.13) are listed in a lexicographical order of the factor levels, often called a *Yates* or *standard* order. One advantage of writing them in such an order is that the matrix $\mathbf{H}$ can be expressed succinctly as

$$\mathbf{H} = \begin{bmatrix} 1 & 1 \\ -1 & 1 \end{bmatrix} \otimes \begin{bmatrix} 1 & 1 \\ -1 & 1 \end{bmatrix} \otimes \begin{bmatrix} 1 & 1 \\ -1 & 1 \end{bmatrix},$$

where $\otimes$ is the Kronecker product. Recall that for any $p \times q$ matrix $\mathbf{A} = [a_{ij}]$ and $r \times s$ matrix $\mathbf{B} = [b_{kl}]$, the Kronecker product $\mathbf{A} \otimes \mathbf{B}$ is defined as the $pr \times qs$ matrix $[a_{ij}\mathbf{B}]$.

It was pointed out earlier that the coefficients of the main-effect and interaction contrasts in (6.5)–(6.11) depend only on the levels of the factors that are involved. In general we can define main-effect and interaction contrasts inductively as follows.

Definition 6.2. A treatment contrast is said to represent the main effect of factor $A_i, i = 1, \ldots, n$, if its coefficients depend only on the levels of A_i. A treatment contrast is said to represent the interaction of factors $A_{i_1}, \ldots, A_{i_k}, k > 1$, if its coefficients depend only on the levels of $A_{i_1}, \ldots, A_{i_k}$, and it is orthogonal to every treatment contrast that represents a factorial effect involving a proper subset of $A_{i_1}, \ldots, A_{i_k}$.

We refer to the main-effect and interaction contrasts as *factorial effects*. By a simple extension of (6.12), we show in the next two sections how to construct a set of mutually orthogonal contrasts with the properties described in Definition 6.2.

6.2 Factorial effects for more than three two-level factors

Suppose there are n two-level treatment factors. Then (6.12) suggests that a $2^n \times 1$ vector $\boldsymbol{\beta}$ whose components consist of the mean and various factorial effects of the n factors can be defined as

$$\boldsymbol{\beta} = \frac{1}{2^n} \mathbf{M} \boldsymbol{\alpha},$$

where $\mathbf{M}$ contains a row of 1's corresponding to the mean and, for each treatment factor A_i, a row whose entries are 1 (or -1) if A_i is at the high (or low) level in the corresponding treatment combinations in $\boldsymbol{\alpha}$. The latter clearly defines a treatment contrast whose coefficients depend only on the levels of A_i and, according to Definition 6.2, represents the main effect of A_i. Taking the Hadamard product of k distinct rows, each of which defines the main effect of a factor, we obtain another row of $\mathbf{M}$ whose entries depend only on the levels of the k factors in the corresponding treatment combinations in $\boldsymbol{\alpha}$. This construction produces a total of $\sum_{k=1}^{n} \binom{n}{k} = 2^n - 1$ treatment contrasts, which, as we will see in Section 6.3, are mutually orthogonal. Then by Definition 6.2, the $2^n - 1$ treatment contrasts so constructed together represent the main effects and interactions of the n factors.

For a 2^n factorial with $x_i = 1$ or -1, the construction described in the previous paragraph produces an orthogonal basis of the $(2^n - 1)$-dimensional space of treatment contrasts, with one contrast corresponding to each nonempty subset of $\{1,\ldots,n\}$. We can write

$$\alpha(x_1, x_2, \ldots, x_n) = \beta_0 + \sum_{i=1}^{n} \beta_i x_i + \sum_{1 \leq i < j \leq n} \beta_{ij} x_i x_j + \sum_{1 \leq i < j < k \leq n} \beta_{ijk} x_i x_j x_k$$
$$+ \cdots + \beta_{1\ldots n} x_1 \cdots x_n, \tag{6.15}$$

where

$$\beta_{i_1 \cdots i_k} = \frac{1}{2^n} \left\{ \sum_{(x_1,\ldots,x_n):\, x_{i_1} \cdots x_{i_k} = 1} \alpha(x_1, \ldots, x_n) - \sum_{(x_1,\ldots,x_n):\, x_{i_1} \cdots x_{i_k} = -1} \alpha(x_1, \ldots, x_n) \right\}. \tag{6.16}$$

Suppose the factors are labeled $A, B, C, \ldots$, etc. For two-level factors, we often use a combination of lower case letters to represent a treatment combination and its effect. The rule is that a letter is present if and only if the corresponding factor is at level 1. For example, when there are four factors $A, B, C,$ and D, ac represents both the combination $(1, 0, 1, 0)$ and its effect α_{1010}. The treatment combination with all the factors at level 0 and its effect α_{0000} are denoted by (1). We also use a combination of capital letters to represent a factorial effect. For example, A represents both factor A and its main effect, and ACD represents the interaction of factors $A, C,$ and D. Then we have the following relations between the treatment and factorial effects when they are placed in standard orders:

$$\begin{bmatrix} (1) \\ a \\ b \\ ab \\ c \\ ac \\ bc \\ abc \\ \vdots \end{bmatrix} = \underbrace{\begin{bmatrix} 1 & -1 \\ 1 & 1 \end{bmatrix} \otimes \begin{bmatrix} 1 & -1 \\ 1 & 1 \end{bmatrix} \otimes \cdots \otimes \begin{bmatrix} 1 & -1 \\ 1 & 1 \end{bmatrix}}_{n} \begin{bmatrix} \mu \\ \frac{1}{2}A \\ \frac{1}{2}B \\ \frac{1}{2}AB \\ \frac{1}{2}C \\ \frac{1}{2}AC \\ \frac{1}{2}BC \\ \frac{1}{2}ABC \\ \vdots \end{bmatrix}, \tag{6.17}$$

$$
\begin{bmatrix} \mu \\ \frac{1}{2}A \\ \frac{1}{2}B \\ \frac{1}{2}AB \\ \frac{1}{2}C \\ \frac{1}{2}AC \\ \frac{1}{2}BC \\ \frac{1}{2}ABC \\ \vdots \end{bmatrix} = 2^{-n} \underbrace{ \begin{bmatrix} 1 & 1 \\ -1 & 1 \end{bmatrix} \otimes \begin{bmatrix} 1 & 1 \\ -1 & 1 \end{bmatrix} \otimes \cdots \otimes \begin{bmatrix} 1 & 1 \\ -1 & 1 \end{bmatrix} }_{n} \begin{bmatrix} (1) \\ a \\ b \\ ab \\ c \\ ac \\ bc \\ abc \\ \vdots \end{bmatrix}. \tag{6.18}
$$

The order of the entries on the left side of (6.17) is determined as follows. We start with (1), a. The next two entries b, ab are obtained by symbolically multiplying the first two by b subject to the rule that $x^2 = 1$. We then multiply the four entries that have been generated by c to obtain four more entries c, ac, bc, abc. The process is continued in the same manner if there are more factors. The same construction applies to the vector on the right side of (6.17) and those on both sides of (6.18).

When the factorial effects are represented by strings of capital letters, we also refer to them as *words*.

6.3 The general case

For any k distinct treatment factors $A_{i_1}, \ldots, A_{i_k}$, let $V_{A_{i_1} \wedge \cdots \wedge A_{i_k}}$ be the set of all the linear functions of the $s_1 \cdots s_n$ treatment effects whose coefficients depend only on the levels of $A_{i_1}, \ldots, A_{i_k}$, and let $W_{A_{i_1} \wedge \cdots \wedge A_{i_k}}$ be the set of all the treatment contrasts that represent the factorial effects involving all of $A_{i_1}, \ldots, A_{i_k}$, as defined in Definition 6.2. (Thus the contrasts in $W_{A_{i_1} \wedge \cdots \wedge A_{i_k}}$ represent main effects when $k = 1$ and represent k-factor interactions when $k > 1$.) In Section 6.8, the notations $V_{A_{i_1} \wedge \cdots \wedge A_{i_k}}$ and $W_{A_{i_1} \wedge \cdots \wedge A_{i_k}}$ will be explained in the language of Chapter 4; however, readers who will skip that optional section can simply think of $V_{A_{i_1} \wedge \cdots \wedge A_{i_k}}$ and $W_{A_{i_1} \wedge \cdots \wedge A_{i_k}}$ as somewhat complicated notations for what were defined in the previous sentence.

If each linear function of the treatment effects is identified with its coefficient vector, then $V_{A_{i_1} \wedge \cdots \wedge A_{i_k}}$ can be thought of as an $(s_{i_1} \cdots s_{i_k})$-dimensional subspace of $\mathbb{R}^{s_1 \cdots s_n}$, and $W_{A_{i_1} \wedge \cdots \wedge A_{i_k}}$ is a subspace of $V_{A_{i_1} \wedge \cdots \wedge A_{i_k}}$. According to Definition 6.2,

$$W_{A_i} = V_{A_i} \ominus V_U,$$

where V_U is the one-dimensional space of vectors with constant entries as defined in Section 4.3. Therefore

$$\dim(W_{A_i}) = s_i - 1.$$

We say that the main effect of A_i has $s_i - 1$ degrees of freedom.

For any two factors A_i and A_j, $i \neq j$, by Theorem 2.5, W_{A_i} and W_{A_j} are orthogonal. Then, by Definition 6.2, we have $W_{A_i \wedge A_j} = V_{A_i \wedge A_j} \ominus (V_U \oplus W_{A_i} \oplus W_{A_j})$. It follows that

$$
\begin{aligned}
\dim(W_{A_i \wedge A_j}) &= \dim(V_{A_i \wedge A_j}) - \dim(V_U) - \dim(W_{A_i}) - \dim(W_{A_j}) \\
&= s_i s_j - 1 - (s_i - 1) - (s_j - 1) \\
&= (s_i - 1)(s_j - 1).
\end{aligned}
$$

Thus the interactions of factors A_i and A_j have $(s_i - 1)(s_j - 1)$ degrees of freedom. For any three distinct factors A_i, A_j, and A_k, it can be shown that $W_{A_i \wedge A_j}$, $W_{A_i \wedge A_k}$, and $W_{A_j \wedge A_k}$ are mutually orthogonal; see Exercise 6.6. Thus $W_{A_i \wedge A_j \wedge A_k} = V_{A_i \wedge A_j \wedge A_k} \ominus (V_U \oplus W_{A_i} \oplus W_{A_j} \oplus W_{A_k} \oplus W_{A_i \wedge A_j} \oplus W_{A_i \wedge A_k} \oplus W_{A_j \wedge A_k})$. It follows that

$$
\begin{aligned}
\dim\left(W_{A_i \wedge A_j \wedge A_k}\right) &= s_i s_j s_k - 1 - (s_i - 1) - (s_j - 1) - (s_k - 1) \\
&\quad - (s_i - 1)(s_j - 1) - (s_i - 1)(s_k - 1) - (s_j - 1)(s_k - 1) \\
&= (s_i - 1)(s_j - 1)(s_k - 1).
\end{aligned}
$$

We say that the interactions of factors A_i, A_j, and A_k have $(s_i - 1)(s_j - 1)(s_k - 1)$ degrees of freedom.

For any k factors $A_{i_1}, \ldots, A_{i_k}$,

$$
\dim\left(W_{A_{i_1} \wedge \cdots \wedge A_{i_k}}\right) = (s_{i_1} - 1) \cdots (s_{i_k} - 1). \tag{6.19}
$$

All these W spaces, one for each nonempty subset of $\{1, \ldots, n\}$, together form an orthogonal decomposition of the $(s_1 \cdots s_n - 1)$-dimensional space of treatment contrasts. For $s_1 = \cdots = s_n = 2$, by (6.19), $\dim(A_{i_1} \wedge \cdots \wedge A_{i_k}) = 1$ for all subsets of factors. In this case, each main effect or interaction component has one degree of freedom. We have shown in Section 6.2 how to choose a set of $2^n - 1$ mutually orthogonal contrasts, each of which represents a main effect or interaction. In the general case, we can choose a set of $(s_{i_1} - 1) \cdots (s_{i_k} - 1)$ mutually orthogonal contrasts from each $W_{A_{i_1} \wedge \cdots \wedge A_{i_k}}$. Together we have $s_1 \cdots s_n - 1$ mutually orthogonal contrasts that form an orthogonal basis of the space of all the treatment contrasts. We present one way of constructing such contrasts due to Kurkjian and Zelen (1962).

As in the two-level case discussed in Section 6.2, the first step is to choose, for each $1 \le i \le n$, $s_i - 1$ mutually orthogonal treatment contrasts that represent the main effect of factor A_i. To do this we choose an $s_i \times s_i$ matrix

$$
\mathbf{P}_{s_i} = \begin{bmatrix} \mathbf{p}_0^i & \mathbf{p}_1^i & \cdots & \mathbf{p}_{s_i-1}^i \end{bmatrix} = \begin{bmatrix} 1 & p_{01}^i & \cdots & p_{0,s_i-1}^i \\ 1 & p_{11}^i & \cdots & p_{1,s_i-1}^i \\ \vdots & \vdots & \vdots & \vdots \\ 1 & p_{s_i-1,1}^i & \cdots & p_{s_i-1,s_i-1}^i \end{bmatrix}, \tag{6.20}
$$

with mutually orthogonal columns and $\mathbf{p}_0^i = \mathbf{1}$, the all-one vector. For each $1 \le j \le s_i - 1$, define a treatment contrast in which the coefficient of a treatment combination is $p_{l,j}^i$ if factor A_i is at level l in that treatment combination; so each row of $\mathbf{P}_{s_i}$ is associated with a level of A_i, and each of the last $s_i - 1$ columns of $\mathbf{P}_{s_i}$ is used to define a treatment contrast. Clearly the resulting treatment contrasts have their coefficients depending only on the levels of A_i, and therefore are main-effect contrasts of A_i. Since the columns of $\mathbf{P}_{s_i}$ are mutually orthogonal, the $s_i - 1$ treatment contrasts so constructed are mutually orthogonal and form an orthogonal basis of W_{A_i}.

The Hadamard product of any of the $s_i - 1$ columns defining the main-effect contrasts of A_i and any of the $s_j - 1$ columns defining the main-effect contrasts of A_j,

$i \neq j$, is a column whose entries depend only on the levels of A_i and A_j. Altogether we can construct $(s_i - 1)(s_j - 1)$ such contrasts. Similarly, for any $1 \leq i_1 < \cdots < i_k \leq n$, all possible Hadamard products of k columns, one from each of the $s_{i_h} - 1$ columns that define main-effect contrasts of $A_{i_h}, 1 \leq h \leq k$, have entries depending only on the levels of $A_{i_1}, \ldots, A_{i_k}$. Altogether $(s_{i_1} - 1) \cdots (s_{i_k} - 1)$ such contrasts can be constructed. This construction yields a total of $\sum_{k=1}^{n} \sum_{1 \leq i_1 < \cdots < i_k \leq n} (s_{i_1} - 1) \cdots (s_{i_k} - 1) = s_1 \cdots s_n - 1$ treatment contrasts. We will prove at the end of this section that these contrasts are mutually orthogonal. Then, by Definition 6.2, they represent various treatment factorial effects and form an orthogonal basis of the space of treatment contrasts. This extends the construction for the two-level case in Section 6.2.

For the case with $s_1 = \cdots = s_n = 2$, the columns of $\mathbf{P}_2$ are unique up to multiplicative constants. The construction in Section 6.2 was based on the matrix

$$\mathbf{P}_2 = \begin{bmatrix} 1 & -1 \\ 1 & 1 \end{bmatrix}.$$

For $s_i > 2$, however, there are infinitely many choices. When the factors are quantitative, treatment contrasts defined by orthogonal polynomials are commonly used. One can choose $\mathbf{P}_{s_i}$ so that, for each $1 \leq j \leq s_i - 1$, the column $\mathbf{p}_j^i$ consists of the values of a jth degree orthogonal polynomial at the s_i levels. For three levels, this results in the matrix

$$\mathbf{P}_3 = \begin{bmatrix} 1 & -1 & 1 \\ 1 & 0 & -2 \\ 1 & 1 & 1 \end{bmatrix}, \tag{6.21}$$

where the second and third columns consist of the values of x and $3x^2 - 2$, respectively, at $x = -1, 0$, and 1.

Example 6.1. For $n = 2$, $s_1 = s_2 = 3$, the following columns constitute an orthogonal basis of the eight-dimensional space of treatment contrasts.

	A_l	A_q	B_l	B_q	$A_l \times B_l$	$A_l \times B_q$	$A_q \times B_l$	$A_q \times B_q$
00	-1	1	-1	1	1	-1	-1	1
10	0	-2	-1	1	0	0	2	-2
20	1	1	-1	1	-1	1	-1	1
01	-1	1	0	-2	0	2	0	-2
11	0	-2	0	-2	0	0	0	4
21	1	1	0	-2	0	-2	0	-2
02	-1	1	1	1	-1	-1	1	1
12	0	-2	1	1	0	0	-2	-2
22	1	1	1	1	1	1	1	1

Suppose the two factors are denoted by A and B. The columns labeled A_l and B_l are based on the second column of the matrix in (6.21): an entry is equal to -1, 0, or 1, depending on whether the factor is at level 0, 1, or 2, respectively. On the other hand, the two columns labeled A_q and B_q are based on the third column of the same matrix. If we code the three levels by -1, 0, and 1, then columns A_l, A_q, B_l, and B_q give the values of the orthogonal polynomials x_1, $3x_1^2 - 2$, x_2, and $3x_2^2 - 2$, respectively, where x_1 and x_2 are the levels of A and B, respectively. Here l stands for linear and q stands for quadratic. The main-effect contrasts defined by these columns are referred to as linear A, quadratic A, linear B, and quadratic B contrasts, respectively. The column labeled $A_l \times B_q$ is the Hadamard product of the two columns labeled by A_l and B_q. It consists of the values of $x_1(3x_2^2 - 2)$, and defines a contrast representing the linear $A \times$ quadratic B component of the interaction of factors A and B.

Now we prove the mutual orthogonality of the contrasts constructed above for $s_1 \times \cdots \times s_n$ experiments. For each $\mathbf{z} = (z_1, \ldots, z_n)$, where $z_i \in \{0, 1, \ldots, s_i - 1\}$, we define an $(s_1 \cdots s_n) \times 1$ vector $\mathbf{p}^{\mathbf{z}}$ as

$$\mathbf{p}^{\mathbf{z}} = \mathbf{p}_{z_1}^1 \otimes \cdots \otimes \mathbf{p}_{z_n}^n. \tag{6.22}$$

For $\mathbf{z} = \mathbf{0}$, $\mathbf{p}^{\mathbf{z}}$ is the column of all 1's. It can be seen that if $\mathbf{z}$ contains exactly one nonzero entry, say $z_i \neq 0$, and all the other components of $\mathbf{z}$ are equal to zero, then $\mathbf{p}^{\mathbf{z}}$ defines a main-effect contrast of A_i. Specifically, if $z_i = j$, then $\mathbf{p}^{\mathbf{z}}$ is the main-effect column that is constructed by using the entries of $\mathbf{p}_j^i$. If $z_{i_1}, \ldots, z_{i_k}$ are nonzero and all the other entries of $\mathbf{z}$ are equal to zero, then $\mathbf{p}^{\mathbf{z}}$ is the Hadamard product of k columns, one each from the $s_{i_h} - 1$ columns defining main-effect contrasts of A_{i_h}, $1 \leq h \leq k$. Therefore the $s_1 \cdots s_n - 1$ columns constructed by using the method presented in this section are the $s_1 \cdots s_n - 1$ vectors $\mathbf{p}^{\mathbf{z}}$ with $\mathbf{z} \neq \mathbf{0}$. Together with $\mathbf{p}^0$, the all-one vector, they are the $s_1 \cdots s_n$ columns of $\mathbf{P}_{s_1} \otimes \cdots \otimes \mathbf{P}_{s_n}$. Therefore it suffices to show that the columns of $\mathbf{P}_{s_1} \otimes \cdots \otimes \mathbf{P}_{s_n}$ are mutually orthogonal. Now the mutual orthogonality of the columns of each $\mathbf{P}_{s_i}$ implies that $\mathbf{P}_{s_i}^T \mathbf{P}_{s_i}$ is a diagonal matrix for all $1 \leq i \leq n$. It follows that $(\mathbf{P}_{s_1} \otimes \cdots \otimes \mathbf{P}_{s_n})^T (\mathbf{P}_{s_1} \otimes \cdots \otimes \mathbf{P}_{s_n}) = (\mathbf{P}_{s_1}^T \mathbf{P}_{s_1}) \otimes \cdots \otimes (\mathbf{P}_{s_n}^T \mathbf{P}_{s_n})$ is also a diagonal matrix. This completes the proof.

Thus for each $W_{A_{i_1} \wedge \cdots \wedge A_{i_k}}$, the construction in this section yields an orthogonal basis that consists of the treatment contrasts $(\mathbf{p}^{\mathbf{z}})^T \boldsymbol{\alpha}$ whose coefficient vectors are such that $z_{i_1}, \ldots, z_{i_k}$ are nonzero, and all the other entries of $\mathbf{z}$ are equal to zero. The set of all the $(\mathbf{p}^{\mathbf{z}})^T \boldsymbol{\alpha}$'s with $\mathbf{z} \neq \mathbf{0}$ is an orthogonal basis of all the treatment contrasts.

Since the $\mathbf{p}^{\mathbf{z}}$'s, including that with $\mathbf{z} = \mathbf{0}$, form an orthogonal basis of $\mathbb{R}^{s_1 \cdots s_n}$, we have

$$\boldsymbol{\alpha} = \sum_{\mathbf{z} \in S_1 \times \cdots \times S_n} \beta^{\mathbf{z}} \mathbf{p}^{\mathbf{z}}, \tag{6.23}$$

where

$$\beta^{\mathbf{z}} = \frac{1}{\|\mathbf{p}^{\mathbf{z}}\|^2} (\mathbf{p}^{\mathbf{z}})^T \boldsymbol{\alpha}. \tag{6.24}$$

This provides a parametrization of the treatment effects in terms of the factorial effects. Except for a multiplicative constant, each $\beta^{\mathbf{z}}$ with $\mathbf{z} \neq \mathbf{0}$ is a main effect or interaction contrast defined above. Note that (6.15) and (6.16) are special cases of

(6.23) and (6.24), respectively.

Example 6.2. (Example 6.1 continued) We have

$$
\begin{aligned}
\alpha(x_1, x_2) = {} & \beta_0 + \beta_{A_l} x_1 + \beta_{B_l} x_2 + \beta_{A_q}(3x_1^2 - 2) + \beta_{B_q}(3x_2^2 - 2) \\
& + \beta_{A_l B_l} x_1 x_2 + \beta_{A_l B_q} x_1 (3x_2^2 - 2) + \beta_{A_q B_l}(3x_1^2 - 2)x_2 \\
& + \beta_{A_q B_q}(3x_1^2 - 2)(3x_2^2 - 2),
\end{aligned}
$$

for some constants β_0, β_{A_l}, β_{B_l}, β_{A_q}, β_{B_q}, $\beta_{A_l B_l}$, $\beta_{A_l B_q}$, $\beta_{A_q B_l}$, and $\beta_{A_q B_q}$, where $x_i = -1$, 0, or 1. For example, the main-effect contrast $\beta_{A_l} = \frac{1}{6}[\sum_{x_2} \alpha(1, x_2) - \sum_{x_2} \alpha(-1, x_2)]$ gives an overall comparison of level 1 and level -1 of factor A, and $\beta_{A_q} = \frac{1}{18}[\sum_{x_2} \alpha(1, x_2) + \sum_{x_2} \alpha(-1, x_2) - 2\sum_{x_2} \alpha(0, x_2)] = \frac{1}{18}\{[\sum_{x_2} \alpha(1, x_2) - \sum_{x_2} \alpha(0, x_2)] - [\sum_{x_2} \alpha(0, x_2) - \sum_{x_2} \alpha(-1, x_2)]\}$. If A is a qualitative factor, then β_{A_q} is a comparison between level 0 and the other two levels of A. If A is quantitative, then it is also a measure of curvature.

Let $\mathbf{P}$ be the $(s_1 \cdots s_n) \times (s_1 \cdots s_n)$ matrix $\mathbf{P}_{s_1} \otimes \cdots \otimes \mathbf{P}_{s_n}$, where $\mathbf{P}_{s_i}$ is as in (6.20), and $\boldsymbol{\beta}$ be the $(s_1 \cdots s_n) \times 1$ vector whose components are the $\beta^{\mathbf{z}}$'s ordered lexicographically according to the components of $\mathbf{z}$. If the components of $\boldsymbol{\alpha}$ are also placed in the corresponding standard order, then (6.23) can be written as

$$
\boldsymbol{\alpha} = \mathbf{P}\boldsymbol{\beta} = \left(\mathbf{P}_{s_1} \otimes \cdots \otimes \mathbf{P}_{s_n} \right) \boldsymbol{\beta}, \tag{6.25}
$$

and

$$
\boldsymbol{\beta} = \left(\mathbf{P}_{s_1}^{-1} \otimes \cdots \otimes \mathbf{P}_{s_n}^{-1} \right) \boldsymbol{\alpha}. \tag{6.26}
$$

Note that (6.17) and (6.18) are special cases of (6.25) and (6.26), respectively, for two-level factors.

If each of the matrices $\mathbf{P}_{s_i}$ in (6.20) is an orthogonal matrix (each column is normalized to have unit length), then $\mathbf{P}_{s_1} \otimes \cdots \otimes \mathbf{P}_{s_n}$ is also an orthogonal matrix, and all the vectors $\mathbf{p}^{\mathbf{z}}$ in the orthogonal basis constructed earlier have unit length. Then (6.24) becomes

$$
\beta^{\mathbf{z}} = \left(\mathbf{p}^{\mathbf{z}} \right)^T \boldsymbol{\alpha}. \tag{6.27}
$$

An alternative parametrization that defines the factorial effects with reference to natural baseline levels of the factors, called *baseline parametrization*, was proposed by Yang and Speed (2002) for cDNA microarray experiments when the cell populations have a factorial structure. Banerjee and Mukerjee (2008) studied optimal factorial designs under this parametrization. Unlike the approach discussed here, the contrasts of interest under the baseline parametrization are not orthogonal.

6.4 Analysis of complete factorial designs

Suppose each of the $s_1 \cdots s_n$ treatment combinations is replicated q times, $q \geq 1$, and the experiment is run according to a completely randomized design, a randomized complete block design, or a randomized Latin square design (or more generally a design in which each treatment combination appears the same number of times in each row and the same number of times in each column). Then by Theorem 5.2, each normalized factorial effect β^z in (6.27) is estimated by

$$\widehat{\beta}^z = (\mathbf{p}^z)^T \widehat{\alpha},$$

where $\widehat{\alpha}$ is the vector of treatment means. We have

$$\text{var}\left(\widehat{\beta}^z\right) = \frac{\|\mathbf{p}^z\|^2 \xi}{q} = \frac{\xi}{q}, \tag{6.28}$$

where, depending on whether a completely randomized design, a randomized complete block design, or a randomized Latin square design is used, ξ is the overall variance, intrablock variance, or unit variance adjusted for the row variability and column variability.

With each normalized β^z there is associated a sum of squares

$$\text{SS}(\beta^z) = q\left[(\mathbf{p}^z)^T \widehat{\alpha}\right]^2,$$

carrying one degree of freedom. The treatment sum of squares is equal to the sum of all these single-degree-of-freedom sums of squares.

For $q > 1$, the residual sum of squares has $(q-1)s_1 \cdots s_n$ degrees of freedom under a completely randomized design, $(q-1)(s_1 \cdots s_n - 1)$ degrees of freedom under a randomized complete block design, and $t^2 - 3t + 2$ degrees of freedom, where $t = s_1 \cdots s_n$, under a randomized Latin-square design.

If it is known a priori that some of the factorial effects are negligible, then we have a reduced treatment model after the negligible effects are dropped. Under the full model,

$$E(\mathbf{y}) = \mu \mathbf{1}_N + \mathbf{X}_T \alpha = \mu \mathbf{1}_N + \mathbf{X}_T \mathbf{P} \beta.$$

Suppose $\beta = \begin{bmatrix} \beta_1^T & \beta_2^T \end{bmatrix}^T$ and $\beta_2 = \mathbf{0}$. Then the estimates of the parameters in β_1 and their sums of squares are computed in the usual way, but the sums of squares associated with the negligible factorial effects in β_2 are pooled into the residual sum of squares.

For two-level designs, the estimates of various factorial effects can be obtained from (6.16) or (6.18) by replacing the effect of each treatment combination with the average of all the observations on that treatment combination. Note that the factorial effects in (6.16) are not normalized. Each of these contrasts is estimated with variance $\xi/(q2^n)$, and the corresponding sum of squares is equal to the squared estimate multiplied by $q2^n$. The fact that the matrix relating the vector of factorial effects to that of the treatment effects is a repeated Kronecker product of $\begin{bmatrix} 1 & 1 \\ -1 & 1 \end{bmatrix}$ with itself leads to the following Yates algorithm (Yates, 1937), a convenient method for calculating estimates of factorial effects in the two-level case, though nowadays these estimates can easily be obtained through computer software.

Yates algorithm for two-level experiments:

Step 1. Set $i = 1$. Put the treatment means in column 1, arranged in the standard order of the corresponding factor-level combinations.

Step 2. Generate column $i + 1$ from column i. For each $1 \leq m \leq 2^{n-1}$, let the mth entry of column $i + 1$ be the sum of the $(2m)$th and $(2m - 1)$th entries of column i. This gives the first half of column $i + 1$. For the last half, subtract the $(2m - 1)$th entry of column i from the $(2m)$th entry, and let the difference be the $(2^{n-1} + m)$th entry of column $i + 1$.

Step 3. Increase i by 1. Stop if $i = n + 1$; otherwise go back to step 2 and continue.

Then the first entry of column $n + 1$ divided by 2^n is the grand mean. Dividing each of the remaining entries by 2^{n-1}, we obtain the estimates of various factorial effects arranged in the standard order.

6.5   Analysis of unreplicated experiments

In this section, some methods for the analysis of unreplicated complete factorial experiments are presented. When $q > 1$, the stratum variance ξ in (6.28) can be estimated by the residual mean square in the ANOVA table. In a completely randomized single-replicate complete factorial experiment, there is no degree of freedom for residual and ξ cannot be estimated under the full treatment model. Tests of significance cannot be carried out in this case unless some interactions are negligible. Alternatively, an informal graphical method based on a normal or half-normal probability plot due to Daniel (1959, 1976) can be used. Let $m = s_1 \cdots s_n - 1$, and $\widehat{\beta}_{[1]} \leq \widehat{\beta}_{[2]} \leq \cdots \leq \widehat{\beta}_{[m]}$ be the ordered values of the factorial-effect estimates. A normal probability plot is constructed by plotting the order statistics $\widehat{\beta}_{[i]}$ versus the normal quantiles $\Phi^{-1}((i - 0.5)/m)$ for $i = 1, \ldots, m$, where Φ is the cumulative distribution function of $N(0, 1)$. Under the normality assumption for the observations, if all the normalized factorial effects are zero, then their estimates would behave like a random sample from $N(0, \xi)$, and the points on the plot would fall roughly on a straight line. Thus the effects whose corresponding points fall off the line are declared significant. An alternative procedure is to use a half-normal probability plot, in which $|\widehat{\beta}|_{[i]}$ versus $\Phi^{-1}(0.5 + 0.5(i - 0.5)/m)$ is plotted.

Using normal or half-normal probability plots requires subjective judgment. Lenth (1989) proposed a statistical test based on the assumption of *effect sparsity* that only a small number of the β_i's are nonzero. Given estimates $\widehat{\beta}_1, \ldots, \widehat{\beta}_m$ of the factorial effects, one needs an estimate of the standard error of $\widehat{\beta}_i$ to carry out a test. Lenth proposed a robust estimator, called *pseudo standard error* (PSE):

$$\text{PSE} = 1.5 \cdot \underset{\{|\widehat{\beta}_i| < 2.5 \cdot s_0\}}{\text{median}} \left| \widehat{\beta}_i \right|, \tag{6.29}$$

where the median is computed among the $\widehat{\beta}_i$'s with $|\widehat{\beta}_i| < 2.5 \cdot s_0$, and

$$s_0 = 1.5 \cdot \operatorname{median} \left|\widehat{\beta}_i\right|. \tag{6.30}$$

This estimator can be justified as follows. Under effect sparsity, only a small number of the β_i's are nonzero. Tentatively, assume that all of them are zero. Then under the normality assumption, $\widehat{\beta}_1, \ldots, \widehat{\beta}_m$ can be regarded as a random sample from $N(0, \xi)$. Since $P(|Z| \geq 0.675) = 0.5$ for a standard normal random variable Z, $\operatorname{median}|\widehat{\beta}_i|$ is an estimate of $0.675\sqrt{\xi}$. It follows that if all the β_i's are zero, then $s_0 = 1.5 \cdot \operatorname{median}|\widehat{\beta}_i|$ is an estimate of $(1.5)(0.675)\sqrt{\xi} \approx \sqrt{\xi}$. Since a small number of β_i's are nonzero, as an attempt to remove the estimates that are associated with nonzero β_i's, (6.30) is repeated, but with the median taken over only those with $|\widehat{\beta}_i| < 2.5 \cdot s_0$. This results in the PSE in (6.29). Since $P\{|Z| \geq 2.57\} = 0.01$, this amounts to trimming about 1% of the $\widehat{\beta}_i$'s if all the β_i's were zero. The reason the median is used is due to its robustness against extreme $|\widehat{\beta}_i|$ values.

After the pseudo standard error is computed, one can use the statistic $\widehat{\beta}_i/\text{PSE}$ to perform the test. The null distribution of this statistic was studied by Ye and Hamada (2000) by simulation. Critical values for IER (individual error rate) and EER (experimentwise error rate) were provided. The former controls the probability of type I error for individual tests, while the latter accounts for multiple testing since more than one null hypothesis are tested simultaneously.

Box and Meyer (1986) proposed a Bayesian analysis of unreplicated factorial designs. Suppose each factorial effect has probability $1 - \alpha$ to be inert in the sense that its usual estimator is normally distributed with zero mean and variance σ^2, and probability α to be active in the sense that its estimator is normally distributed with zero mean and a much larger variance $k^2\sigma^2$. The prior information can be combined with the data to produce posterior probabilities that the effects are active.

6.6   Defining factorial effects via finite geometries

When $s_1 = \cdots = s_n = s$ is a prime number or power of a prime number, a one-to-one correspondence can be established between the treatment combinations and the points in the n-dimensional Euclidean geometry $EG(n, s)$ with s points per line. This leads to an alternative choice of mutually orthogonal contrasts representing various factorial effects (Bose and Kishen, 1940; Bose, 1947), useful for design construction. For a brief introduction to finite Euclidean geometry, see Section A.4.

In the usual Euclidean geometry $\mathbb{R}^n$, real numbers are used to set up the coordinates $(x_1, \ldots, x_n)$ of each point. In $EG(n, s)$, the set of real numbers $\mathbb{R}$ is replaced by a finite set F of size s. While there are infinitely many points in $\mathbb{R}^n$, the total number of points in $EG(n, s)$ is s^n since there are s possible values for each x_i. As in $\mathbb{R}$, we also need two operations + (addition) and $\cdot$ (multiplication) to perform the arithmetic in F, and we require F to be a *field* under these two operations. (See Section A.2.) A field with s elements is denoted as $GF(s)$. The arithmetic rules in the definition of a field allow one to solve linear equations. Then the notions of linear independence, basis, etc., are defined in the same way as in the linear algebra based on real numbers. Therefore the usual results in linear algebra still apply. In the special case where s is a prime number, we take F to be $\mathbb{Z}_s = \{0, 1, \ldots, s-1\}$ with modulo s addition and multiplication.

Each nonzero vector $\mathbf{a} \in EG(n, s)$ defines a partition of the s^n points (vectors) $\mathbf{x}$ in $EG(n, s)$ into s disjoint sets according to the values of $\mathbf{a}^T\mathbf{x}$. Each such set, called an $(n-1)$-flat, consists of the s^{n-1} solutions to the equation $\mathbf{a}^T\mathbf{x} = b$, where b is one of the s elements of $GF(s)$. The set of the s disjoint $(n-1)$-flats defined by the same $\mathbf{a}$, denoted by $P(\mathbf{a})$, is called a *pencil* of $(n-1)$-flats.

If $\mathbf{a}_1, \ldots, \mathbf{a}_k$ are k linearly independent vectors in $EG(n, s)$, $k < n$, then we can partition the s^n points $\mathbf{x}$ in $EG(n, s)$ into s^k disjoint sets according to the values of $\mathbf{a}_1^T\mathbf{x}, \ldots, \mathbf{a}_k^T\mathbf{x}$. Each set contains s^{n-k} points and is called an $(n-k)$-flat. The set of these s^k disjoint $(n-k)$-flats, denoted by $P(\mathbf{a}_1, \ldots, \mathbf{a}_k)$, is called a pencil of $(n-k)$-flats.

We identify each of the s^n points in $EG(n, s)$ with a treatment combination in an s^n experiment. For any nonzero vector $\mathbf{a} \in EG(n, s)$, let $H_0, \ldots, H_{s-1}$ be the s disjoint $(n-1)$-flats in $P(\mathbf{a})$. Consider a function of the treatment effects of the form

$$\tau = \sum_{i=0}^{s-1} c_i \sum_{\mathbf{x} \in H_i} \alpha(\mathbf{x}), \tag{6.31}$$

where $c_i \in \mathbb{R}$ and $\sum_{i=0}^{s-1} c_i = 0$. Since all the $(n-1)$-flats contain s^{n-1} points, the sum of the coefficients of τ in (6.31) is equal to zero; therefore τ is a treatment contrast. We call τ a treatment contrast defined by $\mathbf{a}$ (or by the pencil $P(\mathbf{a})$).

The same argument shows that if $\sum_{i=0}^{s-1} d_i = 0$ and $\sum_{i=0}^{s-1} c_i d_i = 0$, then the two contrasts $\sum_{i=0}^{s-1} c_i \sum_{\mathbf{x} \in H_i} \alpha(\mathbf{x})$ and $\sum_{i=0}^{s-1} d_i \sum_{\mathbf{x} \in H_i} \alpha(\mathbf{x})$ are orthogonal. Suppose $\{\mathbf{p}_1, \ldots, \mathbf{p}_{s-1}\}$ is a set of mutually orthogonal $s \times 1$ vectors that are orthogonal to the vector of 1's, say $\mathbf{p}_j = (p_{0j}, \ldots, p_{s-1,j})^T$. For each $1 \le j \le s-1$, define a treatment contrast τ_j as $\sum_{i=0}^{s-1} p_{ij} \sum_{\mathbf{x} \in H_i} \alpha(\mathbf{x})$. Then $\tau_1, \ldots, \tau_{s-1}$ are mutually orthogonal. Thus given each pencil of $(n-1)$-flats (or each nonzero $\mathbf{a}$), we can construct $s-1$ mutually orthogonal treatment contrasts based on a set of $s-1$ mutually orthogonal vectors in $\mathbb{R}^s$ that are orthogonal to the vector of 1's.

Example 6.3. Suppose $n = 3$ and $s_1 = s_2 = s_3 = 3$. Denote the three factors by A, B, and C, respectively. Then the vector $(1, 0, 0)^T$ defines a partition of the 27 treatment combinations into three sets of size 9 according to the values of x_1. The three subsets $\{(x_1, x_2, x_3)^T : x_1 = 0\}$, $\{(x_1, x_2, x_3)^T : x_1 = 1\}$, and $\{(x_1, x_2, x_3)^T : x_1 = 2\}$ can be used to construct two orthogonal contrasts of the form $c_0 \sum_{x_2, x_3} \alpha(0, x_2, x_3) + c_1 \sum_{x_2, x_3} \alpha(1, x_2, x_3) + c_2 \sum_{x_2, x_3} \alpha(2, x_2, x_3)$. They are contrasts of the three levels of factor A, therefore representing its main effect. We denote the two degrees of freedom defined in this way by A. Likewise the two vectors $(0, 1, 0)$ and $(0, 0, 1)$ partition the 27 treatment combinations according to the values of x_2 and x_3, respectively, each defining two degrees of freedom for the main effects of factor B and factor C.

If $\mathbf{a}$ and $\mathbf{b}$ are two nonzero vectors in $EG(n,s)$ such that $\mathbf{a} = \lambda\mathbf{b}$ for some $\lambda \in GF(s)$, then $\mathbf{a}^T\mathbf{x} = \mathbf{a}^T\mathbf{y} \Leftrightarrow \mathbf{b}^T\mathbf{x} = \mathbf{b}^T\mathbf{y}$. This implies that $\mathbf{a}$ and $\mathbf{b}$ yield the same partition of the points of $EG(n,s)$. Therefore they define the same pencil of $(n-1)$-flats and the same treatment contrasts. On the other hand, suppose $\mathbf{a} \neq \lambda\mathbf{b}$ for all $\lambda \in GF(s)$. Let $H_0,\ldots,H_{s-1}$ be the s disjoint $(n-1)$-flats in $P(\mathbf{a})$ and $K_0,\ldots,K_{s-1}$ be the $(n-1)$-flats in $P(\mathbf{b})$. Then for each $0 \leq i, j \leq s-1$, $H_i \cap K_j$ is an $(n-2)$-flat, and therefore contains s^{n-2} points. Let

$$\tau = \sum_{i=0}^{s-1} c_i \sum_{\mathbf{x}\in H_i} \alpha(\mathbf{x}) \text{ and } \varphi = \sum_{j=0}^{s-1} d_j \sum_{\mathbf{x}\in K_j} \alpha(\mathbf{x}),$$

where $\sum_{i=0}^{s-1} c_i = 0$ and $\sum_{j=0}^{s-1} d_j = 0$. Then the inner product of the coefficient vectors of τ and φ is equal to $s^{n-2}\sum_{i=0}^{s-1}\sum_{j=0}^{s-1} c_i d_j = 0$. This shows that contrasts defined by different pencils are orthogonal to each other.

Each nonzero vector in $EG(n,s)$ defines one pencil of $(n-1)$-flats, and two vectors $\mathbf{a}$ and $\mathbf{b}$ define the same pencil if and only if $\mathbf{a} = \lambda\mathbf{b}$ for some nonzero $\lambda \in GF(s)$. Therefore there are a total of $(s^n-1)/(s-1)$ pencils of $(n-1)$-flats in $EG(n,s)$. From each pencil, we can construct $s-1$ mutually orthogonal contrasts. Altogether, we obtain $[(s^n-1)/(s-1)](s-1) = s^n - 1$ mutually orthogonal contrasts, which form an orthogonal basis of the space of all the treatment contrasts. By definition, if $\mathbf{a}$ has k nonzero entries $a_{i_1},\ldots,a_{i_k}$, then the coefficients of the contrasts defined by $\mathbf{a}$ depend only on the levels of factors $i_1,\ldots,i_k$. Therefore these contrasts are interaction contrasts of factors $i_1,\ldots,i_k$ if $k > 1$ and are main-effect contrasts if $k = 1$.

When $s_1 = \cdots = s_n = 2$, this yields contrasts that are equivalent to those given in (6.16). Suppose $\mathbf{a}$ is a vector in $EG(n,2)$ with k nonzero entries $a_{i_1},\ldots,a_{i_k}$. Then $a_{i_1} = \cdots = a_{i_k} = 1$, and all the other entries are 0. A contrast as in (6.31), except for a multiplicative constant, must be of the form

$$\sum_{(x_1,\ldots,x_n):\, x_{i_1}+\cdots+x_{i_k}=0} \alpha(x_1,\ldots,x_n) - \sum_{(x_1,\ldots,x_n):\, x_{i_1}+\cdots+x_{i_k}=1} \alpha(x_1,\ldots,x_n).$$

If we replace level 0 with -1, then the above contrast reduces to that in (6.16) except for a possible sign change.

For the case of more than two levels, the contrasts constructed by using finite geometry as discussed here are different from those based on orthogonal polynomials and might not have interpretable physical meanings, but they are convenient and useful for design construction and other purposes, as we will see in later chapters.

Example 6.4. (Example 6.3 continued) The vector $(1,1,0)^T$ partitions the treatment combinations into three sets of size nine according to $x_1 + x_2 \equiv 0$, 1, or 2 (mod 3). Two orthogonal contrasts of the three sets can be constructed, and the two degrees of freedom obtained are denoted by AB. The coefficients of these contrasts depend only on the levels of A and B; therefore they represent the interaction of factors A

and B. With three levels for each factor, there are four degrees of freedom for a two-factor interaction. The other two degrees of freedom, obtained by using $(1,2,0)^T$ and denoted by AB^2, are based on the partition of the treatment combinations according to the values of $x_1 + 2x_2$. For each nonzero vector $\mathbf{a}$, $\mathbf{a}$ and $2\mathbf{a}$ yield the same partition of the treatment combinations. Without loss of generality, we only have to consider the nonzero vectors of $EG(3,3)$ in which the first nonzero entry is 1. There are 13 such vectors: $(1,0,0)^T$, $(0,1,0)^T$, $(0,0,1)^T$, $(1,1,0)^T$, $(1,2,0)^T$, $(1,0,1)^T$, $(1,0,2)^T$, $(0,1,1)^T$, $(0,1,2)^T$, $(1,1,1)^T$, $(1,1,2)^T$, $(1,2,1)^T$, $(1,2,2)^T$, leading to a partition of the 26 degrees of freedom for treatment contrasts into 13 sets of two degrees of freedom each, denoted by A, B, C, AB, AB^2, AC, AC^2, BC, BC^2, ABC, ABC^2, AB^2C, AB^2C^2, respectively.

6.7 Defining factorial effects via Abelian groups

The geometric approach presented in Section 6.6 applies only to the symmetric case $s_1 = \cdots = s_n = s$, where s is a prime number or power of a prime number. Bailey, Gilchrist, and Patterson (1977) developed an algebraic method that can be applied to the asymmetrical case as well; furthermore the numbers of levels of the treatment factors are not necessarily prime numbers or powers of prime numbers.

Suppose there are n treatment factors $A_1, \ldots, A_n$ with $s_1, \ldots, s_n$ levels, respectively. The levels of A_i can be identified with the elements of $\mathbb{Z}_{s_i} = \{0, \ldots, s_i - 1\}$. Then the treatment combinations can be identified with the elements of $\mathbb{Z}_{s_1} \times \cdots \times \mathbb{Z}_{s_n}$, which is an Abelian group with respect to the operation $(x_1, \ldots, x_n)^T + (y_1, \ldots, y_n)^T = (z_1, \ldots, z_n)^T$, where $z_i = x_i + y_i \pmod{s_i}$. Since the operation of addition is used, hereafter we write $\mathbb{Z}_{s_1} \times \cdots \times \mathbb{Z}_{s_n}$ as $\mathbb{Z}_{s_1} \oplus \cdots \oplus \mathbb{Z}_{s_n}$. For any $\mathbf{a}$, $\mathbf{x} \in \mathbb{Z}_{s_1} \oplus \cdots \oplus \mathbb{Z}_{s_n}$, define

$$[\mathbf{a}, \mathbf{x}] = \sum_{i=1}^{n} a_i \frac{l}{s_i} x_i \pmod{l}, \tag{6.32}$$

where l is the least common multiple of $s_1, \ldots, s_n$. Note that when $s_1 = \cdots = s_n = s$, $[\mathbf{a}, \mathbf{x}] - \mathbf{a}^T \mathbf{x} \pmod{s}$.

Suppose the order of a nonzero $\mathbf{a}$ is p; that is, p is the smallest positive integer such that $p\mathbf{a} = \mathbf{0}$. Then it can be shown that $[\mathbf{a}, \mathbf{x}]$ has p distinct values as $\mathbf{x}$ varies over $\mathbb{Z}_{s_1} \oplus \cdots \oplus \mathbb{Z}_{s_n}$. Thus the $s_1 \cdots s_n$ treatment combinations can be partitioned into p disjoint classes $H_0, \ldots, H_{p-1}$, with the treatment combinations $\mathbf{x}$ in the same H_i having the same value of $[\mathbf{a}, \mathbf{x}]$. It can also be shown that all of $H_0, \ldots, H_{p-1}$ are of equal size. Using the same notation as in Section 6.6, we denote this partition induced by $\mathbf{a}$ as $P(\mathbf{a})$. Then, as in (6.31), functions of the form

$$\tau = \sum_{i=0}^{p-1} c_i \sum_{\mathbf{x} \in H_i} \alpha(\mathbf{x}),$$

where $c_i \in \mathbb{R}$ and $\sum_{i=0}^{p-1} c_i = 0$ constitute a $(p-1)$-dimensional space of treatment contrasts. We denote this space by $C_{\mathbf{a}}$.

Therefore each nonzero $\mathbf{a}$ in $\mathbb{Z}_{s_1} \oplus \cdots \oplus \mathbb{Z}_{s_n}$ of order p can be used to define $p-1$ degrees of freedom of treatment contrasts. However, the example below shows that, unlike in Section 6.6, the spaces $C_{\mathbf{a}}$ defined by different $\mathbf{a}$'s can be neither equal nor orthogonal, and not all the contrasts in the same $C_{\mathbf{a}}$ belong to the same main effect or interaction.

Example 6.5. Suppose $n = 3, s_1 = 2, s_2 = 2, s_3 = 3$. Let $\mathbf{a} = (1,1,1)^T$. Then the order of $\mathbf{a}$ is 6. Thus it can be used to partition the 12 treatment combinations $\mathbf{x}$ into six disjoint classes based on the values of $[\mathbf{a}, \mathbf{x}] = 3x_1 + 3x_2 + 2x_3 \pmod{6}$:

$$
\begin{aligned}
H_0\ ([\mathbf{a}, \mathbf{x}] = 0): & \quad \mathbf{x} = (0,0,0)^T, (1,1,0)^T, \\
H_1\ ([\mathbf{a}, \mathbf{x}] = 1): & \quad \mathbf{x} = (0,1,2)^T, (1,0,2)^T, \\
H_2\ ([\mathbf{a}, \mathbf{x}] = 2): & \quad \mathbf{x} = (0,0,1)^T, (1,1,1)^T, \\
H_3\ ([\mathbf{a}, \mathbf{x}] = 3): & \quad \mathbf{x} = (0,1,0)^T, (1,0,0)^T, \\
H_4\ ([\mathbf{a}, \mathbf{x}] = 4): & \quad \mathbf{x} = (0,0,2)^T, (1,1,2)^T, \\
H_5\ ([\mathbf{a}, \mathbf{x}] = 5): & \quad \mathbf{x} = (0,1,1)^T, (1,0,1)^T.
\end{aligned}
$$

Based on this partition, five degrees of freedom for treatment contrasts can be defined. In Section 6.6, each nonzero $\mathbf{x} \in EG(n,s)$ and all its nonzero multiples define the same degrees of freedom. This does not hold here. For example, since $2\mathbf{a} = (0,0,2)^T$, the partition induced by $2\mathbf{a}$ coincides with that according to the values of x_3. It consists of $H_0 \cup H_3$ ($x_3 = 0$), $H_1 \cup H_4$ ($x_3 = 2$), and $H_2 \cup H_5$ ($x_3 = 1$). Thus $2\mathbf{a}$ defines the main effect of A_3. This partition is coarser than that induced by $\mathbf{a}$ and accounts for two of the five degrees of freedom defined by $\mathbf{a}$. Since $3\mathbf{a} = (1,1,0)^T$, the partition induced by $3\mathbf{a}$ coincides with that based on $x_1 + x_2 \pmod{2}$. It consists of $H_0 \cup H_2 \cup H_4$ ($x_1 + x_2 = 0$) and $H_1 \cup H_3 \cup H_5$ ($x_1 + x_2 = 1$) and defines the interaction of A_1 and A_2. For the other multiples of $\mathbf{a}$, $4\mathbf{a}$ produces the same partition as $2\mathbf{a}$, and $5\mathbf{a}$ produces the same partition as $\mathbf{a}$. Therefore $4\mathbf{a}$ and $2\mathbf{a}$ define the same treatment contrasts, and $5\mathbf{a}$ and $\mathbf{a}$ also define the same treatment contrasts. Removing the two degrees of freedom associated with the main effect of A_3 defined by $2\mathbf{a}$ (or $4\mathbf{a}$) and the one degree of freedom associated with the interaction of A_1 and A_2 defined by $3\mathbf{a}$, we have that the two remaining degrees of freedom defined by $\mathbf{a}$ belong to the three-factor interaction. Thus $C_{\mathbf{a}}$ contains not only some three-factor-interaction contrasts, but also main-effect and two-factor interaction contrasts.

In general, if $\mathbf{a}$ has order p, then the subgroup generated by $\mathbf{a}$ is the cyclic group $\{\mathbf{0}, \mathbf{a}, \ldots, (p-1)\mathbf{a}\}$. For any positive integer r, $1 \le r \le p-1$, it follows from $[r\mathbf{a}, \mathbf{x}] = r[\mathbf{a}, \mathbf{x}] \pmod{l}$ that if $[\mathbf{a}, \mathbf{x}_1] = [\mathbf{a}, \mathbf{x}_2]$, then we also have $[r\mathbf{a}, \mathbf{x}_1] = [r\mathbf{a}, \mathbf{x}_2]$. This implies that the partition induced by $\mathbf{a}$ is nested in that induced by $r\mathbf{a}$. Therefore $C_{r\mathbf{a}} \subseteq C_{\mathbf{a}}$, and $C_{r\mathbf{a}} = C_{\mathbf{a}}$ if and only if $\mathbf{a}$ and $r\mathbf{a}$ generate the same cyclic group; otherwise, $C_{r\mathbf{a}}$ is a proper subspace of $C_{\mathbf{a}}$. Note that in Example 6.5, $C_{5\mathbf{a}} = C_{\mathbf{a}}$ and $C_{4\mathbf{a}} = C_{2\mathbf{a}}$; this is because $5\mathbf{a}$ and $\mathbf{a}$ generate the same group, and $4\mathbf{a}$ and $2\mathbf{a}$ also generate the same group. Since $\mathbf{a}$ and $r\mathbf{a}$ generate the same cyclic group if and only if r is coprime to p (r and p have no common prime divisor), the number of $r\mathbf{a}$'s such that $C_{r\mathbf{a}} = C_{\mathbf{a}}$ is equal to the Euler function $\phi(p)$, the number of integers between 1 and

p that are coprime to p. Because these $\phi(p)$ generators of the same cyclic group define the same treatment degrees of freedom, only one of them is needed for defining factorial effects. On the other hand, when r is not coprime to p, since C_{ra} is a proper subspace of C_a, to obtain an orthogonal decomposition of the space of treatment contrasts, the contrasts in C_{ra} need to be removed from C_a. Now r is not coprime to p if and only if they have a common prime divisor, say q. Then, since $C_{ra} \subseteq C_{qa}$, removing C_{qa} automatically removes C_{ra}. Therefore it is sufficient to remove C_{qa} for prime divisors q of p only. (These are C_{2a} and C_{3a} in Example 6.5; note that 6 has two prime divisors, 2 and 3.) We have the following result from Bailey, Gilchrist, and Patterson (1977) on the decomposition of the space of treatment contrasts into various orthogonal factorial components.

Theorem 6.1. *For each nonzero* $a \in \mathbb{Z}_{s_1} \oplus \cdots \oplus \mathbb{Z}_{s_n}$ *of order* p, *let*

$$W_a = \{v \in C_a : v \perp C_{ra} \text{ for all proper prime divisors } r \text{ of } p\}.$$

That is,

$$W_a = C_a \ominus \left(\sum_{r:\, r \text{ is a proper prime divisor of } p} C_{ra} \right).$$

Then

(i) *For any nonzero* $a, b \in \mathbb{Z}_{s_1} \oplus \cdots \oplus \mathbb{Z}_{s_n}$, W_a *and* W_b *are orthogonal unless* a *and* b *generate the same cyclic group, in which case* $W_a = W_b$.

(ii) *If* a *has order* p, *then* $\dim(W_a) = \phi(p)$, *and if* a *has exactly* k *nonzero components, say* $a_{i_1}, \ldots, a_{i_k}$, *then* $W_a \subseteq W_{A_{i_1} \wedge \cdots \wedge A_{i_k}}$; *that is,* W_a *represents* $\phi(p)$ *degrees of freedom of the factorial effects of* $A_{i_1}, \ldots, A_{i_k}$.

(iii) *The* $(s_1 \cdots s_n - 1)$-*dimensional space of treatment contrasts can be decomposed as* $\oplus_a W_a$, *where the sum is over nonzero* $a \in \mathbb{Z}_{s_1} \oplus \cdots \oplus \mathbb{Z}_{s_n}$, *but with only one such* a *from those that generate the same cyclic group.*

This result is based on the character theory of finite groups. We refer readers to Bailey (1985) and Bailey (1990) for the algebraic theory. Chakravarti (1976) was the first to use characters in the case of two-level factors.

When $s_1 = \cdots = s_n = s$ is a prime number, each nonzero a has order s, $\phi(s) = s - 1$, and all the nonzero multiples of a generate the same cyclic group. In this case, Theorem 6.1 gives the same breakdown of treatment degrees of freedom as the method presented in Section 6.6.

Example 6.6. (Example 6.5 continued) The factorial effects in a $2 \times 2 \times 3$ experiment can be defined by $(1,0,0)^T$, $(0,1,0)^T$, $(0,0,1)^T$, $(1,1,0)^T$, $(1,0,1)^T$, $(0,1,1)^T$, and $(1,1,1)^T$. Each of the other nonzero a's in $\mathbb{Z}_2 \oplus \mathbb{Z}_2 \oplus \mathbb{Z}_3$ generates the same group as one of those listed here. The orders of $(1,0,0)^T$, $(0,1,0)^T$, and $(1,1,0)^T$ are equal to 2; so each of them defines one degree of freedom. They are, respectively, main effect of A_1, main effect of A_2, and the interaction of A_1 and A_2. The order of $(0,0,1)^T$ is three. Hence it defines the two degrees of freedom for the main effect of A_3. The orders of $(1,0,1)$, $(0,1,1)^T$, and $(1,1,1)^T$ are all equal to 6. It follows that each defines $\phi(6) = 2$ degrees of freedom of factorial effects. They are, respectively, interaction of A_1 and A_3, interaction of A_2 and A_3, and interaction of A_1, A_2, and A_3. This accounts for all the 11 degrees of freedom.

6.8   More on factorial treatment structure*

We explain here the notations $V_{A_{i_1} \wedge \cdots \wedge A_{i_k}}$ and $W_{A_{i_1} \wedge \cdots \wedge A_{i_k}}$ in the language of Chapter 4. Each of the n treatment factors A_i, $i = 1, \ldots, n$, can be regarded as a partition of the $\prod_{i=1}^{n} s_i$ treatment combinations $(x_1, \ldots, x_n)^T$, $x_i \in \{0, 1, \ldots, s_i - 1\}$, into s_i disjoint classes according to the values of x_i. By the discussion in the paragraph preceding (4.15), for any pair of factors A_i and A_j, $i \neq j$, $A_i \wedge A_j$ is the factor that partitions the treatment combinations into $s_i s_j$ disjoint classes according to the levels of A_i and A_j. In general, for any $1 \le i_1 < \cdots < i_k \le n$, $1 \le k \le n$, $A_{i_1} \wedge \cdots \wedge A_{i_k}$ is the factor whose levels are all the level combinations of $A_{i_1}, \ldots, A_{i_k}$. The factorial treatment structure $s_1 \times \cdots \times s_n$ can be thought to consist of the 2^n factors $\{\wedge_{i \in Z} A_i : Z \subseteq \{1, \ldots, n\}\}$, where $\wedge_{i \in Z} A_i$ is the universal factor U when Z is the empty set and is the equality factor E when $Z = \{1, \ldots, n\}$. We have

$$\wedge_{i \in Z_2} A_i \preceq \wedge_{i \in Z_1} A_i \text{ if and only if } Z_1 \subseteq Z_2. \tag{6.33}$$

Figure 6.1 shows the Hasse diagram for the factorial treatment structure $s_1 \times s_2 \times s_3$.

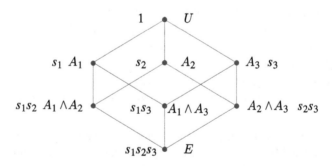

Figure 6.1 *Hasse diagram for the factorial treatment structure $s_1 \times s_2 \times s_3$*

For each subset of factors $\{A_{i_1}, \ldots, A_{i_k}\}$, the space $V_{A_{i_1} \wedge \cdots \wedge A_{i_k}}$ associated with $A_{i_1} \wedge \cdots \wedge A_{i_k}$, as defined in Section 4.3, consists of all the $s_1 \cdots s_n \times 1$ vectors whose entries depend only on the levels of $A_{i_1}, \ldots, A_{i_k}$. Thus $\dim(V_{A_{i_1} \wedge \cdots \wedge A_{i_k}}) = s_{i_1} \cdots s_{i_k}$. In particular, $V_E = \mathbb{R}^{s_1 \cdots s_n}$ and V_U is the one-dimensional space of vectors with constant entries.

By (6.33) and (4.2),

$$Z_1 \subseteq Z_2 \Rightarrow V_{\wedge_{i \in Z_1} A_i} \subseteq V_{\wedge_{i \in Z_2} A_i}. \tag{6.34}$$

For each set $Z \subseteq \{1, \ldots, n\}$, define $W_{\wedge_{i \in Z} A_i}$ as

$$W_{\wedge_{i \in Z} A_i} = V_{\wedge_{i \in Z} A_i} \ominus \left(\sum_{Z' : Z' \subset Z} V_{\wedge_{i \in Z'} A_i} \right). \tag{6.35}$$

In view of (6.34), the definition in (6.35) makes sense. Then by Definition 6.2, (6.35), and the discussion in the paragraph preceding (6.34),

a treatment contrast $\mathbf{c}^T \boldsymbol{\alpha}$ represents the main effect of factor $i \Leftrightarrow \mathbf{c} \in W_{A_i}$,

and

$\mathbf{c}^T \boldsymbol{\alpha}$ represents the interaction of factors $A_{i_1}, \ldots, A_{i_k} \Leftrightarrow \mathbf{c} \in W_{A_{i_1} \wedge \cdots \wedge A_{i_k}}$.

The following is a consequence of the definition of the W spaces in (6.35).

Theorem 6.2. *The 2^n spaces $W_{\wedge_{i \in Z} A_i}$, $Z \subseteq \{1, \ldots, n\}$, are mutually orthogonal. Furthermore, for each $Z \subseteq \{1, \ldots, n\}$,*

$$V_{\wedge_{i \in Z} A_i} = \bigoplus_{Z' : Z' \subseteq Z} W_{\wedge_{i \in Z'} A_i}. \tag{6.36}$$

In particular

$$V_E = \mathbb{R}^{s_1 \cdots s_n} = \bigoplus_{Z : Z \subseteq \{1, \ldots, n\}} W_{\wedge_{i \in Z} A_i}. \tag{6.37}$$

It follows from (6.36) that

$$W_{\wedge_{i \in Z} A_i} = V_{\wedge_{i \in Z} A_i} \ominus \left(\bigoplus_{Z' : Z' \subset Z} W_{\wedge_{i \in Z'} A_i} \right). \tag{6.38}$$

Equation (6.37) gives an orthogonal decomposition of $\mathbb{R}^{s_1 \cdots s_n}$ into 2^n components, one for each subset of the n treatment factors. All the spaces $W_{\wedge_{i \in Z} A_i}$ can be obtained by repeatedly using (6.38), working up from the smallest subset of $\{1, \ldots, n\}$, i.e., the empty set (or, down the Hasse diagram from the top node), as follows. Since $\wedge_{i \in Z} A_i$ is the universal factor U when Z is the empty set, and the empty set has no proper subset, by (6.38), $W_U = V_U$. For each $1 \leq i \leq n$, since the empty set is the only proper subset of $\{i\}$, again by (6.38), we have $W_{A_i} = V_{A_i} \ominus W_U = V_{A_i} \ominus V_U$; note that in Figure 6.1, U is the only node that is above each A_i. Then, since there are three nodes U, A_i, and A_j above $A_i \wedge A_j$, by (6.38), $W_{A_i \wedge A_j} = V_{A_i \wedge A_j} \ominus (V_U \oplus W_{A_i} \oplus W_{A_j})$. All the other W spaces can be obtained by continuing this process. This construction is essentially what was used in Section 6.3. We point out that Theorem 6.2 is a special case of Theorem 12.6, to be presented in Section 12.7 as a tool for constructing the strata associated with an orthogonal block structure.

Exercises

6.1 Verify that under a factorial experiment with three two-level factors A, B, and C, the factorial effects can be obtained from the treatment effects by symbolically expanding the following:

$$A = \frac{1}{4}(a-1)(b+1)(c+1), \quad B = \frac{1}{4}(a+1)(b-1)(c+1),$$

$$C = \frac{1}{4}(a+1)(b+1)(c-1), \quad AB = \frac{1}{4}(a-1)(b-1)(c+1),$$

$$AC = \frac{1}{4}(a-1)(b+1)(c-1), \quad BC = \frac{1}{4}(a+1)(b-1)(c-1),$$

$$ABC = \frac{1}{4}(a-1)(b-1)(c-1).$$

Extend this to the case of more than three factors and justify your answer.

6.2 Verify the details in Examples 6.5 and 6.6.

6.3 A vector $\mathbf{a} = (a_1, \ldots, a_N)$ is said to be k-trend free, where k is a nonnegative integer, if $\mathbf{a}$ is orthogonal to the vector $(1^t, 2^t, \ldots, N^t)$ for all integers t such that $0 \leq t \leq k$. Let $\mathbf{K}$ be the matrix on the right-hand side of (6.18). Show that each row of $\mathbf{K}$ that is the coefficient vector of a k-factor interaction contrast is $(k-1)$-trend free. An implication of this result is that if the treatment combinations in a complete factorial experiment are observed in the Yates order in an equally spaced time sequence, then the estimates of k-factor interaction contrasts are not affected by a $(k-1)$th degree polynomial time trend.

[Cheng and Jacroux (1988)]

6.4 Extend the result in Exercise 6.3 to complete $s_1 \times \cdots \times s_n$ experiments.

[Bailey, Cheng, and Kipnis (1992)]

6.5* It was shown in Sections 6.3 and 6.8 how factorial effects can be constructed for real-valued functions α defined on $S_1 \times \cdots \times S_n$, where each $S_i = \{0, \ldots, s_i - 1\}$ is a finite set. Suggest how Definition 6.2, results such as Theorem 6.2, and the construction of the W spaces for various factorial effects can be extended to the case where each S_i is an interval. This leads to a functional ANOVA.

6.6* Prove Theorem 6.2.

Hint: Use Exercise 4.7 to prove the mutual orthogonality of the spaces $W_{\wedge_{i \in Z} A_i}$, $Z \subseteq \{1, \ldots, n\}$.

Chapter 7

Blocked, Split-Plot, and Strip-Plot Complete Factorial Designs

Blocking is one of the most important principles in experimental design. In this chapter we address the issue of designing complete factorial experiments with incomplete blocks. As the number of factors increases, the total number of treatment combinations soon becomes very large. Two challenges to the experimenter arise. There may not be enough resources or time to run a complete factorial experiment. Second, even if a complete factorial experiment is feasible, incomplete blocks may have to be used since a block accommodating all the treatment combinations may be too large to have acceptable within-block variability, or not all the treatment combinations can be run in the same setting. We focus on the latter issue here and address the former in later chapters. If some factorial effects (for example, the higher-order effects) are deemed less important or are negligible, then one can design an experiment so that these effects are estimated by less precise interblock contrasts. They are said to be *confounded* with blocks and are sacrificed to achieve better precision for the other more important effects. We also discuss similar issues for row-column designs where some factorial effects are confounded with rows or columns. In split-plot and strip-plot experiments, due to practical considerations, certain main effects must be confounded with blocks, rows, or columns.

7.1 An example

Suppose an experiment with two two-level treatment factors A and B is to be performed using a randomized block design with six blocks of size two. In this case there exists a balanced incomplete block design (BIBD) (see Section A.8 for a definition) in which each treatment appears in three blocks and every pair of treatments appears together in one block:

(1)	a	(1)	b	(1)	a
b	ab	a	ab	ab	b

$$(7.1)$$

A BIBD is known to have some optimality property when the treatments are unstructured and the interest is in, e.g., making all pairwise comparisons (Kiefer, 1958,

1975); see Exercise 2.5. Under an incomplete block design, there is information for treatment contrasts in both the inter- and intrablock strata; it can be shown that under a BIBD, all the normalized treatment contrasts $l = \mathbf{c}^T \boldsymbol{\alpha}$ ($\|\mathbf{c}\| = 1$) are estimated with the same precision in each of the two strata. (See Exercises 2.5 and 5.5.) Let $\widehat{l}^{\mathcal{B}}$ and $\widehat{l}^{\mathcal{E}}$ be the inter- and intrablock estimators of l, respectively. Then it can be shown that under design (7.1), $\mathrm{var}(\widehat{l}^{\mathcal{B}}) = \xi_B$ and $\mathrm{var}(\widehat{l}^{\mathcal{E}}) = \frac{1}{2}\xi_{\mathcal{E}}$. It follows that all the factorial effects A, B, and AB are estimated with the same precision. If the interaction AB is negligible, then it is not necessary to estimate AB with the same precision as the main effects. The objective is to construct a design under which the main effects are estimated more efficiently. Ideally we would want to have them estimated entirely in the intrablock stratum.

Now consider the following alternative design:

$$
\begin{array}{|c|}\hline (1) \\ \hline ab \\ \hline \end{array} \quad
\begin{array}{|c|}\hline a \\ \hline b \\ \hline \end{array} \quad
\begin{array}{|c|}\hline (1) \\ \hline ab \\ \hline \end{array} \quad
\begin{array}{|c|}\hline a \\ \hline b \\ \hline \end{array} \quad
\begin{array}{|c|}\hline (1) \\ \hline ab \\ \hline \end{array} \quad
\begin{array}{|c|}\hline a \\ \hline b \\ \hline \end{array} \qquad (7.2)
$$

Note that AB is defined as $\frac{1}{2}((1) - a - b + ab))$, and design (7.2) is constructed in such a way that in each of the three replicates, the treatment combinations with coefficients $+1$ in the interaction contrast ((1) and ab) are in the same block, and those with coefficients -1 (a and b) are also in the same block. Consequently AB is estimated by a block contrast. We say that it is *confounded* with blocks. When AB is written in the form $\mathbf{c}_{AB}^T \boldsymbol{\alpha}$, the vector $\mathbf{c}_{AB}^*$ as defined in (5.2) has constant entries corresponding to the observations in the same block. Therefore $\mathbf{c}_{AB}^* \in V_B \ominus V_U = W_B$. For the main effects, both $\mathbf{c}_A^*$ and $\mathbf{c}_B^*$ are in $V_B^{\perp} = W_{\mathcal{E}}$. We say that they are *orthogonal* to block contrasts (or orthogonal to blocks). It follows from Theorem 5.1 that the best linear unbiased estimators of A, B, and AB are obtained by replacing the treatment effects in the contrasts defining these factorial effects with the respective treatment means. The variance of the estimator of AB is $\xi_B/3$, and both estimators of A and B have variances $\xi_{\mathcal{E}}/3$. Furthermore, the estimators of A, B, and AB are uncorrelated. This is a simple example of orthogonal designs in the sense of Section 5.8.

Under design (7.2), estimators of the main effects are 50% more efficient than their intrablock estimators under design (7.1). This is achieved by sacrificing the interaction, which can only be estimated by a between-block contrast. Under a model with fixed block effects, AB is not even estimable under design (7.2). This design, for which the condition in Theorem 2.4 does not hold, is not connected and hence is not suitable when one is interested in all the treatment comparisons.

Under design (7.2), the sum of squares associated with each factorial effect can be computed easily. Let l be any of the three effects A, B, and AB. Then, by (5.30), $SS(l) = 3\left(\widehat{l}\right)^2$. As discussed in Section 5.8, the treatment sum of squares is split into two parts. We have that $SS(A)$ and $SS(B)$ appear in the intrablock stratum ANOVA, leaving four degrees of freedom for residual, and $SS(AB)$ appears in the interblock stratum ANOVA, also with four degrees of freedom for residual; see Table 7.1.

Table 7.1 *ANOVA for the design in (7.2)*

source	sum of squares	d.f.	E(MS)
interblock	$\sum_{i=1}^{6} 2(y_{i\cdot} - y_{\cdot\cdot})^2$	5	
AB	$3\left(\widehat{AB}\right)^2$	1	$\xi_B + 3(AB)^2$
residual	By subtraction	4	ξ_B
intrablock	$\sum_{i=1}^{6}\sum_{j=1}^{2}(y_{ij} - y_{i\cdot})^2$	6	
A	$3\left(\widehat{A}\right)^2$	1	$\xi_\varepsilon + 3A^2$
B	$3\left(\widehat{B}\right)^2$	1	$\xi_\varepsilon + 3B^2$
residual	By subtraction	4	ξ_ε
Total	$\sum_{i=1}^{6}\sum_{j=1}^{2}(y_{ij} - y_{\cdot\cdot})^2$	11	

7.2   Construction of blocked complete factorial designs

When all the factors have s levels, where s is a prime number or power of a prime number, as in Section 6.6, each treatment combination can be considered as a point in EG(n,s), and each nonzero $\mathbf{a} = (a_1,\ldots,a_n)^T \in$ EG(n,s) defines a partition of the s^n points $\mathbf{x}$ into s disjoint $(n-1)$-flats according to the values of $\mathbf{a}^T\mathbf{x}$. If we consider each of these $(n-1)$-flats as a block, then we have a design with s blocks each of size s^{n-1}. Since all the treatment combinations $\mathbf{x}$ in the same block have the same value of $\mathbf{a}^T\mathbf{x}$, they have the same coefficient in any treatment contrast defined by the pencil $P(\mathbf{a})$. It follows that these contrasts are confounded with blocks, and all the other factorial effects are orthogonal to blocks.

In general, to construct an s^n design in s^q blocks of size s^{n-q}, $q < n$, we choose q linearly independent vectors $\mathbf{a}_1,\ldots,\mathbf{a}_q$ in EG(n,s). The s^n treatment combinations are partitioned into s^q blocks according to the values of $\mathbf{a}_1^T\mathbf{x},\ldots,\mathbf{a}_q^T\mathbf{x}$. Each block consists of all the treatment combinations $\mathbf{x}$ satisfying the q equations

$$\mathbf{a}_i^T\mathbf{x} = b_i, \ i = 1,\ldots,q,$$

where each $b_i \in$ GF(s). The s^q choices of $b_1,\ldots,b_q$ define the s^q blocks, which are all the disjoint $(n-q)$-flats in the pencil $P(\mathbf{a}_1,\ldots,\mathbf{a}_q)$. Then for each $i = 1,\ldots,q$, since all the treatment combinations in the same block have the same value of $\mathbf{a}_i^T\mathbf{x}$, they have the same coefficient in each of the contrasts defined by $\mathbf{a}_i$. Therefore the contrasts defined by $\mathbf{a}_1,\ldots,\mathbf{a}_q$ are confounded with blocks.

Furthermore, for any nonzero vector $\mathbf{a}$ that is a linear combination of $\mathbf{a}_1,\ldots,\mathbf{a}_q$, say $\mathbf{a} = \sum_{i=1}^{q}\lambda_i\mathbf{a}_i$, all the treatment combinations $\mathbf{x}$ in the same block also have the same value of $\sum_{i=1}^{q}\lambda_i\mathbf{a}_i^T\mathbf{x} = \mathbf{a}^T\mathbf{x}$. It follows that the factorial effects defined by all the nonzero vectors in the q-dimensional space generated by $\mathbf{a}_1,\ldots,\mathbf{a}_q$ are also confounded with blocks. There are a total of $s^q - 1$ nonzero linear combinations $\sum_{i=1}^{q}\lambda_i\mathbf{a}_i$, where $\lambda_i \in \mathrm{GF}(s)$. Each nonzero vector $\mathbf{a}$ defines $s - 1$ degrees of freedom; on the other hand, all the $s - 1$ nonzero multiples of $\mathbf{a}$ ($\lambda\mathbf{a}$ with $\lambda \in \mathrm{GF}(s)$, $\lambda \neq 0$) define the same pencil of $(n - 1)$-flats and hence the same treatment contrasts. So the total number of degrees of freedom that are confounded with blocks is equal to $[(s^q - 1)/(s - 1)](s - 1) = s^q - 1$, accounting for the $s^q - 1$ degrees of freedom in the interblock stratum. We call $\mathbf{a}_1, \ldots, \mathbf{a}_q$ *independent blocking words*. Nonzero linear combinations of $\mathbf{a}_1, \ldots, \mathbf{a}_q$ are called *blocking words*, and the factorial effects they define are called blocking effects.

We call the factorial effect defined by a nonzero linear combination of $\mathbf{a}_1, \ldots, \mathbf{a}_q$ a *generalized interaction* of the q factorial effects defined by $\mathbf{a}_1, \ldots, \mathbf{a}_q$. When we choose to confound certain interactions with blocks, all their generalized interactions are also confounded with blocks. We should make sure that these generalized interactions can be safely confounded.

The block consisting of the solutions to

$$\mathbf{a}_i^T \mathbf{x} = 0, \ i = 1, \ldots, q, \tag{7.3}$$

is called the *principal block*. Let this block be denoted by H_1. Then H_1 is a subgroup of $\mathrm{EG}(n, s)$, and the other blocks are its cosets. Therefore the blocks other than the principal block are of the form $H_1 + \mathbf{x}$, where $\mathbf{x}$ is a nonzero vector in $\mathrm{EG}(n, s)$ (see Section A.4). One can use the following procedure to generate the s^q blocks:

Step 1: Set $k = 0$. Solve the simultaneous equations $\mathbf{a}_i^T \mathbf{x} = 0$, $i = 1, \ldots, q$, to obtain the principal block H_1.

Step 2: Pick a treatment combination $\mathbf{x}$ that is not in $\bigcup_{i=1}^{s^k} H_i$. Add each of the $s - 1$ nonzero multiples of $\mathbf{x}$ to all the treatment combinations in $H_1, \ldots, H_{s^k}$ to form $(s - 1)s^k$ additional blocks $H_{s^k+1}, \ldots, H_{s^{k+1}}$.

Step 3: Increase k by 1. Stop if $k = q$; otherwise go back to step 2 and continue.

For two-level designs, if the shorthand notation introduced in the paragraph following (6.16) is used, then the string of capital letters (word) representing the factorial effect defined by $\mathbf{a} + \mathbf{b}$, where $\mathbf{a}, \mathbf{b} \in \mathrm{EG}(n, 2)$, can be obtained by symbolically multiplying those representing the factorial effects defined by $\mathbf{a}$ and $\mathbf{b}$, subject to the rule that the square of any letter is removed; for example, $(ACD)(CDE) = AE$. Then the linear independence of $\mathbf{a}_1, \ldots, \mathbf{a}_q$ is equivalent to that none of the corresponding factorial effects, when written as a word, can be expressed as a product of some of the other $q - 1$ words. We say that the q factorial effects (or words) are linearly independent (or independent for short). For example, AB, ACD, and BCD are dependent since $BCD = (AB)(ACD)$. For the case of four factors, this corresponds to that $(0, 1, 1, 1) = (1, 1, 0, 0) + (1, 0, 1, 1)$. Similarly, the treatment combination $\mathbf{x} + \mathbf{y}$, when expressed as a string of lower case letters, is the product of those representing $\mathbf{x}$ and $\mathbf{y}$, again subject to the rule that the square of any letter is removed. If the i_1th, $\ldots$, and i_kth entries of the vector $\mathbf{a}_i$ in (7.3) are nonzero and all the other entries are zero, then the equation $\mathbf{a}_i^T \mathbf{x} = 0$ means that an even number of $x_{i_1}, \ldots, x_{i_k}$ are equal to 1. When the shorthand notation is used, this implies that the principal block consists of all the treatment combinations that have even numbers of letters in common with all the independent blocking words. We then multiply all the treatment combinations in the blocks that have been generated by a treatment combination that has not appeared yet, until all the blocks are constructed. We also multiply the independent blocking words in all possible ways to obtain their generalized interactions, which are confounded with blocks.

Example 7.1. To construct a 2^5 design in eight blocks of size four, suppose we pick the three independent interactions AC, BD, and ABE to be confounded with blocks. Then we will also confound $(AC)(BD) = ABCD$, $(AC)(ABE) = BCE$, $(BD)(ABE) = ADE$, and $(AC)(BD)(ABE) = CDE$. The four treatment combinations that have even numbers of letters in common with all of AC, BD, and ABE are (1), ace, bde, and $abcd$. With these four treatment combinations in the principal block, the other seven blocks can easily be constructed:

(1)	c	d	cd	e	ce	de	cde
ace	ae	acde	ade	ac	a	acd	ad
bde	bcde	be	bce	bd	bcd	b	bc
abcd	abd	abc	ab	abcde	abde	abce	abe

The second block is obtained by multiplying all the treatment combinations in the first block by c, which does not appear in the principal block. The third and fourth blocks are obtained by multiplying all the treatment combinations in the first two blocks by d, and the last four blocks are obtained by multiplying all the treatment combinations in the first four blocks by e. If we choose to confound the highest-order interactions $ABCD$ and $ABCDE$ with blocks, then the main effect E would also be confounded.

Example 7.2. To construct a 3^3 design in nine blocks of size three, suppose we choose the two blocking words AB^2 and AC^2. Then $\mathbf{a}_1 = (1,2,0)^T$ and $\mathbf{a}_2 = (1,0,2)^T$. The principal block consists of the solutions (x_1,x_2,x_3) to the two equations $x_1 + 2x_2 \equiv 0 \pmod 3$ and $x_1 + 2x_3 \equiv 0 \pmod 3$: $\{(0,0,0), (1,1,1), (2,2,2)\}$. The nine blocks are

0 0 0	0 1 0	0 2 0	0 0 1	0 1 1	0 2 1	0 0 2	0 1 2	0 2 2
1 1 1	1 2 1	1 0 1	1 1 2	1 2 2	1 0 2	1 1 0	1 2 0	1 0 0
2 2 2	2 0 2	2 1 2	2 2 0	2 0 0	2 1 0	2 2 1	2 0 1	2 1 1

The second and third blocks are obtained by adding $(0,1,0)$ and $2 \cdot (0,1,0) = (0,2,0)$ to all the treatment combinations in the first block, and blocks 4–6 and 7–9 are obtained by adding $(0,0,1)$ and $2 \cdot (0,0,1) = (0,0,2)$, respectively, to all the treatment combinations in the first three blocks. The linear combinations of $\mathbf{a}_1$ and $\mathbf{a}_2$ that define different contrasts can be obtained by calculating $\mathbf{a}_1 + \mathbf{a}_2 = (2,2,2)^T$ and $\mathbf{a}_1 + 2\mathbf{a}_2 = (0,2,1)^T$. The former defines the same two degrees of freedom of treatment contrasts as $2 \cdot (2,2,2)^T = (1,1,1)^T$, which are denoted as ABC, and the latter defines the same two degrees of freedom of treatment contrasts as $2 \cdot (0, 2, 1)^T = (0, 1, 2)^T$, which are denoted as BC^2. Therefore the eight degrees of freedom that are confounded with blocks are AB^2, AC^2, ABC, and BC^2.

A blocking scheme of a complete factorial design is said to have *estimability of order e* if e is the largest integer such that all the factorial effects involving e or fewer factors are not confounded with blocks (Sun, Wu, and Chen, 1997). For a given positive integer e and block size, there is a limit on the number of factors one can accommodate to achieve estimability of order e. Specifically, let $g(e; r, s)$ be the maximum number of s-level factors that can be accommodated in a blocked complete factorial in blocks of size s^r such that no factorial effect involving e or fewer factors is confounded with blocks. Fisher (1942, 1945) showed that

$$g(2; r, s) = \frac{s^r - 1}{s - 1}. \tag{7.4}$$

Bose (1947) determined $g(e; r, s)$ for $e = 3$ in the case of two-level designs and a few other cases with $e > 3$. In particular, he showed that

$$g(3; r, 2) = 2^{r-1}. \tag{7.5}$$

We will revisit these bounds in Remark 9.7 in Section 9.6.

The idea of confounding was introduced in Fisher (1926). The construction in this section was mainly due to Fisher (1942), Yates (1935, 1937), Bose and Kishen (1940), Bose (1947), and Kempthorne (1947).

The selection of independent blocking words will be addressed in Sections 9.7 and 10.9.

7.3   Analysis

When a design is constructed by the method described in Section 7.2, for each factorial effect $\mathbf{c}^T \boldsymbol{\alpha}$ that is confounded with blocks, the vector $\mathbf{c}^*$ as defined in (5.2) belongs to $V_B \ominus V_{\mathcal{U}}$, while for all the other factorial effects, $\mathbf{c}^* \in V_B^{\perp}$. Theorem 5.1 implies that estimates of all the factorial effects and their associated sums of squares can be computed in exactly the same way as when the experiment is not blocked. In particular, the Yates algorithm as described in Section 6.4 can be used for two-level designs. In a single-replicate design, normalized factorial effects that are confounded with blocks are estimated with variance ξ_B (but are not estimable under models with fixed block effects), and those that are orthogonal to blocks are estimated with variance $\xi_{\mathcal{E}}$. Compared with when the experiment is run in complete blocks, the only difference in the ANOVA tables is that the sums of squares of the factorial effects that are confounded with blocks appear in the interblock stratum instead. For the second design in Section 7.1, the two main effect contrasts A and B are tested against the intrablock error, whereas the interaction contrast AB is tested against the interblock error. In general, if normal or half-normal probability plots are used to judge the significance of effects in single-replicate experiments, then two separate plots should be used, one for the effects estimated in the intrablock stratum, and the other for those estimated in the interblock stratum. This is due to different precisions of the estimates in the two strata.

For the designs in Section 7.1, if randomization is carried out in each replicate separately, that is, the block structure is (3 replicates)/(2 blocks)/(2 units), then, as we will see in Chapter 12, there is an additional inter-replicate stratum with two degrees of freedom. The ANOVA in the intrablock stratum remains the same, but two degrees of freedom for residual in the interblock stratum move up to the inter-replicate stratum, leaving no residual degree of freedom in the interblock stratum for the design in (7.1) and two degrees of freedom for the design in (7.2).

7.4 Pseudo factors

In Section 7.2, when s is a prime number, addition and multiplication are carried out as usual except that the results are reduced mod s. For $s = p^r$, where p is a prime number and $r > 1$, as discussed in Section A.2, one cannot use the usual arithmetic operations. An alternative approach is to represent each of the p^r levels as a combination of r *pseudo factors* with p levels each. For example, in an experiment with three four-level factors A, B, and C, suppose each level of A (B and C, respectively) is represented by a combination of two two-level pseudo factors A_1 and A_2 (B_1 and B_2, C_1 and C_2, respectively). Then A_1, A_2, and A_1A_2 correspond to the three degrees of freedom of the main effect of factor A; all the nine interactions A_1B_1, A_1B_2, A_2B_1, A_2B_2, $A_1B_1B_2$, $A_2B_1B_2$, $A_1A_2B_1$, $A_1A_2B_2$, and $A_1A_2B_1B_2$ correspond to the interaction of factors A and B. So, for example, if we do not want the main effect of factor A to be confounded with blocks, then all the three contrasts A_1, A_2, and A_1A_2 need to be estimated in the intrablock stratum.

7.5 Partial confounding

When more than one replicate of the complete factorial is to be conducted, one has the option of confounding different factorial effects in different replicates. This is useful when confounding some important effects with blocks is inevitable in a single replicate. Both designs in Section 7.1 consist of three replicates of a complete 2^2. If the interaction AB cannot be ignored, then one may not want to confound it with blocks in all three replicates, as is the case under the design in (7.2). The balanced incomplete block design in (7.1), on the other hand, confounds A with blocks in the first replicate (the first two blocks), confounds B in the second replicate (the third and fourth blocks), and confounds AB in the third replicate (the fifth and sixth blocks). This is called *partial confounding*.

With partial confounding, the condition in Theorem 5.1 no longer holds, but one can still take advantage of the factorial structure to compute the estimates of factorial effects in a simple manner. The interblock estimate of a factorial effect and the associated sum of squares can be computed from the replicates in which the factorial effect is confounded with blocks, as if there were no blocking. Likewise, the intrablock estimate and the associated sum of squares can be computed from the replicates in which the factorial effect is orthogonal to blocks, also as if there were no blocking. For the design in (7.1), this results in a variance of ξ_B for the interblock estimates and $\xi_{\mathcal{E}}/2$ for the intrablock estimates. Since there is information for all the factorial effects in both strata, this is not an orthogonal design in the sense of Section 5.8.

One can perform recovery of interblock information to combine the inter- and intrablock estimates of the factorial effects. Under a model with fixed block effects, each factorial effect can only be estimated in the two replicates in which it is not confounded with blocks.

7.6 Design keys

We present here an equivalent construction of single-replicate factorial designs with incomplete blocks based on *design keys* due to H. D. Patterson; see Patterson (1965, 1976) and Patterson and Bailey (1978). Instead of picking independent interactions to be confounded with blocks, using them to construct the blocking scheme, and checking that none of the important factorial effects are confounded with blocks, the method of design key first designates the strata where the treatment main effects are to be estimated. If we want to have the treatment main effects estimated in the intrablock stratum, then we can designate a set of n linearly independent contrasts (of the units) in the intrablock stratum to be *aliases* of the treatment main effects, in a way to be described more precisely later. This guarantees that the treatment main effects are estimated in the intrablock stratum. Furthermore, we will show that the specified aliasing of treatment main-effect contrasts with intrablock contrasts can be used to construct the blocking scheme in a systematic manner and can also be used to obtain unit aliases of all the other treatment factorial effects to determine in which stratum each of them is estimated.

The method of design key can be applied more generally to designs with simple block structures and also to asymmetric (mixed-level) factorial designs in which the factors may have different numbers of levels and the numbers of levels may not be prime numbers or powers of prime numbers. In this book we mainly consider s^n experiments where s is a prime number or power of a prime number. The construction of single-replicate complete factorial designs with incomplete blocks is treated in this section. Applications to experiments with more complicated block structures will be discussed in Chapter 13 (for single-replicate designs) and Chapter 14 (for fractional factorial designs).

Suppose the s^n treatment combinations are to be assigned to s^n units partitioned into s^q blocks of size s^{n-q}. The s^n experimental units can be thought of as all the combinations of an s^q-level factor $\mathcal{B}$ and an s^{n-q}-level factor $\mathcal{P}$. The combination where $\mathcal{B}$ is at level i and $\mathcal{P}$ is at level j corresponds to the jth unit in the ith block. Ignore the nesting structure and define the "factorial-effect" ("main-effect" and "interaction") contrasts of $\mathcal{B}$ and $\mathcal{P}$ as in Chapter 6 as if $\mathcal{B}$ and $\mathcal{P}$ were crossed. Then the "main-effect" contrasts of $\mathcal{B}$, which have $s^q - 1$ degrees of freedom, are the interblock contrasts. On the other hand, since the "main-effect" contrasts of $\mathcal{P}$ and the "interaction" contrasts of $\mathcal{B}$ and $\mathcal{P}$ are orthogonal to the "main-effect" contrasts of $\mathcal{B}$, they together generate the intrablock contrasts. If a design is constructed in such a way that each treatment main-effect contrast coincides with a contrast representing either the main effect of $\mathcal{P}$ or the interaction of $\mathcal{B}$ and $\mathcal{P}$, then all the treatment main effects are estimated in the intrablock stratum. For ease of construction, we further consider each level of $\mathcal{B}$ as a combination of q pseudo factors $\mathcal{B}_1, \ldots, \mathcal{B}_q$ with s levels each, and each level of $\mathcal{P}$ as a combination of $n - q$ pseudo factors $\mathcal{P}_1, \ldots, \mathcal{P}_{n-q}$, also with s levels each. Then a main-effect or interaction contrast of the n factors $\mathcal{B}_1, \ldots, \mathcal{B}_q, \mathcal{P}_1, \ldots, \mathcal{P}_{n-q}$ represents an interblock (respectively, intrablock) contrast if and only if it involves none (respectively, at least one) of the $\mathcal{P}_j$'s. We call $\mathcal{B}_1, \ldots, \mathcal{B}_q, \mathcal{P}_1, \ldots, \mathcal{P}_{n-q}$ unit factors. For a reason to be explained later, we place $\mathcal{P}_1, \ldots, \mathcal{P}_{n-q}$ before $\mathcal{B}_1, \ldots, \mathcal{B}_q$ and label each experimental unit by $(p_1, \ldots, p_{n-q}, b_1, \ldots, b_q)^T$, where $p_i, b_j \in \mathrm{GF}(s)$.

Since each experimental unit is a combination of the levels of $\mathcal{P}_1, \ldots, \mathcal{P}_{n-q}$, $\mathcal{B}_1, \ldots, \mathcal{B}_q$, we can describe the relation between the experimental units and the unit factors by an $n \times s^n$ matrix $\mathbf{Y}$ such that for each j, $1 \leq j \leq s^n$, the jth column of $\mathbf{Y}$ is the level combination of the unit factors corresponding to the jth unit. What we need is an $n \times s^n$ matrix $\mathbf{X}$ whose jth column gives the treatment combination assigned to the jth unit. The matrix $\mathbf{X}$ then produces the design. We obtain $\mathbf{X}$ from $\mathbf{Y}$ via a matrix multiplication

$$\mathbf{X} = \mathbf{KY}, \qquad (7.6)$$

where $\mathbf{K}$ is an $n \times n$ matrix with entries from $\mathrm{GF}(s)$. The matrix $\mathbf{K}$ connects the treatment factors to the unit factors and is called a *design key* matrix. By (7.6), the design key requires that the treatment combination $(x_1, \ldots, x_n)^T$ assigned to the experimental unit $(p_1, \ldots, p_{n-q}, b_1, \ldots, b_q)^T$ satisfy

$$x_i = \sum_{j=1}^{n-q} k_{ij} p_j + \sum_{l=1}^{q} k_{i,n-q+l} b_l. \qquad (7.7)$$

Equation (7.7) indicates that under the constructed design, the main effect of the ith treatment factor coincides with the "factorial" effect of the unit factors defined by $(k_{i1}, \ldots, k_{i,n-q}, k_{i,n-q+1}, \ldots, k_{in})$, the ith row of $\mathbf{K}$. We say that the latter is the *unit alias* of the former. The main effect of the ith treatment factor is estimated in the intrablock stratum if its unit alias involves at least one of the $\mathcal{P}_j$'s, which means that at least one of $k_{i,1}, \ldots, k_{i,n-q}$ is nonzero. In order to generate all the s^n treatment combinations, the n columns of $\mathbf{K}$ must be linearly independent.

To obtain the design layout, however, it is not necessary to carry out the matrix multiplication in (7.6). We describe below a simple systematic way of generating the design.

We call n linearly independent vectors $\mathbf{x}_1, \ldots, \mathbf{x}_n$ in $\mathrm{EG}(n, s)$ a set of *independent generators*. Given a set of independent generators and a sequence $a_1, \ldots, a_{s-1}$ of the nonzero elements of $\mathrm{GF}(s)$ with $a_1 = 1$, we can generate all the s^n treatment combinations in the following order:

1. Start with $\mathbf{0}$, which has all the components equal to zero. Set $k = 0$.
2. For each $0 \leq k \leq n - 1$, follow the s^k treatment combinations that have been generated by their sums with $\mathbf{x}_{k+1}$ in the same order, then the sums with $a_2 \mathbf{x}_{k+1}$ in

the same order, ... , up to the sums with $a_{s-1}\mathbf{x}_{k+1}$ in the same order. We have now a sequence of s^{k+1} treatment combinations.

3. Increase k by 1. Stop if $k = n$; otherwise go back to step 2 and continue.

We call the order determined by this procedure the *Yates order* with respect to $\mathbf{x}_1, \ldots, \mathbf{x}_n$. When s is a prime number, use $a_i = i$ and mod s operations.

Let $\mathbf{e}_i$ be the vector with the ith component equal to 1 and all the other components equal to zero. Suppose we arrange the experimental units in the Yates order with respect to $\mathbf{e}_1, \ldots, \mathbf{e}_n$. Since $\mathcal{P}_1, \ldots, \mathcal{P}_{n-q}$ are placed before $\mathcal{B}_1, \ldots, \mathcal{B}_q$, the first s^{n-q} units in the generated sequence form one block, and each of the succeeding set of s^{n-q} units is also a block. Since $\mathbf{K}$ is nonsingular, $\mathbf{Ke}_1, \ldots, \mathbf{Ke}_n$ form a set of independent generators for the treatment combinations. Then the treatment combinations assigned to the experimental units arranged in the Yates order with respect to $\mathbf{e}_1, \ldots, \mathbf{e}_n$ are themselves in the Yates order with respect to the independent generators $\mathbf{Ke}_1, \ldots, \mathbf{Ke}_n$. Since $\mathbf{Ke}_j$ is the jth column of the design key matrix, the treatment combinations can be generated and arranged in the Yates order with respect to the columns of $\mathbf{K}$. The first s^{n-q} treatment combinations are in the first block, and each subsequent set of s^{n-q} treatment combinations is also in the same block.

We note that in Patterson (1965) and Patterson and Bailey (1978), $\mathbf{K}^T$, instead of $\mathbf{K}$, is called the key matrix.

Example 7.3. We revisit Example 7.1 and consider the construction of a 2^5 design in eight blocks of size four. Let the five treatment factors be A, B, C, D, E, and the unit factors be $\mathcal{P}_1$, $\mathcal{P}_2$ and $\mathcal{B}_1$, $\mathcal{B}_2$, $\mathcal{B}_3$. To avoid confounding treatment main effects with blocks, we need to choose five linearly independent factorial effects of $\mathcal{P}_1$, $\mathcal{P}_2$, and $\mathcal{B}_1$, $\mathcal{B}_2$, $\mathcal{B}_3$, each involving at least one of $\mathcal{P}_1$ and $\mathcal{P}_2$, to be unit aliases of the five treatment main effects. Suppose we choose the unit aliases of A, B, C, D, E to be $\mathcal{P}_1$, $\mathcal{P}_2$, $\mathcal{P}_1\mathcal{B}_1$, $\mathcal{P}_2\mathcal{B}_2$, and $\mathcal{P}_1\mathcal{P}_2\mathcal{B}_3$, respectively. Then the design key matrix is

$$\mathbf{K} = \begin{matrix} & \begin{matrix} \mathcal{P}_1 & \mathcal{P}_2 & \mathcal{B}_1 & \mathcal{B}_2 & \mathcal{B}_3 \end{matrix} & \\ \begin{bmatrix} 1 & 0 & 0 & 0 & 0 \\ 0 & 1 & 0 & 0 & 0 \\ 1 & 0 & 1 & 0 & 0 \\ 0 & 1 & 0 & 1 & 0 \\ 1 & 1 & 0 & 0 & 1 \end{bmatrix} & \begin{matrix} A \\ B \\ C \\ D \\ E \end{matrix} \end{matrix} \qquad (7.8)$$

In (7.8), the correspondence of the treatment and unit factors to the rows and columns is indicated. For example, since the unit alias of C is $\mathcal{P}_1\mathcal{B}_1$, in the third row, we have 1's at the two entries corresponding to $\mathcal{P}_1$ and $\mathcal{B}_1$. Under the resulting design, the unit aliases of all the treatment interactions can be obtained from those of the treatment main effects. For example, the unit aliases of the three interactions AC, BD, and ABE, chosen to be confounded with blocks in Example 7.1, are, respectively, $\mathcal{P}_1(\mathcal{P}_1\mathcal{B}_1) = \mathcal{B}_1$, $\mathcal{P}_2(\mathcal{P}_2\mathcal{B}_2) = \mathcal{B}_2$, and $\mathcal{P}_1\mathcal{P}_2(\mathcal{P}_1\mathcal{P}_2\mathcal{B}_3) = \mathcal{B}_3$, which indeed represent interblock contrasts. The unit aliases of the other four interactions confounded with blocks, $ABCD$, BCE, ADE, and CDE, are, respectively, $\mathcal{B}_1\mathcal{B}_2$, $\mathcal{B}_1\mathcal{B}_3$, $\mathcal{B}_2\mathcal{B}_3$, and $\mathcal{B}_1\mathcal{B}_2\mathcal{B}_3$, all of which involve only the $\mathcal{B}$'s and are interblock contrasts.

The five columns of the design key matrix $\mathbf{K}$ in (7.8) give five independent generators ace, bde, c, d, and e. The first four treatment combinations in the Yates order of all the 32 treatment combinations with respect to these generators are (1), ace, bde, and $abcd$. These are the four treatment combinations in the first block. The second block can be obtained by multiplying all the treatment combinations in the first block by the third generator c. The fourth generator d generates two more blocks. Finally, four additional blocks are constructed by using the last generator e. This results in the same design as constructed in Example 7.1.

Thus unit aliases of the treatment main effects are used to write down the rows of $\mathbf{K}$. This determines $\mathbf{K}$, and then the columns of $\mathbf{K}$ are used to generate the design layout. We will explain later how to choose unit aliases of the treatment main effects. In particular, a template for the design key matrix will be provided in Section 7.7 to facilitate the construction.

One can determine the treatment interactions that are confounded with blocks by identifying those whose unit aliases involve $\mathcal{B}_1, \ldots, \mathcal{B}_q$ only. This information can also be obtained from the inverse relation $\mathbf{Y} = \mathbf{K}^{-1}\mathbf{X}$. In $\mathbf{K}^{-1}$, which is called the *inverse design key* matrix, the roles of treatment and unit factors are reversed, with its rows corresponding to unit factors and its columns corresponding to treatment factors. Thus the rows of $\mathbf{K}^{-1}$ determine the treatment factorial effects that are confounded with main effects of the unit factors and can be used to determine the treatment factorial effects that are estimated in each stratum.

Example 7.4. (Example 7.3 revisited) The inverse of the design key matrix in (7.8) is

$$
\mathbf{K}^{-1} = \begin{array}{ccccc} A & B & C & D & E \\ \begin{bmatrix} 1 & 0 & 0 & 0 & 0 \\ 0 & 1 & 0 & 0 & 0 \\ 1 & 0 & 1 & 0 & 0 \\ 0 & 1 & 0 & 1 & 0 \\ 1 & 1 & 0 & 0 & 1 \end{bmatrix} & \begin{array}{c} \mathcal{P}_1 \\ \mathcal{P}_2 \\ \mathcal{B}_1 \\ \mathcal{B}_2 \\ \mathcal{B}_3 \end{array} \end{array}
$$

The last three rows of $\mathbf{K}^{-1}$ correspond to $\mathcal{B}_1$, $\mathcal{B}_2$, and $\mathcal{B}_3$. Three treatment interactions AC, BD, and ABE can be read from these rows. Therefore the between-block contrasts coincide with AC, BD, ABE, and their generalized interactions. These are the treatment factorial effects confounded with blocks under the design constructed in Examples 7.1 and 7.3. We note that $\mathbf{K}^{-1} = \mathbf{K}$. Therefore the information about the treatment factorial effects that are confounded with blocks can also be determined directly from $\mathbf{K}$. When this is done, we need to label the columns of $\mathbf{K}$ by the treatment factors that originally labeled the rows. For example, the last three rows $(1, 0, 1, 0, 0)$, $(0, 1, 0, 1, 0)$, and $(1, 1, 0, 0, 1)$ of the design key matrix in (7.8) now correspond to the independent blocking words AC, BD, and ABE. We will return to a more general discussion of this in the next section where a template for design keys is presented.

7.7   A template for design keys

A major problem in searching/selecting design keys is that many different design keys produce exactly the same design. Cheng and Tsai (2013) presented templates to facilitate design key construction and to remove redundancies.

It will be shown in Remark 9.6 in Section 9.5 that without loss of generality (up to factor relabeling), an s^n design in s^q blocks of size s^{n-q} can be constructed by using a design key $\mathbf{K}$ of the form

$$\mathbf{K} = \begin{bmatrix} \mathbf{I}_{n-q} & \mathbf{0}_{n-q,q} \\ \mathbf{b}_1^T & \\ \vdots & \mathbf{I}_q \\ \mathbf{b}_q^T & \end{bmatrix}, \tag{7.9}$$

where $\mathbf{0}_{n-q,q}$ is the $(n-q) \times q$ matrix with all the entries equal to zero, and each $\mathbf{b}_i$ is an $(n-q) \times 1$ vector with entries from GF(s). The construction based on (7.9) is equivalent to that given by Das (1964) and Cotter (1974). This explicit design key template was given by Cheng and Tsai (2013).

Only the q vectors $\mathbf{b}_1, \ldots, \mathbf{b}_q$ in (7.9) need to be chosen. If it is required that none of the treatment main effects be confounded with blocks, then all of $\mathbf{b}_1, \ldots, \mathbf{b}_q$ must be nonzero, which guarantees that all the unit aliases of treatment main effects are intrablock contrasts.

It is easy to verify that

$$\mathbf{K}^{-1} = \begin{bmatrix} \mathbf{I}_{n-q} & \mathbf{0}_{n-q,q} \\ -\mathbf{b}_1^T & \\ \vdots & \mathbf{I}_q \\ -\mathbf{b}_q^T & \end{bmatrix}. \tag{7.10}$$

In particular, for two-level designs,

$$\mathbf{K}^{-1} = \mathbf{K};$$

this is because $-\mathbf{b}_i = \mathbf{b}_i$ when $s = 2$.

We already know that each of the last q rows of $\mathbf{K}^{-1}$ determines a blocking word (a treatment factorial effect that is confounded with blocks). It is clear that the last q rows of $\mathbf{K}^{-1}$ in (7.10) are linearly independent. Therefore choosing the q nonzero vectors $\mathbf{b}_1, \ldots, \mathbf{b}_q$ is equivalent to choosing q independent blocking words (but notice that $\mathbf{b}_1, \ldots, \mathbf{b}_q$ themselves do not have to be linearly independent). Furthermore, by (7.10), a set of independent blocking words can readily be identified from the last q rows of the design key matrix. Specifically, each of the last q rows of $\mathbf{K}$ with the first $n-q$ components replaced by their additive inverses (unchanged when $s = 2$) defines a treatment factorial effect that is confounded with blocks; also see Das (1964).

Example 7.5. (Examples 7.3 and 7.4 revisited) To construct a 2^5 design in eight blocks of size four, we can use the design key template

$$
\mathbf{K} =
\begin{array}{ccccc}
\mathcal{P}_1 & \mathcal{P}_2 & \mathcal{B}_1 & \mathcal{B}_2 & \mathcal{B}_3 \\
\end{array}
\begin{bmatrix}
1 & 0 & 0 & 0 & 0 \\
0 & 1 & 0 & 0 & 0 \\
* & * & 1 & 0 & 0 \\
* & * & 0 & 1 & 0 \\
* & * & 0 & 0 & 1 \\
\end{bmatrix}
\begin{array}{c}
A \\ B \\ C \\ D \\ E
\end{array}
$$

We need to fill in the six vacant entries. The last three rows of $\mathbf{K}$ in (7.8) result from the choices of $\mathbf{b}_1 = (1,0)^T$, $\mathbf{b}_2 = (0,1)^T$, and $\mathbf{b}_3 = (1,1)^T$, respectively. We have seen in Example 7.4 that indeed $\mathbf{K}^{-1} = \mathbf{K}$. Thus three independent blocking words AC, BD, and ABE can be identified from the last three rows of $\mathbf{K}$, or equivalently, from what are chosen for the unfilled entries in the design key template.

The discussion above shows that the construction based on design keys is equivalent to the method described in Section 7.2. Several things are built into the design key template: a set of independent generators, independence of blocking words, and constraints imposed by the block structure. This leads to the following advantages:

1. There is no need to solve equations to determine the principal block.
2. The design can be constructed in a simple systematic manner.
3. There is no need to check the independence of blocking words.
4. It is ensured that the treatment main effects are orthogonal to blocks. There is no need to check whether the choice of independent blocking words leads to the confounding of some treatment main effects with blocks.

Example 7.6. We revisit Example 7.2 and consider the construction of a 3^3 design in nine blocks of size three. Let the treatment factors be A, B, C and the unit factors be $\mathcal{P}$, $\mathcal{B}_1$ and $\mathcal{B}_2$. The template in (7.9) reduces to

$$
\mathbf{K} =
\begin{array}{ccc}
\mathcal{P} & \mathcal{B}_1 & \mathcal{B}_2 \\
\end{array}
\begin{bmatrix}
1 & 0 & 0 \\
* & 1 & 0 \\
* & 0 & 1 \\
\end{bmatrix}
\begin{array}{c}
A \\ B \\ C
\end{array}
$$

where the two $*$'s must be nonzero. If we choose both to be 1, then

$$
\mathbf{K} =
\begin{array}{ccc}
\mathcal{P} & \mathcal{B}_1 & \mathcal{B}_2 \\
\end{array}
\begin{bmatrix}
1 & 0 & 0 \\
1 & 1 & 0 \\
1 & 0 & 1 \\
\end{bmatrix}
\begin{array}{c}
A \\ B \\ C
\end{array}
$$

This gives the unit aliases of A, B, C as $\mathcal{P}$, $\mathcal{PB}_1$, and $\mathcal{PB}_2$, respectively. The three columns then give the independent generators $(1,1,1)^T$, $(0,1,0)^T$, and $(0,0,1)^T$. The first generator determines the first block $\{(0,0,0)^T, (1,1,1)^T, (2,2,2)^T\}$. The second generator determines two additional blocks, and the last generator produces six more, resulting in the same design as constructed in Example 7.2. A set of independent treatment factorial effects that are confounded with blocks can be identified from the last two rows of the design key matrix: A^2B and A^2C, which are the same as AB^2 and AC^2, respectively; note that we need to replace the first entries of these two rows by -1, which is congruent to 2 mod 3.

7.8   Construction of blocking schemes via Abelian groups

If s is a prime power but not a prime number, then pseudo factors, briefly discussed in Section 7.4, can be used for the construction of blocking schemes. An alternative method is to define the factorial effects as in Section 6.7 instead of using finite fields. This approach can also be applied to asymmetrical designs where the numbers of levels are not prime numbers or powers of prime numbers. The results presented in Section 6.7 can be used to help determine the treatment factorial effects that are confounded with blocks.

We have seen in Section 7.2 that each set of q linearly independent vectors $\mathbf{a}_1, \dots, \mathbf{a}_q \in \mathrm{EG}(n, s)$, where s is a prime number or power of a prime number, can be used to divide s^n treatment combinations into s^q blocks of size s^{n-q} by placing solutions to the simultaneous equations $\mathbf{a}_i^T \mathbf{x} = b_i$, $i = 1, \dots, q$, where $b_i \in \mathrm{EG}(n, s)$, in the same block. Let H be the subspace of $\mathrm{EG}(n, s)$ generated by $\mathbf{a}_1, \dots, \mathbf{a}_q$. Then the factorial effects defined by nonzero elements of H are confounded with blocks, and the principal block is the subspace of $\mathrm{EG}(n, s)$ consisting of all the $\mathbf{x}$'s satisfying $\mathbf{a}^T \mathbf{x} = 0$ for all $\mathbf{a} \in H$. Denote the principal block by H^0. Then the s^q blocks are H^0 and all the disjoint $(n-q)$-flats in the same pencil. This construction can be extended naturally to the case where the set of treatment combinations is an Abelian group (Bailey, 1977, 1985).

As in Section 6.7, suppose the treatment combinations are identified with the elements of $\mathbb{Z}_{s_1} \oplus \cdots \oplus \mathbb{Z}_{s_n}$, where $\mathbb{Z}_{s_i} = \{0, \dots, s_i - 1\}$. Given a subgroup H of $\mathbb{Z}_{s_1} \oplus \cdots \oplus \mathbb{Z}_{s_n}$, let $H^0 = \{\mathbf{x} \in \mathbb{Z}_{s_1} \oplus \cdots \oplus \mathbb{Z}_{s_n} : [\mathbf{a}, \mathbf{x}] = 0 \text{ for all } \mathbf{a} \in H\}$, where $[\mathbf{a}, \mathbf{x}]$ is as defined in (6.32). Then H^0, called the *annihilator* of H, is a subgroup of $\mathbb{Z}_{s_1} \oplus \cdots \oplus \mathbb{Z}_{s_n}$. Thus H^0 and all its cosets partition $\mathbb{Z}_{s_1} \oplus \cdots \oplus \mathbb{Z}_{s_n}$ into disjoint blocks of equal size. If $\mathbf{a}_1, \dots, \mathbf{a}_k$ are generators of H, then the principal block H^0 can be obtained by solving the equations $[\mathbf{a}_i, \mathbf{x}] = 0$ for $i = 1, \dots, k$. Since each coset of H^0 is of the form $H^0 + \mathbf{z}$ for some $\mathbf{z} \in \mathbb{Z}_{s_1} \oplus \cdots \oplus \mathbb{Z}_{s_n}$, for any $\mathbf{a} \in H$, we have $[\mathbf{a}, \mathbf{x}] = [\mathbf{a}, \mathbf{z}]$ for all $\mathbf{x} \in H^0 + \mathbf{z}$. That is, $[\mathbf{a}, \mathbf{x}]$ is a constant for all $\mathbf{x}$ in the same block. It follows that the factorial effects defined by nonzero elements of H are confounded with blocks. Furthermore, by results in linear algebra, we have $(H^0)^0 = H$, that is, H contains all the $\mathbf{a}$'s with $[\mathbf{a}, \mathbf{x}] = 0$ for all $\mathbf{x} \in H^0$. Therefore the treatment factorial effects that are confounded with blocks are precisely those defined by nonzero elements of H. Thus by part (ii) of Theorem 6.1, if all nonzero $\mathbf{a} \in H$ have more than e nonzero entries, then all the factorial effects involving no more than e factors are not confounded with blocks.

In general, any subgroup G of $\mathbb{Z}_{s_1} \oplus \cdots \oplus \mathbb{Z}_{s_n}$ and its cosets provide a blocking scheme. The factorial effects confounded with blocks are those defined by the nonzero elements of G^0. Readers are referred to Bailey (1977) for details. More general results can be stated and studied in terms of characters and dual groups; see Bailey (1985) and Kobilinsky (1985).

Example 7.7. Consider Example 3 in John and Dean (1975), also discussed in Bailey, Gilchrist, and Patterson (1977). This is a single-replicate 6^3 design. The six levels of each factor are represented by the elements of $\mathbb{Z}_6$. Let H be the cyclic group generated by $\mathbf{a} = (1, 2, 5)^T$. Then H^0 consists of the solutions to $x_1 + 2x_2 + 5x_3 \equiv 0 \pmod 6$. Since $\mathbf{a}$ has order six, H^0 and its five cosets give a partition of the treatment combinations into six blocks of size 36. The 36 treatment combinations in the principal block H^0 can be generated by $(1, 0, 1)^T$ and $(0, 1, 2)^T$, which are two solutions to $x_1 + 2x_2 + 5x_3 \equiv 0 \pmod 6$, each generating a distinct cyclic group of order six. The treatment factorial effects confounded with blocks are the five degrees of freedom of treatment contrasts defined by the partition of the treatment combinations into six disjoint classes induced by $(1, 2, 5)^T$. To determine what these contrasts are, by the discussions in Section 6.7, we only need to consider the multiples $r\mathbf{a}$ where r is a prime divisor of 6, i.e., $\mathbf{a}$, $2\mathbf{a}$, and $3\mathbf{a}$. Now $2\mathbf{a} = (2, 4, 4)^T$ has order 3 and $3\mathbf{a} = (3, 0, 3)^T$ has order 2. Since $\phi(2) = 1$, and $\phi(3) = 2$, the treatment factorial effects that are confounded with blocks are the interaction of A_1 and A_3 defined by $(3, 0, 3)^T$ (one degree of freedom), the interaction of A_1, A_2, and A_3 defined by $(1, 2, 5)^T$ (two degrees of freedom), and the interaction of A_1, A_2, and A_3 defined by $(2, 4, 4)^T$ (two degrees of freedom).

The method of design key can also be applied to the current setting. We illustrate its application to the construction of the design in Example 7.7, and refer readers to Bailey, Gilchrist, and Patterson (1977) and Bailey (1977) for more general discussions.

Example 7.8. (Example 7.7 continued) Denote the three treatment factors by A_1, A_2, and A_3, and identify each unit with a combination of three six-level factors $\mathcal{B}$, $\mathcal{P}_1$, and $\mathcal{P}_2$, where the levels of $\mathcal{B}$ represent the six blocks, and $\mathcal{P}_1$, $\mathcal{P}_2$ are pseudo factors whose level combinations represent the 36 units in each block. Then the design in Example 7.7 can be constructed by using the design key

$$\mathbf{K} = \begin{matrix} & \mathcal{P}_1 & \mathcal{P}_2 & \mathcal{B} & \\ & \begin{bmatrix} 1 & 0 & 0 \\ 0 & 1 & 0 \\ 1 & 2 & 1 \end{bmatrix} & & & \begin{matrix} A_1 \\ A_2 \\ A_3 \end{matrix} \end{matrix}$$

with the modulo 6 arithmetic. One can see that this design key matrix is obtained by using the template in (7.9). Replacing the first two entries of the last row with their additive inverses, we obtain $(5,4,1)^T = 5(1,2,5)^T$, which is a blocking word that yields the same partition of the treatment combinations into six blocks of size 36 as the blocking word $(1,2,5)^T$ in Example 7.7. The design can be obtained by generating and arranging the treatment combinations in the Yates order with respect to the three columns $(1,0,1)^T$, $(0,1,2)^T$, and $(0,0,1)^T$ of $\mathbf{K}$. The first 36 treatment combinations are in the same block, and each subsequent set of 36 consecutive treatment combinations is also in the same block. We have seen in Example 7.7 that indeed $(1,0,1)^T$ and $(0,1,2)^T$ are generators of the principal block.

For the case where the numbers of levels are not prime numbers, John and Dean (1975) and Dean and John (1975) introduced *generalized cyclic designs*, which are blocked symmetric or asymmetric factorial designs constructed by using independent generators of a group. They also showed how the confounding pattern can be determined from generators of the initial (principal) block. Such designs can be constructed by the method of design key. We refer readers to Dean (1990) for a survey of the use of generalized cyclic designs.

7.9 Complete factorial experiments in row-column designs

Suppose a complete factorial experiment is to be conducted by using a row-column design. If both the numbers of rows and columns are multiples of the number of treatment combinations, then one should use a design under which all the treatment combinations appear equally often in each row and each column. Such a design allows for simple analysis, but requires a large number of experimental units. If the numbers of rows and columns are less than the number of treatment combinations, then, as in the case of incomplete blocks, there may be treatment information in the row and column strata. We discuss here the construction of row-column designs under which some factorial effects are confounded with row contrasts or column contrasts.

Suppose there are n treatment factors with s levels each, where s is a prime number or power of a prime number. The s^n treatment combinations are to be observed on s^n experimental units arranged in s^p rows and s^q columns with $p+q=n$. Choose a set of $p+q$ linearly independent vectors $\mathbf{a}_1, \ldots, \mathbf{a}_p, \mathbf{b}_1, \ldots, \mathbf{b}_q$ in EG(n,s). Divide the s^n treatment combinations $\mathbf{x}$ into s^p rows according to the values of $\mathbf{a}_i^T \mathbf{x}$, $i = 1, \ldots, p$, and s^q columns according to the values of $\mathbf{b}_j^T \mathbf{x}$, $j = 1, \ldots, q$. Then there is a single treatment combination at each row-column intersection. Similar to Section 7.2, under the resulting design the $s^p - 1$ degrees of freedom defined by $\mathbf{a}_1, \ldots, \mathbf{a}_p$ and all their nonzero linear combinations are confounded with rows, the $s^q - 1$ degrees of freedom defined by $\mathbf{b}_1, \ldots, \mathbf{b}_q$ and all their nonzero linear combinations are confounded with columns, and the other factorial effects are estimated in the bottom stratum. To construct the design, as in Section 7.2, we start with $\mathbf{x} = \mathbf{0}$, which is the only solution to the $p+q$ equations $\mathbf{a}_i^T \mathbf{x} = 0$, $i = 1, \ldots, p$, and $\mathbf{b}_j^T \mathbf{x} = 0$, $j = 1, \ldots, q$. The first column consists of $\mathbf{0}$ and the treatment combinations obtained by repeatedly adding solutions to the equations $\mathbf{b}_j^T \mathbf{x} = 0$, $j = 1, \ldots, q$, and their nonzero multiples to the treatment combinations that have been generated. All the other columns can be obtained by repeatedly adding solutions to the equations $\mathbf{a}_i^T \mathbf{x} = 0$, $i = 1, \ldots, p$, and their nonzero multiples to the treatment combinations in the columns that have been constructed, while keeping the sequential positions of the treatment combinations in each column.

One can also apply the method of design key. Represent the s^p rows by p pseudo factors $\mathcal{R}_1, \ldots, \mathcal{R}_p$ and the s^q columns by q pseudo factors $\mathcal{C}_1, \ldots, \mathcal{C}_q$, all with s levels. Then the "main-effect" and "interaction" contrasts of $\mathcal{R}_1, \ldots, \mathcal{R}_p$ are in the row-stratum, and the "main-effect" and "interaction" contrasts of $\mathcal{C}_1, \ldots, \mathcal{C}_q$ are in the column-stratum. It can be shown (see Cheng and Tsai (2013) and Exercise 9.5) that if the columns corresponding to $\mathcal{R}_1, \ldots, \mathcal{R}_p$ are placed before those corresponding to $\mathcal{C}_1, \ldots, \mathcal{C}_q$, then, subject to factor relabeling, without loss of generality, we can use a design key matrix of the form

$$\mathbf{K} = \begin{bmatrix} \mathbf{I}_p & \mathbf{B} \\ \mathbf{C} & \mathbf{I}_q \end{bmatrix}, \tag{7.11}$$

where $\mathbf{B}$ and $\mathbf{C}$ are such that $\mathbf{K}$ is nonsingular. The design key construction requires choosing the p rows of $\mathbf{B}$ and the q rows of $\mathbf{C}$. As in Section 7.7, this is equivalent to selecting $p+q$ independent treatment factorial effects with p of which confounded with rows and the other q confounded with columns. Each of the last q (respectively, first p) rows of $\mathbf{K}$ with its first p (respectively, last q) components replaced by their additive inverses (unchanged when $s = 2$) defines a treatment factorial effect that is confounded with columns (respectively, rows). If the treatment main effects are not to be confounded with rows or columns, then we must choose factorial effects of $\mathcal{R}_1, \ldots, \mathcal{R}_p, \mathcal{C}_1, \ldots, \mathcal{C}_q$ that involve *at least one $\mathcal{R}_i$ and at least one $\mathcal{C}_j$* as unit aliases of the treatment main effects; that is, all the rows of $\mathbf{B}$ and $\mathbf{C}$ are nonzero.

Example 7.9. To construct a 2^4 design in four rows and four columns, suppose the treatment factors are A, B, C, D, and we would like to confound AD and BC with rows and confound ABC and BD with columns. (Note that the four effects AD, BC, ABC, and BD are independent.) Then we will also confound $ABCD$ (the generalized interaction of AD and BC) with rows and ACD (the generalized interaction of ABC and BD) with columns, neither of which is a main effect. To construct the design, we start with (1), followed by a treatment combination that has even numbers of letters in common with both ABC and BD, say ac. Then we multiply both (1) and ac by another treatment combination that has even numbers of letters in common with both ABC and BD, say bcd. This gives the treatment combinations in the first column: (1), ac, bcd, and abd. The second column can be obtained by multiplying all the treatment combinations in the first column by a treatment combination that has even numbers of letters in common with both AD and BC, say bc. The design is completed by multiplying all the treatment combinations in the first two columns by a treatment combination that has not appeared yet and has even numbers of letters in common with both AD and BC, say ad.

(1)	bc	ad	$abcd$
ac	ab	cd	bd
bcd	d	abc	a
abd	acd	b	c

The same design can be constructed by using the design key

$$
\mathbf{K} = \begin{array}{c} \\ \\ \\ \\ \end{array}
\begin{array}{cccc}
\mathcal{R}_1 & \mathcal{R}_2 & \mathcal{C}_1 & \mathcal{C}_2 \\
\left[\begin{array}{cccc}
1 & 0 & 0 & 1 \\
0 & 1 & 1 & 0 \\
1 & 1 & 1 & 0 \\
0 & 1 & 0 & 1
\end{array}\right] &
\begin{array}{c}
A \\ B \\ C \\ D
\end{array}
\end{array}
$$

with $\mathcal{R}_1\mathcal{C}_2$, $\mathcal{R}_2\mathcal{C}_1$, $\mathcal{R}_1\mathcal{R}_2\mathcal{C}_1$, and $\mathcal{R}_2\mathcal{C}_2$ as unit aliases of the main effects of A, B, C, and D, respectively. Then the columns of $\mathbf{K}$ give four independent generators ac, bcd, bc, and ad. The first four treatment combinations (1), ac, bcd, and abd in the Yates order of all the treatment combinations with respect to the four generators are those that appear in the first column, and each succeeding set of four treatment combinations is also in the same column. This produces the same design constructed above. The unit aliases of the treatment interactions are $AB = \mathcal{R}_1\mathcal{R}_2\mathcal{C}_1\mathcal{C}_2$, $AC = \mathcal{R}_2\mathcal{C}_1\mathcal{C}_2$, $AD = \mathcal{R}_1\mathcal{R}_2$, $BC = \mathcal{R}_1$, $BD = \mathcal{C}_1\mathcal{C}_2$, $CD = \mathcal{R}_1\mathcal{C}_1\mathcal{C}_2$, $ABC = \mathcal{C}_2$, $ABD = \mathcal{R}_1\mathcal{C}_1$, $ACD = \mathcal{C}_1$, $BCD = \mathcal{R}_1\mathcal{R}_2\mathcal{C}_2$, $ABCD = \mathcal{R}_2$. These unit aliases confirm that AD, BC, and $ABCD$ are confounded with rows, and ABC, BD, and ACD are confounded with columns. One can also identify AD and BC from the first two rows of $\mathbf{K}$, and ABC and BD from the last two rows of $\mathbf{K}$.

7.10   Split-plot designs

In the previous sections, we discussed complete factorial experiments conducted in incomplete blocks, in which certain factorial effects are confounded with blocks. Typically we prefer not to confound lower-order treatment factorial effects with blocks if it is possible. Under split-plot designs discussed in this section and strip-plot designs to be studied in the next section, some treatment main effects, however, cannot be estimated in the bottom stratum.

In agricultural experiments, sometimes certain treatment factors require larger plots than others. For example, it may be that the machines for sowing seeds require larger plots for the seeds, but each fertilizer can be applied to a small plot. The seeds are randomly assigned to plots of a suitable size, which are called *whole-plots*. Each whole-plot is then divided into smaller *subplots*, and different fertilizers are randomly assigned to the subplots within each whole-plot. This leads to the same block structure as a block design with each whole-plot as a block and each subplot as a unit in a block. The counterparts of inter- and intrablock strata are called whole-plot and subplot strata, respectively. In this case, the seed factor is called a *whole-plot treatment factor* and the fertilizer factor is called a *subplot treatment factor*. The main-effect contrasts of the seed factor are confounded with between-whole-plot contrasts since the same seed is applied to all the subplots in the same whole-plot. As a result, the main-effect contrasts of this treatment factor may be estimated with lower precision. Many industrial experiments also have a split-plot structure. It may be difficult to change the levels of certain treatment factors, which have to be kept the same on all the experimental runs throughout the day, whereas the levels of the other factors can be varied from run to run. (See Example 1.2.) Then each run is a subplot and the experimental runs on the same day constitute a whole-plot. Many experiments involve several processing stages, with the treatment factor levels assigned at different stages. Suppose batches of material are treated by several different methods, and each batch is divided into several samples to receive different levels of another factor. Then each batch is a whole-plot and the samples are subplots.

Sometimes certain treatment factors are included in the experiment mainly to study their possible interactions with the other treatment factors, and the main effects of such factors are of less interest. For example, in experiments for robust parameter designs in quality improvement (see Chapter 11 of Wu and Hamada (2009)), it is desirable to have higher precision for estimates of the main effects of control factors and interactions of control and noise factors. We refer readers to Box and Jones (1992) for a split-plot experiment with noise factors as whole-plot treatment factors.

Suppose there are a whole-plots, with each whole-plot split into b subplots, and there are two treatment factors A and B, where A has a levels and B has b levels. Each level of factor A is assigned randomly to one whole-plot, and each level of B is assigned randomly to one subplot in each whole-plot. The ab experimental units have the structure (a whole-plots)/(b subplots). It is clear that the main-effect contrasts of A coincide with between-whole-plot contrasts. The main-effect contrasts of B and the interaction contrasts of A and B are estimated in the subplot stratum since they are orthogonal to the main-effect contrasts of A and therefore are orthogonal to whole-plot contrasts. Estimates of these contrasts and the associated sums of squares can be computed in the same simple manner as under complete randomization. The following is a skeleton of the ANOVA.

Source of variation	d.f.
Whole-plot stratum	
A	$a-1$
Subplot stratum	
B	$b-1$
AB	$(a-1)(b-1)$
Total	$ab-1$

The design and analysis discussed above can be extended to the case where there are more than two treatment factors. Consider the design of a complete s^n factorial experiment in s^q whole-plots each containing s^{n-q} subplots, where s is a prime number or power of a prime number. Suppose n_1 of the n treatment factors are whole-plot factors. In order to have a complete factorial design, we must have $n_1 \leq q$. We allow the possibility that $n_1 < q$, so the number of whole-plots can be greater than the number of level combinations of whole-plot treatment factors. Bingham, Schoen, and Sitter (2004) used a cheese-making experiment to illustrate the practical need of having a sufficiently large number of whole-plots. For example, there may be physical limitations on the number of subplots per whole-plot; therefore more whole-plots are needed. Furthermore, when normal or half-normal probability plots are used to assess the significance of the effects, two separate plots should be constructed for the factorial effects estimated in the whole-plot and subplot strata. If there are too few whole-plots, then on the probability plot for the effects estimated in the whole-plot stratum, there might not be enough points to separate active effects from inactive ones. Under these circumstances, replications of the complete factorial of whole-plot treatment factors at the whole-plot level may be necessary.

Technically, the construction of such split-plot designs is the same as that of complete s^n factorial designs in s^q blocks of size s^{n-q} described in Section 7.2, except that since the main effect of each of the n_1 whole-plot treatment factors must be confounded with whole-plots, the q linearly independent vectors $\mathbf{a}_1, \ldots, \mathbf{a}_q$ used for partitioning the s^n treatment combinations into s^q disjoint groups can be chosen in such a way that n_1 of them, say $\mathbf{a}_1, \ldots, \mathbf{a}_{n_1}$, define the main effects of the whole-plot treatment factors. For each $1 \leq i \leq n$, let $\mathbf{e}_i$ be the $n \times 1$ vector whose ith entry is equal to 1 and all the other entries are zero. Without loss of generality, let the first n_1 treatment factors be whole-plot factors; then we have $\mathbf{a}_i = \mathbf{e}_i$ for $1 \leq i \leq n_1$. Under the resulting design all the factorial effects defined by nonzero linear combinations of $\mathbf{a}_1, \ldots, \mathbf{a}_q$ are estimated in the whole-plot stratum, and the other factorial effects are estimated in the subplot stratum.

In the construction procedure described above, effectively we only need to choose $q - n_1$ words $\mathbf{a}_{n_1+1}, \ldots, \mathbf{a}_q$ that involve at least some subplot treatment factors. In particular, when $n_1 = q$, no such word is needed, and we simply assign the s^{n_1} level combinations of the whole-plot treatment factors randomly to the s^{n_1} whole-plots, and assign the s^{n-n_1} level combinations of the subplot treatment factors randomly to the s^{n-n_1} subplots in each whole-plot. In this case, factorial effects of the whole-plot treatment factors are estimated in the whole-plot stratum, and all the other factorial effects are estimated in the subplot stratum.

When $n_1 < q$, the construction procedure is equivalent to the following. Start with the design with s^{n_1} whole-plots each containing s^{n-n_1} subplots as constructed in the previous paragraph. Then we use the $q - n_1$ independent words $\mathbf{a}_{n_1+1}, \ldots, \mathbf{a}_q$ to partition the s^{n-n_1} treatment combinations in each whole-plot into s^{q-n_1} disjoint groups of size $s^{n-n_1}/s^{q-n_1} = s^{n-q}$. Now treat each of these groups as a new whole-plot. In other words, each of the s^{n_1} original whole-plots of size s^{n-n_1} is split into s^{q-n_1} smaller whole-plots, creating a total of s^q whole plots of size s^{n-q}, as desired. The $q - n_1$ independent words $\mathbf{a}_{n_1+1}, \ldots, \mathbf{a}_q$ are called *independent splitting words*. The factorial effects defined by nonzero linear combinations of $\mathbf{a}_{n_1+1}, \ldots, \mathbf{a}_q$ and their generalized interactions with the factorial effects of whole-plot treatment factors are called *splitting effects*. Under the resulting design, all the factorial effects of whole-plot treatment factors and all the splitting effects are estimated in the whole-plot stratum, and all the other treatment factorial effects are estimated in the subplot stratum. Note that not only $\mathbf{a}_{n_1+1}, \ldots, \mathbf{a}_q$ are linearly independent, all the q vectors $\mathbf{e}_1, \ldots, \mathbf{e}_{n_1}, \mathbf{a}_{n_1+1}, \ldots, \mathbf{a}_q$ need to be linearly independent. We also need to check that no main effect of a subplot treatment factor ends up being estimated in the whole-plot stratum. This is equivalent to that no subplot treatment main effect is a splitting effect.

Example 7.10. Suppose the block structure is (8 whole-plots)/(4 subplots) and there are five two-level treatment factors A, B, C, S, T, where A, B, and C are whole-plot factors, and S, T are subplot factors. In this case $n = 5$, $n_1 = 3$, and $q = 3$. Since $n_1 = q$, we assign each level combination of the whole-plot treatment factors to all the subplots in one whole-plot, and each level combination of the subplot treatment factors is assigned to one subplot in each whole-plot:

(1)	a	b	ab	c	ac	bc	abc
st	ast	bst	$abst$	cst	$acst$	$bcst$	$abcst$
t	at	bt	abt	ct	act	bct	$abct$
s	as	bs	abs	cs	acs	bcs	$abcs$

The factorial effects of A, B, and C are estimated in the whole-plot stratum, and the other factorial effects are estimated in the subplot stratum.

Example 7.11. Suppose the block structure is (16 whole-plots)/(2 subplots) and the treatment factors are as in Example 7.10. In this case $n = 5$, $n_1 = 3$, and $q = 4$. We need one splitting effect which, together with the factorial effects of A, B, and C, is confounded with whole-plots. Suppose we choose ST to be a splitting effect. To construct the design, we begin by identifying the treatment combinations that have even numbers of letters in common with each of the four independent words A, B, C, and ST: (1) and st. These two treatment combinations constitute the first (principal) whole-plot. Repeatedly multiplying the treatment combinations in the whole-plots that have been generated by a treatment combination that has not appeared yet completes the design:

(1)	a	b	ab	c	ac	bc	abc
st	ast	bst	$abst$	cst	$acst$	$bcst$	$abcst$

t	at	bt	abt	ct	act	bct	$abct$
s	as	bs	abs	cs	acs	bcs	$abcs$

Under this design, A, B, C, AB, AC, BC, ABC, ST, AST, BST, CST, $ABST$, $ACST$, $BCST$, and $ABCST$ are estimated in the whole-plot stratum. These are all the factorial effects of whole-plot treatment factors, their generalized interactions with ST, and ST. All the other factorial effects are estimated in the subplot stratum. We can also start with the design constructed in Example 7.10, and split each whole-plot of size four into two whole-plots of size two by placing the treatment combinations with even numbers of letters in common with the splitting effect ST in the same smaller whole-plot, and those with odd numbers of letters in common in the other. For example, the first whole-plot $\{(1), s, t, st\}$ of the design in Example 7.10 is split into $\{(1), st\}$ and $\{s, t\}$. As a third option, we can construct the design by using the design key matrix

$$
\begin{array}{ccccc}
\mathcal{S} & \mathcal{P}_1 & \mathcal{P}_2 & \mathcal{P}_3 & \mathcal{P}_4 \\
\begin{bmatrix}
1 & 0 & 0 & 0 & 0 \\
0 & 1 & 0 & 0 & 0 \\
0 & 0 & 1 & 0 & 0 \\
0 & 0 & 0 & 1 & 0 \\
1 & 0 & 0 & 0 & 1
\end{bmatrix}
\begin{matrix}
S \\ A \\ B \\ C \\ T
\end{matrix}
\end{array}
\tag{7.12}
$$

where the two-level unit factors $\mathcal{P}_1$, $\mathcal{P}_2$, $\mathcal{P}_3$, and $\mathcal{P}_4$ define the 16 whole-plots and $\mathcal{S}$ defines the two subplots in each whole plot. We let the first row correspond to S, the next three rows correspond to the whole-plot treatment factors, and the last row correspond to T; the reason for this arrangement will be explained when a general template is presented in the next paragraph. Since the unit alias of each of A, B, and C does not involve $\mathcal{S}$, and the unit aliases of both S and T involve $\mathcal{S}$, this guarantees that A, B, C are whole-plot treatment factors and S, T are subplot treatment factors. The design can be constructed by generating and arranging the 32 treatment combinations in the Yates order with respect to the five generators st, a, b, c, and t, identified from the columns of the design key matrix. Each consecutive pair of treatment combinations starting from the beginning constitutes a whole-plot. That ST is a splitting effect can be seen from the fact that its unit alias is $\mathcal{S}(\mathcal{S}\mathcal{P}_4) = \mathcal{P}_4$, which does not involve $\mathcal{S}$ and hence is a whole-plot contrast. This splitting effect can also be identified from the last row of the design key.

We now present a template for the design key construction of a complete factorial split-plot design. Because the block structure of a split-plot design is the same as that of a block design, the results in Section 7.7 can be applied, and we can use a design key $\mathbf{K}$ of the form

$$
\mathbf{K} = \begin{bmatrix} \mathbf{I}_{n-q} & \mathbf{0}_{n-q,q} \\ \mathbf{B} & \mathbf{I}_q \end{bmatrix},
\tag{7.13}
$$

where $\mathbf{B}$ is $q \times (n-q)$. Since all the first $n-q$ components of any row of $\mathbf{K}$ corresponding to a whole-plot treatment factor must be zero, the first $n-q$ rows of $\mathbf{K}$ in (7.13) can only correspond to subplot treatment factors. Without loss of generality, we can let the next n_1 rows correspond to the whole-plot treatment factors, and the last $q-n_1$ rows of $\mathbf{K}$ correspond to subplot treatment factors if $n_1 < q$. Thus, subject to factor relabeling and the aforementioned constraints on the rows of $\mathbf{K}$, a complete s^n factorial split-plot design with n_1 whole-plot treatment factors can be constructed by using the design key template in (7.13), where the first n_1 rows of $\mathbf{B}$ are zero, and the remaining rows, if there are any, are nonzero (Cheng and Tsai, 2013).

For the case $n_1 = q$, (7.13) reduces to

$$
\mathbf{K} = \begin{bmatrix} \mathbf{I}_{n-q} & \mathbf{0} \\ \mathbf{0} & \mathbf{I}_q \end{bmatrix}.
\tag{7.14}
$$

For $n_1 < q$, the design key construction requires the selection of $q-n_1$ nonzero rows of $\mathbf{B}$. The construction presented in the paragraph preceding Example 7.10 requires $q-n_1$ independent splitting words. The same argument as in Section 7.7 shows that the selection of the last $q - n_1$ rows of **B** is equivalent to that of independent splitting words. Each of the last $q - n_1$ rows of **K** with the first $n - q$ components replaced by their additive inverses (unchanged when $s = 2$) defines a splitting effect. This establishes the equivalence between the construction/search of split-plot designs based on the method presented in the paragraph preceding Example 7.10 and that based on the design key. In addition to the advantages mentioned in the paragraph preceding Example 7.6 (built-in independent generators for design construction, no need to check independence of splitting words, and no need to check that no main effects of subplot treatment factors are confounded with whole-plot contrasts), the design key construction does not need to go through the splitting process.

Example 7.12. (Example 7.11 revisited) We have $n = 5$, $n_1 = 3$, and $q = 4$. Since $q - n_1 = 1$, we only have to choose the last row of **B**, which is a number. The only choice is $b_1 = 1$. Then we have the design key matrix in (7.12), and the last row $(1\,0\,0\,0\,1)$ confirms that ST is a splitting effect.

7.11   Strip-plot designs

Strip-plot designs also originated from agricultural experiments. Suppose there are two treatment factors A and B, both of which require large plots. If A has a levels and B has b levels, then one replicate of a complete factorial requires ab large plots, which may not be practical. An alternative is to divide the experimental area into a horizontal strips and b vertical strips, resulting in a total of ab plots with the structure (a rows)$\times$(b columns). Each level of factor A is assigned to all the plots in one row, and each level of B is assigned to all the plots in one column. This is a more economic way to run the experiment, but the main effects of the two treatment factors are confounded with rows and columns, respectively. All their interaction contrasts are estimated in the bottom stratum. We call A a *row treatment factor* and B a *column treatment factor*.

Many industrial experiments are also run in strip-plots to reduce cost. Experiments involving multiple processing stages are increasingly common in industrial applications. For example, the fabrication of integrated circuits involves a sequence of processing stages. At each stage several wafers may be processed together and they are all assigned the same levels of some treatment factors. A strip-plot design is useful for experiments with two processing stages. The experiment discussed in Example 1.3 is a blocked strip-plot experiment. There are two processing stages within each block: washing and drying.

Suppose there are s^p rows, s^q columns, p row treatment factors with s levels each, and q column treatment factors also with s levels each, where s is a prime number or power of a prime number. Each level combination of the row treatment factors is assigned to all the units in one row, and each level combination of the column treatment factors is assigned to all the units in one column. Then all the factorial effects of the row treatment factors are estimated in the row stratum, all the factorial effects of the column treatment factors are estimated in the column stratum, and all the other factorial effects are estimated in the bottom stratum. The design key (7.11) reduces to

$$\mathbf{K} = \begin{bmatrix} \mathbf{I}_p & \mathbf{0} \\ \mathbf{0} & \mathbf{I}_q \end{bmatrix}, \tag{7.15}$$

where the first p rows of $\mathbf{K}$ correspond to row treatment factors and the last q rows correspond to column treatment factors.

We note that strip-plot designs are also called strip-block designs.

Exercises

7.1 Show that under design (7.1), the interblock estimator of any treatment contrast $\mathbf{c}^T\boldsymbol{\alpha}$ has variance $\|\mathbf{c}\|^2 \xi_{\mathcal{B}}$, and the intrablock estimator has variance $\frac{1}{2}\|\mathbf{c}\|^2 \xi_{\mathcal{E}}$.

7.2 Use a design key template to show the following fact observed in Kerr (2006): Under a complete 2^n factorial design in 2^{n-1} blocks of size two, if no main effect is confounded with blocks, then all the interactions involving even numbers of factors are confounded with blocks.

7.3 Use a design key template to show (7.4).

7.4 Construct a design for a complete 3^4 factorial in nine blocks of size 9 without confounding main effects and two-factor interactions with blocks.

7.5 Construct a design for a complete $2^3 \times 3^2$ factorial in 12 blocks of size 6 without confounding main effects with blocks.

7.6 Construct a design for a complete $2^3 \times 4$ factorial in four blocks of size 8 without confounding main effects and two-factor interactions with blocks.

7.7 Construct a design for a complete $2^3 \times 6$ factorial in four blocks of size 12 and determine the treatment factorial effects that are confounded with blocks.

7.8 In Example 7.9, use the inverse of the design key matrix to verify that AD, BC, $ABCD$ are confounded with rows and ABC, ACD, BD are confounded with columns.

Chapter 8

Fractional Factorial Designs and Orthogonal Arrays

When the number of factors is large, cost and other practical considerations often call for observing a fraction of all the treatment combinations, in particular if only a subset of the factorial effects is expected to be active. Under such a *fractional factorial design*, first introduced by Finney (1945), not all the factorial effects can be estimated, and follow-up experiments are often needed to resolve ambiguity. In this and the next three chapters, we assume that the experimental units are unstructured and the experiment is to be conducted with complete randomization. Designs of factorial experiments with more complicated block structures will be discussed in Chapters 13 and 14. We introduce in this chapter an important combinatorial structure called orthogonal arrays and describe how they can be used to run factorial experiments. The estimability of factorial effects under an orthogonal array is discussed. It is shown that when some effects are assumed to be negligible, certain other effects can be uncorrelatedly estimated. Several examples are given to illustrate different types of orthogonal arrays and the distinction between regular and nonregular designs. While the emphasis of this book is on regular designs, we briefly discuss a class of nonregular designs derived from Hadamard matrices. We also present the methods of foldover and difference matrices for constructing fractional factorial designs. The chapter ends with a survey of some variants of orthogonal arrays that were introduced in recent years for applications to the design of computer experiments.

8.1 Treatment models for fractional factorial designs

When the experimental units are unstructured, a factorial design d can be specified by the number of observations to be taken on each treatment combination. Thus it can be thought of as a *multiset* of treatment combinations, allowing repeated observations on the same treatment combination.

Suppose the run size is N, and let $y_1, \ldots, y_N$ be the N observed responses. In view of the analysis of completely randomized designs derived in Section 5.2, without loss of generality, we may assume that $y_1, \ldots, y_N$ are uncorrelated with constant variance, and if y_i is an observation on treatment combination $(x_1, \ldots, x_n)^T$, then

$$E(y_i) = \alpha(x_1, \ldots, x_n). \tag{8.1}$$

We have seen in Chapter 6 how to parametrize $\alpha(x_1, \ldots, x_n)$ in terms of various factorial effects. We can write $\boldsymbol{\alpha}$ as

$$\boldsymbol{\alpha} = \mathbf{P}\boldsymbol{\beta}, \tag{8.2}$$

where $\mathbf{P}$ is as in (6.25). A full model with all the factorial effects present can be expressed as

$$E(\mathbf{y}) = \mathbf{X}_{\mathcal{T}}\boldsymbol{\alpha} = \mathbf{X}_{\mathcal{T}}\mathbf{P}\boldsymbol{\beta}, \tag{8.3}$$

where $\mathbf{y}$ is the $N \times 1$ vector of observed responses, and $\mathbf{X}_{\mathcal{T}}$ is the unit-treatment incidence matrix defined in Section 4.3.

Often some of the factorial effects are assumed to be negligible; then the associated terms are dropped from (8.3). This leads to a reduced linear model

$$E(\mathbf{y}) = \mathbf{X}_{\mathcal{T}}\mathbf{Q}\tilde{\boldsymbol{\beta}}, \tag{8.4}$$

where $\mathbf{Q}$ (respectively, $\tilde{\boldsymbol{\beta}}$) is obtained by deleting the columns of $\mathbf{P}$ (respectively, entries of $\boldsymbol{\beta}$) that correspond to negligible factorial effects

We follow the notation in (6.24) to write each component of $\boldsymbol{\beta}$ as $\beta^{\mathbf{z}}$, where $\mathbf{z} \in S_1 \times \cdots \times S_n$, and denote the column of the full model matrix $\mathbf{X}_{\mathcal{T}}\mathbf{P}$ corresponding to $\beta^{\mathbf{z}}$ by $\mathbf{x}^{\mathbf{z}}$. The reduced model matrix $\mathbf{X}_{\mathcal{T}}\mathbf{Q}$ for $\tilde{\boldsymbol{\beta}}$ is obtained from the full model matrix by dropping the columns $\mathbf{x}^{\mathbf{z}}$ corresponding to negligible factorial effects.

To determine the model matrix $\mathbf{X}_{\mathcal{T}}\mathbf{P}$, as in Section 6.3, we can first write down the columns $\mathbf{x}^{\mathbf{z}}$ that correspond to main effects. For example, suppose orthogonal polynomials are used to construct $\mathbf{P}$. If $\mathbf{z}$ has exactly one nonzero entry, say $z_i = j$ and all the other entries are equal to zero, then the entries of the $N \times 1$ vector $\mathbf{x}^{\mathbf{z}}$ consist of the values of a jth degree orthogonal polynomial at the levels of factor i in the N experimental runs. (For two-level factors, an entry of such an $\mathbf{x}^{\mathbf{z}}$ is 1 or -1 depending on whether the ith factor is at the high or low level in the corresponding experimental run.) If $\mathbf{z}$ has k nonzero entries, then $\mathbf{x}^{\mathbf{z}}$ is the Hadamard product of the k corresponding main-effect columns in the full model matrix.

If each treatment combination is observed exactly once, and none of the factorial effects is negligible, then by Theorem 5.2, each factorial effect $\beta^{\mathbf{z}} = \sum c_{x_1 \cdots x_n} \alpha_{x_1 \cdots x_n}$ is estimated by $\sum c_{x_1 \cdots x_n} y_{x_1 \cdots x_n}$, where $y_{x_1 \cdots x_n}$ is the observation on treatment combination $(x_1, \ldots, x_n)^T$, and the estimators of all the $\beta^{\mathbf{z}}$'s are uncorrelated.

Under a fractional factorial design, not all the factorial effects are estimable. Two issues arise here. Given a model in which some of the factorial effects are assumed negligible, under what designs are the nonnegligible effects estimable? Given a fractional factorial design, what models are estimable? In the former question, one may also ask, under what designs are the estimators of nonnegligible effects uncorrelated? In the next section, we introduce a combinatorial structure called orthogonal arrays and show that they provide some answers to these questions.

8.2   Orthogonal arrays

Definition 8.1. An orthogonal array $OA(N, s_1 \times \cdots \times s_n, t)$ is an $N \times n$ matrix with s_i distinct symbols in the ith column, $1 \le i \le n$, such that in each $N \times t$ submatrix, all combinations of the symbols appear equally often as row vectors. The positive integers N and t are called, respectively, the size and strength of the orthogonal array. If $s_1 = \cdots = s_n = s$, then it is called a symmetric orthogonal array and is denoted as $OA(N, s^n, t)$. Otherwise, it is called an asymmetrical (mixed-level) orthogonal array.

Symmetric orthogonal arrays were first introduced by Rao (1946, 1947) as designs for fractional factorial experiments. The definition was extended to cover asymmetrical orthogonal arrays in Rao (1973). There is an extensive literature on orthogonal arrays; we refer readers to Hedayat, Sloane, and Stufken (1999) for a comprehensive account.

An $OA(N, s_1 \times \cdots \times s_n, t)$ can be used to define a factorial design of size N for n factors with $s_1, \ldots, s_n$ levels: each column corresponds to one factor and each row represents a treatment combination to be observed.

Given a factorial design d with n factors, we call the design obtained by restricting d to a subset of m factors, where $m \leq n$, a *projection* of d onto m factors, or an m-dimensional projection of d. We allow the possibility $m = n$ so that a design is considered as its own projection onto all the factors. It follows from Definition 8.1 that, for a factorial design defined by an orthogonal array of strength t, all its projections onto t factors, say factors $i_1, \ldots, i_t$, are one or more replicates of a complete $s_{i_1} \times \cdots \times s_{i_t}$ factorial. Thus if an $OA(N, s_1 \times \cdots \times s_n, t)$ exists, then for any $1 \leq i_1 < \cdots < i_t \leq n$, N is a multiple of $s_{i_1} \cdots s_{i_t}$. In particular, if an $OA(N, s^n, t)$ exists, then N must be a multiple of s^t.

The following result explains the role of the strength and the utility of orthogonal arrays.

Theorem 8.1. *An orthogonal array of strength $t = 2k$ ($k \geq 1$) can be used to estimate all the main-effect contrasts and all the interaction contrasts involving up to k factors, assuming that all the interactions involving more than k factors are negligible. An orthogonal array of strength $t = 2k - 1$ ($k \geq 2$) can be used to estimate all the main-effect contrasts and all the interaction contrasts involving up to $k - 1$ factors, assuming that all the interactions involving more than k factors are negligible.*

Proof. Suppose all the interactions involving more than k factors are negligible. Let $\mathbf{X}_T \mathbf{Q}$ be the model matrix as in (8.4). Then each column of $\mathbf{X}_T \mathbf{Q}$ is an $\mathbf{x}^\mathbf{z}$, where $\mathbf{z}$ contains at most k nonzero entries.

Case 1 $t = 2k$:

We claim that the information matrix $(\mathbf{X}_T \mathbf{Q})^T (\mathbf{X}_T \mathbf{Q})$ is a diagonal matrix; then $(\mathbf{X}_T \mathbf{Q})^T (\mathbf{X}_T \mathbf{Q})$ is invertible and all the unknown parameters in the model are estimable with uncorrelated estimators.

If the nonzero entries of $\mathbf{z}$ are $z_{i_1}, \ldots, z_{i_r}$, then formally we can write

$$\mathbf{x}^\mathbf{z} = \mathbf{x}^{z_{i_1} \mathbf{e}_{i_1}} \odot \cdots \odot \mathbf{x}^{z_{i_r} \mathbf{e}_{i_r}}, \tag{8.5}$$

where the ith entry of $\mathbf{e}_i$ is 1, and all its other entries are zero. Notice that in this notation the i_lth entry of each $z_{i_l}\mathbf{e}_{i_l}$ is z_{i_l} and all its other entries are zero, so each $\mathbf{x}^{z_{i_l}\mathbf{e}_{i_l}}$ is one of the $s_{i_l}-1$ main-effect columns associated with factor i_l in the full model matrix.

Let $\mathbf{x}^{\mathbf{u}}$ and $\mathbf{x}^{\mathbf{v}}$ be two different columns of $\mathbf{X}_T\mathbf{Q}$. We want to show that $(\mathbf{x}^{\mathbf{u}})^T\mathbf{x}^{\mathbf{v}} = 0$. Let $u_{i_1},\ldots,u_{i_g}$ and $v_{j_1},\ldots,v_{j_h}$ be the nonzero entries of $\mathbf{u}$ and $\mathbf{v}$, respectively. As in (8.5), write $\mathbf{x}^{\mathbf{u}}$ and $\mathbf{x}^{\mathbf{v}}$ as

$$\mathbf{x}^{\mathbf{u}} = \mathbf{x}^{u_{i_1}\mathbf{e}_{i_1}} \odot \cdots \odot \mathbf{x}^{u_{i_g}\mathbf{e}_{i_g}},$$

$$\mathbf{x}^{\mathbf{v}} = \mathbf{x}^{v_{j_1}\mathbf{e}_{j_1}} \odot \cdots \odot \mathbf{x}^{v_{j_h}\mathbf{e}_{j_h}}.$$

Then $(\mathbf{x}^{\mathbf{u}})^T\mathbf{x}^{\mathbf{v}}$ is the sum of all the entries of $\mathbf{x}^{\mathbf{u}} \odot \mathbf{x}^{\mathbf{v}} = (\mathbf{x}^{u_{i_1}\mathbf{e}_{i_1}} \odot \cdots \odot \mathbf{x}^{u_{i_g}\mathbf{e}_{i_g}})\odot$ $(\mathbf{x}^{v_{j_1}\mathbf{e}_{j_1}} \odot \cdots \odot \mathbf{x}^{v_{j_h}\mathbf{e}_{j_h}})$. The total number of columns involved in this Hadamard product is $g + h$, with possibly some repeated columns. Since g, $h \le k$, $g + h$ is no more than the strength $(2k)$ of the orthogonal array. Therefore each row of the $N \times (g+h)$ matrix $[\mathbf{x}^{u_{i_1}\mathbf{e}_{i_1}} \cdots \mathbf{x}^{u_{i_g}\mathbf{e}_{i_g}} \mathbf{x}^{v_{j_1}\mathbf{e}_{j_1}} \cdots \mathbf{x}^{v_{j_h}\mathbf{e}_{j_h}}]$ is replicated the same number of times ($s_1 s_2 \cdots s_n/N$ times) as row vectors of the $(s_1 s_2 \cdots s_n) \times (g+h)$ matrix $[\mathbf{p}^{u_{i_1}\mathbf{e}_{i_1}} \cdots \mathbf{p}^{u_{i_g}\mathbf{e}_{i_g}} \mathbf{p}^{v_{j_1}\mathbf{e}_{j_1}} \cdots \mathbf{p}^{v_{j_h}\mathbf{e}_{j_h}}]$. It follows that $(\mathbf{x}^{\mathbf{u}})^T\mathbf{x}^{\mathbf{v}} = \frac{N}{s_1 s_2 \cdots s_n}(\mathbf{p}^{\mathbf{u}})^T\mathbf{p}^{\mathbf{v}} = 0$.

Case 2 $t = 2k - 1$:

Partition $\mathbf{X}_T\mathbf{Q}$ as $[\mathbf{U}_1 \ \mathbf{U}_2]$ where $\mathbf{U}_1$ consists of all the $\mathbf{x}^{\mathbf{u}}$'s such that $\mathbf{u}$ contains at most $k-1$ nonzero entries, and $\mathbf{U}_2$ consists of all the $\mathbf{x}^{\mathbf{v}}$'s such that $\mathbf{v}$ contains k nonzero entries. Then, by Theorem 2.3, the information matrix for the effects involving up to $k-1$ factors is equal to $\mathbf{U}_1^T\mathbf{U}_1 - \mathbf{U}_1^T\mathbf{U}_2(\mathbf{U}_2^T\mathbf{U}_2)^-\mathbf{U}_2^T\mathbf{U}_1$. Similar to the proof of case 1, $\mathbf{U}_1^T\mathbf{U}_1$ is a diagonal matrix and $\mathbf{U}_1^T\mathbf{U}_2 = \mathbf{0}$, since each off-diagonal entry of $\mathbf{U}_1^T\mathbf{U}_1$ and each entry of $\mathbf{U}_1^T\mathbf{U}_2$ is the sum of all the entries of the Hadamard product of no more than $2k - 1 = t$ main-effect columns in the model matrix $\mathbf{X}_T\mathbf{Q}$. Thus $\mathbf{U}_1^T\mathbf{U}_1 - \mathbf{U}_1^T\mathbf{U}_2(\mathbf{U}_2^T\mathbf{U}_2)^-\mathbf{U}_2^T\mathbf{U}_1 = \mathbf{U}_1^T\mathbf{U}_1$ is a diagonal matrix. $\qquad\square$

In either case, the least squares estimators of the factorial effects of interest in Theorem 8.1 can be calculated easily: $\widehat{\beta^{\mathbf{z}}} = \frac{1}{\|\mathbf{x}^{\mathbf{z}}\|^2}(\mathbf{x}^{\mathbf{z}})^T\mathbf{y}$.

For orthogonal arrays of odd strengths, one can prove a result somewhat stronger than that in Theorem 8.1. Under an orthogonal array of strength $2k - 1$, we may not be able to estimate all the k-factor interaction contrasts. However, consider the model that contains all the main-effect contrasts, all the interaction contrasts involving up to $k-1$ factors, and all the k-factor interaction contrasts involving a given factor. In this case, the information matrix $(\mathbf{X}_T\mathbf{Q})^T(\mathbf{X}_T\mathbf{Q})$ for all these nonnegligible effects is also diagonal. This can be proved as follows. For any two columns $\mathbf{x}^{\mathbf{u}}$ and $\mathbf{x}^{\mathbf{v}}$ of $\mathbf{X}_T\mathbf{Q}$, unless both $\mathbf{x}^{\mathbf{u}}$ and $\mathbf{x}^{\mathbf{v}}$ correspond to k-factor interactions, $(\mathbf{x}^{\mathbf{u}})^T\mathbf{x}^{\mathbf{v}}$ is the sum of all the entries of the Hadamard product of no more than $2k - 1$ main-effect columns of $\mathbf{X}_T\mathbf{Q}$. Since the strength of the array is $2k - 1$, as in the proof of Theorem 8.1, we have $(\mathbf{x}^{\mathbf{u}})^T\mathbf{x}^{\mathbf{v}} = 0$. On the other hand, if both $\mathbf{x}^{\mathbf{u}}$ and $\mathbf{x}^{\mathbf{v}}$ correspond to k-factor interactions, then by assumption, one common factor is involved in both of these k-factor interactions. Then the actual number of factors involved is no more than $2k - 1$. Again, $(\mathbf{x}^{\mathbf{u}})^T\mathbf{x}^{\mathbf{v}} = 0$. We state this result in the following theorem.

Theorem 8.2. *An orthogonal array of strength $t = 2k - 1$ $(k > 1)$ can be used to estimate all the main-effect contrasts, all the interaction contrasts involving up to $k - 1$ factors, and all the k-factor interaction contrasts involving a given factor, assuming that all the other interaction contrasts are negligible.*

For the treatment models described in Theorems 8.1 and 8.2, orthogonal arrays of suitable strengths not only provide uncorrelated estimators of the factorial effects as specified, but also have strong optimality properties. See Exercise 8.9 for a simple result that under two-level orthogonal arrays of suitable strengths, the variances of the estimators of all the factorial effects specified in Theorems 8.1 and 8.2 are minimized. We refer readers to Sections 2.5–2.6 and Chapter 6 of Dey and Mukerjee (1999) for more results on the optimality of orthogonal arrays and other fractional factorial designs for models containing specific factorial effects.

Some lower bounds on the sizes of orthogonal arrays due to Rao (1947) can be obtained by counting the degrees of freedom for the estimable orthogonal treatment contrasts in Theorems 8.1 and 8.2.

Theorem 8.3. *If there exists an $OA(N, s^n, t)$, then*

(i) for $t = 2k$, $N \geq \sum_{i=0}^{k} \binom{n}{i}(s-1)^i$;

(ii) for $t = 2k - 1$, $N \geq \sum_{i=0}^{k-1} \binom{n}{i}(s-1)^i + \binom{n-1}{k-1}(s-1)^k$.

Two results follow easily from Theorem 8.3.

Corollary 8.4. *If there exists an $OA(N, s^n, 2)$, then $n \leq (N-1)/(s-1)$. In particular, if there exists an $OA(N, 2^n, 2)$, then $n \leq N - 1$.*

Corollary 8.5. *If there exists an $OA(N, 2^n, 3)$, then $n \leq N/2$.*

We refer readers to Hedayat, Sloane, and Stufken (1999) for an extension of Theorem 8.3 to mixed level orthogonal arrays and other more refined bounds.

Orthogonal arrays achieving the bounds in Corollaries 8.4 and 8.5 will be presented in Sections 8.5, 8.7, 9.5, and 9.6. The counterpart of the inequality in Corollary 8.4 for mixed-level orthogonal arrays is that if there exists an $OA(N, s_1^{n_1} \times \cdots \times s_h^{n_h}, 2)$, then $N - 1 \geq \sum_{j=1}^{h}(s_j - 1)n_j$. An orthogonal array achieving this bound is called *saturated*. In particular, an N-run design for n two-level factors is saturated if $N = n + 1$. Such a design has the minimum number of runs for estimating all the main effects of given factors (or can accommodate the maximum number of factors for a given run size) when the interactions are negligible.

Mukerjee and Wu (1995) studied the number of coincidences between different rows of a saturated orthogonal array of strength two and used the result to derive necessary conditions for the existence of such designs. This result is also useful in Sections 10.6, 15.8, and 15.9.

Theorem 8.6. *For any two rows of a saturated $OA(N, s_1^{n_1} \times \cdots \times s_h^{n_h}, 2)$ and for each $m = 1, \ldots, h$, let Δ_m be number of s_m-level factors for which the two rows have the same level. Then $\sum_{m=1}^{h} s_m \Delta_m = \sum_{m=1}^{h} n_m - 1$. In particular, for a saturated $OA(N, s^n, 2)$, the number of coincidences between any two rows is a constant $\Delta = (n-1)/s$.*

Proof. We only prove the theorem for symmetric arrays. The asymmetrical case can be proved similarly. Consider an $OA(N, s^n, 2)$. Let $\mathbf{U}$ be the model matrix under the assumption that all the interactions are negligible. Then $\mathbf{U} = [\mathbf{x}_0, \mathbf{A}_1, \ldots, \mathbf{A}_n]$, where $\mathbf{x}_0$ is an $N \times 1$ vector with constant entries and, for $i = 1, \ldots, n$, $\mathbf{A}_i$ is an $N \times (s-1)$ matrix corresponding to the main effect of the ith factor. Let $\mathbf{P}$ be the matrix $\mathbf{P}_{s_i}$ in (6.20) with $s_i = s$. Suppose the columns of $\mathbf{P}$ are normalized to have unit lengths. Then

$$\mathbf{P}^T \mathbf{P} = \mathbf{P}\mathbf{P}^T = \mathbf{I}_s. \tag{8.6}$$

Denote the (v, j)th entry of $\mathbf{A}_i$ by a_{vj}^i. Without loss of generality, we may assume that $\mathbf{x}_0$ has all the entries equal to $(1/N)^{1/2}$ and, for $j = 1, \ldots, s-1$, $a_{vj}^i = (s/N)^{1/2} p_{lj}$ if factor i appears at level l on the vth experimental run, where p_{lj} is the (l, j)th entry of $\mathbf{P}$. Then the columns of $\mathbf{U}$ also have unit lengths. By the proof of Theorem 8.1, $\mathbf{U}^T \mathbf{U} = \mathbf{I}$. Since $\mathbf{U}$ is a square matrix, we also have $\mathbf{U}\mathbf{U}^T = \mathbf{I}$. Then $\frac{1}{N}\mathbf{J} + \sum_{i=1}^{n} \mathbf{A}_i \mathbf{A}_i^T = \mathbf{I}$. Thus, for any $u \neq v$,

$$\sum_{i=1}^{n} \sum_{j=1}^{s-1} a_{uj}^i a_{vj}^i = -\frac{1}{N}. \tag{8.7}$$

If factor i has the same level, say level l, on the uth and vth experimental runs, then by (8.6), $\sum_{j=1}^{s-1} a_{uj}^i a_{vj}^i = (s/N)\sum_{j=1}^{s-1} p_{lj} p_{lj} = (s/N)(1 - 1/s)$; otherwise it is equal to $(s/N)(-1/s)$. Let Δ be the number of coincidences between these two experimental runs. Then

$$\sum_{i=1}^{n} \sum_{j=1}^{s-1} a_{uj}^i a_{vj}^i = (s/N)(\Delta - n/s). \tag{8.8}$$

By (8.7) and (8.8), $s\Delta = n - 1$. It follows that $\Delta = (n-1)/s$ is a constant. $\qquad\square$

Chakravarti (1956) extended symmetric orthogonal arrays to *partially balanced arrays*. Let $\mathbf{X}$ be an $N \times n$ matrix with entries from $S = \{0, 1, \ldots, s-1\}$. Then $\mathbf{X}$ is called a partially balanced array of strength t if, for any $\mathbf{x} = (x_1, \ldots, x_t)$, where $x_i \in S$, there is an integer $n(\mathbf{x})$ such that $\mathbf{x}$ appears $n(\mathbf{x})$ times as row vectors of any $N \times t$ submatrix of $\mathbf{X}$, and $n(\mathbf{x})$ does not depend on the permutations of the entries of $\mathbf{x}$. Srivastava and Chopra (1971) renamed such arrays *balanced arrays*. In a series of papers, they studied the properties and existence of balanced arrays as well as optimal designs among balanced arrays.

8.3   Examples of orthogonal arrays

We list here a few examples of orthogonal arrays. The first three are symmetric two-level arrays in which the two levels are denoted by 1 and -1. The fourth is asymmetrical with seven three-level factors and one two-level factor. The first two designs are examples of the so-called *regular* fractional factorial designs, while the last two designs are nonregular. The construction of the designs in Examples 8.1, 8.2, 8.3, and 8.4 will be discussed in Sections 8.4, 8.7, 8.5, and 8.8, respectively.

Example 8.1. An OA$(8, 2^6, 2)$:

$$
\begin{array}{rrrrrr}
-1 & -1 & -1 & -1 & -1 & -1 \\
1 & -1 & -1 & 1 & -1 & 1 \\
-1 & 1 & -1 & 1 & 1 & 1 \\
1 & 1 & -1 & -1 & 1 & -1 \\
-1 & -1 & 1 & -1 & 1 & 1 \\
1 & -1 & 1 & 1 & 1 & -1 \\
-1 & 1 & 1 & 1 & -1 & -1 \\
1 & 1 & 1 & -1 & -1 & 1
\end{array}
\tag{8.9}
$$

Example 8.2. An OA$(16, 2^8, 3)$:

$$
\begin{array}{rrrrrrrr}
-1 & -1 & -1 & -1 & -1 & -1 & -1 & -1 \\
-1 & 1 & -1 & 1 & -1 & 1 & -1 & 1 \\
-1 & -1 & 1 & 1 & -1 & -1 & 1 & 1 \\
-1 & 1 & 1 & -1 & -1 & 1 & 1 & -1 \\
-1 & -1 & -1 & -1 & 1 & 1 & 1 & 1 \\
-1 & 1 & -1 & 1 & 1 & -1 & 1 & -1 \\
-1 & -1 & 1 & 1 & 1 & 1 & -1 & -1 \\
-1 & 1 & 1 & -1 & 1 & -1 & -1 & 1 \\
1 & 1 & 1 & 1 & 1 & 1 & 1 & 1 \\
1 & -1 & 1 & -1 & 1 & -1 & 1 & -1 \\
1 & 1 & -1 & -1 & 1 & 1 & -1 & -1 \\
1 & -1 & -1 & 1 & 1 & -1 & -1 & 1 \\
1 & 1 & 1 & 1 & -1 & -1 & -1 & -1 \\
1 & -1 & 1 & -1 & -1 & 1 & -1 & 1 \\
1 & 1 & -1 & -1 & -1 & -1 & 1 & 1 \\
1 & -1 & -1 & 1 & -1 & 1 & 1 & -1
\end{array}
\tag{8.10}
$$

Example 8.3. An OA$(12, 2^{11}, 2)$:

$$
\begin{array}{rrrrrrrrrrr}
1 & -1 & -1 & 1 & -1 & -1 & -1 & 1 & 1 & 1 & -1 \\
-1 & 1 & -1 & -1 & 1 & -1 & -1 & -1 & 1 & 1 & 1 \\
1 & -1 & 1 & -1 & -1 & 1 & -1 & -1 & -1 & 1 & 1 \\
1 & 1 & -1 & 1 & -1 & -1 & 1 & -1 & -1 & -1 & 1 \\
1 & 1 & 1 & -1 & 1 & -1 & -1 & 1 & -1 & -1 & -1 \\
-1 & 1 & 1 & 1 & -1 & 1 & -1 & -1 & 1 & -1 & -1 \\
-1 & -1 & 1 & 1 & 1 & -1 & 1 & -1 & -1 & 1 & -1 \\
-1 & -1 & -1 & 1 & 1 & 1 & -1 & 1 & -1 & -1 & 1 \\
1 & -1 & -1 & -1 & 1 & 1 & 1 & -1 & 1 & -1 & -1 \\
-1 & 1 & -1 & -1 & -1 & 1 & 1 & 1 & -1 & 1 & -1 \\
-1 & -1 & 1 & -1 & -1 & -1 & 1 & 1 & 1 & -1 & 1 \\
1 & 1 & 1 & 1 & 1 & 1 & 1 & 1 & 1 & 1 & 1
\end{array}
\tag{8.11}
$$

Example 8.4. An OA($18, 2 \times 3^7, 2$):

$$
\begin{matrix}
0 & 0 & 0 & 1 & 1 & 0 & 2 & 2 \\
0 & 1 & 2 & 2 & 1 & 0 & 0 & 1 \\
0 & 2 & 2 & 1 & 2 & 0 & 1 & 0 \\
1 & 0 & 0 & 0 & 0 & 0 & 0 & 0 \\
1 & 1 & 1 & 2 & 0 & 0 & 1 & 2 \\
1 & 2 & 1 & 0 & 2 & 0 & 2 & 1 \\
0 & 0 & 1 & 2 & 2 & 1 & 0 & 0 \\
0 & 1 & 0 & 0 & 2 & 1 & 1 & 2 \\
0 & 2 & 0 & 2 & 0 & 1 & 2 & 1 \\
1 & 0 & 1 & 1 & 1 & 1 & 1 & 1 \\
1 & 1 & 2 & 0 & 1 & 1 & 2 & 0 \\
1 & 2 & 2 & 1 & 0 & 1 & 0 & 2 \\
0 & 0 & 2 & 0 & 0 & 2 & 1 & 1 \\
0 & 1 & 1 & 1 & 0 & 2 & 2 & 0 \\
0 & 2 & 1 & 0 & 1 & 2 & 0 & 2 \\
1 & 0 & 2 & 2 & 2 & 2 & 2 & 2 \\
1 & 1 & 0 & 1 & 2 & 2 & 0 & 1 \\
1 & 2 & 0 & 2 & 1 & 2 & 1 & 0
\end{matrix}
\tag{8.12}
$$

8.4 Regular fractional factorial designs

One major difference between regular and nonregular designs is their construction. The design in Example 8.1 is constructed in the following simple manner. The first three columns of this design consist of all the eight treatment combinations of the first three factors. Column 6 is the Hadamard product of the first three columns, and except for sign changes, column 4 is the Hadamard product of the first two columns, and column 5 is the Hadamard product of the second and third columns. If each row (treatment combination) is denoted by $(x_1, \ldots, x_6)$, where $x_i = 1$ or -1, then the eight treatment combinations in the design satisfy

$$
x_4 = -x_1 x_2, \quad x_5 = -x_2 x_3, \quad x_6 = x_1 x_2 x_3. \tag{8.13}
$$

In other words, since the design has $8 = 2^3$ runs, we first write down all the combinations of three factors (in this case, factors 1, 2, and 3); these factors are called *basic factors*. Then we use (8.13) to define three additional factors, called *added factors*.

Note that (8.13) is equivalent to

$$
-x_1 x_2 x_4 = 1, \quad -x_2 x_3 x_5 = 1, \quad x_1 x_2 x_3 x_6 = 1. \tag{8.14}
$$

Then all the products $(-x_1 x_2 x_4)(-x_2 x_3 x_5) = x_1 x_3 x_4 x_5$, $(-x_1 x_2 x_4)(x_1 x_2 x_3 x_6) = -x_3 x_4 x_6$, $(-x_2 x_3 x_5)(x_1 x_2 x_3 x_6) = -x_1 x_5 x_6$, and $(-x_1 x_2 x_4)(-x_2 x_3 x_5)(x_1 x_2 x_3 x_6) = x_2 x_4 x_5 x_6$ are also equal to 1. Thus the eight treatment combinations in the fraction are solutions to the equations

$$
1 = -x_1 x_2 x_4 = -x_2 x_3 x_5 = x_1 x_2 x_3 x_6 = x_1 x_3 x_4 x_5 = -x_3 x_4 x_6 = -x_1 x_5 x_6 = x_2 x_4 x_5 x_6. \tag{8.15}
$$

Suppose the two levels are represented by the two elements 0 and 1 of $\mathbb{Z}_2$, with -1 replaced by 0. Then (8.14) is equivalent to

$$x_1 + x_2 + x_4 = 0, \ x_2 + x_3 + x_5 = 0, \ x_1 + x_2 + x_3 + x_6 = 0. \tag{8.16}$$

Thus the eight treatment combinations in the fraction are those in the principal block when the treatment combinations in a complete 2^6 factorial are divided into eight blocks of size eight by confounding the interaction of factors 1, 2, 4, that of factors 2, 3, 5, and that of factors 1, 2, 3, 6 with blocks; see Section 7.2. These eight treatment combinations constitute a three-dimensional subspace of $EG(6,2)$ when each treatment combination is considered as a point in $EG(6,2)$. In general, when s is a prime or prime power, subspaces (or flats) of $EG(n,s)$ are called *regular* fractional factorial designs. This implies that the run size of a regular design must be a power of s. The construction and properties of regular designs will be discussed in more detail in Chapter 9. In particular, we show in Theorem 9.5 that regular fractional factorial designs are orthogonal arrays.

Remark 8.1. When converting $(-1, 1)$ to $(0, 1)$ coding, algebraically it is simpler to use the correspondence $1 \leftrightarrow 0$ and $-1 \leftrightarrow 1$. This is because 1 is the multiplicative identity and 0 is the additive identity; furthermore $(-1)(-1) = 1$, which corresponds to $1 + 1 = 0$ in $\mathbb{Z}_2$. Then the equation $x_1 + x_2 + x_4 = 0$ with $x_i = 0$, 1 would be converted to $x_1 x_2 x_4 = 1$ without the negative sign. However, in practice, often the two levels are called high and low levels, respectively, especially when they are quantitative. It is common to denote the high level by 1 and the low level by 0 (or -1). This results in the algebraically less convenient conversion $1 \leftrightarrow 1$ and $-1 \leftrightarrow 0$.

8.5 Designs derived from Hadamard matrices

The design in Example 8.3 can be constructed from a Hadamard matrix of order 12. If we multiply all the entries in the same row or the same column of a Hadamard matrix by -1, then the resulting matrix is still a Hadamard matrix. Therefore, without loss of generality, we may assume that all the entries in the first row and/or the first column of a Hadamard matrix are equal to 1.

We show the equivalence of a Hadamard matrix of order N to an $OA(N, 2^{N-1}, 2)$.

Theorem 8.7. *Suppose $\mathbf{H}$ is a Hadamard matrix of order $N > 2$ such that all the entries in the first column are equal to 1. Then the matrix obtained by deleting the first column of $\mathbf{H}$ is an $OA(N, 2^{N-1}, 2)$. Conversely, adding a column of 1's to an $OA(N, 2^{N-1}, 2)$ results in a Hadamard matrix of order N.*

Proof. It is trivial that an $OA(N, 2^{N-1}, 2)$ supplemented by a column of 1's is a Hadamard matrix. Thus it is enough to prove the converse. Suppose $\mathbf{H}$ is a Hadamard matrix such that all the entries in the first column are equal to 1. Let $\mathbf{h}_i = [h_{1i}, \dots, h_{Ni}]^T$ and $\mathbf{h}_j = [h_{1j}, \dots, h_{Nj}]^T$ be the ith and jth columns of $\mathbf{H}$, where $i \neq 1$, $j \neq 1$, and $i \neq j$. We need to show that each of the four pairs $(1,1)$, $(1,-1)$,

$(-1, 1)$, and $(-1, -1)$ appears $N/4$ times among the N pairs $(h_{1i}, h_{1j}), \ldots, (h_{Ni}, h_{Nj})$. Suppose $(1, 1)$ appears a times, $(1, -1)$ appears b times, $(-1, 1)$ appears c times, and $(-1, -1)$ appears d times. Then, since $\mathbf{h}_i$ is orthogonal to $\mathbf{h}_j$, and they are both orthogonal to the column of 1's, we have

$$a + b - c - d = a - b + c - d = a - b - c + d = 0.$$

Solving these equations, we conclude that $a = b = c = d$. $\square$

We have actually proved the following result.

Corollary 8.8. *Let* $\mathbf{X}$ *be an* $N \times n$ *matrix of 1's and* -1*'s, with all the entries of its first column equal to 1 and* $n \geq 3$*. Then the columns of* $\mathbf{X}$ *are pairwise orthogonal if and only if the array obtained by deleting the first column of* $\mathbf{X}$ *is an orthogonal array of strength two.*

The design in Example 8.3 can be obtained by deleting a column of 1's from a Hadamard matrix of order 12. We call an $OA(N, 2^{N-1}, 2)$ constructed from a Hadamard matrix of order N as described in Theorem 8.7 a Hadamard design. Hadamard designs are saturated orthogonal arrays of strength two.

An immediate consequence of Theorem 8.7 is that if there exists a Hadamard matrix of order $N > 2$, then N must be a multiple of 4. It has been conjectured that a Hadamard matrix of order N exists for every N that is a multiple of 4. This conjecture has not been proved, and no counterexample has been found. Hadamard matrices have been constructed for many orders that are multiples of four, and the smallest order for which no Hadamard matrix has been found is 668. Hadamard designs provide more flexibility than regular designs in terms of run sizes. Plackett and Burman (1946) proposed the use of Hadamard designs in factorial experiments. The Hadamard designs presented in their paper are referred to as *Plackett–Burman designs*.

The matrix below is a Hadamard matrix of order 2:

$$\mathbf{P}_2 = \begin{bmatrix} 1 & -1 \\ 1 & 1 \end{bmatrix}.$$

It is easy to see that if $\mathbf{H}$ and $\mathbf{K}$ are Hadamard matrices of orders m and n, respectively, then the Kronecker product $\mathbf{H} \otimes \mathbf{K}$ is a Hadamard matrix of order mn. Applying this result to $\mathbf{P}_2$ repeatedly, we conclude that there exists a Hadamard matrix of order 2^n for every positive integer n.

Corollary 8.9. *For* $N = 2^k$*, the design obtained by deleting the first column of 1's from the k-fold Kronecker product*

$$\begin{bmatrix} 1 & -1 \\ 1 & 1 \end{bmatrix} \otimes \cdots \otimes \begin{bmatrix} 1 & -1 \\ 1 & 1 \end{bmatrix}$$

is a saturated $OA(N, 2^{N-1}, 2)$*.*

Except for possible sign changes and row permutations, the regular fractional factorial designs in Examples 8.1 and 8.2. can be obtained by selecting columns from,
respectively, the 8-run and 16-run saturated orthogonal arrays constructed in Corollary 8.9. The saturated orthogonal arrays constructed in Corollary 8.9 are regular
designs; see Remark 9.5 in Section 9.5. Hadamard designs are not regular when N is
not a power of 2; when N is a power of 2, they may or may not be regular.

An important method for constructing Hadamard matrices used by Plackett and
Burman (1946) was due to Paley (1933). Suppose N is a multiple of 4 such that
$N-1$ is an odd prime power. Let $q = N-1$; then a finite field GF(q) exists. Denote
its elements by $\alpha_1 = 0, \alpha_2, \ldots, \alpha_q$, and define a function χ: GF(q) $\to \{0, 1, -1\}$ by

$$\chi(\beta) = \begin{cases} 1, & \text{if } \beta = y^2 \text{ for some } y \in \text{GF}(q); \\ 0, & \text{if } \beta = 0; \\ -1, & \text{otherwise.} \end{cases}$$

Let $\mathbf{A}$ be the $q \times q$ matrix $[a_{ij}]$ with $a_{ij} = \chi(\alpha_i - \alpha_j)$ for $i, j = 1, 2, \ldots, q$, and

$$\mathbf{Q}_N = \begin{bmatrix} 1 & -\mathbf{1}_q^T \\ \mathbf{1}_q & \mathbf{A} + \mathbf{I}_q \end{bmatrix}. \tag{8.17}$$

Then it can be shown that $\mathbf{Q}_N$ is a Hadamard matrix; see Hedayat, Sloane, and
Stufken (1999). We call the OA($N, 2^{N-1}, 2$) obtained by deleting the first column
of (8.17) the *Paley design* of order N.

There is also a connection between Hadamard matrices and balanced incomplete
block designs (BIBD). Suppose $\mathbf{H}$ is a Hadamard matrix of order N. Without loss of
generality, assume that all the entries in the first row and first column of $\mathbf{H}$ are equal
to 1. Deleting the first row and first column from $\mathbf{H}$, we obtain an $(N-1) \times (N-1)$
matrix $\mathbf{H}^*$ of 1's and -1's. Define a block design d with $N-1$ treatments and $N-1$
blocks such that the ith treatment appears in the jth block once if the (i, j)th entry
of $\mathbf{H}^*$ is equal to 1; otherwise, it does not appear. Then d is a balanced incomplete
block design. The block size is $N/2-1$ since there are $N/2-1$ ones in each column
of $\mathbf{H}^*$. Conversely, given a balanced incomplete block design d with $N-1$ treatments
and $N-1$ blocks of size $N/2-1$, write down an $(N-1) \times (N-1)$ incidence matrix
$\mathbf{H}^*$ of 1's and -1's such that the (i, j)th entry of $\mathbf{H}^*$ is equal to 1 if and only if the
ith treatment appears in the jth block of d. Supplement $\mathbf{H}^*$ by a row and column of
1's; then we obtain an $N \times N$ Hadamard matrix. For a proof of this result, readers are
referred to Hall (1986).

The OA($12, 2^{11}, 2$) in (8.11) can be constructed by applying the method described
in the previous paragraph to a balanced incomplete block design with 11 treatments
and 11 blocks of size 5. The first 11 rows of the array come from the incidence matrix
of the BIBD which can be developed from the initial block $\{1, 3, 4, 5, 9\}$ in a cyclic
manner (see Section A.8). Thus the associated incidence matrix is a circulant matrix.
One row of 1's is added at the bottom to produce an orthogonal array. If we also add
a column of 1's, then a Hadamard matrix of order 12 is obtained.

Hadamard matrices can also be used to construct mixed-level orthogonal arrays;
see Section 4.3 of Dey and Mukerjee (1999).

8.6   Mutually orthogonal Latin squares and orthogonal arrays

Latin squares were defined in Section 5.5. We say that two $s \times s$ Latin squares with entries $1, \ldots, s$ are orthogonal if, when one is superimposed on the other, each of the s^2 pairs $(i, j), 1 \leq i, j \leq s$, appears in exactly one cell.

Theorem 8.10. *A set of k mutually orthogonal $s \times s$ Latin squares is equivalent to an $OA(s^2, s^{k+2}, 2)$.*

Proof. Given a set of k mutually orthogonal $s \times s$ Latin squares, an $s^2 \times (k+2)$ orthogonal array can be constructed. Start with an $s^2 \times 2$ matrix whose rows are all the 1×2 vectors (i, j), where i and j are the positive integers $1, \ldots, s$; the pairs (i, j) are used to index the entries of an $s^2 \times 1$ column. For each of the k Latin squares, we write down an additional $s^2 \times 1$ column by placing the entry that appears at the ith row and jth column of the Latin square at the entry of the $s^2 \times 1$ column indexed by (i, j). Then it can easily be verified that the resulting $s^2 \times (k+2)$ array is an orthogonal array of strength two. Conversely, given an $OA(s^2, s^{k+2}, 2)$, we can write down an $s \times s$ Latin square from each of the last k columns. Specifically, if the $n^2 \times 3$ subarray consisting of the first two and the $(l+2)$th columns has a row (i, j, t), then we put t at the ith row and jth column of the lth square array. It can easily be verified that each of the k square arrays so constructed is a Latin square, and the k Latin squares constructed are mutually orthogonal. □

Example 8.5. The two orthogonal Latin squares in Example 1.4 are equivalent to the following OA(16, 4^4, 2):

$$
\begin{bmatrix}
1 & 1 & 1 & 1 & 2 & 2 & 2 & 2 & 3 & 3 & 3 & 3 & 4 & 4 & 4 & 4 \\
1 & 2 & 3 & 4 & 1 & 2 & 3 & 4 & 1 & 2 & 3 & 4 & 1 & 2 & 3 & 4 \\
1 & 2 & 3 & 4 & 2 & 1 & 4 & 3 & 3 & 4 & 1 & 2 & 4 & 3 & 2 & 1 \\
1 & 2 & 3 & 4 & 4 & 3 & 2 & 1 & 2 & 1 & 4 & 3 & 3 & 4 & 1 & 2
\end{bmatrix}^T
$$

By Corollary 8.4, $(k+2)(s-1) \leq s^2 - 1$. Therefore $k \leq s - 1$. This provides an upper bound on the number of mutually orthogonal $s \times s$ Latin squares. We will see in Section 8.8 that this upper bound can be achieved when s is a prime or a prime power. A set of $s - 1$ mutually orthogonal $s \times s$ Latin squares is called complete. A complete set of mutually orthogonal $s \times s$ Latin squares is equivalent to a saturated s-level orthogonal array of size s^2 and strength two.

8.7   Foldover designs

The orthogonal array in Example 8.2 has strength three. One can see that the first eight rows of this design constitute a Hadamard matrix, which is an OA(8, 2^7, 2) supplemented by a column of -1's. The last eight rows of array (8.10) are obtained from the first eight rows by interchanging the two levels. For any two-level design **X**, where the two levels are represented by 1 and -1, we call $-$**X** the *foldover* design of **X**. Then the array in (8.10) is obtained by combining a Hadamard matrix with its foldover. In general, a two-level design $\mathbf{X}$ of size N can be combined with $-\mathbf{X}$ to form a design of size $2N$. Often a column of -1's is added before the design is folded over to accommodate one extra factor. Then we obtain the following design of size $2N$ and $n+1$ two-level factors:

$$\tilde{\mathbf{X}} = \begin{bmatrix} -\mathbf{1} & \mathbf{X} \\ \mathbf{1} & -\mathbf{X} \end{bmatrix}, \tag{8.18}$$

where $\mathbf{1}$ is the $N \times 1$ vector of 1's.

The method of foldover was proposed by Box and Wilson (1951) for regular fractional factorial designs. The following result for general two-level orthogonal arrays is due to Seiden and Zemach (1966) (the case $t = 2$ was treated in Seiden (1954)).

Theorem 8.11. *If* $\mathbf{X}$ *is a two-level orthogonal array of even strength* t, *then the design* $\tilde{\mathbf{X}}$ *constructed in (8.18) is an orthogonal array of strength* $t + 1$.

Theorem 8.11 follows from a lemma that is useful again in Section 15.3.

Lemma 8.12. *Suppose* $\mathbf{X}$ *is an* $OA(N, 2^n, t)$ *with* $n \geq t + 1$ *in which the two levels are denoted by* 1 *and* -1. *Then for any* $N \times (t + 1)$ *submatrix* $\mathbf{Y}$ *of* $\mathbf{X}$, *there are nonnegative integers* α *and* β *with* $\alpha + \beta = N/2^t$ *such that each of the vectors* $\mathbf{x} = (x_1, x_2, \ldots, x_{t+1})$ *with* $x_1 x_2 \cdots x_{t+1} = 1$, *where* $x_i = 1$ *or* -1, *appears* α *times as row vectors of* $\mathbf{Y}$, *and each of those with* $x_1 x_2 \cdots x_{t+1} = -1$ *appears* β *times.*

Proof. For each $\mathbf{x} = (x_1, \ldots, x_{t+1})$ with $x_i = 1$ or -1, let $f(\mathbf{x})$ be the number of times $\mathbf{x}$ appears as row vectors of $\mathbf{Y}$. Then since $\mathbf{X}$ has strength t, we have

$$f(x_1, \ldots, x_t, x_{t+1}) + f(x_1, \ldots, x_t, -x_{t+1}) = N/2^t. \tag{8.19}$$

Also,

$$f(x_1, \ldots, -x_t, -x_{t+1}) + f(x_1, \ldots, x_t, -x_{t+1}) = N/2^t. \tag{8.20}$$

It follows from (8.19) and (8.20) that $f(x_1, \ldots, x_t, x_{t+1}) = f(x_1, \ldots, -x_t, -x_{t+1})$. Repeating the same argument, we see that any two rows $\mathbf{x}$ and $\mathbf{y}$ differing in an even number of components appear the same number of times as row vectors of $\mathbf{Y}$. Thus all the vectors $\mathbf{x} = (x_1, x_2, \ldots, x_{t+1})$ with $x_1 x_2 \cdots x_{t+1} = 1$ appear the same number of times, say α times, and those with $x_1 x_2 \cdots x_{t+1} = -1$ also appear the same number of times, say β times. By (8.19), we have $\alpha + \beta = N/2^t$. $\square$

Now we prove Theorem 8.11. Let $\mathbf{X}$ be an $OA(N, 2^n, t)$, where t is even, and let $\tilde{\mathbf{Y}}$ be a $2N \times (t + 1)$ submatrix of $\tilde{\mathbf{X}}$. If $\tilde{\mathbf{Y}}$ contains the first column of $\tilde{\mathbf{X}}$ as displayed in (8.18), then, since $\mathbf{X}$ has strength t, it is clear that all the $(t + 1)$-tuples of 1's and -1's appear the same number of times as row vectors of $\tilde{\mathbf{Y}}$. On the other hand, suppose

$$\tilde{\mathbf{Y}} = \begin{bmatrix} \mathbf{Y} \\ -\mathbf{Y} \end{bmatrix},$$

where $\mathbf{Y}$ is an $N \times (t+1)$ submatrix of $\mathbf{X}$. Then by Lemma 8.12, there exists a non-negative integer α such that each $\mathbf{x} = (x_1, x_2, \ldots, x_{t+1})$ appears either α or $N/2^t - \alpha$ times as row vectors of $\mathbf{Y}$. Since $t+1$ is odd, if $\mathbf{x}$ appears α times, then $-\mathbf{x}$ must appear $N/2^t - \alpha$ times. It follows that $\mathbf{x}$ appears $N/2^t - \alpha$ times in $-\mathbf{Y}$, and thus appears $\alpha + (N/2^t - \alpha) = N/2^t$ times in $\tilde{\mathbf{Y}}$. $\qquad \square$

It is also clear from the above proof that the foldover method does not increase the strength when it is applied to an orthogonal array of odd strength.

Remark 8.2. By Corollary 8.5, an $OA(N, 2^n, 3)$ must have $n \leq N/2$. It follows from Theorem 8.11 that this upper bound is achieved by the design obtained by applying the foldover construction (8.18) to a saturated two-level orthogonal array of strength two and size $N/2$. It can be shown that the converse is also true; that is, every $OA(N, 2^{N/2}, 3)$ can be constructed from an $OA(N/2, 2^{N/2-1}, 2)$ according to (8.18). We will revisit this in Sections 9.6 and 15.11, where two more general results are stated.

Foldover is a popular method for constructing two-level orthogonal arrays of strength three. However, there are two-level orthogonal arrays of strength three that cannot be constructed by the method of foldover. Several methods of constructing such orthogonal arrays can be found in the exercises of this chapter.

8.8   Difference matrices

The orthogonal array in Example 8.4, which was referred to as L_{18} by Genichi Taguchi and used extensively in industrial experiments for off-line quality control (Taguchi, 1987), can be constructed by using the method of difference matrices introduced by Bose and Bush (1952). Let $F = \{a_1, \ldots, a_s\}$ be an Abelian group containing s elements. Denote the group operation by $+$. A $\lambda s \times k$ matrix $\mathbf{D}$ of elements from F is called a difference matrix $D(\lambda s, k, s)$ if each element of F appears λ times in the componentwise difference of any two columns of $\mathbf{D}$. We start with a basic result that provides a simple tool for using difference matrices to construct orthogonal arrays of strength two.

Proposition 8.13. *The existence of a difference matrix $D(\lambda s, k, s)$ implies that of an $OA(\lambda s^2, s^k, 2)$.*

Proof. Let $\mathbf{D}$ be a difference matrix $D(\lambda s, k, s)$. For each $a \in F$, let $\mathbf{D} + a$ be the matrix obtained by adding a to all the entries of $\mathbf{D}$. Then it is easy to see that the $\lambda s^2 \times k$ matrix

$$\mathbf{M} = \begin{bmatrix} \mathbf{D} + a_1 \\ \vdots \\ \mathbf{D} + a_s \end{bmatrix} \qquad (8.21)$$

is an $OA(\lambda s^2, s^k, 2)$.   □

The orthogonal array constructed in (8.21) is *resolvable* in the sense that it can be partitioned into $s \times k$ subarrays of strength one, λs in total. Here each subarray consists of the s rows $\mathbf{d} + a_1, \ldots, \mathbf{d} + a_s$, where $\mathbf{d}$ is a row of $\mathbf{D}$. If we supplement $\mathbf{M}$ by a column in which each element of F appears λs times and all its entries corresponding to the s rows $\mathbf{d} + a_1, \ldots, \mathbf{d} + a_s$ are equal, then the resulting array is an $OA(\lambda s^2, s^{k+1}, 2)$. In particular, let $\mathbf{e}$ be a $\lambda s \times 1$ vector in which each element of F appears λ times. Then

$$\begin{bmatrix} \mathbf{e} & \mathbf{D} + a_1 \\ \vdots & \vdots \\ \mathbf{e} & \mathbf{D} + a_s \end{bmatrix} \tag{8.22}$$

is an $OA(\lambda s^2, s^{k+1}, 2)$.

If we choose $\mathbf{e}$ to be a $\lambda s \times 1$ vector in which each of the integers $0, 1, \ldots, \lambda s - 1$ appears once, then we have a mixed-level $OA(\lambda s^2, \lambda s \times s^k, 2)$.

Proposition 8.14. *The existence of a difference matrix $D(\lambda s, k, s)$ implies that of an $OA(\lambda s^2, s^{k+1}, 2)$ and an $OA(\lambda s^2, \lambda s \times s^k, 2)$.*

If s is a prime number or a prime power, then it is easy to see that the multiplication table of $GF(s)$ is a difference matrix $D(s, s, s)$. Applying Proposition 8.14 to it, we obtain a saturated $OA(s^2, s^{s+1}, 2)$ that, by Theorem 8.10, is equivalent to a complete set of mutually orthogonal $s \times s$ Latin squares. This establishes the existence of a complete set of mutually orthogonal $s \times s$ Latin squares when s is a prime or a prime power, a result mentioned at the end of Section 8.6.

Example 8.6. Masuyama (1957) constructed a difference matrix $D(6, 6, 3)$.

$$\begin{bmatrix} 0 & 1 & 1 & 0 & 2 & 2 \\ 2 & 2 & 1 & 0 & 0 & 1 \\ 2 & 1 & 2 & 0 & 1 & 0 \\ 0 & 0 & 0 & 0 & 0 & 0 \\ 1 & 2 & 0 & 0 & 1 & 2 \\ 1 & 0 & 2 & 0 & 2 & 1 \end{bmatrix}$$

Applying the construction in Proposition 8.14 to this difference matrix with modulo 3 arithmetic, we obtain an $OA(18, 3^7, 2)$ and an $OA(18, 6 \times 3^6, 2)$. The latter is saturated.

The construction in (8.21) and (8.22) can be extended further. Let $\mathbf{a} = (a_1, \ldots, a_s)^T$. Then the orthogonal array in (8.21) can be expressed as a Kronecker sum $\mathbf{a} \oplus \mathbf{D}$, where the Kronecker sum is defined in the same way as the Kronecker product, except that multiplication is replaced by addition. We can replace $\mathbf{a}$ with an

s-level orthogonal array of strength 2.

Theorem 8.15. *If* $\mathbf{M}$ *is an OA* $(N, s^n, 2)$ *and* $\mathbf{D}$ *is a difference matrix* $D(M, k, s)$, *where the factor levels in* $\mathbf{M}$ *and the entries of* $\mathbf{D}$ *are elements of an Abelian group* F, *then* $\mathbf{M} \oplus \mathbf{D}$ *is an* $OA(NM, s^{nk}, 2)$.

On the other hand, the $\lambda s \times 1$ vector $\mathbf{e}$ in (8.22) can be replaced by an $OA(\lambda s, s_1^{r_1} \times \cdots \times s_h^{r_h}, 2)$ if it exists. Denote such an OA by $\mathbf{L}$. Then it is easy to see that

$$\begin{bmatrix} \mathbf{L} & \mathbf{D} + a_1 \\ \vdots & \vdots \\ \mathbf{L} & \mathbf{D} + a_s \end{bmatrix} = [\mathbf{0}_s \oplus \mathbf{L} \quad \mathbf{a} \oplus \mathbf{D}]$$

is an $OA(\lambda s^2, s^k \times s_1^{r_1} \times \cdots \times s_h^{r_h}, 2)$, where $\mathbf{0}_s$ is the $s \times 1$ vector of zeros. This useful technique of adding extra factors to resolvable orthogonal arrays for constructing mixed-level orthogonal arrays has been used, e.g., by Chacko and Dey (1981), Wang and Wu (1991), Hedayat, Pu, and Stufken (1992), and Dey and Midha (1996). In particular, it can be applied to the orthogonal arrays constructed in Theorem 8.15, which are also resolvable.

Theorem 8.16. *If* $\mathbf{M}$ *is an* $OA(N, s^n, 2)$, $\mathbf{D}$ *is a difference matrix* $D(M, k, s)$, *where the factor levels in* $\mathbf{M}$ *and the entries of* $\mathbf{D}$ *are elements of an Abelian group* F, *and* $\mathbf{L}$ *is an arbitrary symmetric or asymmetrical strength-two orthogonal array of size* M, *then* $[\mathbf{0}_N \oplus \mathbf{L} \quad \mathbf{M} \oplus \mathbf{D}]$ *is an orthogonal array of strength two.*

More generally, we have the following result from Wang and Wu (1991).

Theorem 8.17. *For* $i = 1, \ldots, h$, *let* $\mathbf{M}_i$ *be an* $OA(N, s_i^{n_i}, 2)$ *and* $\mathbf{D}_i$ *be a difference matrix* $D(M, k_i, s_i)$ *such that the levels of the factors in* $\mathbf{M}_i$ *and the entries of* $\mathbf{D}_i$ *are elements of an Abelian group* F_i. *Suppose* $\mathbf{M} = [\mathbf{M}_1 \cdots \mathbf{M}_h]$ *is a mixed-level strength-two orthogonal array and* $\mathbf{L}$ *is an arbitrary symmetric or mixed-level strength-two orthogonal array of size* M. *Then* $[\mathbf{0}_N \oplus \mathbf{L} \quad \mathbf{M}_1 \oplus \mathbf{D}_1 \cdots \mathbf{M}_h \oplus \mathbf{D}_h]$ *is an orthogonal array of strength two.*

Wang and Wu (1991) used Theorem 8.17 to construct many mixed-level orthogonal arrays of strength two. Hedayat, Stufken, and Su (1996) extended difference matrices from strength two to higher strengths.

Another useful technique of constructing mixed-level orthogonal arrays is the method of replacement (Addelman, 1962a). Suppose an orthogonal array $\mathbf{O}_1$ of strength two involves a q-level factor A, and $\mathbf{O}_2$ is an $OA(q, s_1 \times \cdots \times s_h, 2)$. A one-to-one correspondence can be established between the q levels of A and the q rows of $\mathbf{O}_2$. If we replace each level of A in $\mathbf{O}_1$ with the corresponding row of $\mathbf{O}_2$, then we obtain an orthogonal array of strength two in which A is replaced by h factors with $s_1, \ldots, s_h$ levels, respectively. Hedayat, Sloane, and Stufken (1999) called this method *expansive replacement*. Conversely, suppose an orthogonal array $\mathbf{O}_3$ of size N and strength two contains an $N \times h$ submatrix $\mathbf{L}$ such that all the rows of $\mathbf{O}_2$ appear the same number of times as rows of $\mathbf{L}$. If we replace each row of $\mathbf{L}$ with the corresponding level of A, then we obtain an orthogonal array of strength two in which h factors in $\mathbf{O}_3$ are replaced by a single q-level factor. This is called *contractive replacement*.

Example 8.7. (Examples 8.4 and 8.6 revisited) In the OA$(18, 6 \times 3^6, 2)$ constructed in Example 8.6, if we replace each level of the six-level factor with a combination of a three-level factor and a two-level factor, then we obtain an OA$(18, 2 \times 3^7, 2)$. This is the orthogonal array in (8.12).

A more detailed discussion of contractive replacement will be given in Section 9.8. We refer readers to Chapter 4 of Dey and Mukerjee (1999) and Chapter 9 of Hedayat, Sloane, and Stufken (1999) for more discussions of the construction of mixed-level orthogonal arrays.

8.9 Enumeration of orthogonal arrays

Compared to the work on the construction of orthogonal arrays, there has been little done on their enumeration and classification into isomorphism sets, which are useful for selecting efficient designs with respect to various statistical criteria. Two orthogonal arrays are said to be isomorphic if they can be obtained from each other by row/column permutations and/or symbol (level) relabeling.

Schoen, Eendebak, and Nguyen (2010) proposed a general algorithm for complete enumeration of nonisomorphic symmetric and mixed-level orthogonal arrays for given run size, number of factors, numbers of factor levels, and strength. Before this work, there were some sporadic results on the enumeration of orthogonal arrays with specific parameters. We refer readers to Schoen, Eendebak, and Nguyen (2010) for a review and references to earlier works. For example, Hedayat, Seiden, and Stufken (1997) showed analytically that there are exactly four nonisomorphic OA$(54, 3^5, 3)$'s.

Most of the existing algorithms for enumerations of orthogonal arrays are sequential in nature, extending nonisomorphic orthogonal arrays with a given number of factors by adding one column (factor) at a time. One crucial step in these algorithms is isomorphism checking. A survey and evaluation of methods for the determination of isomorphism of factorial designs can be found in Katsaounis and Dean (2008).

Instead of adding one whole-column, Schoen, Eendebak, and Nguyen (2010) used element-wise addition to facilitate earlier abortion when the strength requirement is violated. They retained only arrays of a specific form so that pairwise isomorphism testing is replaced by a fast test to determine whether a single array is of some special form. By implementing the proposed algorithm, Schoen, Eendebak, and Nguyen (2010) enumerated nonisomorphic orthogonal arrays in the following cases:

all mixed-level strength two OA's with $N \leq 28$, all symmetric strength two OA's with $N \leq 27$, all $OA(28, 2^a, 2)$ with $a \leq 7$, all strength three OA's with $N \leq 64$ except $OA(56, 2^a, 3)$, $OA(64, 2^a, 3)$, and $OA(64, 4^1 \times 2^a, 3)$, and all strength four OA's with $N \leq 168$ except $OA(160, 2^a, 4)$. For $OA(56, 2^a, 3)$, $OA(64, 2^a, 3)$, $OA(64, 4^1 \times 2^a, 3)$, and $OA(160, 2^a, 4)$, complete enumerations were accomplished for $a \leq 8$, 6, 6, and 8, respectively. In many cases, previously unknown results on the maximum number of factors that can be accommodated were established.

Bulutoglu and Margot (2008) showed that the problem of classifying all isomorphism classes of orthogonal arrays is equivalent to finding isomorphism classes of nonnegative integer solutions to a system of linear equations. They then used integer linear programming techniques to find all isomorphism classes of solutions to the system of equations generated by an orthogonal array.

We mention here some general analytical results on the enumeration of whole families of orthogonal arrays. Seiden and Zemach (1966) determined all nonisomorphic $OA(\lambda 2^d, 2^{d+1}, d)$'s. Fujii, Namikawa, and Yamamoto (1989) did the same for all $OA(2^{d+1}, 2^{d+2}, d)$'s and $OA(2^{d+2}, 2^{d+3}, d)$'s. Stufken and Tang (2007) provided a complete solution to the enumeration and systematic construction of nonisomorphic $OA(\lambda 2^d, 2^{d+2}, d)$'s.

Enumeration of regular designs will be surveyed in Section 10.6.

Remark 8.3. When the treatment factors are qualitative, the labeling of factor levels is not important, and isomorphic designs have the same statistical properties. For quantitative factors with more than two levels, since the factor levels have a definite order, isomorphic designs may not be equivalent under some statistical models. For example, if orthogonal polynomials are used to represent factorial effects of quantitative factors, then relabeling the factor levels may result in designs with different properties. We refer readers to Cheng and Ye (2004) for a discussion of this issue. In this case, more strict relabeling that preserves the order needs to be considered for checking design equivalence.

8.10   Some variants of orthogonal arrays*

Several variants of orthogonal arrays have been proposed for different applications. For example, compound orthogonal arrays were introduced by Rosenbaum (1994) to study dispersion effects in factorial experiments for quality improvement. Orthogonal multi-arrays introduced by Brickell (1984) (whose connection with a class of designs called semi-Latin squares was noted in Bailey (1992)) were generalized by Sitter (1993) to balanced orthogonal multi-arrays for applications in sample surveys. Both of these classes of arrays were discussed in Hedayat, Sloane, and Stufken (1999). We concentrate here on the applications of orthogonal arrays to computer experiments and discuss three generalizations (strong, nested, and sliced orthogonal arrays) introduced in recent years.

In computer experiments, complex mathematical models are implemented in computer codes, from which outputs are generated in lieu of physical experiments that may be time consuming, expensive, or unfeasible. The design issue is to choose combinations of the input variables to run the computer code for building a surrogate model. We refer readers to Santner, Williams, and Notz (2003) and Fang, Li, and Sudjianto (2006) for comprehensive discussions of the design and analysis of computer experiments.

For simplicity, suppose there are n input variables (factors), and the experimental region is the unit cube $[0, 1)^n$; the other cases can often be dealt with by appropriate transformations. One major difference between computer experiments and physical experiments is that, since the computer code is deterministic, the same response is obtained when the computer code is repeatedly run with the same values of the input variables. Typically, the response functions in computer experiments are modeled by Gaussian stochastic processes. An optimal design approach can be used to find good model-based designs using appropriate optimality criteria.

Alternatively, since the relation between the response variable and the input variables is unknown and is usually very complex, the design points should be spread evenly to fill the design region well. This leads to the so-called *space filling designs*. Johnson, Moore and Ylvisaker (1990) used some distance-based criteria to produce space-filling designs.

Another approach, surveyed in Fang, Lin, Winker, and Zhang (2000), is based on the *uniform designs* introduced by Fang (1980). The idea is to obtain a set of points that is uniformly scattered in the design region by minimizing a measure of the discrepancy between the empirical distribution of the design points and the uniform distribution on the design region. The (w, m, n)-nets (Niederreiter, 1987), defined as follows, are examples of sets of points with good uniformity. Let an elementary interval in base s be an interval in $[0, 1)^n$ of the form $\prod_{j=1}^{n} [c_j/s^{d_j}, (c_j + 1)/s^{d_j})$, where c_j and d_j are integers such that $0 \leq c_j < s^{d_j}$. Each of these intervals has volume $s^{-(d_1 + \cdots + d_n)}$. Then a (w, m, n)-net is a set of s^m points in $[0, 1)^n$ that has the same number (s^w) of points in all the elementary intervals in base s of volume s^{w-m}. For $w = 0$, this gives exactly one point in each elementary interval in base s of volume s^{-m}.

A third approach, where orthogonal arrays play an important role, is based on a class of designs called *Latin hypercubes*. We first discuss the related problem of numerical evaluation of the integral of a real-valued function f over $[0, 1)^n$. A simple unbiased estimator of $\int_{[0,1)^n} f(\mathbf{x}) d\mathbf{x}$ can be obtained by ordinary Monte Carlo, that is, estimate $\int_{[0,1)^n} f(\mathbf{x}) d\mathbf{x}$ by the mean of $f(\mathbf{x}_i)$ over a simple random sample $\mathbf{x}_1, \ldots, \mathbf{x}_n$ drawn from the uniform distribution on $[0, 1)^n$. McKay, Beckman, and Conover (1979) proposed Latin hypercube sampling as an alternative to ordinary Monte Carlo. Under a Latin hypercube sample of size N, for each $i = 1, \ldots, n$, the ith coordinates of the sampled points are scattered in the N intervals $[0, 1/N)$, $[1/N, 2/N)$, $\ldots, [(N-1)/N, 1)$, with exactly one value in each interval. In other words, the sampled points are spread "evenly" in all the univariate projections. As in Chapter 6, we can decompose f into main-effect and interaction components (see Exercise 6.5). Stein (1987) showed that the variance of the sample mean under Latin hypercube sampling is asymptotically smaller than that under simple random sampling unless the main effects of all the variables are zero. Thus the stratification of each univariate margin in a Latin hypercube sample results in variance reduction by filtering out the main effects. Owen (1992) and Tang (1993) independently proposed the use of orthogonal arrays of strength t to construct Latin hypercube samples that achieve stratification in all the t and lower dimensional margins. This leads to further improvement by filtering out interactions. Owen (1995) introduced randomized (w, m, n)-nets, which are obtained by scrambling the digits in (w, m, n)-nets. The ratio of the variance under a randomized (w, m, n)-net to that under ordinary Monte Carlo tends to zero as $N = s^m \to \infty$.

Space-filling properties in low dimensional projections are also desirable for designs of computer experiments under factor sparsity. An N-point Latin hypercube with n factors is defined as an $N \times n$ matrix $\mathbf{D} = [d_{ij}]$ such that each column of $\mathbf{D}$ is a permutation of $\{1, \ldots, N\}$; that is, $\mathbf{D}$ is an $OA(N, N^n, 1)$. A design $\mathbf{X} = [x_{ij}]$ on $[0, 1)^n$ for computer experiments can be constructed by letting $x_{ij} = (d_{ij} - 0.5)/N$. Then the projections of the N points (rows of $\mathbf{X}$) onto any variable have exactly one point in each of the intervals $[0, 1/N), [1/N, 2/N), \ldots, [(N-1)/N, 1)$. Such a design achieves "uniformity" in all one-dimensional projections.

Uniformity in all two-dimensional projections is achieved by a Latin hypercube based on an orthogonal array of strength two. In general, to construct an OA-based Latin hypercube proposed by Tang (1993), we start with an $OA(N, s^n, t)$ with entries $0, \ldots, s - 1$, where $N = \lambda s^t$. In each column, the λs^{t-1} cells where k, $k = 0, \ldots, s - 1$, appears are replaced by a permutation of $k\lambda s^{t-1}, k\lambda s^{t-1} + 1, \ldots, k\lambda s^{t-1} + \lambda s^{t-1} - 1$. Then we have an N-point Latin hypercube that is space-filling in low dimensional projections in the sense that there is an equal number of points in h dimensional $s \times \cdots \times s$ grids for all $h \le t$.

He and Tang (2013) proposed *strong orthogonal arrays* for the construction of Latin hypercubes that better fill low dimensional projections than the usual OA-based Latin hypercubes in the sense that they provide stratification on finer grids.

Definition 8.2. An $N \times n$ matrix with entries from $0, \ldots, s^t - 1$ is called a strong orthogonal array of size N, n factors, s^t levels, and strength t, denoted by $SOA(N, (s^t)^n, t)$, if, for any $1 \le g \le t$ and any positive integers $u_1, \ldots, u_g$ with $u_1 + \cdots + u_g = t$, any subarray of g columns becomes an $OA(N, s_1 \times \cdots \times s_g, g)$, where $s_j = s^{u_j}$, after the s^t levels in the jth column, $1 \le j \le g$, are collapsed into s^{u_j} levels, with each level a replaced by $\lfloor a/s^{t-u_j} \rfloor$, where $\lfloor x \rfloor$ is the integral part of x.

By taking $g = t$ and $u_1 = \cdots = u_t = 1$, we can collapse an $SOA(N, (s^t)^n, t)$ into an $OA(N, s^n, t)$. Therefore $N = \lambda s^t$ for some positive integer λ. If $\lambda = 1$, then the $SOA(N, (s^t)^n, t)$ is itself a Latin hypercube. Otherwise, it can be converted into a Latin hypercube in the same way as a usual OA is converted: in each column, the λ cells where k, $k = 0, \ldots, s^t - 1$, appears are replaced by a permutation of $k\lambda, k\lambda + 1, \ldots, (k+1)\lambda - 1$. The resulting Latin hypercube achieves stratification on $s^{u_1} \times \cdots \times s^{u_g}$ grids, for all positive integers $u_1, \cdots, u_g$ with $u_1 + \cdots + u_g = t$, in all g-dimensional projections with $g \le t$. For example, for two-dimensional projections of the Latin hypercube constructed from an $OA(N, s^n, 3)$, in general we can only conclude that it has the same number of points in $s \times s$ grids, but for one constructed from an $SOA(N, (s^3)^n, 3)$, this projection property holds in $s^2 \times s$ and $s \times s^2$ grids as well.

The concept of strong orthogonal arrays was motivated by nets. He and Tang

(2013) showed that the existence of a (w,m,n)-net in base s is equivalent to that of an SOA($s^{w+t}, (s^t)^n, t$), where $t = m - w$. Such arrays have $\lambda = s^w$. Strong orthogonal arrays are more general than nets since λ is not necessarily a power of s. He and Tang (2013) also presented some results on the construction of strong orthogonal arrays.

Computer experiments can be conducted at different levels of sophistication. To model and calibrate the differences between experiments with different levels of accuracy, it is desirable that the design points for the more expensive high-accuracy experiment are a subset of those of the less expensive low-accuracy experiment. Qian, Tang, and Wu (2009) proposed and constructed such nested designs in which both designs are Latin hypercubes and are space-filling in their low dimensional projections. This is achieved by constructing an OA(N, s^n, t) $\mathbf{A}_1$ and an $M \times n$ subarray $\mathbf{A}_2$ such that $\mathbf{A}_2$ becomes an OA(M, r^n, t), $r < s$, after a certain mapping collapses the s levels into r levels. Qian, Ai, and Wu (2009) further studied the construction of such *nested orthogonal arrays*.

Qian and Wu (2009) introduced *sliced orthogonal arrays* to construct sliced space-filling designs for computer experiments in which there are qualitative as well as quantitative factors. Let $\mathbf{B}$ be an OA(N, s^n, t). If the rows of $\mathbf{B}$ can be partitioned into subarrays $\mathbf{B}_1, \ldots, \mathbf{B}_h$, such that each $\mathbf{B}_i$ becomes an OA($N/h, r^n, t$), $r < s$, after a certain mapping collapses the s levels into r levels, then $(\mathbf{B}_1, \ldots, \mathbf{B}_h)$ is called a sliced orthogonal array. The orthogonal array $\mathbf{B}$ is used to obtain a Latin hypercube design $\mathbf{D}$ with space-filling low dimensional projections. The points corresponding to each $\mathbf{B}_i$ are assigned the same level combination of the qualitative factors. This gives a partition of $\mathbf{D}$ into h subdesigns, where each subdesign consisting of the points associated with the same level combination of the qualitative factors also has space-filling low dimensional projections.

Exercises

8.1   In (8.18), let $\mathbf{X}$ be a saturated OA($N/4, 2^{N/4-1}, 2$). Then $\tilde{\mathbf{X}}$ is an OA($N/2, 2^{N/4}, 3$). Write $\tilde{\mathbf{X}}$ as [$\mathbf{B}$ $\mathbf{C}$]. Show that

$$\begin{bmatrix} -1 & \mathbf{B} & \mathbf{C} \\ 1 & -\mathbf{B} & \mathbf{C} \end{bmatrix}$$

is an OA($N, 2^{N/4+1}, 3$). Note that this array cannot be constructed by the method of foldover.   [Cheng, Mee, and Yee (2008)]

8.2   Show that if $\mathbf{X}$ is a two-level orthogonal array of strength three, then

$$\begin{bmatrix} \mathbf{X} & \mathbf{X} \\ \mathbf{X} & -\mathbf{X} \end{bmatrix},$$

called the double of $\mathbf{X}$, is an orthogonal array of strength three. Even if $\mathbf{X}$ has strength greater than three, its double cannot have strength greater than three.
                                                              [Cheng, Mee, and Yee (2008)]

8.3   The double of $\mathbf{X}$ in Exercise 8.2 can be written as $\mathbf{H} \otimes \mathbf{X}$, where $\mathbf{H}$ is the 2×2

Hadamard matrix

$$\begin{bmatrix} 1 & 1 \\ 1 & -1 \end{bmatrix}.$$

Show that if $\mathbf{X}$ is a two-level orthogonal array of strength three and $\mathbf{H}$ is a Hadamard matrix, then $\mathbf{H} \otimes \mathbf{X}$ is an orthogonal array of strength three.

[Cheng, Mee, and Yee (2008)]

8.4   Construct an OA$(20, 2^{19}, 2)$.

8.5   Construct a balanced incomplete block design with 15 treatments in 15 blocks of size 7.

8.6   Prove Theorem 8.15.

8.7   (a) Verify that the multiplication table of GF(s) is a difference matrix $D(s, s, s)$.

   (b) The discussion following Proposition 8.14 shows how the difference matrix in part (a) can be used to construct an OA$(s^2, s^{s+1}, 2)$. Show that by Theorem 8.15, an OA$(s^3, s^{s^2+s+1}, 2)$ can be constructed by adding a suitable column to the Kronecker sum of an OA$(s^2, s^{s+1}, 2)$ and a difference matrix $D(s, s, s)$.

   (c) Show that by applying the construction in part (b) repeatedly, one obtains saturated OA$(s^k, s^{(s^k-1)/(s-1)}, 2)$ for all k.

8.8   Use a difference matrix to construct an OA$(25, 5^6, 2)$ and four mutually orthogonal 5×5 Latin squares.

8.9   Use Exercise 2.8 to show the following.

   (a) An orthogonal array of strength $2k$ ($k \geq 1$) minimizes the variances of the estimators of all the main-effect contrasts and all the interaction contrasts involving up to k factors under the assumption that the interactions involving more than k factors are negligible.

   (b) An orthogonal array of strength $2k - 1$ ($k \geq 2$) minimizes the variances of the estimators of all the main-effect contrasts and all the interaction contrasts involving up to $k - 1$ factors under the assumption that the interactions involving more than k factors are negligible. It also minimizes the variances of the estimators of all the main-effect contrasts, all the interaction contrasts involving up to $k - 1$ factors, and all the k-factor interaction contrasts involving a given factor under the assumption that all the other interaction contrasts are negligible.

8.10   Justify the methods of expansive and contractive replacement described in the paragraph preceding Example 8.7.

Chapter 9

Regular Fractional Factorial Designs

An example of regular fractional factorial designs was discussed in Section 8.4. Some basic properties of regular fractional factorial designs are presented in this chapter. When only a fraction of the treatment combinations are observed, the factorial effects are mixed up (aliased). Under a regular fractional factorial design, aliasing of the factorial effects can be determined in a simple manner. Such designs are easy to construct, have nice structures, and are relatively straightforward to analyze, but the number of levels must be a prime number or power of a prime number. We show that all regular fractional factorial designs are orthogonal arrays. An algorithm for constructing a regular fractional factorial design under which certain required effects are estimable is presented. We also discuss connections of regular fractional factorial designs with finite projective geometries and linear codes. Results from finite projective geometry and coding theory provide useful tools for studying the structures and construction of regular fractional factorial designs.

9.1   Construction and defining relation

The discussion in Section 8.4 explains how the design in (8.9) can be obtained by solving three independent linear equations and therefore can be viewed as a three-dimensional subspace of a six-dimensional finite Euclidean geometry. This can be extended to the construction of general regular fractional factorial designs.

As seen in Section 6.6, in an s^n experiment where s is a prime number or power of a prime number, the s^n treatment combinations can be identified with the points (or vectors) in an n-dimensional Euclidean geometry EG(n,s), and pencils of $(n-1)$-flats can be used to define contrasts representing factorial effects. For any positive integer p such that $p < n$, the design consisting of all the s^{n-p} treatment combinations in an $(n-p)$-flat is an s^{-p} fraction of the complete s^n factorial and is called a *regular s^{n-p} fractional factorial design*. Such a design is the solution set of equations

$$\mathbf{a}_i^T \mathbf{x} = b_i, \ i = 1, \ldots, p, \tag{9.1}$$

where $\mathbf{a}_1, \ldots, \mathbf{a}_p$ are linearly independent vectors in EG(n,s), and $b_1, \ldots, b_p \in$ GF(s); see Section A.4. The equations in (9.1) define an $(n-p)$-dimensional subspace when $b_1 = \cdots = b_p = 0$, and parallel $(n-p)$-flats in the same pencil for the other b_i's. Any of the s^p disjoint $(n-p)$-flats in the same pencil can be used as an s^{n-p} fractional factorial design. Without loss of generality and for notational convenience, unless otherwise stated, we take $b_1 = \cdots = b_p = 0$ and write (9.1) as

$$\mathbf{A}^T \mathbf{x} = \mathbf{0}, \tag{9.2}$$

where $\mathbf{A} = [\mathbf{a}_1 \cdots \mathbf{a}_p]$. In this case, the design contains the treatment combination with all the factors at level 0.

Let $R(\mathbf{A})$ be the space generated by $\mathbf{a}_1, \ldots, \mathbf{a}_p$. Then the regular fractional factorial design defined by (9.2) consists of all the treatment combinations $\mathbf{x}$ that satisfy $\mathbf{a}^T \mathbf{x} = 0$ for all $\mathbf{a} \in R(\mathbf{A})$. The p-dimensional space $R(\mathbf{A})$ is called the *defining contrast subgroup* of the design. The factorial effects (contrasts) defined by the nonzero vectors in $R(\mathbf{A})$ are called *defining effects* or *defining contrasts* of the fractional factorial design. When these factorial effects are denoted by strings of capital letters as described in the paragraph following (6.16) and Example 6.4, they are called *defining words*. Sometimes we also call the nonzero vectors in $R(\mathbf{A})$ themselves defining words. Two defining words $\mathbf{x}$ and $\mathbf{y}$ are said to be equivalent if $\mathbf{x} = \lambda \mathbf{y}$ for some nonzero $\lambda \in GF(s)$. In general, there are a total of $(s^p - 1)/(s - 1)$ nonequivalent defining words.

Remark 9.1. The construction of regular s^p fractional factorial designs presented above amounts to using the method in Section 7.2 to divide the treatment combinations in an s^n complete factorial into s^p blocks and taking any of the blocks as an s^{n-p} fraction. The factorial effects in the defining contrast subgroup are exactly those that are confounded with blocks in the associated confounding scheme.

Example 9.1. (Example 8.1 revisited) It follows from (8.16) that the design in Example 8.1 is a regular 2^{6-3} design defined by $\mathbf{a}_1 = (1,1,0,1,0,0)^T$, $\mathbf{a}_2 = (0,1,1,0,1,0)^T$, and $\mathbf{a}_3 = (1,1,1,0,0,1)^T$. The defining contrast subgroup consists of $\mathbf{a}_1, \mathbf{a}_2, \mathbf{a}_3$ and their linear combinations. It contains seven nonzero vectors: $\mathbf{a}_1$, $\mathbf{a}_2, \mathbf{a}_3, \mathbf{a}_1 + \mathbf{a}_2 = (1,0,1,1,1,0)^T$, $\mathbf{a}_1 + \mathbf{a}_3 = (0,0,1,1,0,1)^T$, $\mathbf{a}_2 + \mathbf{a}_3 = (1,0,0,0,1,1)^T$, and $\mathbf{a}_1 + \mathbf{a}_2 + \mathbf{a}_3 = (0,1,0,1,1,1)^T$. The seven defining words are ABD, BCE, $ABCF$, $ACDE$, CDF, AEF, and $BDEF$.

When 0 is replaced by -1, the seven equations $\mathbf{a}^T \mathbf{x} = 0$ for all $\mathbf{a} \in R(\mathbf{A}), \mathbf{a} \neq \mathbf{0}$, determined in Example 9.1, are equivalent to (8.15). We call (8.15) the *defining relation* of the design, which is usually written as

$$I = -ABD = -BCE = ABCF = ACDE = -CDF = -AEF = BDEF. \tag{9.3}$$

All the defining words that contain an odd number of letters (factors) carry a negative sign. One easy way to see this is that the design contains the treatment combination with all the factors at level -1.

In Example 9.1, if we choose the design defined by $\mathbf{A}^T \mathbf{x} = (1,0,0)^T$ instead of $\mathbf{A}^T \mathbf{x} = (0,0,0)^T$, then (8.13) becomes $x_4 = x_1 x_2$, $x_5 = -x_2 x_3$, and $x_6 = x_1 x_2 x_3$. The corresponding defining relation is written as

$$I = ABD = -BCE = ABCF = -ACDE = CDF = -AEF = -BDEF.$$

In general, for regular 2^{n-p} fractional factorial designs, the 2^p combinations of $\pm$ signs for the p independent defining words correspond to the 2^p flats in the same pencil, or equivalently, the 2^p choices of the b_i's in (9.1). One can first write down the p independent defining words with appropriate signs and then form all the possible products to complete the defining relation. When multiplying the words, we follow the rule that $X^2 = I$ and $XI = I$ for any letter X.

The design in Example 8.2 can be constructed by making $x_4 = x_1 x_2 x_3$, $x_6 = x_1 x_2 x_5$, $x_7 = x_1 x_3 x_5$, and $x_8 = x_2 x_3 x_5$, so $x_1 x_2 x_3 x_4 = 1$, $x_1 x_2 x_5 x_6 = 1$, $x_1 x_3 x_5 x_7 = 1$, and $x_2 x_3 x_5 x_8 = 1$. Therefore it is a regular 2^{8-4} design. If the eight factors are denoted by A, B, C, D, E, F, G, and H, respectively, then $ABCD, ABEF, ACEG$, and $BCEH$ are four independent defining words, generating a total of 15 defining words. The design can be constructed by using A, B, C, and E as basic factors.

Example 9.2. For an example of three-level designs, consider the 2-flat in $EG(4,3)$ defined by $x_1 + x_2 + 2x_3 = 0$ and $x_1 + 2x_2 + 2x_4 = 0$. Writing each treatment combination as a row, we have the following regular 3^{4-2} fractional factorial design.

$$
\begin{array}{cccc}
0 & 0 & 0 & 0 \\
0 & 1 & 1 & 2 \\
0 & 2 & 2 & 1 \\
1 & 0 & 1 & 1 \\
1 & 1 & 2 & 0 \\
1 & 2 & 0 & 2 \\
2 & 0 & 2 & 2 \\
2 & 1 & 0 & 1 \\
2 & 2 & 1 & 0
\end{array}
\tag{9.4}
$$

The defining contrast subgroup, consisting of $(1,1,2,0)^T$, $(1,2,0,2)^T$, and their nonzero linear combinations, contains eight nonzero vectors. Since each nonzero vector $\mathbf{x}$ and its nonzero multiples $\lambda \mathbf{x}$ define the same factorial effect, we only need to consider the two linear combinations $(1,1,2,0)^T + (1,2,0,2)^T = (2,0,2,2)^T$ and $(1,1,2,0)^T + 2(1,2,0,2)^T = (0,2,2,1)^T$. The nine treatment combinations in (9.4) are solutions to the four linear equations $x_1 + x_2 + 2x_3 = x_1 + 2x_2 + 2x_4 = 2x_1 + 2x_3 + 2x_4 = 2x_2 + 2x_3 + x_4 = 0$. To make the leading nonzero coefficients equal to 1, as discussed in Example 6.4, we replace the last two equations with $2(2x_1 + 2x_3 + 2x_4) = 2(2x_2 + 2x_3 + x_4) = 0$. Then $x_1 + x_3 + x_4 = x_2 + x_3 + 2x_4 = 0$, and the defining equations of the design can be written as

$$
x_1 + x_2 + 2x_3 = x_1 + 2x_2 + 2x_4 = x_1 + x_3 + x_4 = x_2 + x_3 + 2x_4 = 0. \tag{9.5}
$$

If the factors are labeled by A, B, C, and D, then these equations are written as

$$
I = ABC^2 = AB^2 D^2 = ACD = BCD^2. \tag{9.6}
$$

Again, (9.6) is called the defining relation of the design. Each word that appears in the defining relation represents two degrees of freedom of factorial effects. It follows from (9.5) that $x_3 = x_1 + x_2$ and $x_4 = x_1 + 2x_2$. Therefore, this design can also be constructed by writing down two columns that consist of all the nine level combinations of the first two factors (basic factors) and defining the levels of the last two factors (added factors) as linear functions of those of the basic factors.

Remark 9.2. In the literature, treatment factors are sometimes labeled $1, 2, 3, \ldots$, etc., instead of capital letters. Then, e.g., the interaction of factors 1, 2, and 4 is denoted by **124**, and the defining relation in (9.3) is written as

$$I = -124 = -235 = 1236 = 1345 = -346 = -156 = 2456.$$

9.2 Aliasing and estimability

With the two levels represented by 1 and -1, the 2^{6-3} fractional factorial design in Example 8.1 is defined by $x_4 = -x_1 x_2$, $x_5 = -x_2 x_3$, and $x_6 = x_1 x_2 x_3$. In the full model matrix, the column corresponding to the main effect of factor D, except for a sign change, is identical to that corresponding to the two-factor interaction AB. It implies that these two effects are completely mixed up and cannot be estimated at the same time. We say that they are *aliases* of each other. Likewise the main effect of factor E is an alias of the two-factor interaction BC, and the main effect of F is an alias of the three-factor interaction ABC.

Complete information about the aliasing of factorial effects can be determined from the defining relation as follows. The defining relation (8.15) or (9.3) implies that the columns of the model matrix corresponding to the seven defining effects $-ABD$, $-BCE$, $ABCF$, $ACDE$, $-CDF$, $-AEF$, and $BDEF$ are the vector of 1's. Thus these effects are aliases of the mean and cannot be estimated.

If each term in (8.15) is multiplied by x_1, then we obtain

$$x_1 = -x_2 x_4 = -x_1 x_2 x_3 x_5 = x_2 x_3 x_6 = x_3 x_4 x_5 = -x_1 x_3 x_4 x_6 = -x_5 x_6 = x_1 x_2 x_4 x_5 x_6.$$

This implies that the columns of the model matrix corresponding to the eight factorial effects A, $-BD$, $-ABCE$, BCF, CDE, $-ACDF$, $-EF$, and $ABDEF$ are identical, and these effects are therefore completely mixed up. We say that they constitute an *alias set*, and write

$$A = -BD = -ABCE = BCF = CDE = -ACDF = -EF = ABDEF.$$

These aliased effects can be obtained by multiplying each term in (9.3) by A. Other alias sets can be determined similarly. Altogether there are seven alias sets.

$$
\begin{aligned}
A &= -BD = -ABCE = BCF = CDE = -ACDF = -EF = ABDEF \\
B &= -AD = -CE = ACF = ABCDE = -BCDF = -ABEF = DEF \\
C &= -ABCD = -BE = ABF = ADE = -DF = -ACEF = BCDEF \\
D &= -AB = -BCDE = ABCDF = ACE = -CF = -ADEF = BEF \qquad (9.7) \\
E &= -ABDE = -BC = ABCEF = ACD = -CDEF = -AF = BDF \\
F &= -ABDF = -BCEF = ABC = ACDEF = -CD = -AE = BDE \\
AC &= -BCD = -ABE = BF = DE = -ADF = -CEF = ABCDEF
\end{aligned}
$$

In general, in a regular 2^{n-p} fractional factorial design, the $2^p - 1$ defining effects are aliased with the mean and cannot be estimated. The other $2^n - 2^p$ effects are divided into $2^{n-p} - 1$ alias sets, each containing 2^p effects. In algebraic terms, the alias sets are cosets of the defining contrast subgroup. It will be shown later in this section that if all but one effect in the same alias set are negligible, then the nonnegligible effect is estimable. Therefore we can estimate a total of $2^{n-p} - 1$ effects. With 2^{n-p} observations, we do have $2^{n-p} - 1$ degrees of freedom excluding the mean.

Example 9.3. (**Example 9.2 revisited**) Under the 3^{4-2} design in (9.4), adding x_1 and $2x_1$ to each term in the four defining equations in (9.5), we obtain

$$x_1 = 2x_1 + x_2 + 2x_3 = 2x_1 + 2x_2 + 2x_4 = 2x_1 + x_3 + x_4 = x_1 + x_2 + x_3 + 2x_4 \qquad (9.8)$$

and

$$2x_1 = x_2 + 2x_3 = 2x_2 + 2x_4 = x_3 + x_4 = 2x_1 + x_2 + x_3 + 2x_4. \qquad (9.9)$$

Multiplying each term in (9.9) by 2, we have

$$x_1 = 2x_2 + x_3 = x_2 + x_4 = 2x_3 + 2x_4 = x_1 + 2x_2 + 2x_3 + x_4. \qquad (9.10)$$

It follows from (9.8) and (9.10) that coefficients of the contrasts represented by A, AB^2C, ABD, AC^2D^2, $ABCD^2$, BC^2, BD, CD, and AB^2C^2D coincide on all the treatment combinations in the chosen fraction; here we follow the convention of making the leading exponent equal to 1. This determines the aliases of the main effect of factor A. The other alias sets can be determined similarly. There are a total of four alias sets:

$$\{A, AB^2C, ABD, AC^2D^2, ABCD^2, BC^2, BD, CD, AB^2C^2D\},$$

$$\{B, AB^2C^2, AD^2, ABCD, BC^2D, AC^2, ABD^2, AB^2CD, CD^2\},$$

$$\{C, AB, AB^2CD^2, AC^2D, BC^2D^2, ABC, AB^2C^2D^2, AD, BD^2\},$$

$$\{D, ABC^2D, AB^2, ACD^2, BC, ABC^2D^2, AB^2D, AC, BCD\}.$$

Let d be the regular s^{n-p} fractional factorial design consisting of the treatment combinations that satisfy (9.2). For each nonzero vector $\mathbf{g}$ in $R(\mathbf{A})$, $\mathbf{g}^T\mathbf{x} = 0$ for all $\mathbf{x} \in d$. Thus all the treatment combinations in d fall entirely in the same $(n-1)$-flat defined by $\mathbf{g}$ and therefore have the same coefficient in any contrast defined by the pencil $P(\mathbf{g})$. Then the corresponding column in the model matrix has constant entries. It follows that the contrasts defined by $P(\mathbf{g})$ are completely aliased with the grand mean and cannot be estimated.

For any two vectors $\mathbf{g}_1, \mathbf{g}_2 \notin R(\mathbf{A})$, if $\mathbf{g}_1 - \lambda\mathbf{g}_2 \in R(\mathbf{A})$ for some nonzero $\lambda \in GF(s)$, then $\mathbf{g}_1^T\mathbf{x} = \lambda\mathbf{g}_2^T\mathbf{x}$ for all $\mathbf{x} \in d$. This implies that $\mathbf{g}_1$ and $\mathbf{g}_2$ yield identical partitions of d into disjoint $(n-p-1)$-flats. Each treatment contrast defined by $P(\mathbf{g}_1)$, when restricted to the treatment combinations in d, completely coincides with a treatment contrast defined by $P(\mathbf{g}_2)$. Hereafter we abuse the language by saying that $P(\mathbf{g}_1)$

and $P(\mathbf{g}_2)$ are aliased when we actually mean that the effects defined by them are aliased. The set of all the pencils of $(n-1)$-flats that are aliased with one another is called an alias set.

For any $\mathbf{g} \notin R(\mathbf{A})$, the alias set containing $P(\mathbf{g})$ consists of all the pencils defined by nonzero vectors of the form $\lambda \mathbf{g} + \mathbf{h}$, where $\lambda \in GF(s)$, $\lambda \neq 0$, and $\mathbf{h} \in R(\mathbf{A})$. Since $R(\mathbf{A})$ contains s^p vectors, there are $s^p(s-1)$ vectors of this form. Each vector and its nonzero multiples define the same partition of the treatment combinations. Therefore each alias set contains $s^p(s-1)/(s-1) = s^p$ pencils of $(n-1)$-flats. On the other hand, the total number of vectors not in $R(\mathbf{A})$ is equal to $s^n - s^p$. It follows that the total number of alias sets is equal to

$$(s^n - s^p)/[s^p(s-1)] = (s^{n-p} - 1)/(s-1).$$

We summarize these results in a theorem.

Theorem 9.1. *For any regular s^{n-p} fractional factorial design,*

(i) *the treatment contrasts defined by each of the $s^p - 1$ nonzero vectors in the defining contrast subgroup $R(\mathbf{A})$ cannot be estimated;*

(ii) *For any two vectors $\mathbf{g}_1$, $\mathbf{g}_2 \notin R(\mathbf{A})$, $P(\mathbf{g}_1)$ and $P(\mathbf{g}_2)$ are in the same alias set if and only if there is a nonzero $\lambda \in GF(s)$ such that $\mathbf{g}_1 - \lambda \mathbf{g}_2 \in R(\mathbf{A})$;*

(iii) *there are a total of $(s^{n-p} - 1)/(s-1)$ alias sets, each containing s^p pencils of $(n-1)$-flats.*

For the design in Example 9.3 ($s = 3, n = 4, p = 2$), there are four alias sets, with nine pencils of 3-flats defining a total of 18 degrees of freedom in each alias set.

We now present a result on the estimability of the factorial effects.

Theorem 9.2. *For any regular s^{n-p} fractional factorial design, if, in each alias set, the treatment contrasts defined by all but one pencil are negligible, then the nonnegligible treatment contrasts are estimable, and the least squares estimators of orthogonal nonnegligible treatment contrasts are uncorrelated.*

Proof. As in the discussion preceding Example 6.3, each set of $s - 1$ mutually orthogonal $s \times 1$ vectors that are orthogonal to the vector of 1's can be used to construct $s - 1$ mutually orthogonal treatment contrasts representing the factorial effects defined by each pencil of $n - 1$-flats in $EG(n, s)$. We show that under the assumption of the theorem, the columns of the model matrix corresponding to such nonnegligible contrasts are mutually orthogonal and are orthogonal to the vector of all 1's.

Suppose $\mathbf{g} \notin R(\mathbf{A})$ is such that the factorial effects defined by $P(\mathbf{g})$ are the only effects in the alias set containing $P(\mathbf{g})$ that are not negligible. Let $H_0, \ldots, H_{s-1}$ be the s disjoint $(n-1)$-flats in $P(\mathbf{g})$. Then each treatment contrast defined by $P(\mathbf{g})$ is of the form $\tau = \sum_{i=0}^{s-1} c_i \sum_{\mathbf{x} \in H_i} \alpha(\mathbf{x})$, where $\sum_{i=0}^{s-1} c_i = 0$. Since $\mathbf{g} \notin R(\mathbf{A})$, the $p+1$ vectors $\mathbf{a}_1, \ldots, \mathbf{a}_p$ and $\mathbf{g}$ are linearly independent. Therefore each $d \cap H_i$ is an $(n-p-1)$-flat. It follows that each $d \cap H_i$ contains s^{n-p-1} treatment combinations. The sum of the coefficients of τ over the treatment combinations in d is equal to $s^{n-p-1} \sum_{i=0}^{s-1} c_i = 0$. In other words, the column corresponding to τ in the model matrix for the fractional factorial design is orthogonal to the vector of 1's. Similarly, if $\sum_{i=0}^{s-1} e_i = 0$ and $\sum_{i=0}^{s-1} c_i e_i = 0$, then for the two orthogonal contrasts $\sum_{i=0}^{s-1} c_i \sum_{\mathbf{x} \in H_i} \alpha(\mathbf{x})$ and $\sum_{i=0}^{s-1} e_i \sum_{\mathbf{x} \in H_i} \alpha(\mathbf{x})$ defined by $P(\mathbf{g})$, the inner product of their corresponding columns in the model matrix is equal to $s^{n-p-1} \sum_{i=0}^{s-1} c_i e_i$, which is also 0.

If $P(\mathbf{g}_1)$ and $P(\mathbf{g}_2)$ are in different alias sets, then $\mathbf{g}_1 - \lambda \mathbf{g}_2 \notin R(\mathbf{A})$ for all nonzero $\lambda \in GF(s)$. It follows that $\mathbf{g}_1$, $\mathbf{g}_2$, and $\mathbf{a}_1, \ldots, \mathbf{a}_p$ are linearly independent. Let $E_0, \ldots, E_{s-1}$ be the partition of d according to the values of $\mathbf{g}_1^T \mathbf{x}$, and $F_0, \ldots, F_{s-1}$ be the partition of d according to the values of $\mathbf{g}_2^T \mathbf{x}$. Then each $E_i \cap F_j$ contains s^{n-p-2} treatment combinations since it is an $(n-p-2)$-flat. Let τ and ϕ be two treatment contrasts defined by $P(\mathbf{g}_1)$ and $P(\mathbf{g}_2)$, respectively. Suppose τ has coefficient c_i for each treatment combination in E_i and ϕ has coefficient e_j for each treatment combination in F_j. Then the inner product of the coefficient vectors of τ and ϕ, restricted to the treatment combinations in d, is $s^{n-p-2} \sum_{i=0}^{s-1} \sum_{j=0}^{s-1} c_i e_j$, which is equal to zero. $\qquad\square$

Remark 9.3. As in Sections 6.7 and 7.8, suppose the treatment combinations are identified with the elements of $\mathbb{Z}_{s_1} \oplus \cdots \oplus \mathbb{Z}_{s_n}$, where $\mathbb{Z}_{s_i} = \{0, \ldots, s_i - 1\}$. Each subgroup G of $\mathbb{Z}_{s_1} \oplus \cdots \oplus \mathbb{Z}_{s_n}$ can be regarded as a fractional factorial design. The treatment combinations in G are the solutions $\mathbf{x}$ to the equations $[\mathbf{a}, \mathbf{x}] = 0$ for all $\mathbf{a} \in G^0$, where G^0 is the annihilator of G. For any two nonzero $\mathbf{a}$ and $\mathbf{b}$ in the same coset of G^0, we have $[\mathbf{a}, \mathbf{x}] = [\mathbf{b}, \mathbf{x}]$ for all $\mathbf{x} \in G$. Thus the factorial effects defined by $\mathbf{a}$ and $\mathbf{b}$ are aliased. This extends the construction of regular fractional factorial designs to $s_1 \times \cdots \times s_n$ experiments.

9.3 Analysis

We will show in Section 9.5 that there is a complete factorial of $n - p$ factors embedded in each regular s^{n-p} fractional factorial design, and each regular s^{n-p} design can be constructed by using the interactions of $n - p$ basic factors to define p additional factors. We have seen that this holds for the designs in Examples 8.1, 8.2, and 9.2. An algorithm for identifying basic factors from the defining relation will also be presented in Section 9.5. Once we have found a set of basic factors, we can calculate estimates of their factorial effects in the usual way, tentatively ignoring the added factors. However, what we get are actually estimates of linear combinations of factorial effects of the basic factors and their aliases.

Suppose we have $E(\mathbf{y}) = \mathbf{X}_1 \boldsymbol{\beta}_1 + \mathbf{X}_2 \boldsymbol{\beta}_2$, where $\boldsymbol{\beta}_1$ consists of factorial effects of the basic factors, and $\boldsymbol{\beta}_2$ consists of all the other factorial effects. The least squares estimator of $\boldsymbol{\beta}_1$ when $\boldsymbol{\beta}_2$ is ignored is $\hat{\boldsymbol{\beta}}_1 = (\mathbf{X}_1^T \mathbf{X}_1)^{-1} \mathbf{X}_1^T \mathbf{y}$. Under the full model,

$$
\begin{aligned}
E\left(\hat{\boldsymbol{\beta}}_1\right) &= E\left[(\mathbf{X}_1^T \mathbf{X}_1)^{-1} \mathbf{X}_1^T \mathbf{y}\right] \\
&= (\mathbf{X}_1^T \mathbf{X}_1)^{-1} \mathbf{X}_1^T [\mathbf{X}_1 \boldsymbol{\beta}_1 + \mathbf{X}_2 \boldsymbol{\beta}_2] \\
&= \boldsymbol{\beta}_1 + \left(\mathbf{X}_1^T \mathbf{X}_1\right)^{-1} \mathbf{X}_1^T \mathbf{X}_2 \boldsymbol{\beta}_2.
\end{aligned}
$$

So what $\widehat{\boldsymbol{\beta}}_1$ actually estimates is $\boldsymbol{\beta}_1 + (\mathbf{X}_1^T\mathbf{X}_1)^{-1}\mathbf{X}_1^T\mathbf{X}_2\boldsymbol{\beta}_2$. The matrix $(\mathbf{X}_1^T\mathbf{X}_1)^{-1}\mathbf{X}_1^T\mathbf{X}_2$ is called the *alias matrix*.

Consider, for example, two-level designs. Define the factorial effects as in (6.18). Then for any regular 2^{n-p} design, $\mathbf{X}_1^T\mathbf{X}_1 = 2^{n-p}\mathbf{I}$. Therefore we have

$$\mathrm{E}\left(\widehat{\boldsymbol{\beta}}_1\right) = \boldsymbol{\beta}_1 + \frac{1}{2^{n-p}}\mathbf{X}_1^T\mathbf{X}_2\boldsymbol{\beta}_2.$$

Each entry of $\mathbf{X}_1^T\mathbf{X}_2$ is the inner product of a column of $\mathbf{X}_1$ and a column of $\mathbf{X}_2$. Since the main-effect column of each added factor is the Hadamard product of some columns of $\mathbf{X}_1$, possibly with a sign change, all the columns corresponding to the factorial effects of the added factors and the interactions between basic and added factors are Hadamard products of certain columns of $\mathbf{X}_1$, possibly with sign changes. Furthermore, the Hadamard product of any two columns of $\mathbf{X}_1$ is also a column of $\mathbf{X}_1$. Therefore for each column $\mathbf{u}$ of $\mathbf{X}_2$, either $\mathbf{u}$ or $-\mathbf{u}$ is a column of $\mathbf{X}_1$. It follows that for any column $\mathbf{v}$ of $\mathbf{X}_1$ and any column $\mathbf{u}$ of $\mathbf{X}_2$,

$$\mathbf{u}^T\mathbf{v} = \begin{cases} \pm 2^{n-p}, & \text{if the corresponding effects in } \boldsymbol{\beta}_1 \text{ and } \boldsymbol{\beta}_2 \text{ are in the same alias set;} \\ 0, & \text{otherwise.} \end{cases}$$

This implies that for each factorial effect in $\boldsymbol{\beta}_1$, say β,

$$\mathrm{E}\left(\widehat{\beta}\right) = \beta + \sum_{\tilde{\beta}: \, \tilde{\beta} \text{ is in the same alias set as } \beta} \pm\tilde{\beta}. \qquad (9.11)$$

Example 9.4. Under the design in (8.9), to compute estimates of the factorial effects, we can run the Yates algorithm with the data arranged in the Yates order of the treatment combinations of the basic factors A, B, and C. Each estimate obtained from the Yates algorithm estimates the sum of a string of effects aliased with the corresponding effect of the basic factors. For example, it follows from the aliasing relation exhibited in (9.7) that the estimate of AB computed by ignoring the added factors is an estimate of $AB - D - ACE + CF + BCDE - ABCDF - BEF + ADEF$. If $AB, ACE, CF, BCDE, ABCDF, BEF$, and $ADEF$ are negligible, then it is an estimate of $-D$.

For data collected from fractional factorial designs, half-normal probability plots and Lenth's (1989) or other methods can be used to identify significant effects. When some contrasts are found significant but cannot be attributed to specific factorial effects due to the presence of nonnegligible aliases, follow-up experiments can be performed to resolve the ambiguity. Another fraction (e.g., foldover; see Section 9.6) can be added for de-aliasing, or an optimal design criterion (such as the D- or D_s-criterion) can be used to choose extra experimental runs so that the aliased effects can be disentangled and efficiently estimated. Meyer, Steinberg, and Box (1996) proposed a Bayesian method for choosing follow-up runs. We refer readers to Box, Hunter, and Hunter (2005), Mee (2009), and Wu and Hamada (2009) for detailed discussions on strategies for designing follow-up experiments.

9.4 Resolution

Box and Hunter (1961) defined the *resolution* of a regular fractional factorial design as the length of the shortest defining word, where the length of a defining word is the number of letters it contains, or equivalently, the number of nonzero entries in the corresponding vector in the defining contrast subgroup. According to this definition, the design in Example 8.1 is of resolution three. The design in Example 8.2 is defined by the four independent defining words $ABCD$, $ABEF$, $ACEG$, and $BCEH$. All their products contain at least four letters. Therefore this design is of resolution four.

If the resolution is R, then there is at least one defining word of length R. This implies that, for each a such that $1 \leq a < R$, at least one factorial effect involving a factors is aliased with an effect involving $R - a$ factors. However, since there are no defining words of lengths shorter than R, none of the effects involving a factors is aliased with effects involving fewer than $R - a$ factors. Thus, under a design of resolution three, although some main effects are aliased with two-factor interactions, no main effect is aliased with another main effect; therefore all the main effects are estimable if the interactions are assumed negligible. Similarly, under a design of resolution four, not only the main effects are not aliased with one another, they are also not aliased with two-factor interactions; therefore all the main effects are estimable if the three-factor and higher-order interactions are negligible. Under a design of resolution five, none of the main effects and two-factor interactions are aliased with one another; therefore all the main effects and two-factor interactions are estimable if the three-factor and higher-order interactions are negligible.

Let $A_i(d)$ be the number of (nonequivalent) defining words of length i of a regular s^{n-p} fractional factorial design d. Then the resolution of d is the smallest integer i such that $A_i(d) > 0$. The sequence $(A_1(d), \ldots, A_n(d))$ is called the *wordlength pattern* of d. The wordlength pattern plays an important role in design selection, which will be discussed in Chapter 10.

We say that a design is of resolution $R+$ if it is of resolution R or higher. By an s_R^{n-p} design, we mean an s^{n-p} design of resolution R.

9.5 Regular fractional factorial designs are orthogonal arrays

We will show that there is a complete factorial of $n - p$ basic factors embedded in each regular s^{n-p} fractional factorial design. A corollary of this result is that all regular fractional factorial designs are orthogonal arrays.

Theorem 9.3. *For each regular s^{n-p} fractional factorial design defined by (9.2), there exist $n - p$ factors, say the first $n - p$, such that the projection of the design onto these factors contains each of their level combinations exactly once. There are p vectors $\mathbf{b}_1, \ldots, \mathbf{b}_p \in EG(n-p, s)$ such that all the s^{n-p} treatment combinations $\mathbf{x} = (x_1, \ldots, x_n)^T$ in the design satisfy $x_{n-p+j} = \mathbf{b}_j^T \mathbf{z}$, where $\mathbf{z} = (x_1, \ldots, x_{n-p})^T, 1 \leq j \leq p$. So the levels of the last p factors are determined by those of the first $n - p$ factors.*

Proof. This theorem follows from a result in linear algebra for solving linear equations. We present the constructive proof given in Franklin and Bailey (1977) that, essentially the same as Gaussian elimination, is an algorithm for identifying basic factors from a set of independent defining words. An illustration of the algorithm and the main idea of the proof can be found in Example 9.5.

We note first that if there are p factors, say factors $i_1, \ldots, i_p$, and a basis $\{c_1, \ldots, c_p\}$ of $R(\mathbf{A})$ such that for each $1 \leq j \leq p$, among the i_1th, $\ldots$, and the i_pth entries of $\mathbf{c}_j$, only the i_jth entry is nonzero, then the conclusion of the theorem holds. This is because if $\mathbf{c}_j = (c_{1j}, \ldots, c_{nj})^T$ and $c_{i_j j}$ is the only one among $c_{i_1 j}, \ldots, c_{i_p j}$ that is nonzero, then, since $\mathbf{c}_j^T \mathbf{x} = 0$ for all $\mathbf{x}$ in the design, we have $c_{i_j j} x_{i_j} + \sum_{u \notin \{i_1, \ldots, i_p\}} c_{uj} x_u = 0$, and hence $x_{i_j} = -(c_{i_j j})^{-1} \sum_{u \notin \{i_1, \ldots, i_p\}} c_{uj} x_u$ for all $1 \leq j \leq p$. It follows that factors $i_1, \ldots, i_p$ can be defined by the other factors; so they are added factors and the other $n - p$ factors form a set of basic factors. Then since the run size is s^{n-p}, each level combination of the $n - p$ basic factors must appear exactly once.

If $\{c_1, \ldots, c_p\}$ is a set of independent defining words, then after replacing a $\mathbf{c}_i$ with $\mathbf{c}_i + b\mathbf{c}_j$, where $j \neq i$ and b is a nonzero element of $GF(s)$, we still have a set of independent defining words; this is a simple change of bases of a vector space. We now describe an algorithm that uses this fact to convert any set of independent defining words to a new set of independent defining words that satisfies the condition described in the previous paragraph if it is not already satisfied by the original set of independent defining words. Then the proof is complete.

We first set $j = 1$ and $\mathbf{c}_1 = \mathbf{a}_1, \ldots, \mathbf{c}_p = \mathbf{a}_p$, where $\mathbf{a}_1, \ldots, \mathbf{a}_p$ are a set of independent defining words. At each step $j = 1, \ldots, p$, we select an arbitrary integer $i_j \in \{1, \ldots, n\}$ such that the i_jth entry of $\mathbf{c}_j$ is nonzero. For each $l = j + 1, \ldots, p$, if the i_jth entry of $\mathbf{c}_l$ is nonzero, then we reset $\mathbf{c}_l$ to be the sum of the original $\mathbf{c}_l$ and $b\mathbf{c}_j$, where b is an element of $GF(s)$ chosen so that the i_jth entry of the new $\mathbf{c}_l$ is zero. This is possible since the i_jth entry of $\mathbf{c}_j$ is nonzero. When this is done, we reset j to be $j + 1$, and continue until $j = p + 1$.

We have now chosen p factors $i_1, \ldots, i_p$ and a set of independent defining words $\{c_1, \ldots, c_p\}$ such that for all $1 \leq j \leq p$, the i_jth entry of $\mathbf{c}_j$ is nonzero, and all the i_jth entries of $\mathbf{c}_l$ with $l > j$ are zero.

Next we start with $j = p$ and go backwards. At each step $j = p, p - 1, \ldots, 2$, for each l such that $j - 1 \geq l \geq 1$, if the i_jth entry of $\mathbf{c}_l$ is nonzero, then we reset $\mathbf{c}_l$ to be the sum of the original $\mathbf{c}_l$ and $b'\mathbf{c}_j$, where b' is an element of $GF(s)$ chosen so that the i_jth entry of the new $\mathbf{c}_l$ is zero. Again this is possible since the i_jth entry of $\mathbf{c}_j$ is nonzero. After this is done, we reset j to be $j - 1$ and continue until $j = 1$.

The algorithm thus selects a set of independent defining words $\{c_1, \ldots, c_p\}$ with the desired property that, for each $1 \leq j \leq p$, among the i_1th, $\ldots$, and the i_pth entries of $\mathbf{c}_j$, only the i_jth entry is nonzero. $\qquad\square$

The algorithm described here can be used to identify a set of basic factors from the defining relation. The factors that are not chosen by the algorithm form a set of basic factors, and those that are chosen are added factors. From the proof, we see that at each step, since $\mathbf{c}_j$ may have more than one nonzero entry, the choices of i_j may not be unique. This may lead to different choices of basic factors.

Example 9.5. Consider the 2^{6-3} design in (8.9). Among the three independent defining words $-ABD$, $-BCE$, and $ABCF$, $-ABD$ is the only word that contains D, $-BCE$ is the only word that contains E, and $ABCF$ is the only word that contains F. We can write $D = -AB$, $E = -BC$, and $F = ABC$. Thus the design can be constructed by using A, B, and C as basic factors. Referring to the notations in the proof of Theorem 9.3, we have $p = 3$, D, E, and F correspond to factors i_1, i_2, and i_3, and the defining words $-ABD$, $-BCE$, and $ABCF$ correspond to $\mathbf{c}_1, \mathbf{c}_2$, and $\mathbf{c}_3$. In general, it may not be easy to spot a set of defining words that satisfy the condition described in the beginning of the proof of Theorem 9.3. We now give a step-by-step illustration of the algorithm given in the proof. Starting from $-ABD$, $-BCE$, and $ABCF$, first we choose an arbitrary letter from the first word $-ABD$, say A. The second word $-BCE$ does not contain A, so nothing is done; but since the third word $ABCF$ contains A, we replace it with $(-ABD)(ABCF) = -CDF$ to eliminate A. Now we have a new set of independent defining words $-ABD$, $-BCE$, and $-CDF$ in which only the first word contains A. Next we choose a letter from the second defining word $-BCE$, say B. Then since the third defining word $-CDF$ does not contain B, again nothing is done. Then we choose a letter from $-CDF$, say C. At this point we go backwards to eliminate C from the second defining word $-BCE$ by replacing it with $(-BCE)(-CDF) = BDEF$. Nothing needs to be done to the first word $-ABD$ since it does not contain C. Now we have independent defining words $-ABD$, $BDEF$, and $-CDF$ in which only the first word contains A and only the third word contains C, but B appears in both the first and second words. The last step is to eliminate B from the first word $-ABD$ by replacing it with $(-ABD)(BDEF) = -AEF$. The algorithm ends up with the three independent defining words $-AEF$, $BDEF$, and $-CDF$ in which only the first word contains A, only the second word contains B, and only the third word contains C. Then $A = -EF$, $B = DEF$, and $C = -DF$; thus D, E, and F form a set of basic factors, and A, B, C can be defined by interaction contrasts of D, E, and F.

Remark 9.4. Let $\mathbf{B}$ be the $p \times (n - p)$ matrix with $\mathbf{b}_1^T, \cdots, \mathbf{b}_p^T$ as its rows, where $\mathbf{b}_1, \cdots, \mathbf{b}_p$ are as in Theorem 9.3, and let

$$\mathbf{G} = \begin{bmatrix} \mathbf{I}_{n-p} \\ \mathbf{B} \end{bmatrix}. \tag{9.12}$$

Then subject to a relabeling of the factors, the treatment combinations in the s^{n-p} fractional factorial are all the linear combinations of the columns of $\mathbf{G}$, where the first $n - p$ rows of $\mathbf{G}$ correspond to basic factors. One can choose the matrix $\mathbf{A}$ in (9.2) to be

$$\begin{bmatrix} -\mathbf{B}^T \\ \mathbf{I}_p \end{bmatrix}. \tag{9.13}$$

Then $\mathbf{A}^T \mathbf{G} = \mathbf{0}$, which is equivalent to (9.2).

In the proof of Theorem 9.3, at each step we can choose any factor that appears in the defining word c_j. Suppose $\{F_1,\ldots,F_q\}$ is a set of factors such that none of their interactions appears in the defining relation. Then when we apply the algorithm, at each step, each of the independent defining words $c_1,\ldots,c_p$ is not an interaction of $F_1,\ldots,F_q$; therefore it must contain at least one factor other than $F_1,\ldots,F_q$. It follows that at each step we can choose a factor other than $F_1,\ldots,F_q$. Since the chosen factors are added factors, we have proved the following result.

Theorem 9.4. *Suppose $F_1,\ldots,F_q$ are a set of factors such that none of their interactions appears in the defining relation of a regular s^{n-p} fractional factorial design. Then there must be a set of basic factors that contains $F_1,\ldots,F_q$, and the projection of the design onto these q factors contains each of their level combination s^{n-p-q} times.*

Theorem 9.5. *A regular s^{n-p} fractional factorial design of resolution R is an orthogonal array of strength $R-1$.*

Proof. Since the design has resolution R, there is no defining word of length shorter than R. Thus by Theorem 9.4, the projection of the design onto any $R-1$ factors must contain each of their level combinations the same number of times. □

It follows from Theorem 9.5 that for any $k \leq R-1$, every k-factor projection of a regular s_R^{n-p} design consists of s^{n-p-k} copies of a complete s^k factorial. We note that for a design of resolution R, $R-1$ is also its maximum strength. It is clear that the projection of the design onto R factors that form a defining word consists of $s^{n-p-R+1}$ copies of a $1/s$-fraction of complete s^R factorial; all the other R-factor projections, by Theorem 9.4, consist of s^{n-p-R} copies of a complete s^R factorial. Chen (1998) completely characterized k-factor projections of 2_R^{n-p} designs for all $R+1 \leq k \leq R+\lfloor(R-1)/2\rfloor$, where $\lfloor x \rfloor$ is the integer part of x; see Exercise 9.2.

By Corollaries 8.4, 8.5, and Theorem 9.5, we have the following results.

Corollary 9.6. *If there is a regular s_{III+}^{n-p} design, then $n \leq (s^{n-p}-1)/(s-1)$. For regular two-level designs of resolution IV+, we have $n \leq 2^{n-p-1}$.*

The upper bounds in Corollary 9.6 are sharp.

Theorem 9.7. *For any $N = s^k$, where $k < n$ and s is a prime number or prime power, there exists a regular $s_{III+}^{n-(n-k)}$ design with $n = (s^k-1)/(s-1)$. Such a design can be constructed by first writing down all the level combinations of k treatment factors with s-levels each and using each of the $(s^k-1)/(s-1)-k$ interaction pencils to define a new factor.*

A regular s^{n-p} design with $n = (s^{n-p}-1)/(s-1)$ is called a *saturated* regular fractional factorial design. By Theorem 9.3 all saturated regular designs can be constructed by the method described in Theorem 9.7 and hence are isomorphic. The sharpness of the bound for resolution IV two-level designs in Corollary 9.6 will be proved in Section 9.6.

Example 9.6. The regular 3^{4-2} design constructed in Example 9.2 is saturated. By Theorem 9.5, it is a saturated $OA(9,3^4,2)$ and, by Theorem 8.10, it is equivalent to two orthogonal 3×3 Latin squares.

Specializing Theorem 9.7 to the case $k = n - p = 2$, we obtain the existence of a complete set of mutually orthogonal $s \times s$ Latin squares when s is a prime or prime power. This was stated at the end of Section 8.6 and was proved in Section 8.8 by using difference matrices.

Remark 9.5. As shown in Section 6.3, the $2^k - 1$ columns of the saturated orthogonal array constructed in Corollary 8.9 constitute an orthogonal basis of $\mathbb{R}^{2^k} \ominus V_U$, representing all the main effects and interactions of a 2^k complete factorial. The construction of this design amounts to using interaction contrasts of k basic factors to define $2^k - k - 1$ added factors. Therefore the design constructed in Corollary 8.9 is a saturated *regular* fractional factorial design of size 2^k for $2^k - 1$ factors.

Remark 9.6. Theorem 9.3 has an application to the design key construction of blocked complete factorial designs discussed in Sections 7.6 and 7.7. By Remark 9.1, the principal block is a regular fractional factorial design. If, as in Section 7.6, the first $n - q$ columns of the design key matrix $\mathbf{K}$ correspond to $\mathcal{P}_1,\ldots,\mathcal{P}_{n-q}$, then these columns are generators of the principal block. By (9.12), without loss of generality, we may assume that the first $n - q$ columns of $\mathbf{K}$ are of the form

$$\begin{bmatrix} \mathbf{I}_{n-q} \\ \mathbf{B} \end{bmatrix}.$$

By the construction in Section 7.2, once the principal block is given, the whole blocking scheme is determined. In other words, the blocking scheme does not depend on the last q columns of $\mathbf{K}$. On the other hand, the columns of $\mathbf{K}$ must be linearly independent in order to generate a complete factorial. A simple choice of $\mathbf{K}$ that fulfills this requirement is

$$\begin{bmatrix} \mathbf{I}_{n-q} & \mathbf{0} \\ \mathbf{B} & \mathbf{I}_q \end{bmatrix}.$$

Clearly all the columns of this matrix are linearly independent. No treatment main effect is confounded with blocks if all the rows of $\mathbf{B}$ are nonzero. This is the template given in (7.9).

9.6 Foldovers of regular fractional factorial designs

Theorem 8.11 showed that the design obtained by applying the method of foldover to a two-level orthogonal array of even strength t is an orthogonal array of strength

$t+1$. The following is its counterpart for regular designs.

Theorem 9.8. *The design obtained by applying the foldover construction in (8.18) to a regular 2^{n-p} fractional factorial design of odd resolution R is a regular $2^{(n+1)-p}$ design of resolution $R+1$.*

Proof. Replace the two levels -1 and 1 with 0 and 1, respectively. Suppose d is the regular 2^{n-p} design that consists of all the solutions $\mathbf{x} = (x_1,\ldots,x_n)^T$ to (9.2), where $\mathbf{a}_i \in EG(n,2)$. Apply the construction in (8.18) to d, and let $\bar{d}$ be the resulting design. Each treatment combination in $\bar{d}$ is of the form $\bar{\mathbf{x}} = (x_0,x_1,\ldots,x_n)^T$, where $(x_1,\ldots,x_n)^T \in d$ and $x_0 = 0$, or $(1-x_1,\ldots,1-x_n)^T \in d$ and $x_0 = 1$. For each defining word $\mathbf{a} = (a_1,\ldots,a_n)^T$ of d, we have $\sum_{i=1}^{n} a_i x_i = 0$ for all $\mathbf{x}$ in d. Let

$$\bar{\mathbf{a}} = \begin{cases} (0,a_1,\ldots,a_n)^T, & \text{if } \mathbf{a} \text{ has an even number of nonzero entries;} \\ (1,a_1,\ldots,a_n)^T, & \text{if } \mathbf{a} \text{ has an odd number of nonzero entries.} \end{cases} \tag{9.14}$$

Then it is clear that $\bar{\mathbf{a}}^T \bar{\mathbf{x}} = 0$ for all $\bar{\mathbf{x}}$ in $\bar{d}$. Thus $\bar{d}$ is a regular design whose defining words are those in $\{\bar{\mathbf{a}} : \mathbf{a} \text{ is a defining word of } d\}$. Since R is odd, by (9.14), the resolution of $\bar{d}$ is $R+1$. □

Example 9.7. The regular 2^{6-3} design in Example 8.1 is defined by $I = -ABD = -BCE = ABCF = ACDE = -CDF = -AEF = BDEF$. The foldover construction results in a regular 2^{7-3} design $\bar{d}$ of resolution IV. According to the proof of Theorem 9.8, all the defining words of length four remain as such in $\bar{d}$, and each of the defining words of length three is supplemented by the additional factor (G). This yields the defining relation $I = ABDG = BCEG = ABCF = ACDE = CDFG = AEFG = BDEF$ for $\bar{d}$ when the two levels are represented by 1 and -1.

Corollary 9.9. *For any $N = 2^k, k \geq 3$, the design obtained by applying the foldover construction in (8.18) to a two-level saturated regular design of size $N/2$ is a regular resolution IV design with $n = N/2$.*

By Corollary 9.6, if there is a regular two-level design of resolution IV or higher, then $n \leq N/2$. Corollary 9.9 shows that this upper bound is attained by the design constructed from a saturated regular design by the foldover technique.

Remark 9.7. By Remark 9.1, if s is a prime or prime power, then blocked s^n complete factorials in blocks of size s^r such that no factorial effects involving e or fewer than e factors are confounded with blocks are equivalent to regular $s^{n-(n-r)}$ fractional factorial designs of resolution $e+1$ or higher. Then (7.4) and (7.5) follow from Corollary 9.6, Theorem 9.7, and Corollary 9.9.

The converse of Corollary 9.9, the regular design counterpart of a result stated in Remark 8.2, is also true.

Theorem 9.10. *For any $N = 2^k, k \geq 3$, every two-level regular resolution IV design with $n = N/2$ can be constructed by applying the method of foldover to a saturated regular design of size $N/2$.*

We will prove Theorem 9.10 at the end of Section 9.12. Since all saturated regular designs of the same size are isomorphic, Theorem 9.10 implies that all two-level resolution IV designs with $n = N/2$ are also isomorphic.

Theorem 9.10 is a special case of Theorem 3 of Margolin (1969). Webb (1968) conjectured that for $n = N/2$, every two-level design (not necessarily an orthogonal array) that allows for the estimation of all the main effects under the assumption that all the three-factor and higher-order interactions are negligible can be constructed by the foldover technique. This conjecture was confirmed by Margolin (1969). Theorem 9.14 in this section and Theorem 15.22 in Section 15.11 extend Theorem 9.10 in another direction for orthogonal arrays of strength three.

The result below is obvious from (9.14).

Theorem 9.11. *All the defining words of a regular design constructed by the foldover method have even lengths.*

Draper and Mitchell (1967) called regular designs in which all the defining words have even lengths *even designs*. All the regular designs constructed by the foldover method are even designs. By Theorem 9.10, we have the following result.

Corollary 9.12. *Regular two-level resolution IV designs with $n = N/2$ are even designs.*

The foldover construction (8.18) includes an additional factor that is at a constant level before being folded over. Suppose such a factor is not included. Let

$$\mathbf{X}^* = \begin{bmatrix} \mathbf{X} \\ -\mathbf{X} \end{bmatrix}, \tag{9.15}$$

where $\mathbf{X}$ is a 2^{n-p} design. Then the argument in the proof of Theorem 9.8 shows that the defining words of the $2^{n-(p-1)}$ design $\mathbf{X}^*$ are precisely those defining words of $\mathbf{X}$ that have even lengths. So the conclusion of Theorem 9.11 still holds for $\mathbf{X}^*$. The counterpart of Theorem 9.8 is that if $\mathbf{X}$ is a resolution III design, then $\mathbf{X}^*$ has resolution at least IV. For example, if $\mathbf{X}$ is a resolution III design that has no defining word of length four, then the design $\mathbf{X}^*$ in (9.15) has resolution VI or higher, but the design $\tilde{\mathbf{X}}$ with an extra factor in (8.18) has resolution IV.

Example 9.8. (Example 9.7 revisited) If the 2^{6-3} design in Example 8.1 is folded over without adding an extra factor, then we obtain a 2_{IV}^{6-2} design defined by $I = ABCF = ACDE = BDEF$.

The converse of Theorem 9.11 is also true.

Theorem 9.13. *Every regular even design d of run size N can be constructed by the foldover method in (8.18) or (9.15) and can be obtained by choosing factors (columns) from a resolution IV design with $N/2$ factors.*

The concept of even designs can be extended to nonregular designs, and Theorem 9.13 is a special case of a result that holds more generally for both regular and nonregular designs. We will prove the general result in Section 15.11. In view of Theorem 9.13, we call a two-level resolution IV design with $n = N/2$ a *maximal even design*. This design is also called a minimum resolution IV design in the literature since it has the minimum number of runs among resolution IV designs with n two-level factors.

Theorem 9.8 shows that by adding a mirror-image fraction, we can make all the main effects not aliased with two-factor interactions. This allows for de-aliasing of all the main effects if three-factor and higher-order interactions are negligible. It can also be shown that if the levels of a *single* factor, say factor A, are switched in the added fraction, then the main effect of A and all the two-factor interactions involving A can be estimated if three-factor and higher-order interactions are negligible. This allows for the de-aliasing of the main effect of a specific factor and all the two-factor interactions involving that factor (Box, Hunter, and Hunter, 2005, p. 251). Designs obtained by adding a fraction of the same size as the original design in which the levels of a subset of factors are switched were studied in Montgomery and Runger (1996), Li and Mee (2002), and Li and Lin (2003). Unless specified otherwise, we reserve the term "foldover" (or full foldover) for the case where the levels of all the factors in the added fraction are switched. When the levels of a subset of factors are reversed, it is called a "partial foldover."

While a full foldover cannot increase the resolution when it is applied to a design of even resolution, this is possible for partial foldovers. For example, if we augment the 2_{IV}^{8-2} design defined by $G = ABC$ and $H = DEF$ with 64 runs in which the levels of factors G and H are reversed, and those of the other factors are unchanged, then we obtain the resolution VIII design defined by $I = ABCDEFGH$. Partial foldover designs will be investigated in Section 11.6.

The following result extends Theorem 9.10.

Theorem 9.14. *If $N/2 \geq n > 5N/16$, where n is the number of factors, $N = 2^{n-p}$ is the run size, and $n - p \geq 4$, then a regular 2^{n-p} design of resolution at least IV must be an even design. It can be constructed by applying the foldover construction in (8.18) or (9.15) to a regular $2^{(n-1)-p}$ or $2^{n-(p+1)}$ design of resolution III, respectively.*

The foldover construction and Theorem 9.14 will be revisited from the projective geometric point of view in Section 9.12. It will be explained there how Theorem 9.14 follows from some results in the literature of finite projective geometry.

The bound $5N/16$ in Theorem 9.14 is sharp. The method of foldover is the most well known method of constructing resolution IV (and even) designs. When $n \leq 5N/16$, there are designs of resolution IV that are *not* even designs and hence cannot be constructed by the foldover technique. The structures and construction of resolution IV designs will be discussed in more detail in Chapter 11.

9.7 Construction of designs for estimating required effects

Franklin and Bailey (1977) and Franklin (1985) proposed an algorithm for selecting independent defining contrasts to construct regular designs under which certain required effects can be estimated, improving an earlier algorithm proposed by Greenfield (1976). Suppose the factorial effects are divided into three classes: effects A_i whose estimates are required, effects B_j that are not negligible but whose estimates are not required, and effects C_k that are negligible. The objective is to find a regular design of the smallest run size under which the required effects can be estimated. Franklin and Bailey (1977) considered the two-level case and Franklin (1985) extended the algorithm to the case in which the number of levels is a prime number or power of a prime number.

First we need to determine the effects that cannot appear in the defining relation. These are called *ineligible effects*. It is easy to see that they are all the required effects A_i, all generalized interactions of an A_i and an A_j, $i \neq j$, and all generalized interactions of an A_i and a B_j.

The next step is to determine an initial run size. To avoid wasteful searches, it should be as large as possible without missing possible solutions. A simple lower bound on the run size of a solution can be obtained by counting the total number of degrees of freedom associated with the required effects (Greenfield, 1976). A theoretical lower bound was derived in Franklin and Bailey (1977) and Franklin (1985). Since the defining contrast subgroup for any solution is contained in the set of eligible effects, the defining contrast subgroup for a solution of the smallest run size, say s^{n-p}, is the largest group (of size s^p) contained in the set of eligible effects. Let s^r be the size of the largest group contained in the set of ineligible effects. Then we must have $p + r \leq n$; this is because a group of eligible effects and a group of ineligible effects only share the identity I. It follows that $n - p \geq r$, so $s^{n-p} \geq s^r$.

If the initial size is set to be s^k, then a set of k factors is chosen to serve as basic factors. Independent eligible defining words are selected, one for each of the remaining (added) factors, subject to the constraint that none of the generalized interactions of the independent defining words is ineligible. If the search fails to find a suitable set of defining words, then we change to another set of basic factors. When all choices of basic factors for a given run size are exhausted without producing a suitable design, the run size is increased to the next power of the number of levels.

This is basically an exhaustive search, but Franklin and Bailey (1977) provided a theoretical result to cut unnecessary searches in certain cases. Suppose we run the algorithm without finding a solution by using a set $\mathcal{A}$ of basic factors such that none of their interactions is eligible. Then, since all the interactions of the factors in $\mathcal{A}$ cannot appear in the defining relation of any solution, by Theorem 9.4, if there were solutions of the same size, then we would be able to generate them by using the factors in $\mathcal{A}$ as basic factors. Therefore in this case there is no need to try any other set of basic factors of the same size.

Example 9.9. Suppose there are six two-level factors A, B, C, D, E, and F. We would like to construct a regular fractional factorial design of the smallest run size that allows for the estimation of all six main effects and nine specific two-factor interactions AC, AD, AF, BC, CD, CE, CF, DE, EF, assuming that all the other interactions are negligible. In this case, the set of ineligible effects consists of all 15 required effects and their pairwise products. It can be verified that these are all the main effects, all the two-factor interactions, the three-factor interactions ABC, ACD, ACE, ACF, ADE, AEF, ABD, ABF, BCD, BCE, BCF, BDE, BEF, CDE, CEF, ADF, CDF, DEF, and the four-factor interactions $ACDE$, $ACEF$, $ABCD$, $ACDF$, $ADEF$, $ABCF$, $BCDE$, $BCEF$, $CDEF$. Since there are 15 effects to be estimated, we need at least 16 runs. Alternatively, since all the factorial effects of A, B, C, and D are ineligible, the size of the largest group in the set of ineligible effects is at least 16. This also suggests 16 as an initial run size. We start with A, B, C, and D as the basic factors and set up a table of eligible effects for defining E and F, with the ineligible effects crossed out, as in Table 9.1.

Table 9.1 *Search table for Example 9.9*

	E	F
I	–	–
A	–	–
B	–	–
AB	ABE	–
C	–	–
AC	–	–
BC	–	–
ABC	$ABCE$	–
D	–	–
AD	–	–
BD	–	BDF
ABD	$ABDE$	$ABDF$
CD	–	–
ACD	–	–
BCD	–	$BCDF$
$ABCD$	$ABCDE$	$ABCDF$

We need to choose one effect from each column to define the two added factors E and F. We start with ABE and BDF, the effects at the top of the two columns. If both ABE and BDF are defining effects, then their generalized interaction $ADEF$ must also be a defining effect. Since $ADEF$ is ineligible, this choice is ruled out. Then we keep ABE and go down the second column to choose $ABDF$, which still does not lead to a solution since $(ABE)(ABDF) = DEF$ is also ineligible. The next choice, $BCDF$, however, provides a solution because $(ABE)(BCDF) = ACDEF$ is eligible. If we continue the search, we can see that there is only one other solution with A, B, C, D as basic factors, defined by $ABCE$ from the first column and BDF from the second column. Since all the interactions of A, B, C, and D are ineligible, we can stop and conclude that there are two 16-run solutions. On the other hand, if we were not able to find a solution using A, B, C, D as basic factors, then we could immediately jump to 32-run designs without searching through other choices of four basic factors.

This algorithm can also be used to search for appropriate blocking schemes of complete factorial designs to avoid confounding certain factorial effects with blocks. Constructing a blocked s^n complete factorial in s^q blocks of size s^{n-q} also requires q independent factorial effects. These effects and their generalized interactions, which constitute the defining contrast subgroup of the principal block, are confounded with blocks. The algorithm of Franklin and Bailey (1977) and Franklin (1985) can be applied in the same way, except that for constructing blocking schemes, the ineligible effects consist of those whose confounding with blocks are to be avoided. We pointed out in Section 7.7 that choosing defining effects of the principal block is equivalent to selecting $\mathbf{b}_1, \ldots, \mathbf{b}_q$ in (7.9). Therefore the algorithm presented in this section can also be applied to the search of design keys. In this case basic factors for the principal block correspond to the first $n - q$ rows of the template in (7.9).

Example 9.10. Suppose we would like to divide the 64 treatment combinations in a 2^6 complete factorial into four blocks of size 16 without confounding all the main effects and the two-factor interactions $AC, AD, AF, BC, CD, CE, CF, DE, EF$. This is equivalent to selecting a 2^{6-2} fractional factorial design such that the factorial effects $A, B, C, D, E, F, AC, AD, AF, BC, CD, CE, CF, DE$, and EF do not appear in the defining relation. There are fewer ineligible effects here than in Example 9.9, so there are many solutions. Also, using A, B, C, D as basic factors does not guarantee that all the solutions can be generated since, unlike in Example 9.9, not all the factorial effects of A, B, C, and D are ineligible.

9.8 Grouping and replacement

When $s = p^r$ where p is a prime number and $r > 1$, in constructing the saturated design in Theorem 9.7, one cannot use the usual modulo s arithmetic. In this case, a convenient alternative method of constructing an $OA(s^k, s^{(s^k-1)/(s-1)}, 2)$ is to use the ideas of *grouping* and *replacement* due to Addelman (1962a).

By Theorem 9.7, there is a saturated regular design with p^r runs and $(p^r - 1)/(p - 1)$ factors each with p levels that can be constructed by using modulo p arithmetic. Write such a design as a matrix $\mathbf{M}$ where each row represents a treatment combination, and the first r columns, say $\mathbf{x}_1, \ldots, \mathbf{x}_r$, correspond to basic factors. The other columns, corresponding to the added factors, are nonequivalent linear combinations of $\mathbf{x}_1, \ldots, \mathbf{x}_r$ representing all the interaction pencils of the basic factors. Recall that two vectors $\mathbf{x}$ and $\mathbf{y}$ are said to be nonequivalent if $\mathbf{x} \neq \lambda \mathbf{y}$ for all $\lambda \in \mathrm{GF}(p)$. Then $\mathbf{M}$ is an $\mathrm{OA}(p^r, p^{(p^r-1)/(p-1)}, 2)$. For example, for $p = r = 2$,

$$\mathbf{M} = \begin{bmatrix} 0 & 0 & 0 \\ 0 & 1 & 1 \\ 1 & 0 & 1 \\ 1 & 1 & 0 \end{bmatrix}, \tag{9.16}$$

where the third column is the sum of the first two columns modulo 2. For $p = 3$ and $r = 2$, $\mathbf{M}$ is the array in (9.4), where the four columns $\mathbf{x}_1, \mathbf{x}_2, \mathbf{x}_3, \mathbf{x}_4$ are such that $\mathbf{x}_3 = \mathbf{x}_1 + \mathbf{x}_2 \pmod 3$ and $\mathbf{x}_4 = \mathbf{x}_1 + 2\mathbf{x}_2 \pmod 3$.

One can establish a mapping between the four rows of $\mathbf{M}$ in (9.16) and the levels of a four-level factor:

$$(0,0,0) \leftrightarrow 0, \quad (0,1,1) \leftrightarrow 1, \quad (1,0,1) \leftrightarrow 2, \quad (1,1,0) \leftrightarrow 3. \tag{9.17}$$

Similarly, the nine rows of (9.4) are mapped to 0, 1, 2, 3, 4, 5, 6, 7, and 8, respectively. In general, the p^r rows of $\mathbf{M}$ can be identified with the levels of a p^r-level factor, based on which the methods of expansive and contractive replacement discussed in the paragraph preceding Example 8.7 can be applied. In the following we mainly discuss contractive replacement.

A saturated regular design with $s^k = p^{kr}$ runs and $(p^{kr} - 1)/(p - 1)$ factors each with p levels can also be constructed by using modulo p arithmetic. Display this design as a $p^{kr} \times (p^{kr} - 1)/(p - 1)$ matrix (strength two orthogonal array) $\mathbf{D}$. Suppose a certain set of $(p^r - 1)/(p - 1)$ columns of $\mathbf{D}$ consists of r linearly independent columns and all their nonzero and nonequivalent linear combinations. Then in the submatrix of $\mathbf{D}$ consisting of these columns, we can replace each row with one of the p^r levels $0, 1, \ldots, p^r - 1$ based on the aforementioned mapping. Then the corresponding p-level factors, $(p^r - 1)/(p - 1)$ in total, are replaced by one single p^r-level factor. The resulting array is a strength-two orthogonal array with one p^r-level factor and $(p^{kr} - p^r)/(p - 1)$ factors each with p levels. If there are h disjoint subsets of columns of $\mathbf{D}$ of sizes $(p^{r_1} - 1)/(p - 1), \ldots, (p^{r_h} - 1)/(p - 1)$, respectively, such that each set consists of some linearly independent columns and all their nonzero and nonequivalent linear combinations, then the corresponding p-level factors, $\sum_{i=1}^{h}(p^{r_i} - 1)/(p - 1)$ in total, can be replaced by h factors with $p^{r_1}, \ldots, p^{r_h}$ levels, respectively. These, together with the remaining p-level factors, $[p^{kr} - 1 - \sum_{i=1}^{h}(p^{r_i} - 1)]/(p - 1)$ in total, form an orthogonal array of strength two.

Example 9.11. A saturated regular 2^{15-11} design, shown in Figure 9.1, can be constructed by using four basic factors. Label the four columns corresponding to the basic factors by **1**, **2**, **3**, and **4**, respectively, and label the other 11 columns corresponding to the added factors by the interactions of basic factors $(\mathbf{12}, \mathbf{13}, \ldots, \mathbf{1234})$ used to define them. The 15 columns can be partitioned into five disjoint sets of the form $(\mathbf{a}, \mathbf{b}, \mathbf{ab})$: $(\mathbf{1}, \mathbf{2}, \mathbf{12})$, $(\mathbf{3}, \mathbf{4}, \mathbf{34})$, $(\mathbf{13}, \mathbf{24}, \mathbf{1234})$, $(\mathbf{23}, \mathbf{124}, \mathbf{134})$, $(\mathbf{14}, \mathbf{123}, \mathbf{234})$. Each of the five groups can be used to replace three two-level factors with a four-level factor by making the replacement prescribed in (9.17) within each group.

1	2	12	3	4	34	13	24	1234	23	124	134	14	123	234
0	0	0	0	0	0	0	0	0	0	0	0	0	0	0
1	0	1	0	0	0	1	0	1	0	1	1	1	1	0
0	1	1	0	0	0	0	1	1	1	1	0	0	1	1
1	1	0	0	0	0	1	1	0	1	0	1	1	0	1
0	0	0	1	0	1	1	0	1	1	0	1	0	1	1
1	0	1	1	0	1	0	0	0	1	1	0	1	0	1
0	1	1	1	0	1	1	1	0	0	1	1	0	0	0
1	1	0	1	0	1	0	1	1	0	0	0	1	1	0
0	0	0	0	1	1	0	1	1	0	1	1	1	0	1
1	0	1	0	1	1	1	1	0	0	0	0	0	1	1
0	1	1	0	1	1	0	0	0	1	0	1	1	1	0
1	1	0	0	1	1	1	0	1	1	1	0	0	0	0
0	0	0	1	1	0	1	1	0	1	1	0	1	1	0
1	0	1	1	1	0	0	1	1	1	0	1	0	0	0
0	1	1	1	1	0	1	0	1	0	0	0	1	0	1
1	1	0	1	1	0	0	0	0	0	1	1	0	1	1

Figure 9.1 *A saturated 2^{15-11} design*

This results in the saturated OA$(16, 4^5, 2)$ shown in Figure 9.2. If the replacement is not done in every set, then we can also construct OA$(16, 2^{12} \times 4, 2)$, OA$(16, 2^9 \times 4^2, 2)$, OA$(16, 2^6 \times 4^3, 2)$, and OA$(16, 2^3 \times 4^4, 2)$.

It is interesting and useful to know when the grouping as in Example 9.11 is possible. We have the following result due to André (1954).

Theorem 9.15. *For any $N = p^n$, $s = p^r$, where p is a prime number or prime power and $r < n$, the $(N-1)/(p-1)$ factors in a saturated regular p-level design of size N can be decomposed into $(N-1)/(s-1)$ disjoint groups of $(s-1)/(p-1)$ factors such that the projection of the design onto each group is a replicated saturated regular p-level design of size s if and only if n is a multiple of r.*

André (1954) established this result in the language of finite projective geometry. The projective geometric connection will be discussed in Section 9.11.

In the context of Theorem 9.7, $N = s^k$, $s = p^r$, where p is a prime number. In this case $N = p^{rk}$ and rk is a multiple of r; thus Theorem 9.15 applies, and we can use grouping and replacement to construct an OA$(s^k, s^{(s^k-1)/(s-1)}, 2)$ from the saturated regular p-level design of size s^k. Applying Theorem 9.15 to the case $p = r = 2$, we have that for each even n, the $2^n - 1$ columns of a saturated regular two-level design of size 2^n can be decomposed into $(2^n - 1)/3$ disjoint groups of size three such that any column is the sum of the other two columns in the same group. In this

1, 2, 12	3, 4, 34	13, 24, 1234	23, 124, 134	14, 123, 234
0	0	0	0	0
2	0	2	1	3
1	0	1	3	1
3	0	3	2	2
0	2	2	2	1
2	2	0	3	2
1	2	3	1	0
3	2	1	0	3
0	1	1	1	2
2	1	3	0	1
1	1	0	2	3
3	1	2	3	0
0	3	3	3	3
2	3	1	2	0
1	3	2	0	2
3	3	0	1	1

Figure 9.2 *A saturated* $OA(16, 4^5, 2)$ *constructed from the design in Figure 9.1*

case, one can construct $OA(2^n, 2^{k_1} \times 4^{k_2}, 2)$ for all $k_2 = 0, \ldots, (2^n - 1)/3$ such that $k_1 + 3k_2 = 2^n - 1$. For an odd n, Theorem 9.15 does not apply, and one can only construct up to $(2^n - 5)/3$ disjoint groups, thereby obtaining $OA(2^n, 2^{k_1} \times 4^{k_2}, 2)$ for all $k_2 = 0, \ldots, (2^n - 5)/3$ such that $k_1 + 3k_2 = 2^n - 1$. The upper bound $(2^n - 5)/3$ was mentioned in Addelman (1962a). Wu (1989) proposed an algorithm for grouping the factors to construct $OA(2^n, 2^{k_1} \times 4^{k_2}, 2)$'s with the largest numbers of four-level factors. In general, if n is not a multiple of r, then the full decomposition stated in Theorem 9.15 does not hold, and it is not possible to use up all the p-level factors in the grouping and replacement construction. In this case, we have the following result from Wu, Zhang, and Wang (1992).

Theorem 9.16. *Suppose* $N = p^n$, $s = p^r$, *where* p *is a prime number, and* $n = br + q, 0 < q < r < n$. *Then, under a saturated regular* p-level design of size N, there exist $\frac{p^n - p^{r+q}}{p^r - 1} + 1$ disjoint groups of $(s-1)/(p-1)$ factors such that the projection of the design onto each group is a replicated saturated regular p-level design of size s.

For the case $p = r = 2$ and n is odd, we have $\frac{p^n - p^{r+q}}{p^r - 1} + 1 = (2^n - 5)/3$. Wu, Zhang, and Wang (1992) used Theorem 9.16 to construct orthogonal arrays of the form $OA(s^n, s^u \times (s^{r_1})^{k_1} \times \cdots \times (s^{r_h})^{k_h}, 2)$. The construction of such orthogonal arrays was also studied in Hedayat, Pu, and Stufken (1992) using difference matrices.

The results in this section have applications in the design of factorial experiments with multiple processing stages, to be discussed in Section 13.11.

9.9 Connection with linear codes

Mathematically, fractional factorial designs are equivalent to codes, even though the experimental designers and coding theorists have different objectives and are interested in solving different problems. Such connections were studied by Bose (1961). As we will see later, some results and tools from coding theory are useful for studying the properties, structures, and construction of fractional factorial designs. We review their connections here. Readers are referred to MacWilliams and Sloane (1977) for basic concepts and results of algebraic coding theory.

Given a set S of s symbols, each n-tuple of the symbols is called a *codeword* of length n over an alphabet of size s, and a set C of N codewords is called a *code* of size N. The *Hamming distance* of two codewords $\mathbf{x} = (x_1, \ldots, x_n)$ and $\mathbf{y} = (y_1, \ldots, y_n)$, denoted by dist$(\mathbf{x}, \mathbf{y})$, is the number of components where they differ. The *minimum distance* of a code is defined as $\delta = \min \text{dist}(\mathbf{x}, \mathbf{y})$, where the minimization is over all pairs of distinct codewords. The ability of a code to correct errors is reflected by its minimum distance. For example, a code with minimum distance δ can correct $\lfloor (\delta - 1)/2 \rfloor$ errors. Another useful distance-related quantity is the covering radius. The covering radius of a code C is defined as the smallest integer r such that each codeword is within Hamming distance r of some codeword in C.

For a code C and a codeword $\mathbf{x} \in C$, let $W_i(\mathbf{x})$ be the number of codewords $\mathbf{y} \in C$ such that dist$(\mathbf{x}, \mathbf{y}) = i$, $0 \leq i \leq n$. Let

$$W_i(C) = \frac{1}{N} \sum_{\mathbf{x} \in C} W_i(\mathbf{x}).$$

Then the $(n+1)$-tuple $(W_0(C), \ldots, W_n(C))$ is called the *distance distribution* of C.

A code C is called a *linear code* if S is a finite field and C is a subspace of $S^n = EG(n, s)$. In this case, let the *Hamming weight* $w(\mathbf{x})$ of a codeword $\mathbf{x}$ be the number of nonzero components of $\mathbf{x}$. Then dist$(\mathbf{x}, \mathbf{y}) = w(\mathbf{x} - \mathbf{y})$ for any codewords $\mathbf{x}$ and $\mathbf{y}$. For a linear code, because of its group structure, it is easy to see that the minimum distance is equal to the minimum weight of the nonzero codewords; furthermore let $W_i(C)$ be the number of codewords $\mathbf{x}$ with $w(\mathbf{x}) = i$, then $W_i(\mathbf{x}) = W_i(C)$ for all $\mathbf{x} \in C$. We also refer to the distance distribution of a linear code as its *weight distribution*.

Each codeword can be considered as a treatment combination, and vice versa. Then each fractional factorial design is a code and regular fractional factorial designs are linear codes. The defining contrast subgroup of a regular fractional factorial design d, with the vectors in the group considered as codewords, is also a linear code and is called the *dual code* of d. The dual code of a linear code C is denoted by $C^{\perp}$.

The transposes of the matrices in (9.12) and (9.13) are, respectively, the generator matrix and parity check matrix of a linear code; see (4.5) and (4.8) of Hedayat, Sloane, and Stufken (1999).

We have that the resolution of a regular s^{n-p} fractional factorial design d is equal to the minimum distance of its dual code $d^{\perp}$, and the wordlength pattern of d is essentially the same as the weight distribution of $d^{\perp}$:

$$W_i\left(d^{\perp}\right) = (s - 1)A_i(d). \tag{9.18}$$

Note that the presence of the multiple $s-1$ in (9.18) is due to that all defining words that are multiples of each other are counted as one in the wordlength pattern since they define the same pencil of $(n-1)$-flats.

The *MacWilliams identities* (MacWilliams, 1963) in coding theory connect the weight distribution of a linear code to that of its dual code. We have

$$W_i(d) = s^{-p} \sum_{j=0}^{n} W_j\left(d^{\perp}\right) P_i(j;n,s) \text{ for } 0 \leq i \leq n, \tag{9.19}$$

and

$$W_i\left(d^{\perp}\right) = s^{p-n} \sum_{j=0}^{n} W_j(d) P_i(j;n,s) \text{ for } 0 \leq i \leq n, \tag{9.20}$$

where $P_i(x;n,s) = \sum_{j=0}^{i}(-1)^j(s-1)^{i-j}\binom{x}{j}\binom{n-x}{i-j}$, $0 \leq i \leq n$, are the Krawtchouk polynomials. Note that

$$\binom{x}{k} = \begin{cases} x(x-1)\cdots(x-k+1)/k!, & \text{if } k \text{ is positive;} \\ 1, & \text{if } k = 0; \\ 0, & \text{if } k \text{ is negative.} \end{cases}$$

9.10   Factor representation and labeling

Each regular design of resolution III+ can be obtained by choosing factors from a saturated two-level design. We present a convenient way to label the factors in saturated regular designs that is useful for describing and cataloging designs.

Example 9.12. Represent the two levels by 0 and 1. By Remark 9.4, the eight treatment combinations of a saturated regular 2^{7-4} design can be obtained as linear combinations of the row vectors of the following 3×7 array, which is obtained by rearranging the columns of the transpose of the matrix $\mathbf{G}$ in (9.12). The three rows of the array from a basis of the saturated design.

$$
\begin{array}{ccccccc}
\mathbf{1} & \mathbf{2} & \mathbf{12} & \mathbf{3} & \mathbf{13} & \mathbf{23} & \mathbf{123} \\
1 & 0 & 1 & 0 & 1 & 0 & 1 \\
0 & 1 & 1 & 0 & 0 & 1 & 1 \\
0 & 0 & 0 & 1 & 1 & 1 & 1 \\
1 & 2 & 3 & 4 & 5 & 6 & 7
\end{array}
\tag{9.21}
$$

Therefore it is sufficient to use (9.21) to represent a regular 2^{7-4} design. The seven columns in (9.21) are all the 3×1 nonzero vectors of 0's and 1's and are also binary representations of the integers 1, 2, 3, 4, 5, 6, and 7, shown under the array. The three columns that are binary representations of 1, 2, and 4 (2^0, 2^1, and 2^2, respectively) are linearly independent and hence can be made to correspond to a set of basic factors for constructing eight-run designs. The integers 3, 5, 6, and 7 are all possible sums of 1, 2, and 4 (1+2, 1+4, 2+4, and 1+2+4, respectively). Thus the columns corresponding to 3, 5, 6, and 7 are all possible sums of those corresponding to 1, 2, and 4. If columns 1, 2, 4 are made to correspond to basic factors, then the other columns correspond to added factors. Suppose we label the factors corresponding to columns 1, 2, and 4 by **1**, **2**, and **3**, respectively. Then the columns corresponding to columns 3, 5, 6, and 7 can be labeled **12**, **13**, **23**, and **123**, respectively. These are the labels shown at the top of the array in (9.21). We point out that the seven columns are then in the Yates order **1**, **2**, **12**, **3**, **13**, **23**, and **123**. If we choose columns 3 (**12**), 6 (**23**), and 7 (**123**) in addition to the three "basic" columns **1**, **2**, and **3**, and label the three added factors by **4**, **5**, and **6**, respectively, then their main effects are aliased with interactions **12**, **23**, and **123**, respectively. All the linear combinations of the row vectors of the resulting 3×6 matrix yield the regular 2^{6-3} design in Example 8.1, subject to permutations of rows and columns, and relabeling of factor levels.

In general, the factors in saturated regular designs can be represented in the array

1	**2**	**12**	**3**	**13**	**23**	**123**	**4**	**14**	**24**	**124**	**34**	**134**	**234**	**1234**	**5**
1	0	1	0	1	0	1	0	1	0	1	0	1	0	1	0
0	1	1	0	0	1	1	0	0	1	1	0	0	1	1	0
0	0	0	1	1	1	1	0	0	0	0	1	1	1	1	0	$\cdots$
0	0	0	0	0	0	0	1	1	1	1	1	1	1	1	0
0	0	0	0	0	0	0	0	0	0	0	0	0	0	0	1

$$\vdots \qquad\qquad \vdots \qquad\qquad \vdots$$

| 1 | 2 | 3 | 4 | 5 | 6 | 7 | 8 | 9 | 10 | 11 | 12 | 13 | 14 | 15 | 16 | $\cdots$ |

$$(9.22)$$

The columns here are binary representations of the positive integers, with the column corresponding to 2^{i-1} labeled as **i**. All the other columns are sums of some of these "basic" columns and are labeled accordingly. These labels appear at the top of the array in (9.22), and it can be seen that the columns appear in a Yates order. The saturated regular design of size 2^m is represented by the submatrix consisting of the first m rows and the first $2^m - 1$ columns of the array, and its treatment combinations are all the linear combinations of the m rows. A regular $2^{n-(n-m)}$ design can be constructed by choosing n of the first $2^m - 1$ columns. To generate 2^m distinct treatment combinations, m of the n columns must be linearly independent. This is assured if the m basic columns (those corresponding to $2^0, 2^1, \ldots, 2^{m-1}$) are always included. The factors corresponding to these columns are basic factors, and those corresponding to the other $n - m$ columns are added factors. The defining relation is determined accordingly, with one independent defining word obtained from each added factor. This construction is equivalent to the method discussed in Section 9.5. When all the $2^m - 1$ columns are chosen, the resulting design is a saturated design. Each design can be described by the set of integers whose corresponding columns are chosen. This provides a convenient way to describe designs and is what Chen, Sun and Wu (1993) used in their catalogs of 32- and 64-run two-level designs.

We note that the 2^{m-1} factors in a maximal even design of size 2^m correspond to the last 2^{m-1} of the first $2^m - 1$ columns of the array in (9.22), the binary representations of the integers $2^{m-1}, 2^{m-1} + 1, \ldots, 2^m - 1$. Suppose d is a regular $2^{n-(n-m+1)}$ design with factors corresponding to the integers $a_1, \ldots, a_n, a_i \leq 2^{m-1} - 1$ for all i, then the $n + 1$ factors in the design obtained from d by the foldover method (8.18) correspond to the columns that are binary representations of $2^{m-1}, a_1 + 2^{m-1}, \ldots, a_n + 2^{m-1}$. If the foldover construction is as in (9.15) without the additional factor, then the factors in the resulting design correspond to those that are binary representations of $a_1 + 2^{m-1}, \ldots, a_n + 2^{m-1}$.

We justify the observations made in this section in Sections 9.11 and 9.12, where connections with finite projective geometry are discussed.

9.11   Connection with finite projective geometry[*]

A brief introduction to finite projective geometry is given in Section A.5. A k-dimensional Euclidean geometry $EG(k,s)$ consists of all the $k \times 1$ vectors $(x_1,\ldots,x_k)^T$, with each $x_i \in GF(s)$. A $(k-1)$-dimensional projective geometry $PG(k-1,s)$ can be constructed from $EG(k,s)$ as follows. Take the $s^k - 1$ nonzero vectors in $EG(k,s)$, and consider any two such vectors $\mathbf{x}$ and $\mathbf{y}$ as representing the same point of $PG(k-1,s)$ if and only if $\mathbf{x} = \lambda \mathbf{y}$ for some nonzero $\lambda \in GF(s)$. Then we obtain the $(s^k - 1)/(s-1)$ points in $PG(k-1,s)$. Given any m points in $PG(k-1,s)$ represented by linearly independent vectors $\mathbf{a}_1, \ldots, \mathbf{a}_m$, the points represented by all the nonzero linear combinations of $\mathbf{a}_1, \ldots, \mathbf{a}_m$, $(s^m - 1)/(s-1)$ in total, are said to constitute an $(m-1)$-flat. In particular, for $s = 2$, a $PG(k-1,2)$ has $2^k - 1$ points, which can be identified with all the $k \times 1$ nonzero vectors of 0's and 1's, and each $(m-1)$-flat consists of all the nonzero linear combinations of m linearly independent points. In this case, each 1-flat (a line) consists of three points of the form $\mathbf{a}$, $\mathbf{b}$, and $\mathbf{a}+\mathbf{b}$. We say that the three points are collinear.

Throughout this section we write each treatment combination as a row vector. A regular s^{n-p} fractional factorial design defined by (9.2) can be considered as an $(n-p)$-dimensional subspace of $EG(n,s)$. Pick a basis of this subspace, and let $\mathbf{V}$ be the $(n-p) \times n$ matrix with each treatment combination in the basis as a row vector. Then $\mathbf{V}$ has rank $n-p$ and the rows of $\mathbf{V}$ are orthogonal to $R(\mathbf{A})$. The design given by (9.2) consists of all the treatment combinations $\mathbf{x}$ that are linear combinations of the rows of $\mathbf{V}$. Such a design is of resolution three or higher if and only if there are no two columns $\mathbf{x}$ and $\mathbf{y}$ of $\mathbf{V}$ such that $\mathbf{x} = \lambda \mathbf{y}$ for some nonzero λ in $GF(s)$; this is to ensure that no two main effects are aliased with each other. Then the n columns of $\mathbf{V}$ can be thought of as n distinct points of the finite projective geometry $PG(n-p-1,s)$.

Thus the construction of a regular s^{n-p} design of resolution three or higher is equivalent to the selection of n distinct points of $PG(n-p-1, s)$ such that the $(n-p) \times n$ matrix $\mathbf{V}$ with these points as its columns has full row rank $n-p$. The rank condition is to guarantee that s^{n-p} distinct treatment combinations are generated. This geometric interpretation due to Bose (1947) is very useful for the study of regular fractional factorial designs.

We point out that $g \equiv (s^{n-p} - 1)/(s-1)$, the number of points in $PG(n-p-1,s)$, is equal to the number of factors in a saturated regular s^{n-p} design, and a saturated design is obtained when all the g points in $PG(n-p-1,s)$ are chosen. It is also equal to the number of alias sets under a regular s^{n-p} design. A connection in this regard can be made as follows. A nonzero vector $\mathbf{b} \in EG(n,s)$ appears in the defining contrast subgroup $R(\mathbf{A})$ if and only if $\mathbf{Vb} = \mathbf{0}$, and two distinct pencils $P(\mathbf{b}_1)$ and $P(\mathbf{b}_2)$, neither of which appears in the defining relation, are aliased with each other if and only if

$$\mathbf{V}(\mathbf{b}_1 - \lambda \mathbf{b}_2) = \mathbf{0} \tag{9.23}$$

for some $\lambda (\neq 0) \in GF(s)$. The following one-to-one correspondence between the g alias sets under a regular s^{n-p} design and the g points in $PG(n-p-1, s)$ can be established. For all the pencils $P(\mathbf{b})$ in an alias set, by (9.23), the $\mathbf{Vb}$'s represent the same point in $PG(n-p-1,s)$. We say that the alias set corresponds to the unique point so determined. Each of the n points chosen to construct the design corresponds to one factor and can also be used to represent the alias set containing the main effect of that factor.

For two-level designs, in (9.23), let the ith and jth entries of $\mathbf{b}_1$ and the kth entry of $\mathbf{b}_2$ be 1, and all the other entries be zero. Then (9.23) implies that the interaction of factors i and j is aliased with the main effect of factor k if and only if $\mathbf{a}_i + \mathbf{a}_j = \mathbf{a}_k$, where $\mathbf{a}_i$, $\mathbf{a}_j$, and $\mathbf{a}_k$ are the columns of $\mathbf{V}$ corresponding to factors i, j, and k, respectively. This is equivalent to that the points corresponding to factors i, j, and k are collinear. Thus a regular 2^{n-p} design has resolution four or higher if and only if the set of n distinct points of $PG(n-p-1,2)$ chosen to construct the design contains no lines. A set of points in $PG(n-p-1,2)$ that contains no lines is called a *cap* in the literature of projective geometry.

Example 9.13. (Example 9.12 revisited) Let $n-p = 3$. With each point expressed as a column, the seven points in $PG(2,2)$ are those displayed in (9.21). If, along with the basic columns 1, 2, and 4, columns 3, 6, and 7 are chosen to construct a 2^{6-3} design, then the design does not have resolution four since, e.g., the sum of columns 1 and 2 is equal to column 3. Geometrically, the three corresponding points are collinear.

In Theorem 9.15, the factors in a saturated regular p-level design of size p^n are partitioned into disjoint groups such that the projection of the design onto each group is a replicated saturated regular p-level design of size p^r. The factors of the saturated regular p-level design of size p^n correspond to all the points in a $PG(n-1,p)$, while those in the same group correspond to the points in an $(r-1)$-flat. Thus the partition of the factors in Theorem 9.15 is equivalent to that of a $PG(n-1,p)$ into disjoint $(r-1)$-flats. A set of $(r-1)$-flats that forms a partition of the points in a $PG(n-1,p)$ is called an $(r-1)$-*spread*. Theorem 9.15 shows that an $(r-1)$-spread exists if and only if n is a multiple of r. When n is not a multiple of r, there is no $(r-1)$-spread. In this case, a set of disjoint $(r-1)$-flats is called a *partial* $(r-1)$-*spread*. Theorem 9.16 provides a large partial $(r-1)$-spread. Some upper bounds on the sizes of partial $(r-1)$-spreads can be found in Govaerts (2005).

9.12 Foldover and even designs revisited*

We have shown in Theorem 9.11 that all regular designs constructed by the foldover method are even designs. This can also be seen from a geometric point of view.

The array in (9.22) displays a nested sequence of projective geometries with increasing dimensions. The first three columns, when each is considered as a 2×1 vector with the extra zeros eliminated, are the three points of a PG(1,2). When each of these columns is considered as a 3×1 vector, with one zero coordinate appended at the end, they constitute a 1-flat in a PG(2,2). Similarly, the first seven columns in (9.22) constitute a PG(2,2) if each is considered as a 3×1 vector, and are points of a 2-flat in a PG(3,2) if they are considered as 4×1 vectors.

In general, let Σ_{m-1} (respectively, Σ_m) be the set consisting of the first $2^{m-1} - 1$ (respectively, $2^m - 1$) columns, all of which are considered as $m \times 1$ vectors. Then Σ_m is a PG($m-1, 2$) and Σ_{m-1} is an ($m-2$)-flat in Σ_m. Since Σ_{m-1} is itself a projective geometry, we have

$$\mathbf{x}, \mathbf{y} \in \Sigma_{m-1} \Rightarrow \mathbf{x} + \mathbf{y} \in \Sigma_{m-1}. \tag{9.24}$$

Let S be the complement of Σ_{m-1} in Σ_m (columns 2^{m-1} through $2^m - 1$). The last coordinate of each point in Σ_{m-1} is zero, while the last coordinate of each point in S is 1. Let $\mathbf{b}$ be any point in S. Then for any $\mathbf{x} \in \Sigma_{m-1}$, we must have $\mathbf{b} + \mathbf{x} \in S$; for if $\mathbf{b} + \mathbf{x} \in \Sigma_{m-1}$, then by (9.24), $\mathbf{b} = (\mathbf{b} + \mathbf{x}) + \mathbf{x} \in \Sigma_{m-1}$, a contradiction. Therefore

$$S = \{\mathbf{b}\} \cup \left(\mathbf{b} + \Sigma_{m-1}\right) \text{ for all } \mathbf{b} \in S.$$

In particular, let $\mathbf{a}$ be column 2^{m-1}, which has all its first $m-1$ coordinates equal to zero and the mth coordinate equal to 1. Then

$$S = \{\mathbf{a}\} \cup \left(\mathbf{a} + \Sigma_{m-1}\right). \tag{9.25}$$

It follows from (9.24) and (9.25) that if $\mathbf{x}, \mathbf{y} \in S$, then $\mathbf{x} + \mathbf{y} \in \Sigma_{m-1}$. In other words, S does not contain a line and every subset of S is a cap. In fact, the sum of an even number of points from S must be in Σ_{m-1}. Therefore, if all the n points chosen to construct a fractional factorial design are from S, then the resulting design is an even design and is of resolution at least IV.

Suppose all the points in S are chosen. Consider the $m \times 2^{m-1}$ matrix $\mathbf{M}$ obtained by writing down each point in S as a column. The linear combinations of the first $m-1$ rows of $\mathbf{M}$ are precisely the treatment combinations in a saturated resolution III design of size 2^{m-1} supplemented by a factor always at level 0. Since the last row of $\mathbf{M}$ has all the entries equal to 1, adding this row to all the linear combinations of the first $m-1$ rows of $\mathbf{M}$ amounts to switching the two levels. It follows that the design corresponding to S can be obtained by applying the foldover construction (8.18) to a saturated design, so it is a maximal even design. For instance, the design in (8.10) can be constructed by taking all the linear combinations of the first four rows of columns 8–15 of the array in (9.22).

If $\mathbf{a}$ is chosen but not all the points in $\mathbf{a} + \Sigma_{m-1}$ are chosen, then the design can be constructed from an unsaturated design according to (8.18). If $\mathbf{a}$ is not chosen, then the design is of the form (9.15), where $\mathbf{X}$ is a regular $2^{n-(n-m+1)}$ design.

We summarize the above discussion as follows.

Theorem 9.17. *Choosing n points from the complement of an $(n-p-2)$-flat in $PG(n-p-1,2)$ is equivalent to applying the foldover construction (8.18) or (9.15) to a regular $2^{(n-1)-p}$ or $2^{n-(p+1)}$ design, respectively.*

Subsets of the complement of an $(m-2)$-flat in $PG(m-1,2)$ are caps. But they are not the only caps, and there are resolution IV designs that cannot be constructed by the foldover method. Structures of caps were studied, e.g., in Davydov and Tombak (1990), Bruren, Haddad, and Wehlau (1998), and Bruen and Wehlau (1999). A result due to Davydov and Tombak (1990), quoted as Theorem 2.6 in Bruen, Haddad, and Wehlau (1998), implies that if $n > 5 \cdot 2^m/16$, then a cap in $PG(m-1,2)$ containing n points must be a subset of the complement of an $(m-2)$-flat, that is, a subset of the S in (9.25). This leads to Theorem 9.14.

We give a proof of Theorem 9.10 using the projective geometric approach.

Let $N = 2^{n-p}$ and $m = n - p$. We would like to show that every regular two-level resolution IV design with $n = N/2$ can be obtained by applying the foldover construction (8.18) to a saturated regular design. Constructing the design is equivalent to choosing a subset T of 2^{m-1} points from Σ_m. Since the resolution is four, T is a cap. Thus,

$$\mathbf{a}, \mathbf{b} \in T \Rightarrow \mathbf{a} + \mathbf{b} \notin T. \tag{9.26}$$

Let the complement of T in Σ_m be $\overline{T}$. By the discussions that lead to Theorem 9.17, it suffices to show that $\overline{T}$ is an $(m-2)$-flat. Since $\overline{T}$ has the same number of points, $2^{m-1} - 1$, as an $(m-2)$-flat, we only need to show that if $\mathbf{x}, \mathbf{y} \in \overline{T}$ and $\mathbf{x} \neq \mathbf{y}$, then $\mathbf{x} + \mathbf{y} \in \overline{T}$. Choose an arbitrary point $\mathbf{a} \in T$, and let the other $2^{m-1} - 1$ points in T be $\mathbf{b}_1, \ldots, \mathbf{b}_{2^{m-1}-1}$. By (9.26), $\overline{T} = \{\mathbf{a} + \mathbf{b}_1, \ldots, \mathbf{a} + \mathbf{b}_{2^{m-1}-1}\}$. Now for any two points $\mathbf{a} + \mathbf{b}_i$ and $\mathbf{a} + \mathbf{b}_j$ in $\overline{T}$, since $\mathbf{b}_i, \mathbf{b}_j \in T$, by (9.26), we have $(\mathbf{a} + \mathbf{b}_i) + (\mathbf{a} + \mathbf{b}_j) = \mathbf{b}_i + \mathbf{b}_j \in \overline{T}$. □

Exercises

9.1 Suppose there are eight factors A, B, C, D, E, F, G, and H for which all the interactions involving more than two factors are negligible. Construct a smallest possible regular fractional factorial design such that all the main effects and the following two-factor interactions are estimable: $AE, AH, BE, BH, CE, CH, DE, DH$.

9.2 Let d be a regular 2_R^{n-p} design. Show that for $R+1 \leq k \leq R + \lfloor (R-1)/2 \rfloor$, any k-factor projection of d either has no defining word or exactly one defining word. Such a projection consists of either 2^{n-p-k} copies of a complete 2^k factorial or $2^{n-p-k+1}$ copies of a half-replicate of a 2^k factorial. More specifically, show that for $j = R, R+1, \ldots, k$, $\binom{n-j}{k-j}A_j(d)$ of the k-factor projections of d consist of $2^{n-p-k+1}$ copies of a half-replicate of a 2^k factorial of resolution j, and

$\binom{n}{k} - \sum_{j=R}^{k} \binom{n-j}{k-j} A_j(d)$ projections consist of 2^{n-p-k} copies of a complete 2^k factorial. [Chen (1998)]

9.3 Construct an OA$(16, 2^6 \times 4^3, 2)$.

9.4 Suppose d_1 and d_2 are, respectively, regular s^{n-p} and s^{n-q} fractions of the same treatment factors, and $n = p + q$. Show that if $d_1 \cap d_2 = \{0\}$, then there exists a set A_1 of basic factors for d_1 and a set A_2 of basic factors for d_2 such that $A_1 \cap A_2 = \emptyset$.

9.5 For the design key construction of a complete s^n factorial design in s^p rows and s^{n-p} columns, where s is a prime number or a prime power, use the result in Exercise 9.4 to show that, subject to factor relabeling, we may assume that the design key matrix $\mathbf{K}$ is of the form in (7.11). Furthermore, a set of independent treatment factorial effects that are confounded with rows (respectively, columns) can be identified from the first p (last $n - p$) rows of $\mathbf{K}$.

9.6 Add to a regular design of resolution III a fraction of the same size in which the levels of a specific factor A are switched and those of the other factors are unchanged. Show that under the combined design, all the two-factor interactions involving A are not aliased with any main effect or any other two-factor interaction, and that the main effect of A is not aliased with any two-factor or three-factor interaction. State and prove a similar result for orthogonal arrays of strength two.

9.7 Use Theorem 8.6 to show that in an s-level saturated regular design of size s^k that contains the treatment combination with all the factors at level 0, all the nonzero treatment combinations have constant Hamming weight s^{k-1}.

Chapter 10

Minimum Aberration and Related Criteria

A regular fractional factorial design is determined by its defining effects. Different choices of defining effects may lead to designs with different statistical properties. In this chapter, we address the issue of selecting defining effects. When the experimenter has prior knowledge about the relative importance of certain effects, the algorithm proposed by Franklin and Bailey (1977) and Franklin (1985), as discussed in Section 9.7, can be used to select defining effects for constructing designs under which certain required effects are estimable. Here we concentrate on the situation where prior knowledge is diffuse concerning the possible greater importance of factorial effects. Minimum aberration is an established criterion for design selection under the hierarchical assumption that lower-order effects are more important than higher-order effects and effects of the same order are equally important. We discuss some basic properties of this criterion, including a statistical justification showing that it is a good surrogate for maximum estimation capacity under model uncertainty. We also discuss connections to coding theory and finite projective geometry, and present some results on the characterization and construction of minimum aberration designs. One application of coding theory is the equivalence of minimum aberration to a criterion called minimum moment aberration. This equivalence provides a useful tool for determining minimum aberration designs. We end the chapter by presenting a Bayesian approach to factorial designs.

10.1 Minimum aberration

When little information about the importance of the factorial effects is available a priori, it is desirable to have a design with good all-around (model robust) properties. Under the hierarchical assumption, it is desirable to minimize the aliasing among the more important lower-order effects. From this point of view, one may prefer a design that has maximum possible resolution. However, typically many designs have the same resolution. Fries and Hunter (1980) proposed the criterion of *minimum aberration* for further discriminating between designs of the same resolution.

Consider, e.g., the 2^{7-2} designs with defining relations as indicated:

$$d_1 : I = CDFG = ABDEF = \hat{A}BCEG,$$

$$d_2 : I = ABEF = ACDG = BCDEFG.$$

Both d_1 and d_2 are of resolution IV. So if there are no three-factor or higher-order interactions, then all the main effects are estimable under both designs. In this case, one may prefer a design with as much information about two-factor interactions as possible. Under d_1, the defining word $CDFG$ yields three pairs of aliased two-factor interactions: $\{CD, FG\}, \{CF, DG\}$, and $\{CG, DF\}$. Since $CDFG$ is the only defining word of length four, there is no aliasing among the other two-factor interactions under d_1. On the other hand, d_2 has two defining words of length four, each producing three pairs of aliased two-factor interactions. Therefore the aliasing of two-factor interactions under d_2 is more severe. From this point of view, d_1 is better.

The minimum aberration criterion is based on the wordlength patterns defined in Section 9.4.

Definition 10.1. For any two s^{n-p} fractional factorial designs d_1 and d_2, if r is the smallest positive integer such that $A_r(d_1) \neq A_r(d_2)$, then d_1 is said to have less aberration than d_2 if $A_r(d_1) < A_r(d_2)$. A design d is said to have minimum aberration if no other design has less aberration than d.

According to this definition, a minimum aberration design *sequentially minimizes* $A_1(d), \ldots, A_n(d)$. Intuitively, this attempts to minimize aliasing among the more important lower-order effects. It is clear that a minimum aberration design has maximum resolution.

Chen and Hedayat (1996) introduced a weaker version of minimum aberration. A regular design of resolution R is said to have *weak minimum aberration* if it has maximum resolution and minimizes the number of defining words of length R among all the designs of resolution R.

Extension of the minimum aberration criterion to nonregular designs will be discussed in Chapter 15.

10.2   Clear two-factor interactions

Wu and Chen (1992) defined *clear* main effects and two-factor interactions as those that are not aliased with other main effects and two-factor interactions. They also called the two-factor interactions that are not aliased with main effects *eligible*. (Note that this is a different notion of eligibility from that in Franklin and Bailey's (1977) algorithm; see Section 9.7.) Under the assumption that three-factor and higher-order interactions are negligible, when a contrast associated with an alias set containing a clear main effect or clear two-factor interaction is significant, one can attribute it to the clear effect, and there is no need for de-aliasing.

Under a resolution IV design, all the main effects are clear. In this case, it may be desirable to have as many clear two-factor interactions as possible. Since the criterion of minimum aberration is meant to minimize aliasing among two-factor interactions, it is natural to ask whether there is a connection between having a large number of clear two-factor interactions and low aberration. A counterexample was presented in

Chen, Sun, and Wu (1993). The 2^{9-4} design defined by the four independent defining words $ABCF$, $ACDG$, $ADEFJ$, and $ABDEH$ has minimum aberration, where the nine factors are A, B, C, D, E, F, G, H, and J. It is of resolution IV and has eight clear two-factor interactions, which are all the two-factor interactions involving E. On the other hand, the resolution IV design defined by the independent defining words $ABCF$, $BCDH$, $ACDG$, and $ABDEJ$, although not a minimum aberration design, has 15 clear two-factor interactions: all the two-factor interactions involving one or both of E and J. There is even a resolution *three* 2^{9-4} design that also has 15 clear two-factor interactions.

In fact, there are many examples of this kind. This seems to contradict the expectation that under a minimum aberration design more two-factor interactions can be estimated. In order to address this issue, we need to have a better understanding of the statistical meaning of the combinatorial criterion of minimum aberration, in particular its implications on the aliasing of factorial effects. We study the relation between the wordlength pattern of a design and the aliasing pattern of the factorial effects in the next section, and revisit the issue of clear two-factor interactions in Sections 10.4, 10.8, and 11.7.

10.3   Interpreting minimum aberration

For simplicity, we consider two-level designs and assume that three-factor and higher-order interactions are negligible. We also require the designs to be of resolution III or higher so that no main effect is aliased with another main effect. To evaluate the performance of such a design, it is useful to know how the two-factor interactions are aliased with main effects and how they are aliased among themselves. We obtain this information via counting the number of two-factor interactions in each alias set. These counts tell us how the two-factor interactions are distributed over the alias sets and are useful design characteristics in their own right. The main result in this section shows that A_3 and A_4 can be expressed explicitly in terms of these counts, leading to a simple statistical interpretation of the minimum aberration criterion. Most of the results in this and the next sections are from Cheng, Steinberg, and Sun (1999).

Under a regular 2^{n-p} design d, the defining contrast subgroup contains $2^p - 1$ factorial effects. The remaining effects are partitioned into $g = 2^{n-p} - 1$ alias sets each of size 2^p. If the design is of resolution III or higher, then the main effects appear in n different alias sets. Let $f = g - n$. Without loss of generality, suppose the first f alias sets do not contain main effects, and each of the last n alias sets contains one main effect. For $1 \leq i \leq g$, let $m_i(d)$ be the number of two-factor interactions in the ith alias set. We note that a two-factor interaction is clear if it falls in an alias set that does not contain a main effect or another two-factor interaction. Therefore the number of clear two-factor interactions is equal to the number of $m_i(d)$'s, $1 \leq i \leq f$, that are equal to one.

Example 10.1. Consider the 2^{6-2} design defined by $I = ABE = CDEF = ABCDF$. Here $g = 15$ and $f = 9$. Among the 15 two-factor interactions, 3 are aliased with main effects: AB, AE, and BE. Therefore three of the $m_i(d)$'s, $10 \leq i \leq 15$, are equal to 1, and the other three are 0. Six of the twelve two-factor interactions that are not aliased with main effects are partitioned into three aliased pairs: CD is an alias of EF, CE is an alias of DF, and CF is an alias of DE. Therefore three of the $m_i(d)$'s, $1 \leq i \leq 9$, are equal to 2 and the remaining six are equal to 1, indicating that there are six clear two-factor interactions.

From the definition of the $m_i(d)$'s, for each design d of resolution III or higher, the number of two-factor interactions that are not aliased with main effects is equal to $\sum_{i=1}^{f} m_i(d)$. This number can also be counted as follows. Each defining word of length three, say ABC, produces three two-factor interactions AB, AC, and BC that are aliased with main effects. In total there are $3A_3(d)$ two-factor interactions that are aliased with main effects. Therefore the number of two-factor interactions that are not aliased with main effects is equal to

$$\binom{n}{2} - 3A_3(d) = \sum_{i=1}^{f} m_i(d). \tag{10.1}$$

Thus minimizing $A_3(d)$, the first step of the minimum aberration criterion, is equivalent to maximizing the number of two-factor interactions that are not aliased with main effects and is equivalent to maximizing $\sum_{i=1}^{f} m_i(d)$.

Cheng, Steinberg, and Sun (1999) also showed that

$$A_4(d) = \frac{1}{6} \left\{ \sum_{i=1}^{g} [m_i(d)]^2 - \binom{n}{2} \right\}, \tag{10.2}$$

which can be proved by counting the number of pairs of aliased two-factor interactions in two different ways. Since each alias set containing m two-factor interactions produces $m(m-1)/2$ such pairs, the total number of pairs of aliased two-factor interactions is equal to

$$\sum_{i=1}^{g} (m_i(d)[m_i(d) - 1])/2. \tag{10.3}$$

On the other hand, from each defining word of length four, three pairs of aliased two-factor interactions can be identified. Therefore the total number of pairs of aliased two-factor interactions is also equal to $3A_4(d)$. By equating this with (10.3), we obtain

$$A_4(d) = \frac{1}{6} \left\{ \sum_{i=1}^{g} m_i(d)[m_i(d) - 1] \right\}.$$

Then (10.2) follows from this and the fact that $\sum_{i=1}^{g} m_i(d) = \binom{n}{2}$.

Since a minimum aberration design minimizes $A_3(d)$, and minimizes $A_4(d)$ among those that minimize $A_3(d)$, the theorem below due to Cheng, Steinberg, and Sun (1999) follows from (10.1) and (10.2).

Theorem 10.1. *A minimum aberration two-level design of resolution III+ maximizes* $\sum_{i=1}^{f} m_i(d)$, *the number of two-factor interactions that are not aliased with main effects, and minimizes* $\sum_{i=1}^{g}[m_i(d)]^2$ *among the designs that maximize* $\sum_{i=1}^{f} m_i(d)$.

Theorem 10.1 is crucial for gaining some insight into minimum aberration. If there exists a design d of resolution IV+, then, by (10.1), $\sum_{i=1}^{f} m_i(d) = \binom{n}{2}$; so it clearly maximizes $\sum_{i=1}^{f} m_i(d)$. In this case, all the designs that maximize $\sum_{i=1}^{f} m_i(d)$ must be of resolution IV+ and have $m_i(d) = 0$ for all $f+1 \le i \le g$. Then, by Theorem 10.1, a minimum aberration design is of resolution IV+ and minimizes $\sum_{i=1}^{g}[m_i(d)]^2 = \sum_{i=1}^{f}[m_i(d)]^2$ among the resolution IV+ designs. Since $\sum_{i=1}^{f} m_i(d) = \binom{n}{2}$ is fixed, for a minimum aberration design, $m_1(d), \ldots, m_f(d)$ are expected to be nearly equal. When a minimum aberration design d is of resolution III, we would still expect $\sum_{i=1}^{f}[m_i(d)]^2$ to be small since $\sum_{i=1}^{g}[m_i(d)]^2$ is small. Therefore a minimum aberration design of resolution III or higher maximizes the number of two-factor interactions that are not aliased with main effects, and these interactions tend to be distributed over the alias sets very uniformly. The uniformity or near-uniformity of the distribution of two-factor interactions over the alias sets not containing main effects is evident in Table 10.1, which shows the values of the $m_i(d)$'s, $1 \le i \le f$, for the 32-run minimum aberration designs with $9 \le n \le 29$.

The uniform or nearly uniform distribution of the two-factor interactions that are not aliased with main effects over the alias sets is desirable when one does not know a priori which two-factor interactions are important. This connection between the wordlength pattern and alias pattern is the key to the model robustness of minimum aberration designs that we will discuss in the next section.

The same argument can be used to show that a minimum aberration design of resolution V or higher maximizes the number of three-factor interactions that are not aliased with two-factor interactions, and these interactions also tend to be distributed over the alias sets very uniformly. A similar result holds generally for designs of higher resolutions.

In Table 10.1, for $11 \le n \le 15$, there are zeros among the m_i's, and the number of nonzero m_i's is the same for these n values. While all the m_i's for $n = 10$ are nonzero, a zero value appears again for $n = 9$. We will comment on this phenomenon in the next section and study it further in Chapter 11.

10.4 Estimation capacity

In this section, we use the results derived in the previous section to show that minimum aberration is a good surrogate for the criterion of maximum *estimation capacity* proposed by Sun (1993) for design selection under model uncertainty. We say that a model can be estimated (or entertained) by a design d if all the parameters in the model are estimable under d. Suppose $\mathfrak{M}$ is a collection of potential models. For any design d, define its estimation capacity as the number of models in $\mathfrak{M}$ that can be estimated by d. This is a measure of the capability of d to estimate different potential

Table 10.1 $m_1(d), \ldots, m_f(d)$ for minimum aberration $2^{n-(n-5)}$ designs with $9 \leq n \leq 29$

n	$m_1(d), \ldots, m_f(d)$
9	1,1,1,1,1,1,1,1,2,2,2,2,2,2,2,2,2,2,2,4,0
10	2,2,2,2,2,2,2,2,2,2,2,2,2,2,2,2,2,2,2,5
11	3,3,3,3,3,4,4,4,4,4,4,4,4,4,4,0,0,0,0,0
12	4,4,4,4,4,4,4,4,4,5,5,5,5,5,5,0,0,0,0
13	5,5,5,5,5,5,5,5,5,5,5,5,6,6,6,0,0,0
14	6,6,6,6,6,6,6,6,6,6,6,6,6,6,7,0,0
15	7,7,7,7,7,7,7,7,7,7,7,7,7,7,7,0
16	8,8,8,8,8,8,8,8,8,8,8,8,8,8,8
17	8,8,8,8,8,8,8,8,8,8,8,8,8,8
18	8,8,8,8,8,8,8,8,8,8,8,8,9
19	8,8,8,8,8,8,8,8,8,9,9,9
20	8,8,8,8,8,9,9,9,9,9,9
21	9,9,9,9,9,9,9,9,9,9
22	8,8,10,10,10,10,10,10,11
23	8,11,11,11,11,11,11,11
24	12,12,12,12,12,12,12
25	12,12,12,12,12,12
26	12,12,12,12,13
27	12,13,13,13
28	14,14,14
29	14,14

Source: Chen, H. H. and Cheng, C. S. (2004), *Statist. Sinica*, 14, 203–215. With permission.

models. The objective is to find a design whose estimation capacity is as large as possible.

Suppose $\mathfrak{M}_k$ consists of all the models that contain all the main effects and k two-factor interactions, $1 \leq k \leq \binom{n}{2}$. We define $E_k(d)$ as the estimation capacity of d under $\mathfrak{M}_k$; that is, $E_k(d)$ is the number of models containing all the main effects and k two-factor interactions that can be estimated by d. Here k can be thought of as the number of active two-factor interactions. A design is said to have *maximum estimation capacity* if it maximizes $E_k(d)$ for all k. We say that d_1 *dominates* d_2 if $E_k(d_1) \geq E_k(d_2)$ for all k, with strict inequality for at least one k. Note that this notion of domination is only a partial order and usually does not give a complete ranking of all the designs. A set $\mathcal{D}$ of designs is called an *essentially complete class* if for every design d not in $\mathcal{D}$, there is a design d^* in $\mathcal{D}$ such that $E_k(d^*) \geq E_k(d)$ for all k. Such an essentially complete class is called a *minimal essentially complete class* if no two designs d_1 and d_2 in $\mathcal{D}$ satisfy $E_k(d_1) \geq E_k(d_2)$ for all k. It is sufficient to consider designs in a minimal essentially complete class.

Since there may not exist a design that maximizes $E_k(d)$ for all k, to choose a design, one may have to take other considerations into account. For instance, if it is expected that only a small number of two-factor interactions are active, then only the $E_k(d)$'s with small k's need to be considered.

It turns out that $E_k(d)$ can be expressed explicitly in terms of the two-factor interaction counts $m_i(d)$'s. Since only one factorial effect from each alias set can be estimated, and the two-factor interactions that are not aliased with main effects appear in no more than f alias sets, at most f such interactions can be estimated at the same time. It follows that $E_k(d) = 0$ for $k > f$. On the other hand, in the ith alias set there are $m_i(d)$ choices of two-factor interactions to be included in estimable models. Therefore

$$E_k(d) = \sum_{1 \le i_1 < \cdots < i_k \le f} \prod_{j=1}^{k} m_{i_j}(d), \text{ if } k \le f. \tag{10.4}$$

Let $\mathbf{m}(d) = (m_1(d), \ldots, m_f(d))$. Then $E_k(d)$ is a function of $\mathbf{m}(d)$, known as an elementary symmetric function. It has two key properties:

(i) $E_k(d)$ is strictly increasing in each $m_i(d)$;

(ii) $E_k(d)$ is a Schur concave function of $\mathbf{m}(d)$ as defined below.

Definition 10.2. For any vectors $\mathbf{x} = (x_1, \ldots, x_t)$ and $\mathbf{y} = (y_1, \ldots, y_t)$ in $\mathbb{R}^t$, let $x_{[1]} \le \cdots \le x_{[t]}$ and $y_{[1]} \le \cdots \le y_{[t]}$ be the ordered components of $\mathbf{x}$ and $\mathbf{y}$, respectively. We say that $\mathbf{x}$ is majorized by $\mathbf{y}$ if and only if $\sum_{i=1}^{t} x_i = \sum_{i=1}^{t} y_i$, and $\sum_{i=1}^{k} x_{[i]} \ge \sum_{i=1}^{k} y_{[i]}$ for all $1 \le k \le t-1$. If the equality in the first condition is replaced by $\sum_{i=1}^{t} x_i \ge \sum_{i=1}^{t} y_i$, then $\mathbf{x}$ is said to be *upper weakly majorized* by $\mathbf{y}$. A real-valued function f of $\mathbf{x}$ is called *Schur concave* (respectively, *Schur convex*) if for all $\mathbf{x}$ and $\mathbf{y}$ such that $\mathbf{x}$ is majorized by $\mathbf{y}$, we have $f(\mathbf{x}) \ge$ (respectively, $\le$) $f(\mathbf{y})$.

It is an immediate consequence of this definition that if $\mathbf{x}$ is upper weakly majorized by $\mathbf{y}$, then $f(\mathbf{x}) \ge f(\mathbf{y})$ for all f that are Schur concave and nondecreasing in each component of its argument.

Property (i) of $E_k(d)$ is trivial; a proof of its Schur-concavity can be found in Proposition F.1 on p. 78 of Marshall and Olkin (1979). Therefore we have the following result from Cheng, Steinberg, and Sun (1999).

Theorem 10.2. *Given designs d_1 and d_2, if $\mathbf{m}(d_1)$ is upper weakly majorized by $\mathbf{m}(d_2)$, and $\mathbf{m}(d_1)$ cannot be obtained from $\mathbf{m}(d_2)$ by permuting its components, then d_1 dominates d_2 with respect to the criterion of estimation capacity.*

Theorem 10.2 provides a way to compare designs based on the numbers of two-factor interactions in the alias sets.

Example 10.2. Consider the two 2_{IV}^{7-2} designs discussed in Section 10.1. It is easily determined that $\mathbf{m}(d_1)$ has 6 components equal to 0, 15 components equal to 1, and 3 components equal to 2, and $\mathbf{m}(d_2)$ has 9 components equal to 0, 9 components equal to 1, and 6 components equal to 2. We have that $\mathbf{m}(d_1)$ is majorized by $\mathbf{m}(d_2)$. By Theorem 10.2, d_1 dominates d_2 with respect to the criterion of estimation capacity. There is no need to calculate and compare $E_k(d_1)$ and $E_k(d_2)$. Note that d_1 also has less aberration and more clear two-factor interactions than d_2.

It can be seen that if $\sum_{i=1}^{t} x_i = \sum_{i=1}^{t} y_i$ and $x_1 = \cdots = x_t$, then $\mathbf{x}$ is majorized by $\mathbf{y}$; if $\sum_{i=1}^{t} x_i \geq \sum_{i=1}^{t} y_i$ and $x_1 = \cdots = x_t$, then $\mathbf{x}$ is upper weakly majorized by $\mathbf{y}$. The same result also holds when the x_i's differ by at most 1, provided that the entries must be integers. In other words, to make $E_k(d)$ large, $\sum_{i=1}^{f} m_i(d)$ should be large, and $m_1(d), \ldots, m_f(d)$ should be as equal as possible. This, together with the discussion in the paragraph following Theorem 10.1, implies that minimum aberration is a good surrogate for the criterion of maximum estimation capacity.

Corollary 10.3. *If a design d^* minimizes $A_3(d)$, and $m_1(d^*), \ldots, m_f(d^*)$ differ from one another by at most one, then it has maximum estimation capacity.*

It follows from Corollary 10.3 and Table 10.1 that for $16 \leq n \leq 21$ and $24 \leq n \leq 29$, minimum aberration $2^{n-(n-5)}$ designs maximize $E_k(d)$ for all k.

The sufficient condition in Theorem 10.2 also provides a simple tool for eliminating designs that are dominated by others. For example, the two cases ($n = 22$ and 23) not covered in the previous paragraph can be handled in this way. In both cases, all but two designs can be eliminated by applying Theorem 10.2. The $E_k(d)$ values of the two remaining designs can then be compared directly. It turns out that in both cases the minimum aberration design maximizes $E_k(d)$ for all k. Combining this with the result in the previous paragraph, we conclude that for $n \geq 16$, minimum aberration 32-run designs maximize $E_k(d)$ for all k. The same is also true for minimum aberration 16-run designs with $n \geq 8$.

The case where the number of factors is less than half of the run size, however, is different. In this case minimum aberration designs, which must be of resolution IV+, typically do not maximize $E_k(d)$ for all k. It was observed at the end of Section 10.3 that for each of the minimum aberration $2^{n-(n-5)}$ designs with $11 \leq n \leq 15$ listed in Table 10.1, some m_i's with $1 \leq i \leq f$ are 0. The fact that not all the alias sets that contain no main effects are occupied by two-factor interactions has some negative impact on estimation capacity. For example, the minimum aberration 2^{12-7} design d^* has 4 zero m_i's and 15 nonzero m_i's. It cannot entertain models with more than 15 two-factor interactions, so $E_k(d^*) = 0$ for all $k > 15$. On the other hand, there are 2^{12-7}_{III} designs for which all 19 m_i's are nonzero. For such designs, $E_k(d) > 0$ for all $16 \leq k \leq 19$. We have $E_k(d) > E_k(d^*)$ for $16 \leq k \leq 19$, and the inequality continues to hold at least for the k's that are not much smaller than 16.

We point out that for the two cases $n = 10$ and $n = 16$ where all the m_i's in Table 10.1 are nonzero, the minimum aberration resolution IV designs do maximize $E_k(d)$ for all k.

Although minimum aberration designs of resolution IV may not maximize $E_k(d)$ for all k, they tend to maximize $E_k(d)$'s for the practically more important cases where k is not too large. In all the 16-run cases, minimal essentially complete classes contain no more than two designs, and for 32-runs, they contain at most four designs. In all the cases, there is always a minimum aberration design in a minimal essentially complete class.

For all the resolution III designs in Table 10.1, all the m_i's are nonzero. This difference between designs of resolution III and IV is related to a result that has important implications in the structures and construction of resolution IV designs, to be studied in Chapter 11.

The results in Section 10.3 and the connection between minimum aberration and maximum estimation capacity can be extended to the case where the number of levels is a prime power; see Cheng and Mukerjee (1998).

Recall that a two-factor interaction is clear if it is the only two-factor interaction in an alias set that does not contain main effects. We have shown that minimum aberration designs tend to distribute the two-factor interactions that are not aliased with main effects very uniformly over the alias sets. Unless the number of factors is small, as is apparent in Table 10.1, this will leave many two-factor interactions in each alias set, producing no clear two-factor interactions. Therefore low aberration typically is contradictory to having a large number of clear two-factor interactions unless the number of factors is not too large. The counterexample discussed in Section 10.2 is not surprising once one understands the nature of minimum aberration. On the other hand, when the number of factors is not too large, the number of two-factor interactions is also not very large. In this case a nearly uniform distribution of the two-factor interactions that are not aliased with main effects over the alias sets may result in only one two-factor interaction in many alias sets that do not contain main effects, thereby producing many clear two-factor interactions.

It will be shown in Section 10.8 that when $2^{n-p-2}+1 < n \leq 2^{n-p-1}$, even though the maximum resolution is IV, a design that has at least one clear two-factor interaction must be of resolution III. For $n \leq 2^{n-p-2}+1$, there are resolution IV designs with clear two-factor interactions, but they may not maximize the number of such interactions. Again this supports the observations in the previous paragraph.

The choice between a minimum aberration design and a design with the maximum number of clear two-factor interactions depends on the experimenter's prior knowledge and the goal of the experiment. Some information about which two-factor interactions might be important is needed in order to take advantage of clear two-factor interactions. When the experimenter has a specific set of effects that needs to be estimated, a minimum aberration design may not meet the requirement. One needs to keep in mind that minimum aberration is intended for model uncertainty. Having many clear two-factor interactions calls for many alias sets containing only one two-factor interaction. This tends to put all the remaining two-factor interactions in a few alias sets, with a negative impact on model robustness. As noted by Fries and Hunter (1980), the criterion of minimum aberration was proposed for the "situation in which prior knowledge is diffuse concerning the possible greater importance of certain effects." We cannot overemphasize that no single criterion works in all situations.

10.5 Other justifications of minimum aberration

Tang and Deng (1999) provided another justification of minimum aberration as a criterion for minimizing the bias of the estimates of lower-order effects in the presence of higher-order effects. For simplicity, consider two-level designs. Suppose the estimates of main effects are required; then we need a design of resolution III or higher. Under the full model,

$$E(\mathbf{y}) = \mu\mathbf{1} + \mathbf{X}_1\boldsymbol{\beta}_1 + \sum_{j=2}^{n} \mathbf{X}_j\boldsymbol{\beta}_j, \qquad (10.5)$$

where $\boldsymbol{\beta}_1$ consists of all the main effects and $\boldsymbol{\beta}_j, j \geq 2$, consists of all the j-factor interactions. The estimate of $\boldsymbol{\beta}_1$ under the model

$$E(\mathbf{y}) = \mu\mathbf{1} + \mathbf{X}_1\boldsymbol{\beta}_1 \qquad (10.6)$$

is $\widehat{\boldsymbol{\beta}}_1 = (\mathbf{X}_1^T\mathbf{X}_1)^{-1}\mathbf{X}_1^T\mathbf{y} = \frac{1}{N}\mathbf{X}_1^T\mathbf{y}$. Under (10.5), $E(\widehat{\boldsymbol{\beta}}_1) = \boldsymbol{\beta}_1 + \frac{1}{N}\sum_{j=2}^{n}\mathbf{X}_1^T\mathbf{X}_j\boldsymbol{\beta}_j$. Let $\mathbf{C}_j = \frac{1}{N}\mathbf{X}_1^T\mathbf{X}_j$. Then $\mathbf{C}_j\boldsymbol{\beta}_j$ is the contribution of $\boldsymbol{\beta}_j$ to the bias of the estimates of main effects. Since $\boldsymbol{\beta}_j$ is unknown, $\|\mathbf{C}_j\|^2 = \mathrm{tr}(\mathbf{C}_j^T\mathbf{C}_j)$ can be used to measure the contamination of j-factor interactions on the estimation of main effects. Under the hierarchical assumption that lower-order effects are more important than higher-order effects, Tang and Deng (1999) proposed to minimize $\|\mathbf{C}_2\|^2, \ldots, \|\mathbf{C}_n\|^2$ sequentially.

This can be applied to any design, regular or nonregular. For two-level regular fractional factorial designs, each entry of $\mathbf{C}_j$ is 0, 1, or -1: it is 1 or -1 if the corresponding j-factor interaction is aliased with a main effect; see Section 9.3. Furthermore, each column of $\mathbf{C}_j$ has at most one nonzero entry since the same j-factor interaction cannot be aliased with more than one main effect. Therefore $\|\mathbf{C}_j\|^2$, which is the sum of squares of all the entries of $\mathbf{C}_j$, is equal to the number of j-factor interactions that are aliased with main effects. It is easy to see that

$$\|\mathbf{C}_j\|^2 = (j+1)A_{j+1} + (n-j+1)A_{j-1} \qquad (10.7)$$

for $2 \leq j \leq n-1$, and $\|\mathbf{C}_n\|^2 = A_{n-1}$. Thus sequentially minimizing $\|\mathbf{C}_2\|^2, \ldots, \|\mathbf{C}_n\|^2$ is equivalent to sequentially minimizing A_k for $3 \leq k \leq n$.

This approach can be applied to other situations as well. In general, suppose $\boldsymbol{\beta}_1$ consists of all the effects that we need to estimate, and the other effects are divided into J groups $\boldsymbol{\beta}_2, \ldots, \boldsymbol{\beta}_J$, with the effects in $\boldsymbol{\beta}_j$ being more important than those in $\boldsymbol{\beta}_{j+1}$. Then we can adopt the criterion of sequentially minimizing $\|\mathbf{C}_2\|^2, \ldots, \|\mathbf{C}_J\|^2$ to choose from the designs under which the assumed model (10.6) is estimable; see Cheng and Tang (2005) for details. For example, suppose in addition to the main effects, estimates of a subset of two-factor interactions are also required. Then $\boldsymbol{\beta}_1$ consists of the main effects and the required two-factor interactions, $\boldsymbol{\beta}_2$ consists of the remaining two-factor interactions, and for $j > 2$, $\boldsymbol{\beta}_j$ consists of the j-factor interactions. In this case, one can express the resulting minimum aberration criterion in terms of a modified wordlength pattern. Ke and Tang (2003) applied this criterion to study four classes of compromise plans discussed in Addelman (1962b) and Sun

(1993), each with a specific type of nonnegligible two-factor interactions. The best 16- and 32-run designs were searched and tabulated.

Two justifications of the minimum aberration criterion from the projection point of view were provided by Chen (1998) and Tang (2001). Chen (1998) showed that for $R+1 \leq k \leq R + \lfloor (R-1)/2 \rfloor$ and $R \leq j \leq k$, the number of resolution j projections of a regular design d of resolution R onto k-factors is proportional to $A_j(d)$; see Exercise 9.2. Thus minimum aberration designs achieve the good projection property of having fewer projections of low resolutions by sequentially minimizing the $A_j(d)$'s. On the other hand, consider the projection of a design onto a subset S of s factors. For each combination $\mathbf{a}$ of these factors, let $N_S(\mathbf{a})$ be the number of occurrences of $\mathbf{a}$ in the projected design. Then $\sum_{\mathbf{a}} [N_S(\mathbf{a}) - \frac{N}{2^s}]^2$ measures the nonuniformity of the projection onto the factors in S. An overall measure of the nonuniformity of s-factor projections is provided by

$$V_s = \sum_{S:|S|=s} \sum_{\mathbf{a}} \left[N_S(\mathbf{a}) - \frac{N}{2^s} \right]^2 .$$

Tang (2001) showed that sequentially minimizing $A_1, \ldots, A_n$ is equivalent to sequentially minimizing $V_1, \ldots, V_n$.

Fang and Mukerjee (2000) established a connection between aberration and the centered L_2-discrepancy, a measure of uniformity of the distribution of points in the unit hypercube introduced by Hickernell (1998). This criterion considers not just the uniformity over the entire hypercube, but also the uniformity in all lower-dimensional projections. Fang and Mukerjee (2000) showed that for regular fractional factorial designs, the centered L_2-discrepancy is a linear combination of the A_k's, and the coefficient of A_k decreases exponentially with k. Thus a small centered L_2-discrepancy can be achieved by sequentially minimizing $A_k, k = 1, \ldots, n$.

10.6 Construction and complementary design theory

Construction of minimum aberration designs has been studied, e.g., by Franklin (1984), Chen and Wu (1991), Chen (1992), Chen and Hedayat (1996), Tang and Wu (1996), Butler (2003a, 2005), Chen and Cheng (2006, 2009), and Xu and Cheng (2008). Chen, Sun, and Wu (1993) compiled tables of regular 16- and 32-run two-level designs, 27-run three-level designs, and 64-run two-level designs of resolution IV+. Laycock and Rowley (1995) presented a method for generating and labeling regular fractional factorial designs. Block and Mee (2005) presented the results of enumerations of 128-run two-level resolution IV designs (the searches were not complete in some cases). Catalogs of useful regular three-level designs with 27, 81, 243, and 729 runs can be found in Xu (2005a). These designs were obtained by complete enumeration of all 27- and 81-run designs, 243-run designs of resolution IV+, and 729-run designs of resolution V+. Mukerjee and Wu (2006) provided catalogues of two- and three-level designs.

Larger factorial designs are in increasing demand, especially in applications to large scale computer experiments. For large run sizes and numbers of factors, complete design enumerations and searches for minimum aberration designs quickly become a daunting task. Xu (2009) developed an efficient algorithm by cutting some unnecessary searches (thus reducing the isomorphism checks that need to be done) and incorporating a fast isomorphism-check procedure. This algorithm allows him to successfully perform complete enumerations of all two-level 128-run designs of resolution IV+, 256-run designs of resolution IV+ up to 17 factors, 512-run designs of resolution V+, 1024-run designs of resolution VI+, and 2048- and 4,096-run designs of resolution VII+. He also proposed a method for constructing minimum aberration designs by enumerating a subset of good designs when a complete enumeration is not feasible. This, for example, allows him to obtain minimum aberration designs for $N = 256$ and $n \leq 28$. He further proposed three approaches to the construction of good designs for larger run sizes and numbers of factors. Efficient designs were tabulated for 128–4,096 runs and up to 40–160 factors. Many of the tabulated designs have minimum aberration. Ryan and Bulutoglu (2010) improved Xu's algorithm by using a more efficient isomorphism-check procedure and obtained 36 additional minimum aberration designs.

In the rest of this section we concentrate on a useful technique of constructing minimum aberration designs via complementary designs. A regular s^{n-p} fractional factorial design d can be constructed by choosing n of the $(s^{n-p} - 1)/(s - 1)$ factors in a saturated regular design. The factors that are not chosen form another design $\bar{d}$, called the *complementary design* of d. The number of factors in $\bar{d}$ is equal to $f = (s^{n-p} - 1)/(s - 1) - n$.

We first consider an example from Tang and Wu (1996).

Example 10.3. Consider the selection of four factors from the saturated 2^{7-4} design to construct a 2^{4-1} design. Write the saturated 2^{7-4} design as an 8×7 matrix, with the seven columns corresponding to A_1, A_2, A_3, A_1A_2, A_1A_3, A_2A_3, and $A_1A_2A_3$, where A_1, A_2, and A_3 are basic factors. Choosing the columns A_1, A_2, A_3, and $A_1A_2A_3$ gives a design d_1 with less aberration than the design d_2 obtained by choosing A_1, A_2, A_3, and A_1A_2, since d_1 has resolution IV and d_2 has resolution III. The complementary design of d_1 consists of the columns A_1A_2, A_1A_3, and A_2A_3, which form a defining word of length three. On the other hand, the three columns A_1A_3, A_2A_3, and $A_1A_2A_3$ in $\bar{d}_2$ are linearly independent. This example seems to suggest that there is more dependency among the columns of the complement of a minimum aberration design.

Tang and Wu (1996) derived identities relating the wordlength pattern of a regular two-level design to that of its complementary design. Chen and Hedayat (1996) independently obtained such identities for defining words of lengths three and four since their objective was to find weak minimum aberration designs. When d is nearly saturated, since $\bar{d}$ has few factors, it is much easier to determine the wordlength pattern of $\bar{d}$ than that of d. The identities obtained by Tang and Wu (1996) and Chen and Hedayat (1996) can be used to study minimum aberration designs via their complementary designs. These identities were extended to regular s^{n-p} designs by Suen, Chen, and Wu (1997).

Theorem 10.4. *Let d be a regular s^{n-p} design where s is a prime or a prime power, and $\bar{d}$ be its complementary design. Then*

$$A_k(d) \;=\; (s-1)^{-1}(C_k+C_{k0})+\sum_{j=3}^{k-2} C_{kj}A_j\left(\bar{d}\right)$$

$$+(-1)^k[1+(s-2)(k-1)]A_{k-1}(\bar{d})+(-1)^kA_k\left(\bar{d}\right)$$

for $3 \le k \le n$, with

$$C_k = s^{p-n}\left[P_k(0;n,s)-P_k(s^{n-p-1};n,s)\right],$$

$$C_{kj} = s^{-f}\sum_{i=0}^{f}P_k\left(s^{n-p-1}-i;n,s\right)P_i(j;f,s)\,for\,0 \le j \le f,$$

where $P_k(x;n,s)$ are the Krawtchouk polynomials.

Theorem 10.4 shows that for designs of resolution III or higher,

$$A_3(d) = (s-1)^{-1}(C_3+C_{30})-A_3\left(\bar{d}\right),$$

$$A_4(d) = (s-1)^{-1}(C_4+C_{40})+(3s-5)A_3\left(\bar{d}\right)+A_4\left(\bar{d}\right),$$

and in general $A_k(d)$ is a linear function of the $A_j(\bar{d})$'s with $j \le k$, where the coefficient of $A_k(\bar{d})$ is $(-1)^k$. Therefore sequentially minimizing $A_k(d)$ is equivalent to sequentially minimizing $(-1)^kA_k(\bar{d})$.

In Example 10.3, $A_3(\bar{d}_1) = 1$ and $A_3(\bar{d}_2) = 0$. It follows immediately from Theorem 10.4 that d_1 has less aberration than d_2.

The MacWilliams identities in coding theory play an important role in the proof of Theorem 10.4. We only provide a sketch of the proof here and refer readers to Suen, Chen, and Wu (1997) for the details. Treat d and $\bar{d}$ as linear codes. Then, except for a constant multiple, the wordlength patterns of d and $\bar{d}$ are the weight distributions of the dual codes $d^{\perp}$ and $\bar{d}^{\perp}$, respectively. Theorem 10.4 gives the relation between the weight distributions of $d^{\perp}$ and $\bar{d}^{\perp}$. Since MacWilliams identities relate the weight distributions of $d^{\perp}$ and $\bar{d}^{\perp}$ to those of d and $\bar{d}$, respectively, Theorem 10.4 can be proved by finding the missing link: the relation between the weight distributions of d and $\bar{d}$. Write each treatment combination of the saturated regular design as a $1 \times g$ row vector, where $g = (s^{n-p} - 1)/(s-1)$. Then d consists of n of the g columns, and $\bar{d}$ the remaining $g-n$ columns. This partition of the columns of the saturated regular design also partitions each row into two parts: a $1 \times n$ vector that is a treatment combination in d and a $1 \times (g-n)$ vector that is a treatment combination in $\bar{d}$. Therefore the weight of each treatment combination of the saturated regular design is the sum of the weights of the two treatment combinations it is split into. This and the fact that the saturated regular design has a very simple weight distribution, with one single vector of weight zero and $s^{n-p} - 1$ vectors of weight s^{n-p-1} (see Theorem 8.6 and

Exercise 9.7), can be used to write down the relation between the weight distributions of d and $\bar{d}$.

Chen and Hedayat (1996) showed that complementary designs of two-level minimum aberration designs have the same structure as long as they have the same number of factors. The fact that the structures of complementary designs of minimum aberration designs do not depend on the run size is useful for characterizing and cataloging minimum aberration *nearly saturated* designs. A key to this result is stated below. Write a saturated design of size $N = 2^{n-p}$ as a $2^{n-p} \times \left(2^{n-p} - 1\right)$ matrix $\mathbf{X}$ of 0's and 1's as in the previous paragraph. Then the columns of $\mathbf{X}$ are all the nonzero linear combinations of $n - p$ linearly independent columns, say $\mathbf{a}_1, \ldots, \mathbf{a}_{n-p}$. Constructing a 2^{n-p} design amounts to choosing a set d of n columns of $\mathbf{X}$, or deleting from $\mathbf{X}$ the $f = 2^{n-p} - 1 - n$ columns in $\bar{d}$. We define the rank of $\bar{d}$ as the maximum number of linearly independent columns in $\bar{d}$. Chen and Hedayat (1996) showed that if $\bar{d}$ maximizes $A_3(\bar{d})$, then $\bar{d}$ must have *minimum* rank; that is, if $f = 2^{w-1} + q$, where $0 \leq q < 2^{w-1}$, $w \leq n - p$, then the columns in $\bar{d}$ are linear combinations of a set of w linearly independent columns of $\mathbf{X}$. Without loss of generality, we may assume that $\bar{d} \subseteq < \mathbf{a}_1, \ldots, \mathbf{a}_w >$, where $< \mathbf{a}_1, \ldots, \mathbf{a}_w >$ is the set of all the $2^w - 1$ nonzero linear combinations of $\mathbf{a}_1, \ldots, \mathbf{a}_w$. Thus one can solve the simpler problem of maximizing A_3 over the f-factor projections of $< \mathbf{a}_1, \ldots, \mathbf{a}_w >$ instead of choosing f factors from all the $2^{n-p} - 1$ factors of a saturated design.

Chen and Hedayat (1996) also showed that among f-factor projections of $< \mathbf{a}_1, \ldots, \mathbf{a}_w >$, A_3 is maximized by a design of the form

$$\bar{d} = < \mathbf{a}_1, \ldots, \mathbf{a}_{w-1} > \cup \{\mathbf{a}_w, \mathbf{a}_w + \mathbf{b}_1, \ldots, \mathbf{a}_w + \mathbf{b}_q\}, \qquad (10.8)$$

where $\mathbf{b}_1, \ldots, \mathbf{b}_q \in < \mathbf{a}_1, \ldots, \mathbf{a}_{w-1} >$. To show that such a design maximizes A_3 over all the f-factor projections of $< \mathbf{a}_1, \ldots, \mathbf{a}_w >$, by another application of the complementary design theory, it suffices to show that its complement in $< \mathbf{a}_1, \ldots, \mathbf{a}_w >$ has minimum A_3 among all the $(2^w - 1 - f)$-factor projections of $< \mathbf{a}_1, \ldots, \mathbf{a}_w >$. For a $\bar{d}$ of the form in (10.8), all the columns of its complement in $< \mathbf{a}_1, \ldots, \mathbf{a}_w >$ are of the form $\mathbf{a}_w + \mathbf{b}$, where $\mathbf{b} \in < \mathbf{a}_1, \ldots, \mathbf{a}_{w-1} >$. For any two such columns $\mathbf{a}_w + \mathbf{b}$ and $\mathbf{a}_w + \mathbf{b}'$, $(\mathbf{a}_w + \mathbf{b}) + (\mathbf{a}_w + \mathbf{b}') = \mathbf{b} + \mathbf{b}' \in < \mathbf{a}_1, \ldots, \mathbf{a}_{w-1} >$. Therefore the complement of $\bar{d}$ in $< \mathbf{a}_1, \ldots, \mathbf{a}_w >$ has no defining word of length three and hence has zero A_3. This proves that a $\bar{d}$ of the form in (10.8) maximizes $A_3(\bar{d})$ and also shows that for any $\bar{d}$ that maximizes $A_3(\bar{d})$, the complement of $\bar{d}$ in $< \mathbf{a}_1, \ldots, \mathbf{a}_w >$ must have no defining word of length three.

Therefore we have shown that d minimizes $A_3(d)$ if and only if the complement of $\bar{d}$ in $< \mathbf{a}_1, \ldots, \mathbf{a}_w >$ has no defining word of length three. Using this result, Chen and Hedayat (1996) determined all the minimum aberration 2^{n-p} designs with $f \leq 15$. Suen, Chen, and Wu (1997) determined all the minimum aberration 3^{n-p} designs with $f \leq 13$. We refer readers to these sources for lists of complementary designs of such minimum aberration designs. Earlier, Pu (1989) classified two-level complementary designs with $f \leq 15$ and three-level complementary designs with $f \leq 13$.

Remark 10.1. Although complementary designs of the form in (10.8) maximize $A_3(\bar{d})$, they are not the only ones to have this property. Suppose $f = 10$. Let $\bar{d}_1$ be the complement of $\{\mathbf{a}_1, \mathbf{a}_2, \mathbf{a}_3, \mathbf{a}_4, \mathbf{a}_1 + \mathbf{a}_2 + \mathbf{a}_3 + \mathbf{a}_4\}$ in $< \mathbf{a}_1, \mathbf{a}_2, \mathbf{a}_3, \mathbf{a}_4 >$, where $\mathbf{a}_1$, $\mathbf{a}_2$, $\mathbf{a}_3$, and $\mathbf{a}_4$ are linearly independent. Then since $\{\mathbf{a}_1, \mathbf{a}_2, \mathbf{a}_3, \mathbf{a}_4, \mathbf{a}_1 + \mathbf{a}_2 + \mathbf{a}_3 + \mathbf{a}_4\}$ has no defining word of length three, $\overline{d}_1$ maximizes $A_3(\overline{d})$. It can be shown that $\overline{d}_1$ is the complementary design of a minimum aberration design, but it cannot be expressed in the form of (10.8).

10.7   Maximum estimation capacity: a projective geometric approach*

The partition of a saturated regular s^{n-p} design into two designs with complementary factors discussed in the previous section can also be described geometrically. Let $P = \{\mathbf{x}_1, \ldots, \mathbf{x}_g\}$, where $\mathbf{x}_1, \ldots, \mathbf{x}_g$ are the g distinct points of $PG(n - p - 1, s)$, and let T be any subset of P of size n, defining a regular s_{III+}^{n-p} design d_T. The complementary design $\overline{d}_T$ with $f = g - n$ factors then corresponds to the complement $\overline{T}$ of T in P.

For simplicity, we only discuss two-level designs here and refer readers to Cheng and Mukerjee (1998) for results in the case where the number of levels is a prime number or a prime power. In the two-level case, as pointed out in Section 9.11, a defining word of length three corresponds to three collinear points in T. Therefore minimizing the number of defining words of length three is equivalent to minimizing the number of lines in T.

We have also pointed out in Section 9.11 that each $\mathbf{x}_i$ can be identified with an alias set. If $\mathbf{x}_i \in \overline{T}$, then $\mathbf{x}_i$ corresponds to an alias set not containing main effects. The number of two-factor interactions contained in this alias set is equal to the number of lines that pass through $\mathbf{x}_i$ and two points in T. This number is also the m_i defined in Section 10.3. There are two other types of lines that pass through $\mathbf{x}_i$: those that are entirely in $\overline{T}$ and those with two points in $\overline{T}$ and one point in T. Let ϕ_i and h_i be the numbers of such lines, respectively. Then $\phi_i + h_i + m_i$ is a constant, since it is equal to the total number of lines passing through $\mathbf{x}_i$. By counting the number of lines that pass through $\mathbf{x}_i$ and at least another point in $\overline{T}$, we have $f - 1 = h_i + 2\phi_i$; this is because there is a line passing through $\mathbf{x}_i$ and each of the other $f - 1$ points in $\overline{T}$, and if the third point of this line is also in $\overline{T}$, then the line is counted twice. Thus $h_i = f - 1 - 2\phi_i$. Then $\phi_i + h_i + m_i = \phi_i + (f - 1 - 2\phi_i) + m_i = m_i + f - 1 - \phi_i$. Therefore

$$\phi_i = m_i + \text{constant}. \tag{10.9}$$

In Section 10.4, it was shown that a design is expected to have large estimation capacity if it maximizes $\sum_{i=1}^{f} m_i$ and the m_i's are as equal as possible. From (10.9), we see that this is the case if $\sum_{i=1}^{f} \phi_i$ is maximized and the ϕ_i's are as equal as possible. Now, $\frac{1}{3}\sum_{i=1}^{f} \phi_i$ is equal to the total number of lines contained in $\overline{T}$. Thus a regular 2^{n-p} design with large estimation capacity can be constructed by choosing a set T of n points from $PG(n - p - 1, 2)$ such that $\overline{T}$ contains the maximum number of lines among all subsets of $PG(n - p - 1, 2)$ of size f, and these lines are as uniformly distributed among the f points as possible. In particular, if $\sum_{i=1}^{f} \phi_i$ is maximized and the ϕ_i's, $1 \leq i \leq f$, differ from one another by at most one, then the resulting design has maximum estimation capacity, provided that the $(n - p) \times n$ matrix with the n points of T as its columns has full row rank $n - p$. The same argument above shows that minimizing the number of lines in T is equivalent to maximizing the number of lines in $\overline{T}$. This is the geometric version of the fact that minimizing $A_3(d)$ is equivalent to maximizing $A_3(\overline{d})$.

Chen and Hedayat's (1996) result, discussed at the end of the previous section, can be phrased geometrically.

Theorem 10.5. *Suppose $2^{w-1} \le f < 2^w$, where $w \le n - p$. Then an f-subset $\overline{T}$ of $PG(n - p - 1, 2)$ contains the maximum number of lines if and only if it is a subset of a $(w - 1)$-flat F in $PG(n - p - 1, 2)$ and its complement in F contains no line.*

One choice of $\overline{T}$ with its complement in a $(w - 1)$-flat containing no line was given in (10.8).

By the discussions preceding Theorem 10.5, as in Chen and Hedayat (1996), Tang and Wu (1996), and Suen, Chen, and Wu (1997), it is also convenient to construct a nearly saturated design with maximum estimation capacity through its complementary design. Cheng and Mukerjee (1998) constructed designs such that $\sum_{i=1}^{f} \phi_i$ is maximized and the ϕ_i's, $1 \le i \le f$, differ from one another by at most one.

Theorem 10.6. *Suppose $2^{w-1} \le f < 2^w$, where $w \le n - p$, and T^* is an n-subset of $PG(n - p - 1, 2)$ such that $\overline{T}^*$ is contained in a $(w - 1)$-flat F in $PG(n - p - 1, 2)$. Let H^* be the complement of $\overline{T}^*$ in F. If any three or four distinct points of H^* are linearly independent, then d_{T^*} has maximum estimation capacity. In this case, all the designs with maximum estimation capacity must have the structure given here.*

The condition that any three or four distinct points of H^* are linearly independent implies that H^* contains no line. By Theorem 10.5, T^* maximizes $\sum_{i=1}^{f} \phi_i$. The same condition can also be used to show that the ϕ_i's, $1 \le i \le f$, differ from one another by at most one. This is left as an exercise.

We summarize in Table 10.2 the structures of some designs with maximum estimation capacity that can be obtained from Theorem 10.6; for details, see Cheng and Mukerjee (1998). For each f in the first column, the second column shows the set H^* in Theorem 10.6; that is, the points that need to be deleted from a $(w - 1)$-flat to obtain the complement of a design with maximum estimation capacity. The points $\mathbf{a}_i$'s shown in the table are any set of linearly independent points. For the first five cases in the table, a design maximizes $E_k(d)$ for all k if and only if it has minimum aberration. For the first two cases, all the ϕ_i's (and hence the m_i's) are equal, and for each of the first four cases, the structure of $\overline{T}^*$ as described is the only one that can maximize $\sum_{i=1}^{f} \phi_i$ and therefore is also the only structure for a design to have minimum aberration and weak minimum aberration.

Remark 10.2. The $2^{n-(n-5)}$ designs with $16 \le n \le 21$ and $24 \le n \le 29$ in Table 10.1 can be constructed by using the results summarized in Table 10.2.

Table 10.2 *Structures of designs that have maximum estimation capacity*

f	Complement of $\overline{T}^*$ in a $(w-1)$-flat
$f = 2^w - 1 \ (2 \leq w < n - p)$	$\emptyset$
$f = 2^w - 2 \ (3 \leq w < n - p)$	any point
$f = 2^w - 3 \ (3 \leq w < n - p)$	any two points
$f = 2^w - 4 \ (3 \leq w < n - p)$	any three noncollinear points
$f = 2^w - 5 \ (4 \leq w < n - p)$	any four linearly independent points
$f = 2^w - 6 \ (4 \leq w \leq n - p)$	$\mathbf{a}_1, \mathbf{a}_2, \mathbf{a}_3, \mathbf{a}_4, \mathbf{a}_1 + \mathbf{a}_2 + \mathbf{a}_3 + \mathbf{a}_4$
$f = 2^w - 7 \ (5 \leq w \leq n - p)$	$\mathbf{a}_1, \mathbf{a}_2, \mathbf{a}_3, \mathbf{a}_4, \mathbf{a}_5, \mathbf{a}_1 + \mathbf{a}_2 + \mathbf{a}_3 + \mathbf{a}_4 + \mathbf{a}_5$
$f = 2^w - 8 \ (6 \leq w \leq n - p)$	$\mathbf{a}_1, \mathbf{a}_2, \mathbf{a}_3, \mathbf{a}_4, \mathbf{a}_5, \mathbf{a}_6, \mathbf{a}_1 + \mathbf{a}_2 + \mathbf{a}_3 + \mathbf{a}_4 + \mathbf{a}_5 + \mathbf{a}_6$

Example 10.4. Let $n = 19$, $p = 14$. Then $n - p = 5$, $g = 2^5 - 1 = 31$, and $f = 31 - 19 = 12$, which is of the form $2^4 - 4$ with $w = 4$. To construct a design d^* with maximum estimation capacity (and minimum aberration), according to Table 10.2, we drop three noncollinear points from the 15 points (vectors) in a 3-flat to form the complementary design of d^*. Specifically, we pick four linearly independent 5×1 vectors, say $(1,0,0,0,0)^T, (0,1,0,0,0)^T, (0,0,1,0,0)^T$, and $(0,0,0,1,0)^T$. The 15 nonzero linear combinations of these four vectors form a 3-flat in a PG(4, 2). We then drop any three linearly independent vectors, say $(1,0,0,0,0)^T$, $(0,1,0,0,0)^T$, and $(0,0,1,0,0)^T$. Let $\overline{T}^*$ be the set of the remaining 12 vectors. Delete these 12 vectors from the 31 nonzero 5×1 vectors. Write the remaining 19 vectors as a 5×19 matrix. Then all the 32 linear combinations of the five rows of this matrix give the treatment combinations of a 2^{19-14} design that has maximum estimation capacity and minimum aberration. This design has the maximum number (99) of two-factor interactions that are not aliased with main effects. These 99 two-factor interactions are distributed in the 12 alias sets not containing main effects in the most uniform fashion, with three alias sets each containing nine two-factor interactions, and the remaining nine alias sets each containing eight two-factor interactions, as shown in Table 10.1.

10.8   Clear two-factor interactions revisited

We present two results due to Chen and Hedayat (1998) on the existence of clear two-factor interactions.

Theorem 10.7. *When $n > 2^{n-p-1}$, no regular fractional factorial design of resolution III+ has clear two-factor interactions.*

Proof. Suppose there exists a regular 2^{n-p}_{III+} design that contains at least one clear two-factor interaction. Without loss of generality, assume that it is the interaction of factors A_1 and A_2. Then there is no defining word of the form $A_1 A_2 A_i$ or $A_1 A_2 A_i A_j$, where 1, 2, i, j are distinct. Consider a model containing all the main effects, the two-factor interaction $A_1 A_2$, and all three-factor interactions of the form $A_1 A_2 A_i$, where A_i is one of the remaining $n-2$ factors. Then it is easy to see that none of the effects in this model are aliased with one another. Therefore all these effects, $2n-1$ in total, are estimable if the other effects are assumed negligible. This implies that $2n-1 \leq 2^{n-p}-1$, the total number of available degrees of freedom. It then follows that $n \leq 2^{n-p-1}$. So if $n > 2^{n-p-1}$, then it is not possible to have clear two-factor interactions.   □

Theorem 10.7 shows that $n \leq 2^{n-p-1}$ ($n \leq N/2$, where N is the run size) is a necessary condition for the existence of clear two-factor interactions. Chen and Hedayat (1998) showed that this condition is also sufficient.

Note that $n \leq 2^{n-p-1}$ is also a necessary and sufficient condition for the existence of a resolution IV design. However, a resolution IV design may not have clear two-factor interactions.

Theorem 10.8. *When* $n > 2^{n-p-2}+1$, *there does not exist a resolution IV design with clear two-factor interactions.*

Proof. Suppose there exists a regular 2_{IV}^{n-p} design that has at least one clear-two factor interaction, say A_1A_2. Then there is no defining word of length three and no defining word of the form $A_1A_2A_iA_j$, where $1,2,i,j$ are distinct. Consider a model containing all the main effects, all two-factor interactions of the form A_1A_i or A_2A_j, and all three-factor interactions of the form $A_1A_2A_i$. Then it is easy to see that none of the effects in this model are aliased with one another. Therefore all these effects, $4n-5$ in total, are estimable if the other effects are assumed negligible. This implies that $4n-5 \leq 2^{n-p}-1$, from which it follows that $n \leq 2^{n-p-2}+1$.   □

By Theorem 10.8, $n \leq 2^{n-p-2}+1$ ($n \leq N/4+1$) is a necessary condition for the existence of a regular 2_{IV}^{n-p} design that has at least one clear two-factor interaction. Chen and Hedayat (1998) showed that the same condition is also sufficient for the existence of resolution IV+ designs with clear two-factor interactions.

Tang, Ma, Ingram, and Wang (2002) and Wu and Wu (2002) studied designs with the maximum number of clear two-factor interactions. Tang et al. (2002) derived some lower and upper bounds for the number of clear two-factor interactions. Wu and Wu (2002) developed an approach to showing whether or not a given design has the maximum number of clear two-factor interactions by using graphical representations and combinatorial and group-theoretic arguments.

Tang (2006) extended the concept of clear two-factor interactions to nonregular designs. Two factorial effects are said to be orthogonal if their corresponding columns in the model matrix are orthogonal, and a two-factor interaction is said to be clear if it is orthogonal to all the main effects and all the other two-factor interactions. The following is the counterpart of Theorems 10.7 and 10.8 for orthogonal arrays.

Theorem 10.9. *A necessary condition for the existence of an* $OA(N,2^n,2)$ *with at least one clear two-factor interaction is that* N *is a multiple of* 8 *and* $n \leq N/2$. *If* *there is an $OA(N, 2^n, 3)$ with at least one clear two-factor interaction, then N is a multiple of* 16 *and* $n \leq N/4 + 1$.

Tang (2006) also showed that the conditions in Theorem 10.9 are sufficient provided that the Hadamard conjecture holds.

We present some further results on clear two-factor interactions in Section 11.7.

10.9   Minimum aberration blocking of complete factorial designs

An algorithm for searching blocking schemes of complete factorial designs to avoid confounding certain required treatment factorial effects with blocks was discussed in Section 9.7. On the other hand, if one does not have specific knowledge about the required effects, then under the hierarchical assumption it is desirable to confound as few lower-order effects as possible. Let A_i be the number of i-factor interaction pencils (or main-effect pencils when $i = 1$) that are confounded with blocks. If we do not want to confound the main effects with blocks, then we must have $A_1 = 0$. Then a reasonable criterion for selecting a good blocking scheme is to sequentially minimize $A_2, A_3, \ldots$, etc. The resulting blocking scheme is said to have *minimum aberration*. This criterion was adapted from the minimum aberration criterion for selecting fractional factorial designs by Sun, Wu, and Chen (1997). Using this criterion, they obtained all minimum aberration blocking schemes for 2^n designs with up to 256 runs ($n = 8$).

A set of m linearly independent vectors $\mathbf{a}_1, \ldots, \mathbf{a}_m$ in $EG(n, s)$, $m < n$, is required for constructing an s^n design in s^m blocks of size s^{n-m}. By Remark 9.1, $\mathbf{a}_1, \ldots, \mathbf{a}_m$ produce a minimum aberration blocking scheme if and only if they are independent defining words of a minimum aberration fractional factorial design. Therefore all the results on the construction of minimum aberration fractional factorial designs are applicable to that of minimum aberration blocking of complete factorial designs.

As pointed out in Remark 9.7, blocking schemes that correspond to regular fractional factorial designs of resolution R have estimability of order $R - 1$. The literature on the construction of minimum aberration fractional factorial designs deals exclusively with designs of resolution III+. Such results are useful for constructing minimum aberration blocking schemes that have estimability of order 2 or higher. However, blocking schemes with estimability of order 1 are important since they are useful when we try to protect the main effects, but some two-factor interactions can be confounded with blocks. Furthermore, sometimes blocking schemes with higher-order estimability do not exist. For example, for 8 or 16 blocks, there does not exist a blocking scheme of a complete 2^5 design with estimability of order 2. In general, it follows from (7.4) that for $s = 2$, if $n - \log_2(n+1) < m \leq n - 1$, then blocking schemes with estimability of order $e \geq 2$ do not exist. Construction of minimum aberration blocking schemes of complete factorial designs that have estimability of order 1 is equivalent to that of minimum aberration fractional factorial designs of resolution II. We refer readers to Tang (2007) for a complete solution of the two-level case.

For smaller block sizes (larger m), intrablock variances may be smaller since it may be easier to control the within-block homogeneity. This would increase the efficiencies of the estimates of the effects that are orthogonal to blocks. However, more factorial effects will get confounded with blocks.

10.10 Minimum moment aberration

Roughly speaking the objective of minimum aberration is to minimize aliasing of factorial effects. When the factorial effects are expressed as columns and the experimental runs are represented by rows, minimum aberration is a column-based criterion. Interestingly, it is equivalent to a row-based criterion that minimizes the "similarity" of the experimental runs. For any design d with n factors and N experimental runs $\mathbf{r}_1(d), \ldots, \mathbf{r}_N(d)$, where each $\mathbf{r}_i(d)$ is a $1 \times n$ vector representing a level combination of the n factors, let

$$K_t(d) = [N(N-1)/2]^{-1} \sum_{1 \le g < h \le N} \left[\delta(\mathbf{r}_g(d), \mathbf{r}_h(d))\right]^t, \ t = 1, 2, \ldots,$$

where $\delta(\mathbf{r}_g(d), \mathbf{r}_h(d))$ is the number of components where $\mathbf{r}_g(d)$ and $\mathbf{r}_h(d)$ agree. A design is said to have *minimum moment aberration* if it sequentially minimizes $K_1(d), K_2(d), \ldots,$ (Xu, 2003). The following result is due to Xu (2003).

Theorem 10.10. *A regular fractional factorial design has minimum aberration if and only if it has minimum moment aberration.*

What Xu (2003) showed is more general. The minimum aberration criterion cannot be applied to nonregular designs since it is defined in terms of wordlength patterns, and nonregular designs do not have defining words. However, the minimum moment aberration as defined here can readily be applied to all designs: regular, nonregular, or even those with mixed levels. A nontrivial extension of minimum aberration to a criterion called generalized minimum aberration for nonregular designs by Tang and Deng (1999) and Xu and Wu (2001) will be discussed in Chapter 15. When applied to regular designs, generalized minimum aberration reduces to the usual minimum aberration criterion. Xu (2003) showed the equivalence of generalized minimum aberration and minimum moment aberration for the case where all the factors have the same number of levels. Thus, in particular, for regular designs minimum aberration is equivalent to minimum moment aberration. We defer the proof of this result to Chapter 15, where the general case is treated, and only sketch the main idea here.

For any experimental run $\mathbf{r}(d)$ of $\mathbf{X}(d)$, $\delta(\mathbf{r}(d), \mathbf{r}(d)) = n$ is a constant. It follows that

$$K_t(d) = [N(N-1)]^{-1} \sum_{g=1}^{N} \sum_{h=1}^{N} \left[\delta(\mathbf{r}_g(d), \mathbf{r}_h(d))\right]^t - (N-1)^{-1} n^t. \quad (10.10)$$

Thus minimum moment aberration is equivalent to sequential minimization of $\sum_{g=1}^{N} \sum_{h=1}^{N} [\delta(\mathbf{r}_g(d), \mathbf{r}_h(d))]^t$. Then, since $\delta(\mathbf{r}_g(d), \mathbf{r}_h(d)) = n - \text{dist}(\mathbf{r}_g(d), \mathbf{r}_h(d))$, where dist$(\mathbf{r}_g(d), \mathbf{r}_h(d))$ is the Hamming distance of $\mathbf{r}_g(d)$ and $\mathbf{r}_h(d)$, we have

$$
\sum_{g=1}^{N} \sum_{h=1}^{N} \left[\delta(\mathbf{r}_g(d), \mathbf{r}_h(d))\right]^t
$$

$$
= \sum_{g=1}^{N} \sum_{h=1}^{N} \left[n - \mathrm{dist}(\mathbf{r}_g(d), \mathbf{r}_h(d))\right]^t
$$

$$
= \sum_{l=0}^{t} \sum_{g=1}^{N} \sum_{h=1}^{N} \binom{t}{l} n^{t-l} (-1)^l \left[\mathrm{dist}(\mathbf{r}_g(d), \mathbf{r}_h(d))\right]^l. \tag{10.11}
$$

Let

$$
M_t(d) = N^{-2} \sum_{g=1}^{N} \sum_{h=1}^{N} \left[\mathrm{dist}(\mathbf{r}_g(d), \mathbf{r}_h(d))\right]^t, \text{ for } t = 1, 2, \dots, \tag{10.12}
$$

and $M_0(d) = 1$. Then, by (10.11), minimum moment aberration is equivalent to the sequential minimization of $(-1)^t M_t(d), t = 1, 2, \dots$, etc.

We state a key result connecting such moments of pairwise distances (or dissimilarities) of the experimental runs to wordlength patterns.

Lemma 10.11. *(Xu, 2003) For any N-run regular design d with n factors, each factor having s levels,*

$$
M_t(d) = \sum_{i=0}^{\min(n,t)} (-1)^i (s-1) A_i(d) \left[\sum_{j=0}^{t} j! S(t, j) s^{-j} (s-1)^{j-i} \binom{n-i}{j-i}\right] \text{ for } t \ge 0,
$$

where $S(t, j)$ is a Stirling number of the second kind, the number of ways to partition a set of t elements into j nonempty subsets.

From Lemma 10.11, one can draw the conclusion that sequential minimization of $(-1)^t M_t(d)$ is equivalent to that of $A_t(d)$. Combining this with the observation following (10.12), we conclude that minimum moment aberration and minimum aberration are equivalent. We note that Lemma 10.11 is the regular design version of Lemma 15.20, which will be proved in Section 15.10.

Minimum moment aberration was also studied by Butler (2003a,b) in a slightly different but equivalent form for two-level designs. For each two-level design d with the two levels denoted by 1 and -1, let $\mathbf{X}_d$ be the $N \times n$ matrix with each column corresponding to a factor and each row representing a treatment combination, and let $\phi_{gh}(d)$ be the (g, h)th entry of $\mathbf{X}_d \mathbf{X}_d^T$. Independently of Xu (2003), for two-level designs, Butler (2003a) and Butler (2003b), respectively, showed the equivalence of minimum aberration and generalized minimum aberration to the sequential minimization of

$$
H_t(d) = \frac{1}{N^2} \sum_{g=1}^{N} \sum_{h=1}^{N} \left[\phi_{gh}(d)\right]^t \tag{10.13}
$$

for $t = 1, 2, \ldots$, etc. Since

$$\sum_{g=1}^{N} \sum_{h=1}^{N} \left[\phi_{gh}(d) \right]^t = \sum_{1 \leq g \neq h \leq 1} \left[\phi_{gh}(d) \right]^t + Nn^t$$

and

$$\delta(\mathbf{r}_g(d), \mathbf{r}_h(d)) = \{ \mathbf{r}_g(d)[\mathbf{r}_h(d)]^T + n \}/2 = \{ \phi_{gh}(d) + n \}/2,$$

it is easy to see that for two-level regular designs, sequentially minimizing $H_t(d)$ is equivalent to the minimum moment aberration criterion defined in Xu (2003).

We will use the equivalence of minimum aberration and minimum moment aberration to study minimum aberration even designs in Section 11.9 and to derive a complementary design theory for two-level designs of resolution IV in Section 11.10.

10.11   A Bayesian approach

A framework for Bayesian fractional factorial designs was set up by Mitchell, Morris, and Ylvisaker (1995) and further developed in Kerr (2001). Gaussian processes have been used to model unknown deterministic functions in the literature of computer experiments (Sacks, Welch, Mitchell, and Wynn, 1989). In (8.1), consider α as a realization of a Gaussian process. Let $\boldsymbol{\alpha}$ be the $s_1 \cdots s_n \times 1$ vector of the values of $\alpha(x_1, \ldots, x_n)$ for the full factorial design. By (8.2),

$$\boldsymbol{\beta} = \mathbf{P}^{-1} \boldsymbol{\alpha}. \tag{10.14}$$

Then a prior distribution of the factorial effects $\boldsymbol{\beta}$ can be obtained from that of $\boldsymbol{\alpha}$ via (10.14). This approach of using a functional prior of the response function to induce a prior distribution of the many parameters in a factorial treatment model was used in Joseph (2006), Joseph and Delaney (2007), Ai, Kang, and Joseph (2009), Joseph, Ai, and Wu (2009), and Kang and Joseph (2009) for the design and analysis of factorial experiments in several settings for both regular and nonregular designs. Commonly used functional priors based on Gaussian processes require the specification of just a few hyperparameters. As Joseph (2006) pointed out, this averts the difficulty of assigning a large number of hyperparameters when a prior distribution of the factorial effects is to be specified directly.

Suppose the prior distribution of $\boldsymbol{\alpha}$ is multivariate normal with zero mean and covariance matrix $\sigma_0^2 \mathbf{R}$. Then, by (10.14), we have

$$\boldsymbol{\beta} \sim \mathrm{N}\left(\mathbf{0}, \sigma_0^2 \mathbf{P}^{-1} \mathbf{R} \left(\mathbf{P}^T \right)^{-1} \right).$$

In the rest of this section we restrict to two-level designs and denote the two levels by 1 and -1. By (6.17), when the components of $\boldsymbol{\alpha}$ and $\boldsymbol{\beta}$ are in the standard order, we have

$$\mathbf{P} = \underbrace{\begin{bmatrix} 1 & -1 \\ 1 & 1 \end{bmatrix} \otimes \cdots \otimes \begin{bmatrix} 1 & -1 \\ 1 & 1 \end{bmatrix}}_{n}.$$

Thus $\mathbf{P}^T = 2^n \mathbf{P}^{-1}$, and

$$\text{cov}\left(\boldsymbol{\beta}\right) = 2^{-2n} \sigma_0^2 \mathbf{P}^T \mathbf{R} \mathbf{P}. \qquad (10.15)$$

Each component of $\boldsymbol{\beta}$ corresponds to a subset of $\{1 \ldots, n\}$. For convenience, in this section we denote the component of $\boldsymbol{\beta}$ corresponding to $S \subseteq \{1, \ldots, n\}$ by β_S. Thus $\beta_\emptyset$ is the grand mean, and if $S = \{i_1, \ldots, i_k\}$, then β_S is the interaction of factors $i_1, \ldots, i_k$.

We say that $\boldsymbol{\alpha}$ is stationary if the correlation between $\alpha(\mathbf{x}_1)$ and $\alpha(\mathbf{x}_2)$ depends only on the factors where $\mathbf{x}_1$ and $\mathbf{x}_2$ differ. Mitchell, Morris, and Ylvisaker (1995) showed that $\boldsymbol{\alpha}$ is stationary if and only if the β_S's, $S \subseteq \{1, \ldots, n\}$, are independent. In this case, the distribution of $\boldsymbol{\alpha}$ is determined by 2^n parameters: the variances of β_S, $S \subseteq \{1, \ldots, n\}$. Furthermore, the matrix in (10.15) is diagonal. Then, since the columns of $\mathbf{P}$ are mutually orthogonal and have lengths equal to $2^{n/2}$, $(2^{-n/2} \mathbf{P}^T)(\sigma_0^2 \mathbf{R})(2^{-n/2} \mathbf{P})$ is the spectral decomposition of $\text{cov}(\boldsymbol{\alpha})$. Thus the columns of $\mathbf{P}$, which define the β_S's, $S \subseteq \{1, \ldots, n\}$, are eigenvectors, and the variances of the β_S's are the eigenvalues of $\text{cov}(\boldsymbol{\alpha})$ divided by 2^n.

Example 10.5. The following product correlation function of stationary priors is commonly used in the literature of computer experiments.

$$\text{corr}\left(\alpha(\mathbf{x}_1), \alpha(\mathbf{x}_2)\right) = \prod_{i:1 \leq i \leq n, x_{1i} \neq x_{2i}} \rho_i,$$

where $\mathbf{x}_1 = (x_{11}, \ldots, x_{1n})^T$, $\mathbf{x}_2 = (x_{21}, \ldots, x_{2n})^T$, and $0 < \rho_1, \ldots, \rho_n < 1$. Let $r_i = (1 - \rho_i)/(1 + \rho_i)$ and $\tau^2 = \sigma_0^2 \{\prod_{i=1}^n (1 + r_i)\}^{-1}$. Then one can verify that

$$\text{var}\left(\beta_\emptyset\right) = \tau^2, \text{ and for nonempty } S, \ \text{var}\left(\beta_S\right) = \tau^2 \prod_{i \in S} r_i. \qquad (10.16)$$

By (10.16),

$$S \subseteq S' \Rightarrow \text{var}\left(\beta_S\right) \geq \text{var}\left(\beta_{S'}\right). \qquad (10.17)$$

This fact, observed in Mitchell, Morris, and Ylvisaker (1995), is consistent with the hierarchical principle that lower-order effects are more important than higher-order effects. Kerr (2001) referred to (10.17) as the property of nested decreasing interaction variances. Another implication of (10.16) is that larger values of r_i can be assigned to the factors whose main effects are more important than the others, which will also make all the interactions involving these factors more important (Joseph, 2006).

We say that $\boldsymbol{\alpha}$ is isotropic if $\text{cov}(\alpha(\mathbf{x}_1), \alpha(\mathbf{x}_2))$ depends only on the number of components where $\mathbf{x}_1$ and $\mathbf{x}_2$ differ. Mitchell, Morris, and Ylvisaker (1995) showed that $\boldsymbol{\alpha}$ is isotropic if and only if all the factorial effects involving the same number of factors have the same variance. In this case the distribution of $\boldsymbol{\alpha}$ is determined by $n + 1$ parameters.

Example 10.6. (Example 10.5 continued) In Example 10.5, α is isotropic if $\rho_1 = \cdots = \rho_n = \rho$, so all the main effects are considered as equally important. Let

$$r = (1-\rho)/(1+\rho), \quad \tau^2 = \sigma_0^2(1+r)^{-n}.$$

Then we have

$$\text{var}\left(\beta_S\right) = \tau^2 r^k \text{ if } |S| = k.$$

In this case, all the interactions of the same order are also equally important.

Let $y_1, \ldots, y_N$ be observations resulting from a fractional factorial design, and $\mathbf{y} = (y_1, \ldots, y_N)^T$. Suppose $\mathbf{y} = \mathbf{X}_T \mathbf{P}\boldsymbol{\beta} + \mathbf{e}$, where $\mathbf{X}_T \mathbf{P}$ is the model matrix as in (8.3), $\boldsymbol{\beta}$ and $\mathbf{e}$ are independent, $\mathbf{e} \sim N(\mathbf{0}, \sigma^2 \mathbf{I}_N)$, and α is stationary. Let $\mathbf{U} = \mathbf{X}_T \mathbf{P}$, and write the covariance matrix of $\boldsymbol{\beta}$ as $\boldsymbol{\Lambda}$. Then $\boldsymbol{\Lambda}$ is a diagonal matrix, and the joint distribution of $\mathbf{y}$ and $\boldsymbol{\beta}$ is

$$N\left(\begin{bmatrix} \mathbf{0} \\ \mathbf{0} \end{bmatrix}, \begin{bmatrix} \sigma^2 \mathbf{I}_N + \mathbf{U}\boldsymbol{\Lambda}\mathbf{U}^T & \mathbf{U}\boldsymbol{\Lambda} \\ \boldsymbol{\Lambda}\mathbf{U}^T & \boldsymbol{\Lambda} \end{bmatrix}\right).$$

The posterior distribution of $\boldsymbol{\beta}$ given $\mathbf{y}$ can be obtained by using standard results on multivariate normal distributions. In particular, the posterior mean of $\boldsymbol{\beta}$ given $\mathbf{y}$ is

$$E\left(\boldsymbol{\beta}|\mathbf{y}\right) = \boldsymbol{\Lambda}\mathbf{U}^T \left(\mathbf{U}\boldsymbol{\Lambda}\mathbf{U}^T + \sigma^2 \mathbf{I}_N\right)^{-1} \mathbf{y},$$

with the posterior covariance matrix

$$\text{cov}\left(\boldsymbol{\beta}|\mathbf{y}\right) = \boldsymbol{\Lambda} - \boldsymbol{\Lambda}\mathbf{U}^T \left(\mathbf{U}\boldsymbol{\Lambda}\mathbf{U}^T + \sigma^2 \mathbf{I}_N\right)^{-1} \mathbf{U}\boldsymbol{\Lambda}. \tag{10.18}$$

Suppose $N = 2^{n-p}$ and the observations are taken at the treatment combinations in a regular fractional factorial design d. If α is stationary, then the results on the prior distribution of $\boldsymbol{\beta}$ derived above can be applied to the restriction of the Gaussian process to the treatment combinations in d (or, equivalently, the complete factorial of a set of basic factors), which is also stationary. Without loss of generality, we may assume that the first 2^{n-p} columns of the full model matrix $\mathbf{U}$ correspond to factorial effects of the basic factors. Then the first 2^{n-p} columns of $\mathbf{U}$ are given by

$$\tilde{\mathbf{P}} = \underbrace{\begin{bmatrix} 1 & -1 \\ 1 & 1 \end{bmatrix} \otimes \cdots \otimes \begin{bmatrix} 1 & -1 \\ 1 & 1 \end{bmatrix}}_{n-p}.$$

By (9.11), each contrast $\beta = \sum_{\mathbf{x} \in d} c(\mathbf{x})\alpha(\mathbf{x})$ that defines a factorial effect of the basic factors, called a basic effect, is the sum of the basic effect and all its aliases; here without loss of generality we assume that all the terms on the right side of (9.11) have coefficients $+1$. For each nonempty $S \subseteq \{1, \ldots, n\}$, let $A(\beta_S)$ be the set of β_S and all its aliases, and for $S = \emptyset$, let $A(\beta_S)$ be the set consisting of the mean and the interactions that appear in the defining relation. Kerr (2001) called the 2^{n-p} sums $\sum_{\beta:\beta\in A(\beta_S)}\beta$, one for each S, alias interactions. Let $\tilde{\boldsymbol{\beta}}$ be the $2^{n-p}\times 1$ vector consisting of the alias interactions. Then similar to (10.15), we have

$$\text{cov}\left(\tilde{\boldsymbol{\beta}}\right) = 2^{-2(n-p)}\sigma_0^2\tilde{\mathbf{P}}^T\tilde{\mathbf{R}}\tilde{\mathbf{P}},$$

where $\tilde{\mathbf{R}}$ is the submatrix of $\mathbf{R}$ restricted to the treatment combinations in d, $\text{cov}(\tilde{\boldsymbol{\beta}})$ is diagonal, and $(2^{-(n-p)/2}\tilde{\mathbf{P}}^T)(\sigma_0^2\tilde{\mathbf{R}})(2^{-(n-p)/2}\tilde{\mathbf{P}})$ is the spectral decomposition of $\sigma_0^2\tilde{\mathbf{R}}$. Thus the alias interactions are the eigenvalues of $\tilde{\mathbf{R}}$ divided by 2^{n-p}.

Using these observations, (10.18), and the fact that if two factorial effects are aliased, then the corresponding columns of $\mathbf{U}$ are identical, one can show that for each $\beta_S, S \subseteq \{1,\ldots,n\}$,

$$\text{var}\left(\beta_S|\mathbf{y}\right) = \text{var}\left(\beta_S\right) - \frac{\left[\text{var}\left(\beta_S\right)\right]^2}{\sum_{\beta:\beta\in A(\beta_S)}\text{var}(\beta)+2^{-(n-p)}\sigma^2}. \tag{10.19}$$

If β_S and $\beta_{S'}$ are in different alias sets, then $\text{cov}(\beta_S,\beta_{S'}|\mathbf{y}) = 0$; otherwise,

$$\text{cov}\left(\beta_S,\beta_{S'}|\mathbf{y}\right) = -\frac{\text{var}\left(\beta_S\right)\text{var}\left(\beta_{S'}\right)}{\sum_{\beta:\beta\in A(\beta_S)}\text{var}(\beta)+2^{-(n-p)}\sigma^2}. \tag{10.20}$$

We leave the proofs of these as an exercise.

Mitchell, Morris, and Ylvisaker (1995) and Kerr (2001) studied optimal regular two-level fractional factorial designs with respect to several criteria such as the D-, A-, and E-criteria defined in terms of the posterior covariance matrix of $\boldsymbol{\alpha}$ (for prediction) and that of $\boldsymbol{\beta}$ (for examining the factorial effects) under a variety of priors. Kerr (2001) observed that the minimum aberration designs were often optimal. Since the minimum aberration criterion was proposed under the hierarchical assumption, in view of the observations made in Sections 10.3 and 10.4, we expect minimum aberration designs to perform well under priors that are isotropic and have the property of nested decreasing interaction variances.

Joseph (2006) also studied optimal designs with respect to several A-type criteria that are functions of the posterior variances of the β's. Readers are referred to Joseph (2006) for an analysis strategy, including variable selection. While Joseph (2006) concentrated on two-level designs, Joseph and Delaney (2007) extended the approach to designs with more then two levels, where the factors can be qualitative or quantitative.

The minimum aberration criterion is suitable under the hierarchical assumption. Many existing extensions to other situations are based on ad hoc modifications of the usual wordlength patterns. Such modifications are often without strong justifications, since there is no natural order of the words in terms of their desirability, which is crucial for setting up an appropriate wordlength pattern. The Bayesian approach is more flexible in that it allows for incorporating prior knowledge into the distribution of $\boldsymbol{\alpha}$ and assigning different weights for effects of different importance in the criteria. For example, in experiments where the factors have two and four levels, not all the interactions of the same order should be treated equally. Joseph, Ai, and Wu (2009) used a Bayesian approach to formulate minimum aberration type criteria for choosing the grouping schemes discussed in Section 9.8. Earlier, Wu and Zhang (1993) proposed an ad hoc modification of the usual minimum aberration criterion. In experiments for robust parameter designs, the interactions between noise and control factors are more important than those between noise factors. Kang and Joseph (2009) studied Bayesian optimal designs for this kind of experiment by incorporating the importance of control-by-noise interactions in the criterion.

Exercises

10.1 Prove (10.7).

10.2 Verify the claim in Remark 10.1.

10.3 For $N = 32$, the minimum aberration design constructed in Remark 10.1 has $n = 21$. In this case, Table 10.1 shows that this design has the same number of two-factor interactions in all the alias sets that do not contain main effects. Show that this holds for all the designs with $f = 10$ constructed in Remark 10.1.

10.4* Complete the proof of Theorem 10.6.

10.5* Verify that the designs with $f = 2^w - i, 1 \leq i \leq 8$, given in Table 10.2 have maximum estimation capacity.

10.6 Construct the minimum aberration 2^{18-13} design and verify that it has the m_i values shown in Table 10.1.

10.7 Prove (10.16).

10.8 Prove (10.19) and (10.20), and that if β_i and β_j are in different alias sets, then $\text{cov}(\beta_i, \beta_j | \mathbf{y}) = 0$.

Chapter 11

Structures and Construction of Two-Level Resolution IV Designs

In this chapter we study the structures of two-level regular fractional factorial designs, in particular those of resolution IV. The notion of maximal designs is introduced. Each regular fractional factorial design is the projection of a certain maximal design onto a subset of factors. Thus the characterization of maximal designs is useful for the construction of regular designs. While saturated regular designs are the only maximal designs of resolution III, typically there are many nonisomorphic maximal designs of resolution IV. The study of maximal designs of resolution IV brings forth the methods of partial foldover and doubling, in addition to the familiar method of foldover, for constructing resolution IV designs. The methods of partial foldover and doubling produce many resolution IV designs that cannot be constructed by the method of foldover. The structural theorems presented in this chapter are also useful for the determination of minimum aberration designs of resolution IV. In particular, we present a general complementary design theory for two-level designs that are constructed by the method of doubling.

11.1 Maximal designs

We have seen in Sections 9.5 and 9.6 that for a given run size N, two-level saturated regular designs (with $n = N - 1$) and maximal even designs (with $n - N/2$) attain the maximum possible number of factors among designs of resolution III and IV, respectively. It follows that the resolution of a saturated regular design reduces to two whenever an extra factor is added. Likewise, adding another factor also reduces the resolution of a maximal even design. We call a design *maximal* if its resolution decreases whenever an extra factor is added. Then saturated regular designs are maximal designs of resolution III and maximal even designs are maximal designs of resolution IV.

There is a resolution V design (the 2^{5-1} design defined by $I = ABCDE$) whose resolution reduces to III whenever a factor is added. In the rest of this chapter, by maximal designs of resolution IV+ we mean designs of resolution IV+ that become resolution III designs whenever a factor is added.

Given a design of resolution IV that is not maximal, at least one factor can be added so that the resulting design still has resolution IV. One can keep adding factors until no more factors can be added without reducing the resolution. Then a maximal design of resolution IV is obtained. This implies that every nonmaximal design of resolution IV can be expanded to a maximal design of resolution IV by adding factors. In other words, every design of resolution IV can be constructed by choosing a subset of factors from a maximal design.

Given a design d of resolution R, all of its projections have resolution R or higher. Theoretically speaking, if we are able to characterize all the maximal designs of resolution IV, then we would be able to construct all resolution IV designs by considering all possible projections of the maximal designs.

Likewise, all regular designs of resolution III are projections of maximal designs of resolution III. One important difference between designs of resolution III and IV is that saturated regular designs are the *only* maximal designs of resolution III (up to isomorphism), but for a given run size, typically there are maximal resolution IV designs with different numbers of factors, and other than a small number of exceptions, there is more than one nonisomorphic maximal resolution IV design with the same number of factors. Every resolution III+ design can be constructed by choosing factors from a saturated regular design, but not all two-level designs of resolution IV+ are projections of maximal even designs. The following characterization of the projections of maximal even designs is a corollary of Theorem 9.10.

Proposition 11.1. *A regular design is a projection of a maximal even design if and only if it can be constructed by the foldover method.*

While foldover is a very useful and well known method for constructing designs of resolution IV, it cannot produce all the resolution IV designs. We discuss the characterization of maximal designs and the construction of noneven resolution IV designs (called *even-odd designs* in Mee (2009)) in Section 11.4. Some statistical properties of maximal designs of resolution IV are first presented in the next section. Most of the results in Sections 11.2 and 11.3 are from Chen and Cheng (2004) and Chen and Cheng (2006).

11.2   Second-order saturated designs

Maximal resolution IV designs can be characterized by the statistical property of second-order saturation defined as follows. A 2^{n-p} design can be used to estimate up to $2^{n-p} - 1$ factorial effects. If all the n main effects need to be estimated, then we can estimate at most $f = 2^{n-p} - n - 1$ two-factor interactions. Block and Mee (2003) called a design *second-order saturated* (SOS) if it allows for the estimation of all the n main effects and a certain set of $2^{n-p} - n - 1$ two-factor interactions, provided that the other effects are negligible. Under an SOS design, all the available degrees of freedom can be used to estimate main effects and two-factor interactions. Such a design can be used to entertain a model with the largest number of two-factor interactions. Note that saturated designs have one main effect in every alias set, and $f = 0$. In this case, the condition of second-order saturation is trivially satisfied. We exclude this by only considering the case where $f > 0$.

A simple necessary and sufficient condition for a regular design to be second-order saturated can be formulated in terms of the two-factor interaction counts $m_1,\ldots,m_f$ defined in Section 10.3. Since one effect from each alias set can be estimated when the other effects in the same alias set are assumed to be negligible, the result below holds.

Theorem 11.2. *A regular fractional factorial design of resolution III+ is second-order saturated if and only if none of m_1, ..., m_f is equal to zero.*

Table 10.1 shows the values of $m_1,\ldots,m_f$ for 32-run minimum aberration designs with $9 \leq n \leq 29$. It was pointed out before that for all the resolution III designs in the table, none of the m_i values is equal to zero. Therefore all these designs are SOS. In contrast, among the resolution IV designs in the table, only those with $n = 16$ and 10 are SOS. This demonstrates distinct behaviors between designs of resolution III and IV. We present two general results in this regard, covering resolution III and IV designs, respectively. The first result shows that when the maximum possible resolution is three, all the resolution III designs are SOS.

Theorem 11.3. *If $2^{n-p-1} \leq n < 2^{n-p} - 1$, then all the 2_{III}^{n-p} designs are second-order saturated.*

Proof. Chen and Cheng (2004) proved this theorem for $2^{n-p-1} < n < 2^{n-p} - 1$ by using a result from coding theory. The following simpler proof was suggested by Boxin Tang.

Suppose there is a 2_{III}^{n-p} design d that is not SOS. Let the factors in d be $A_1,\ldots,A_n$. Since d is not SOS, by Theorem 11.2, there is an alias set containing neither a main effect nor a two-factor interaction. The effects in this alias set can be used to define a factor A_{n+1} whose main effect is aliased with neither a main effect nor a two-factor interaction. Then none of the $2n+1$ effects $A_1,\ldots,A_{n+1},A_1A_{n+1},\ldots,A_nA_{n+1}$ is aliased. This implies that $2n+1 \leq 2^{n-p} - 1$, which contradicts the assumption that $2^{n-p-1} \leq n$. $\qquad\square$

Theorem 11.4. *A regular design of resolution IV is second-order saturated if and only if it is maximal.*

Proof. Let d be a regular design of resolution IV. If d is second-order saturated, then each alias set of effects that are not aliased with main effects must contain at least one two-factor interaction. When an additional factor is introduced, its main effect must be aliased with at least one two-factor interaction, creating a defining word of length three and reducing the resolution to three. Therefore d is maximal.

On the other hand, suppose d is not SOS. Then, as in the proof of Theorem 11.3, an extra factor can be added so that its main effect is aliased with neither a main effect nor a two-factor interaction. Since no defining word of length two or three will be introduced, d is not maximal. $\qquad\square$

Corollary 11.5. *All resolution IV designs are projections of second-order saturated designs of resolution IV.*

The values of $m_1, \ldots, m_f$ exhibited in Table 10.1 indicate that the minimum aberration 32-run designs with $n = 16$ and 10 are maximal. The maximal design with $n = 16$ is a maximal even design, which has all the two-factor interactions appear in 15 of the 31 alias sets. We will see in Section 11.4. that for $11 \leq n \leq 15$, all the minimum aberration designs are projections of the maximal even design. Deleting some factors does not change in which alias sets the interactions of the remaining factors appear; therefore the number of nonzero m_i's does not increase even though there are more alias sets of effects that are not aliased with main effects. This results in some zero m_i values for such designs and is why even though additional degrees of freedom become available, they cannot be used to estimate more two-factor interactions. For $11 \leq n \leq 15$, one can still estimate at most 15 two-factor interactions, as in the case of $n = 16$. The presence of one zero m_i value for the minimum aberration design with $n = 9$, as shown in the table, indicates that the minimum aberration 32-run design with 9 factors is not maximal (SOS); it is itself a projection of the maximal design with 10 factors. We will see later that there exists a maximal design with $n = 9$ and $N = 32$, which is the design with more clear two-factor interactions than the minimum aberration design discussed in Section 10.2

Remark 11.1. It was pointed out in Section 10.4 that all the 32-run minimum aberration designs of resolution III have maximum estimation capacity, but the minimum aberration resolution IV designs, except for $n = 10$ and 16, typically do not maximize $E_k(d)$ for larger k's. This is closely related to the fact that all resolution III designs with $n > 16$ are second-order saturated (Theorem 11.3), but resolution IV designs are not second-order saturated unless they are maximal (Theorem 11.4). Since most of the minimum aberration designs for $n < N/2$ are not maximal, some of their $m_i(d)$'s are constrained to be equal to zero. In this case, even though the nonzero $m_i(d)$'s are nearly equal, the presence of some zero $m_i(d)$'s upsets the overall near uniformity of the $m_i(d)$ values and prevents the design from maximizing $E_k(d)$ for all k. On the other hand, for $N/2 < n < N-1$, none of the $m_i(d)$'s of a minimum aberration design is constrained to be zero. In this case the $m_i(d)$'s of a minimum aberration resolution III design are truly nearly equal; as a result, minimum aberration and maximum estimation capacity are highly consistent.

Chen and Cheng (2004) defined a notion of *estimation index* and used it to characterize second-order saturated designs. For a regular 2^{n-p} fractional factorial design, let the $2^{n-p} - 1$ alias sets be S_i, $i = 1, \ldots, 2^{n-p} - 1$, and let ρ_i be the minimum length of the words in S_i. Then the estimation index ρ is defined as $\rho = \max\{\rho_i : i = 1, \ldots, 2^{n-p} - 1\}$.

Theorem 11.6. *A regular fractional factorial design with $f > 0$ is second-order saturated if and only if its estimation index is equal to 2.*

Proof. A design has $f > 0$ if and only if there is at least one alias set not containing main effects, and it is second-order saturated if and only if there is at least one two-factor interaction in each alias set that does not contain main effects. The former is equivalent to $\rho \geq 2$, and the latter is equivalent to $\rho \leq 2$.   □

Thus for two-level fractional factorial designs of resolution IV, the three conditions (a) the estimation index is equal to 2, (b) the design is second-order saturated, and (c) the design is maximal, are equivalent. Although the discussions in this section are focused on two-level designs, the notions of maximal designs and estimation index can be defined for designs with more than two levels, and the equivalence of (a), (b), and (c) holds in general.

Block and Mee (2003) defined the *alias length pattern* as the distribution of the m_i values, $1 \leq i \leq f$:

$$\text{alp} = (a_1, \ldots, a_L), \tag{11.1}$$

where $L = \max\{m_i : 1 \leq i \leq f\}$ and, for $1 \leq j \leq L$, a_j is the number of i's, $1 \leq i \leq f$, with $m_i = j$. Thus a resolution IV design is maximal if and only if $\sum_{j=1}^{L} a_j = f$. Furthermore, a_1 is the number of clear two-factor interactions and, by (10.4), the estimation capacity is a function of the alias length pattern.

Remark 11.2. Jacroux (2004) proposed a two-stage maximum rank minimum aberration criterion as a modification of the usual minimum aberration criterion. The first stage is equivalent to maximizing the number of nonzero m_i's, $1 \leq i \leq f$. Under the assumption that all the main effects must be estimated, this is the same as maximizing the number of two-factor interactions that can be entertained in the model. The second stage is to apply the minimum aberration criterion to the first-stage solutions. By Theorem 11.3, for $2^{n-p-1} < n < 2^{n-p} - 1$, all the resolution III designs have $m_i > 0$ for all $1 \leq i \leq f$. In this case, since the maximum possible resolution is III, maximum rank minimum aberration is equivalent to minimum aberration. For $n = 2^{n-p-1}$, the maximal even design has minimum aberration and maximum rank minimum aberration. For $n < 2^{n-p-1}$, if there is a maximal resolution IV design that also has minimum aberration, then it has maximum rank minimum aberration. In the other cases, minimum aberration designs and maximum rank minimum aberration designs may be different.

11.3   Doubling

To prepare for the characterization of maximal designs of resolution IV in the next section, we discuss doubling, a simple method of constructing two-level fractional factorial designs. Given an $N \times n$ matrix $\mathbf{X}$ with entries 1 and -1, the $2N \times 2n$ matrix

$$\begin{bmatrix} \mathbf{X} & \mathbf{X} \\ \mathbf{X} & -\mathbf{X} \end{bmatrix} \tag{11.2}$$

is called the *double* of $\mathbf{X}$ and is denoted by $D(\mathbf{X})$. That is,

$$D(\mathbf{X}) = \begin{bmatrix} 1 & 1 \\ 1 & -1 \end{bmatrix} \otimes \mathbf{X}.$$

If $\mathbf{X}$ is considered as an N-run design for n two-level factors, then $D(\mathbf{X})$ is a design that doubles both the run size and the number of factors of $\mathbf{X}$. This method was used by Plackett and Burman (1946) in their construction of orthogonal main-effect plans based on Hadamard matrices. The key fact is that if $\mathbf{X}$ is a Hadamard matrix of order N, then $D(\mathbf{X})$ is a Hadamard matrix of order $2N$.

Example 11.1. As discussed in Remark 9.5, a saturated regular design with $n = 2^k - 1$ and $N = 2^k$ can be constructed by doubling the matrix $\begin{bmatrix} 1 & 1 \\ 1 & -1 \end{bmatrix}$ repeatedly $k - 1$ times, followed by the deletion of the column of all 1's.

Maximal even designs can also be constructed by the method of doubling.

Theorem 11.7. *The maximal even design with $n = 2^{k-1}$ and $N = 2^k$ ($k \geq 3$) can be constructed by repeatedly doubling the complete 2^1 design $k - 1$ times.*

Some basic properties of doubling were discussed in Chen and Cheng (2006). The following result (Exercise 8.2) can be proved by directly verifying the condition in the definition of an orthogonal array.

Theorem 11.8. *If $\mathbf{X}$ is a two-level orthogonal array of strength two, then $D(\mathbf{X})$ is also an orthogonal array of strength two. If $\mathbf{X}$ is a two-level orthogonal array of strength three or higher, then $D(\mathbf{X})$ is an orthogonal array of strength three.*

In Theorem 11.8, even if $\mathbf{X}$ has strength higher than three, $D(\mathbf{X})$ only has strength three. This is because $D(\mathbf{X})$ contains four columns of the form

$$\begin{bmatrix} \mathbf{a} & \mathbf{b} & \mathbf{a} & \mathbf{b} \\ \mathbf{a} & \mathbf{b} & -\mathbf{a} & -\mathbf{b} \end{bmatrix}, \tag{11.3}$$

whose Hadamard product has all the entries equal to 1. Therefore $D(\mathbf{X})$ cannot have strength four.

Since all the regular designs are projections of saturated regular designs, and saturated regular designs can be constructed by doubling (see Example 11.1), it follows that if $\mathbf{X}$ is a regular design, then $D(\mathbf{X})$ is itself a projection of a saturated regular design and hence is also regular. We have the following counterpart of Theorem 11.8.

Theorem 11.9. *If $\mathbf{X}$ is a two-level regular fractional factorial design of resolution III, then $D(\mathbf{X})$ is also a regular design of resolution III. If $\mathbf{X}$ is a two-level regular design of resolution IV+, then $D(\mathbf{X})$ is a regular design of resolution IV.*

The aliasing pattern of the two-factor interactions under a regular design can be related to that under its double.

Theorem 11.10. *For each regular 2^{n-p} design* **X**, *let* g, f, *and* m_i, $1 \le i \le g$, *be defined as in Section 10.3, and let* g^*, f^*, m_i^* *be the corresponding quantities for* $D(\mathbf{X})$. *Then* $g^* = 2g + 1$, $f^* = 2f + 1$, $m_{2h-1}^* = m_{2h}^* = 2m_h$, *for all* $1 \le h \le g$, *and* $m_{2g+1}^* = n$.

Proof. Since $D(\mathbf{X})$ is a $2^{2n-(n+p-1)}$ design, we have $g^* = 2^{2n-(n+p-1)} - 1$ and $f^* = 2^{2n-(n+p-1)} - 2n - 1$. The relations $f^* = 2f + 1$ and $g^* = 2g + 1$ are obvious.

By the definition of doubling, each column **a** of **X** generates two columns $\begin{bmatrix} \mathbf{a} \\ \mathbf{a} \end{bmatrix}$ and $\begin{bmatrix} \mathbf{a} \\ -\mathbf{a} \end{bmatrix}$ of $D(\mathbf{X})$. We denote the factor corresponding to the former by A^+ and that corresponding to the latter by A^- if A is the factor corresponding to **a**. Then the following facts can easily be verified:

(a) For any pair of factors A and B in **X**, the two two-factor interactions A^+A^- and B^+B^- are aliased under $D(\mathbf{X})$. This follows from the observation that the componentwise product of the four columns in (11.3) has all the entries equal to 1.

(b) Similar to (a), for any pair of factors A and B in **X**, A^+B^+ and A^-B^- are aliased, and A^+B^- and A^-B^+ are also aliased.

(c) If AB and CD are aliased under **X**, then A^+B^+, A^-B^-, C^+D^+, and C^-D^- are aliased under $D(\mathbf{X})$. Likewise, A^+B^-, A^-B^+, C^+D^-, and C^-D^+ are also aliased.

(d) If the main effect of A and two-factor interaction CD are aliased under **X**, then A^+, C^+D^+, and C^-D^- are aliased under $D(\mathbf{X})$, and A^-, C^+D^-, and C^-D^+ are also aliased.

By (b), (c), and (d), each alias set of **X** containing two-factor interactions generates two such alias sets under $D(\mathbf{X})$, both of which contain twice as many two-factor interactions as in the original alias set under **X**. By (a), under $D(\mathbf{X})$, there is an extra alias set containing all the n two-factor interactions of the form A^+A^-, where A is a factor in **X**. This alias set does not contain main effects. The conclusion of the theorem then follows.   □

It follows from Theorem 11.10 that all the $m_1, \ldots, m_f$ values are nonzero if and only if all the corresponding $m_1^*, \ldots, m_{f^*}^*$ values are nonzero. This, together with Theorems 11.2 and 11.4, implies the result below that reveals the crucial role played by doubling in constructing designs of resolution IV, to be discussed in the next section.

Theorem 11.11. *Let* **X** *be a two-level regular design of resolution IV+, then* **X** *is maximal if and only if* $D(\mathbf{X})$ *is maximal.*

Example 11.2. The 2^1 design has only one alias set consisting of one main effect. Doubling once, we obtain the complete 2^2 design, which has three alias sets; two of them each contains one main effect, and the third one contains a two-factor interaction. Applying Theorem 11.10 repeatedly, we can easily see that under the resulting maximal even design with $N = 2^k$ and $n = 2^{k-1}$, the $\binom{2^{k-1}}{2}$ two-factor interactions are uniformly distributed over the $2^{k-1} - 1$ alias sets that do not contain main effects, with 2^{k-2} two-factor interactions in each of these alias sets. It follows from Corollary 10.3 that the maximal even designs have maximum estimation capacity.

Example 11.3. Consider the 2_V^{5-1} design defined by $I = ABCDE$. This design has 15 alias sets, 10 of which do not contain main effects. Since the resolution is V, each of these 10 alias sets contains exactly one two-factor interaction. It follows that the design is maximal, and by Theorems 11.9 and 11.11, its double is a maximal design of resolution IV. This is the maximal (and minimum aberration) design for $n = 10$ and $N = 32$ referred to at the end of the paragraph following Corollary 11.5. Repeated doubling generates a family of maximal resolution IV designs with $n = 5N/16$, where $N = 16 \cdot 2^k$ ($k \geq 1$). By Theorem 11.10, under these designs, there are $10 \cdot 2^k + (2^k - 1)$ alias sets that do not contain main effects, $10 \cdot 2^k$ of which each contains 2^k two-factor interactions, and $2^k - 1$ of which each contains $5 \cdot 2^{k-1}$ two-factor interactions. We will show in Theorem 11.25 that all the designs in this family have minimum aberration.

11.4   Maximal designs with $N/4 + 1 \leq n \leq N/2$

The determination of maximal designs of resolution IV is a mathematical problem whose solution is of considerable importance to the construction of resolution IV designs. Some results are available in the literature of finite projective geometry. While the projective geometric connection will be presented in the last section of this chapter, we translate two important results into the design language and discuss their applications here.

Structures of two-level resolution IV designs with $N/4 + 1 \leq n \leq N/2$ are essentially determined by two key results.

Theorem 11.12. *(Davydov and Tombak, 1990; Bruen, Haddad, and Wehlau, 1998) Every maximal two-level resolution IV+ design with $N/4 + 2 \leq n \leq N/2$, where $N = 2^k$, $k \geq 5$, is the double of a maximal design of resolution IV+.*

Theorem 11.13. *(Bruen and Wehlau, 1999) For each $N = 2^k$ with $k \geq 4$, there exists at least one maximal two-level resolution IV+ design with $n = N/4 + 1$.*

Theorem 11.11 shows that the double of a maximal design of resolution IV+ is also maximal. Theorem 11.12 complements this result by saying that when the number of factors is greater than a quarter of the run size plus 1, any maximal design must be the double of another maximal design. This allows one to determine the maximal designs from the bottom (smallest possible run sizes) up, and Theorem 11.13 assures the existence of the smallest maximal design to be used as the starting point.

For example, a complete search shows that, up to isomorphism, there are exactly two 16-run maximal two-level designs of resolution IV+: the 2^{8-4} maximal even design and the 2^{5-1} design defined by $I = ABCDE$ whose existence is implied by Theorem 11.13. Repeatedly doubling the former yields all maximal even designs with $n = N/2$. The latter, as discussed in Example 11.3, generates a family of maximal resolution IV designs with $n = 5N/16$, where $N = 2^k$, $k \geq 5$. Now, since it is known that there are no 16-run maximal resolution IV+ designs with $5 < n < 8$, it follows from Theorem 11.12 that there are also no maximal resolution IV designs with $5N/16 < n < N/2$ for all $N = 2^k$ with $k \geq 5$. From this we can draw the following important conclusion, which has been stated as Theorem 9.14 in Section 9.6.

Theorem 11.14. *For $5N/16 < n < N/2$, $N = 2^k$, $k \geq 4$, regular designs of resolution IV+ must be projections of maximal even designs; thus they are even designs and can be constructed by the method of foldover.*

Theorem 11.14 leads to considerable simplification in the construction of resolution IV designs, in particular the search for minimum aberration designs, for the case $5N/16 < n < N/2$. For example, suppose we would like to construct a minimum aberration 32-run design with 14 two-level factors. In this case a minimum aberration design must be of resolution IV. Then, instead of choosing 14 from the 31 factors in a 32-run saturated regular design, we only need to choose 14 from the 16 factors in a maximal even design, or equivalently, drop 2 factors from the maximal even design. The latter is considerably easier to do. Also recall that the 32-run maximal even design can be constructed easily by doubling the 2^1 design four times.

Let us now move up to 32-run designs. Doubling the two 16-run maximal designs of resolution IV once, we obtain the 2^{16-11} maximal even design and a 2^{10-5} design, both of which are maximal. Theorem 11.13 assures the existence of at least one 32-run maximal two-level resolution IV+ design with $n = 9$. A computer search shows that, up to isomorphism, there is exactly one such design, the 2_{IV}^{9-4} design with independent defining words $ABCF$, $BCDH$, $ACDG$, and $ABDEJ$, where the nine factors are A, B, C, D, E, F, G, H, and J. It can also be verified that there are no other 32-run maximal two-level resolution IV+ designs. Therefore there are exactly three 32-run maximal two-level designs of resolution IV: one 2^{16-11}, one 2^{10-5}, and one 2^{9-4} design. Repeatedly doubling the maximal 2^{9-4} design yields a family of maximal two-level resolution IV designs with $n = 9N/32$, where $N = 2^k, k \geq 5$. Again Theorem 11.12 implies that there are no maximal two-level resolution IV designs with $9N/32 < n < 5N/16$.

Theorem 11.15. *For $9N/32 < n \leq 5N/16$, where $N = 2^k, k \geq 5$, all two-level resolution IV designs must be projections of either the maximal even design or the maximal resolution IV design with $5N/16$ factors.*

Doubling the three 32-run maximal designs of resolution IV once, we obtain three 64-run maximal designs: the 2^{32-26} maximal even design, a 2^{20-14} design, and a 2^{18-12} design. Again Theorem 11.13 assures the existence of at least one 64-run maximal resolution IV design with $n = 17$. The complete search by Block and Mee (2003) shows that, up to isomorphism, there are five such designs. There is also a 64-run maximal resolution IV design with $n = 13$. Therefore there are nine nonisomorphic 64-run maximal designs of resolution IV: one 2^{32-26}, one 2^{20-14}, one 2^{18-12}, five 2^{17-11}, and one 2^{13-7}. Repeatedly doubling the five 2^{17-11} and the 2^{13-7} maximal designs leads to five families of maximal resolution IV designs with $n = 17N/64$ and one family of maximal resolution IV designs with $n = 13N/64$, for all $N = 2^k$, $k \geq 6$.

Maximal resolution IV 128-run designs were determined by Khatirinejad and Lisoněk (2006). There are a lot more such designs: five 2^{21-14}, two 2^{22-15}, six 2^{24-17}, thirteen 2^{25-18}, two 2^{26-19}, four 2^{27-20}, one 2^{28-21}, three 2^{29-22}, two 2^{31-24}, forty-two 2^{33-26}, five 2^{34-27}, one 2^{36-29}, one 2^{40-33}, and one 2^{64-57}.

Theorems 11.12 and 11.13 together imply that for $N/4 + 1 \leq n \leq N/2$, $N = 2^k$, $k \geq 4$, a two-level maximal resolution IV+ design must be one of three kinds:

(a) a maximal even design $(n = N/2)$;

(b) a design with $n = N/4 + 1$;

(c) a design obtained by repeatedly doubling an N'-run maximal two-level resolution IV+ design with n' factors, where $n' = N'/4 + 1$ and $N' \geq 16$.

Thus for two-level maximal resolution IV+ designs with $N/4 + 1 \leq n \leq N/2$, we must have $n = N/2$, $N/4 + 1$, or $n = (2^i + 1)N/2^{i+2}$, where i is an integer such that $i \geq 2$, so

$$n \in \{N/2, 5N/16, 9N/32, 17N/64, 33N/128, \ldots\}.$$

We state this result on the structures of maximal two-level resolution IV+ designs formally as a theorem.

Theorem 11.16. *For $n \geq N/4 + 1$, where $N = 2^k$, $k \geq 4$, there exists at least one maximal two-level design of resolution IV+ if and only if $n = N/2$, $N/4 + 1$, or $(2^i + 1)N/2^{i+2}$ for some integer $i \geq 2$. A maximal two-level resolution IV+ design with $n = (2^i + 1)N/2^{i+2}$, where $N = 2^k$, $k > i + 2$, $i \geq 2$, can be constructed by doubling a maximal 2^{i+2}-run resolution IV+ design with $2^i + 1$ factors $k - i - 2$ times.*

11.5   Maximal designs with $n = N/4 + 1$

All the maximal resolution IV+ designs characterized in Theorem 11.16 either have $n = N/4 + 1$, $N \geq 16$, or are repeated doubles of such designs. In this section, we present some results on the construction of maximal resolution IV+ designs with $n = N/4 + 1$.

Let $\mathbf{X}$ be a maximal even design with $N/2$ runs and $N/4$ factors. Partition the columns of $\mathbf{X}$ as $\mathbf{X} = [\mathbf{B}\ \mathbf{C}]$, where each of $\mathbf{B}$ and $\mathbf{C}$ contains at least one column. Let

$$\mathbf{S} = \begin{bmatrix} 1 & \mathbf{B} & \mathbf{C} \\ -1 & -\mathbf{B} & \mathbf{C} \end{bmatrix}. \tag{11.4}$$

Recall that $\mathbf{X}$ can be constructed by folding over the saturated regular design of size $N/4$ according to (8.18). The construction of $\mathbf{S}$ in (11.4), an N-run design with $N/4 + 1$ factors, is essentially to foldover once more, except that when folding over $\mathbf{X}$, some columns of $\mathbf{X}$ (those in $\mathbf{C}$) do not change signs. Therefore this is an example of partial foldover. For convenience of presentation, without loss of generality, we take the first column of $\mathbf{S}$ to be $[\mathbf{1}^T \;\; -\mathbf{1}^T]^T$, instead of $[-\mathbf{1}^T \;\; \mathbf{1}^T]^T$.

Theorem 11.17. *Let* $\mathbf{X} = [\mathbf{B} \;\; \mathbf{C}]$ *be a maximal even design. Then the design* $\mathbf{S}$ *constructed in (11.4) is a regular design of resolution IV+. If* $\mathbf{B}$ *contains an odd number of columns, then* $\mathbf{S}$ *is maximal.*

Proof. Since $\mathbf{S}$ has one more factor than $\mathbf{X}$ and twice as many runs, $\mathbf{X}$ and $\mathbf{S}$ have the same number of defining words. Represent each defining word of $\mathbf{X}$ by a vector $\mathbf{a} = (a_1, \ldots, a_{N/4})^T$, where $a_i = 1$ if the corresponding factor appears in the defining word and is 0 otherwise. Suppose $\mathbf{B}$ contains b columns, and $a_1, \ldots, a_b$ correspond to the factors in $\mathbf{B}$. For each defining word $\mathbf{a}$ of $\mathbf{X}$, let

$$\bar{\mathbf{a}} = \begin{cases} (0, a_1, \ldots, a_{N/4})^T, & \text{if an even number of } a_1, \ldots, a_b \text{ are nonzero;} \\ (1, a_1, \ldots, a_{N/4})^T, & \text{if an odd number of } a_1, \ldots, a_b \text{ are nonzero.} \end{cases} \tag{11.5}$$

Then it can easily be checked that $\bar{\mathbf{a}}$ is a defining word of $\mathbf{S}$. Therefore the defining words of $\mathbf{S}$ consist of those of the form $\bar{\mathbf{a}}$, where $\mathbf{a}$ is a defining word of $\mathbf{X}$. Since $\mathbf{X}$ is a resolution IV design, it follows from (11.5) that $\mathbf{S}$ is of resolution IV+.

Since maximal even designs can be constructed from saturated regular designs by the foldover construction (8.18), we may write $\mathbf{X}$ as

$$\mathbf{X} = \begin{bmatrix} \tilde{\mathbf{B}} & \mathbf{1} & \tilde{\mathbf{C}} \\ -\tilde{\mathbf{B}} & -\mathbf{1} & -\tilde{\mathbf{C}} \end{bmatrix}.$$

We consider the case

$$\mathbf{B} = \begin{bmatrix} \tilde{\mathbf{B}} \\ -\tilde{\mathbf{B}} \end{bmatrix}, \quad \mathbf{C} = \begin{bmatrix} \mathbf{1} & \tilde{\mathbf{C}} \\ -\mathbf{1} & -\tilde{\mathbf{C}} \end{bmatrix}. \tag{11.6}$$

The case where $[\mathbf{1}^T \;\; -\mathbf{1}^T]^T$ is a column of $\mathbf{B}$ can be similarly handled. Then $[\tilde{\mathbf{B}} \;\; \tilde{\mathbf{C}}]$ is a saturated regular design of size $N/4$ and $\tilde{\mathbf{B}}$ has b columns. By assumption, $b \geq 1$ and is odd. We have

$$\mathbf{S} = \begin{bmatrix} 1 & \tilde{\mathbf{B}} & 1 & \tilde{\mathbf{C}} \\ 1 & -\tilde{\mathbf{B}} & -1 & -\tilde{\mathbf{C}} \\ -1 & -\tilde{\mathbf{B}} & 1 & \tilde{\mathbf{C}} \\ -1 & \tilde{\mathbf{B}} & -1 & -\tilde{\mathbf{C}} \end{bmatrix}. \tag{11.7}$$

Also, a saturated regular design of size N can be written as

$$\begin{bmatrix} \tilde{\mathbf{B}} & \tilde{\mathbf{C}} & 1 & \tilde{\mathbf{B}} & \tilde{\mathbf{C}} & 1 & \tilde{\mathbf{B}} & \tilde{\mathbf{C}} & 1 & \tilde{\mathbf{B}} & \tilde{\mathbf{C}} \\ \tilde{\mathbf{B}} & \tilde{\mathbf{C}} & -1 & -\tilde{\mathbf{B}} & -\tilde{\mathbf{C}} & 1 & \tilde{\mathbf{B}} & \tilde{\mathbf{C}} & -1 & -\tilde{\mathbf{B}} & -\tilde{\mathbf{C}} \\ \tilde{\mathbf{B}} & \tilde{\mathbf{C}} & 1 & \tilde{\mathbf{B}} & \tilde{\mathbf{C}} & -1 & -\tilde{\mathbf{B}} & -\tilde{\mathbf{C}} & -1 & -\tilde{\mathbf{B}} & -\tilde{\mathbf{C}} \\ \tilde{\mathbf{B}} & \tilde{\mathbf{C}} & -\tilde{\mathbf{1}} & -\tilde{\mathbf{B}} & -\tilde{\mathbf{C}} & -1 & -\tilde{\mathbf{B}} & -\tilde{\mathbf{C}} & 1 & \tilde{\mathbf{B}} & \tilde{\mathbf{C}} \end{bmatrix}, \tag{11.8}$$

which is obtained by deleting the column of 1's from

$$\begin{bmatrix} 1 & 1 \\ 1 & -1 \end{bmatrix} \otimes \begin{bmatrix} 1 & 1 \\ 1 & -1 \end{bmatrix} \otimes [\mathbf{1} \quad \tilde{\mathbf{B}} \quad \tilde{\mathbf{C}}].$$

Now we show that if b is odd, then $\mathbf{S}$ is maximal. We need to verify that adding to $\mathbf{S}$ a column of (11.8) that is not already in $\mathbf{S}$ will create a defining word of length three; that is, the Hadamard product of some three columns of the expanded array is the column of all 1's or all -1's. This is clear if the column is $(\mathbf{1}^T, -\mathbf{1}^T, -\mathbf{1}^T, \mathbf{1}^T)^T$, one of the columns of $[\tilde{\mathbf{B}}^T -\tilde{\mathbf{B}}^T \ \tilde{\mathbf{B}}^T -\tilde{\mathbf{B}}^T]^T$, one of the columns of $[\tilde{\mathbf{B}}^T \ \tilde{\mathbf{B}}^T -\tilde{\mathbf{B}}^T -\tilde{\mathbf{B}}^T]^T$, one of the columns of $[\tilde{\mathbf{C}}^T \ \tilde{\mathbf{C}}^T \ \tilde{\mathbf{C}}^T \ \tilde{\mathbf{C}}^T]^T$, or one of the columns of $[\tilde{\mathbf{C}}^T -\tilde{\mathbf{C}}^T -\tilde{\mathbf{C}}^T \ \tilde{\mathbf{C}}^T]^T$. For example, if the column $(\mathbf{c}^T, \mathbf{c}^T, \mathbf{c}^T, \mathbf{c}^T)^T$ is chosen, where $\mathbf{c}$ is a column of $\tilde{\mathbf{C}}$, then its Hadamard product with the column $(\mathbf{c}^T, -\mathbf{c}^T, \mathbf{c}, -\mathbf{c}^T)^T$ of $\mathbf{S}$ in (11.7) is equal to $(\mathbf{1}^T, -\mathbf{1}^T, \mathbf{1}^T, -\mathbf{1}^T)^T$, which is already in $\mathbf{S}$.

Now suppose a column $(\mathbf{c}^T, \mathbf{c}^T, -\mathbf{c}^T, -\mathbf{c}^T)^T$ of $[\tilde{\mathbf{C}}^T \ \tilde{\mathbf{C}}^T -\tilde{\mathbf{C}}^T -\tilde{\mathbf{C}}^T]^T$ is added to $\mathbf{S}$. Since b is odd, there is at least one column $\mathbf{b}$ of $\tilde{\mathbf{B}}$ such that $\mathbf{c} \odot \mathbf{b}$ is not a column of $\tilde{\mathbf{B}}$. Let $\mathbf{c}_1 = \mathbf{c} \odot \mathbf{b}$. Then $\mathbf{c}_1$ is a column of $\tilde{\mathbf{C}}$ and $\mathbf{c} = \mathbf{b} \odot \mathbf{c}_1$. Then $(\mathbf{c}^T, \mathbf{c}^T, -\mathbf{c}^T, -\mathbf{c}^T)^T$ is the Hadamard product of $(\mathbf{b}^T, -\mathbf{b}^T, -\mathbf{b}^T, \mathbf{b}^T)^T$ and $(\mathbf{c}_1^T, -\mathbf{c}_1^T, \mathbf{c}_1^T, -\mathbf{c}_1^T)^T$; both of the latter two columns are in $\mathbf{S}$. This results in a defining word of length three.

Last, we consider the case of adding a column of the form $(\mathbf{b}^T, \mathbf{b}^T, \mathbf{b}^T, \mathbf{b}^T)^T$ to $\mathbf{S}$, where $\mathbf{b}$ is a column of $\tilde{\mathbf{B}}$. If there are two columns $\mathbf{b}_1$ and $\mathbf{b}_2$ of $\tilde{\mathbf{B}}$ such that $\mathbf{b} = \mathbf{b}_1 \odot \mathbf{b}_2$, then, since $(\mathbf{b}^T, \mathbf{b}^T, \mathbf{b}^T, \mathbf{b}^T)^T$ is the Hadamard product of $(\mathbf{b}_1^T, -\mathbf{b}_1^T, -\mathbf{b}_1^T, \mathbf{b}_1^T)^T$ and $(\mathbf{b}_2^T, -\mathbf{b}_2^T, -\mathbf{b}_2^T, \mathbf{b}_2^T)^T$, a defining word of length three would be created. On the other hand, if there are two columns $\mathbf{c}_1$ and $\mathbf{c}_2$ of $\tilde{\mathbf{C}}$ such that $\mathbf{b} = \mathbf{c}_1 \odot \mathbf{c}_2$, then $(\mathbf{b}^T, \mathbf{b}^T, \mathbf{b}^T, \mathbf{b}^T)^T$ is the Hadamard product of $(\mathbf{c}_1^T, -\mathbf{c}_1^T, \mathbf{c}_1^T, -\mathbf{c}_1^T)^T$ and $(\mathbf{c}_2^T, -\mathbf{c}_2^T, \mathbf{c}_2^T, -\mathbf{c}_2^T)^T$; a defining word of length three is also formed. Therefore it remains to consider the case where $\mathbf{b}$ is neither of the form $\mathbf{b}_1 \odot \mathbf{b}_2$ nor $\mathbf{c}_1 \odot \mathbf{c}_2$, where $\mathbf{b}_1, \mathbf{b}_2$ are columns of $\tilde{\mathbf{B}}$, and $\mathbf{c}_1, \mathbf{c}_2$ are columns of $\tilde{\mathbf{C}}$. Then, for each column $\mathbf{b}_1$ of $\tilde{\mathbf{B}}$, $\mathbf{b} \odot \mathbf{b}_1$ is a column of $\tilde{\mathbf{C}}$, and for each column $\mathbf{c}_1$ of $\tilde{\mathbf{C}}$, $\mathbf{b} \odot \mathbf{c}_1$ is a column of $\tilde{\mathbf{B}}$. This establishes a one-to-one correspondence between the columns of $\tilde{\mathbf{C}}$ and those of $\tilde{\mathbf{B}}$ except $\mathbf{b}$. Let c be the number of columns of $\tilde{\mathbf{C}}$. Then $b = c + 1$. Since $b + c = N/4 - 1$, we must have $b = N/8$. For $N = 2^k$, $k \geq 4$, b must be even. This is a contradiction. $\quad\square$

When $\mathbf{B}$ contains an even number of columns, $\mathbf{S}$ may or may not be maximal.

Proposition 11.18. *In (11.4), if $\mathbf{B}$ consists of two columns, then $\mathbf{S}$ is not maximal.*

We leave the proof of Proposition 11.18 as an exercise. For example, let $\mathbf{b}_1$ and $\mathbf{b}_2$ be the two columns of $\tilde{\mathbf{B}}$ in (11.6). Then $\mathbf{c} = \mathbf{b}_1 \odot \mathbf{b}_2$ must be a column of $\tilde{\mathbf{C}}$. One can verify that if $(\mathbf{c}^T, \mathbf{c}^T - \mathbf{c}^T, -\mathbf{c}^T)^T$ is added to $\mathbf{S}$, then no defining word of length three is formed. The resulting design has resolution IV+ and hence $\mathbf{S}$ is not maximal.

Example 11.4. Maximal 2^{5-1} and 2^{9-4} designs can be obtained by applying the partial foldover construction (11.4) to the 2^{4-1} and 2^{8-4} maximal even designs, respectively. The five nonisomorphic 2^{17-11} maximal designs can be constructed from the 2^{16-11} maximal even design with different choices of $\mathbf{B}$.

Remark 11.3. The construction in (11.4) can also be applied to nonregular designs. If $\mathbf{X}$ is an $OA(N/2, 2^{N/4}, 3)$ obtained from a saturated $OA(N/4, 2^{N/4-1}, 2)$ by the method of foldover, then $\mathbf{S}$ is an orthogonal array of strength three; see Cheng, Mee, and Yee (2008) and Exercise 8.1. The notion of second-order saturation can also be defined for nonregular designs. Cheng, Mee, and Yee (2008) showed that if $\mathbf{B}$ consists of one single column, then $\mathbf{S}$ is second-order saturated.

11.6 Partial foldover

We consider the problem of supplementing a regular 2_{III}^{n-p} design by 2^{n-p} runs in which the two levels of some or all the factors are switched, and, unlike in (8.18), no extra factor is added. Throughout this section, by foldover we mean partial or full foldover. We have seen that the combined design resulting from a full foldover has resolution IV and is even. Partial foldovers, however, may not increase the resolution of the initial design. Several questions arise:

(a) Which resolution IV designs can be obtained as combined designs of resolution III designs and their partial foldovers?

(b) Which 2_{III}^{n-p} designs have partial foldovers with even-odd combined designs of resolution IV?

(c) How do even-odd combined designs of resolution IV compare with even combined designs obtained by full foldovers?

(d) Given a regular 2_{III}^{n-p} design, which foldover produces the best combined design under the minimum aberration criterion?

(e) Suppose the combined design $\tilde{d}$ of a 2_{III}^{n-p} design d and a foldover of d has minimum aberration among all the combined designs of 2_{III}^{n-p} designs and their foldovers. Such a $\tilde{d}$ is called combined-optimal by Li and Lin (2003). When is a combined-optimal design strong combined-optimal in the sense that it has minimum aberration among all the $2^{n-(p-1)}$ designs?

We first present an answer to question (a) from Liau (2006),

Theorem 11.19. *A regular design of resolution IV is the combined design of a resolution III+ design and a partial or full foldover if and only if it is not maximal.*

Proof. Suppose d^* is a $2_{IV}^{n-(p-1)}$ design that is the combined design of a 2_{III+}^{n-p} design d and a foldover of d. Since d consists of half of the treatment combinations in d^*, it has an extra independent defining word, say X, in addition to those of d^*. Furthermore, since d has resolution III+, X has length at least three, and all its generalized interactions with the defining words of d^* also have lengths at least three. If we use X to define a new treatment factor and add it to d^*, then all the defining words of the resulting $2^{(n+1)-p}$ design have lengths at least four. It follows that d^* is not maximal.

Conversely, if d^* is not maximal, then by Theorem 11.2, it has an alias set in which all the factorial effects (words) involve at least three factors. Pick any of these words, say X, to divide the treatment combinations in d^* into two halves: one is defined by the independent defining words of d^* and X, and the other is defined by the independent defining words of d^* and $-X$. The two fractions are regular 2_{III+}^{n-p} designs with the same defining words, but with different signs attached to some of the defining words. It is easy to see that each fraction can be obtained from the other by reversing the levels of the added factors that are defined by the same interactions of the basic factors with different signs in the two fractions.   □

It follows from Theorem 11.14 that all the $2N$-run resolution IV designs with $5N/8 < n \leq N$, where $N = 2^k$, $k \geq 3$, must be even. Therefore partial foldovers satisfying the property in (b) exist only if $n \leq 5N/8$. This necessary condition, observed in Li and Mee (2002), can be weakened to $n < 5N/8$ by Theorem 11.19. If a foldover satisfying the property in (b) exists for $n = 5N/8$, then there is a $2N$-run even-odd resolution IV combined design with $5N/8$ factors. By Theorem 11.14, such a design is either a projection of the maximal even design of size $2N$ or is a maximal design. The latter possibility is ruled out by Theorem 11.19, and the former possibility contradicts the requirement that the combined design is even-odd.

Li and Mee (2002) proposed an algorithm to answer question (b) for any given resolution III design. They applied the algorithm and found that five of the 32-run designs with $n > 16$ cataloged in Chen, Sun, and Wu (1993) have partial foldovers satisfying the property in (b).

Xu and Cheng (2008) showed that for all $N/2 + 1 \leq n < 5N/8$, where $N = 2^k$, $k \geq 3$, the best n-factor projection of the maximal design with $2N$ runs and $5N/8$ factors has less aberration than the best n-factor projection of the maximal even design of size $2N$ and, by Theorem 11.19, can be obtained as a combined design. Thus for all $N/2 + 1 \leq n < 5N/8$, there are partial foldovers that yield better combined designs than full holdovers under the minimum aberration criterion. Also, as we pointed out before, an even design of size $2N$ has all the two-factor interactions in only $N - 1$ alias sets and so is less effective for de-aliasing. This provides some answers to question (c).

To answer questions (d) and (e), Li and Lin (2003) did a computer search and tabulated combined-optimal designs resulting from the foldovers of all 16- and some 32-run designs. Among the designs they considered, there is only one combined optimal design, a 2^{10-5} design, that is not strong combined-optimal. This can be explained by Theorem 11.19: since the minimum aberration 32-run design with 10 factors is maximal, it is not a combined design. A general result is as follows.

Corollary 11.20. *A combined-optimal design is strong combined optimal if and only if there is a minimum aberration design that is not maximal.*

It follows from Theorem 11.16 and Corollary 11.20 that for $n \geq N/2 + 1$, if $n \notin \{N, 5N/8, 9N/16, 17N/32, 33N/64, \ldots\}$, then strong combined-optimal $2N$-run designs with n factors exist. For $n \in \{N, 5N/8, 9N/16, 17N/32, 33N/64, \ldots\}$, there is a strong combined-optimal $2N$-run design with n factors if all the maximal designs with n factors do not have minimum aberration.

Xu and Cheng (2008) showed that for all $17N/32 \le n < 5N/8$, where $N = 2^k$, $k \ge 4$, a minimum aberration design with $2N$ runs and n two-level factors must be a projection of the maximal design with $5N/8$ factors (see Theorem 11.30 in Section 11.10). It follows that in all these cases, the best n-factor projection of the maximal design with $2N$ runs and $5N/8$ factors is strong combined-optimal.

Remark 11.4. When the foldover fraction is used as a follow-up plan at the second stage, the initial design and the foldover design should be regarded as two blocks. In this case, the two blocks are defined by the word X in the proof of Theorem 11.19.

A follow-up fraction as large as the original design can be wasteful when there is only a small number of effects to be de-aliased. Mee and Peralta (2000) pointed out that for many resolution IV designs of size N, adding a foldover fraction provides fewer than $N/2$ additional degrees of freedom for two-factor interactions. *Semifolding*, in which only half of the foldover runs are added, is a more cost-effective alternative (Daniel, 1962; Barnett, Czitrom, John, and Leon, 1997). Efficiencies of the estimates are lower, and there is a loss of orthogonality since the combined designs resulting from semifolding are no longer regular, but Mee and Peralta (2000)'s study showed that a semifold design can often estimate as many two-factor interactions as a foldover plan does. For semifolding, in addition to choosing the factors whose levels are to be reversed, one also has to decide which half of the foldover fraction to add. We refer readers to Balakrishnan and Yang (2009) and Edwards (2011) for some recent work.

11.7 More on clear two-factor interactions

It was shown in Section 10.8 that $n \le 2^{n-p-2} + 1 = N/4 + 1$ is a necessary condition for the existence of a regular 2_{IV}^{n-p} design that has at least one clear two-factor interaction. We show below that for $n = N/4 + 1$, one of the maximal designs constructed in Section 11.5 by the method of partial foldover has the maximum number of clear two-factor interactions among resolution IV designs.

In (11.4), let **B** consist of one single column. Denote the first two columns of the resulting design **S** by $\mathbf{a}_1$, $\mathbf{a}_2$, and the remaining columns by $\mathbf{a}_3, \ldots, \mathbf{a}_{N/4+1}$. These columns correspond to main effects of the $N/4 + 1$ factors, which we label as $A_1, A_2, A_3, \ldots, A_{N/4+1}$. Then it is straightforward to show that none of the $N/2 - 1$ Hadamard products $\mathbf{a}_1 \odot \mathbf{a}_2$, $\mathbf{a}_1 \odot \mathbf{a}_i$, $3 \le i \le N/4+1$, and $\mathbf{a}_2 \odot \mathbf{a}_j$, $3 \le j \le N/4+1$, can be written as the Hadamard product of another pair of columns of **S**. One simple way to see this is to write **S** in the form of (11.7) and compare the signs in pairwise Hadamard products of the columns of **S**. It follows that the $N/2 - 1$ two-factor interactions $A_1 A_2$, $A_1 A_i$, $3 \le i \le N/4+1$, and $A_2 A_j$, $3 \le j \le N/4+1$, are clear. This was observed by Tang et al. (2002), who also showed that in this case, the number of clear two-factor interactions cannot be more than $N/2 + 1$. Wu and Wu (2002) improved the upper bound to $N/2 - 1$; see Proposition 11.21 below. Therefore the maximal design **S** in (11.4) with **B** consisting of one single column maximizes the number of clear two-factor interactions among resolution IV designs. The 2_{IV}^{9-4} design with

15 clear two-factor interactions mentioned in Section 10.2 is a member of this family.

Proposition 11.21. *All regular 2_{IV}^{n-p} designs with $n = 2^{n-p-2} + 1$, where $n - p \geq 4$, have no more than $2n - 3$ ($= N/2 - 1$) clear two-factor interactions.*

Proof. Let $f = 2^{n-p} - n - 1$, the number of alias sets that do not contain main effects, and define $m_1, \ldots, m_f$ as in Section 10.3. Also, recall the notation $a_j = \#\{i : 1 \leq i \leq f, m_i = j\}$ defined in (11.1). Denote the numbers of clear and unclear two-factor interactions by C and U, respectively. Then $C + U = n(n-1)/2$, and $C = a_1$.

The two-factor interactions in the same alias set cannot share any factor; otherwise, some main effects would be aliased. Thus we must have $m_i \leq n/2$. Let $r = \lfloor n/2 \rfloor$. Then $a_j = 0$ for all $j > r$. Also, $r = (n-1)/2$ since n is odd.

For $2 \leq j \leq r$, if $a_j > 0$, then there is at least one alias set containing j two-factor interactions. These j two-factor interactions together involve $2j$ factors, any two of which form a two-factor interaction that is not clear. This shows that

$$\text{if } 2 \leq j \leq r \text{ and } a_j > 0, \text{ then } U \geq \binom{2j}{2} = j(2j-1). \qquad (11.9)$$

Since $n = 2^{n-p-2} + 1$, $f = 3n - 5$. Suppose $a_r > 0$. Then by (11.9), $U \geq r(2r-1) = (n-1)(n-2)/2$. It follows that

$$C = n(n-1)/2 - U \leq n - 1 < 2n - 3. \qquad (11.10)$$

On the other hand, if $a_r = 0$, then, since the design is of resolution IV,

$$
\begin{aligned}
n(n-1)/2 \quad &= \quad a_1 + \sum_{j=2}^{r-1} j a_j \\
&\leq \quad C + (r-1) \sum_{j=2}^{r-1} a_j \qquad (11.11) \\
&= \quad C + (r-1)(f - a_0 - C) \\
&\leq \quad C + (r-1)(f - C) \qquad (11.12) \\
&= \quad C + [(n-1)/2 - 1](3n - 5 - C).
\end{aligned}
$$

By solving the inequality $n(n-1)/2 \leq C + [(n-1)/2 - 1](3n - 5 - C)$, we obtain $C \leq 2n - 3$. $\qquad\square$

The OA($N, 2^{N/4+1}, 3$) constructed in Remark 11.3 with **B** consisting of one single column is the nonregular counterpart of the resolution IV design with the maximum number of clear two-factor interactions discussed above. Tang (2006) showed that this design also has $2n - 3 = N/2 - 1$ clear two-factor interactions (for a definition of clear two-factor interactions under nonregular designs, see Section 10.8). It is an open question whether this design also maximizes the number of clear two-factor interactions among nonregular orthogonal arrays of strength three.

Remark 11.5. Examining the proof of Proposition 11.21, by (11.10), (11.11), and (11.12), we see that $C = 2n - 3$ if and only if $a_r = 0$, $a_j = 0$ for all $2 \leq j \leq r - 2$, and $a_0 = 0$. Therefore $C = 2n - 3$ if and only if $a_j = 0$ for all $j \neq 1$, $r - 1$. Then since $a_1 = C$ and $\sum_{j=0}^{r} a_j = f = 3n - 5$, we must have $a_{r-1} = n - 2$; that is, the design has $2n - 3$ alias sets each containing one two-factor interaction and $n - 2$ alias sets each containing $(n - 3)/2$ two-factor interactions. Cheng and Zhang (2010) showed that the designs satisfying this condition must be isomorphic to the design **S** in (11.4) with **B** consisting of one single column. Thus, for $n = N/4 + 1$, $N \geq 32$, the designs that maximize the number of clear two-factor interactions among regular resolution IV designs are unique up to isomorphism. Such designs are also the unique optimal designs with respect to the *general minimum lower-order confounding* (GMC) criterion proposed by Zhang, Li, Zhao, and Ai (2008). The GMC criterion works in a way similar to the minimum aberration criterion by sequentially minimizing the terms in the so-called *aliased effect-number pattern* (Zhang et al., 2008). The aliased effect-number pattern is more refined than the usual wordlength pattern in that the latter is a function of the former. However, sequential minimization of the terms in the aliased effect-number pattern can be viewed as a refinement of the criterion of maximizing the number of clear two-factor interactions and thus often leads to different solutions from the minimum aberration criterion.

11.8 Applications to minimum aberration designs

We give an example to illustrate how the structural results obtained earlier in this chapter can be used to help determine minimum aberration designs. Discussions in more depth will be continued in Sections 11.9 and 11.10.

We first prove a result from Chen and Cheng (2006) that connects the wordlength pattern of a design to that of its double.

Theorem 11.22. *Let* $(A_1, \ldots, A_n)$ *and* $(A_1^*, \ldots, A_{2n}^*)$ *be the wordlength patterns of a regular* 2^{n-p} *design* **X** *and its double* $D(\mathbf{X})$, *respectively. Then for each* h,

$$
A_h^* = \begin{cases} \sum_{s=0}^{\min\{(h-1)/2, n\}} A_{h-2s} \cdot \binom{n-h+2s}{s} 2^{h-2s-1}, & \text{if } h \text{ is odd;} \\ \sum_{s=0}^{\min\{h/2-1, n\}} A_{h-2s} \cdot \binom{n-h+2s}{s} 2^{h-2s-1} + \binom{n}{h/2}, & \text{if } h/2 \text{ is even;} \\ \sum_{s=0}^{\min\{h/2-1, n\}} A_{h-2s} \cdot \binom{n-h+2s}{s} 2^{h-2s-1}, & \text{if } h \text{ is even and } h/2 \text{ is odd.} \end{cases}
$$

(11.13)

Proof. For each factor Z of **X**, define the factors Z^+ and Z^- as in the proof of Theorem 11.10. The identities in this theorem can be obtained by relating the defining words of $D(\mathbf{X})$ to those of **X**. We first observe that if $Z_{i_1} \cdots Z_{i_h}$ is a defining word of **X**, then $Z_{i_1}^{j_1} \cdots Z_{i_h}^{j_h}$, where each j_l is either $+$ or $-$, is a defining word of $D(\mathbf{X})$ if and only if an even number of the j_l's are $-$'s. Also, for any s distinct factors $Z_{i_1}, \ldots, Z_{i_s}$ of **X**, $Z_{i_1}^+ Z_{i_1}^- \cdots Z_{i_s}^+ Z_{i_s}^-$ is a defining word of $D(\mathbf{X})$ as long as s is even. It can be seen that defining words of $D(\mathbf{X})$ are of the form $Z_{i_1}^+ Z_{i_1}^- \cdots Z_{i_s}^+ Z_{i_s}^- Z_{i_{s+1}}^{j_{s+1}} \cdots Z_{i_{s+t}}^{j_{s+t}}$, where $Z_{i_1}, \ldots, Z_{i_{s+t}}$ are distinct factors of **X**, and $Z_{i_{s+1}} \cdots Z_{i_{s+t}}$ is a defining word of **X**.

Furthermore, if s is even, then an even number of the j_l's are $-$'s, and if s is odd, then an odd number of the j_l's are $-$'s. The formulas in (11.13) can be obtained by counting the number of such defining words of a given length, in conjunction with the following facts: let $Z_{i_1}^+ Z_{i_1}^- \cdots Z_{i_s}^+ Z_{i_s}^- Z_{i_{s+1}}^{j_{s+1}} \cdots Z_{i_{s+t}}^{j_{s+t}}$ be of length h, then $t = h - 2s$, and for given s,

(i) if $t \geq 1$, then there are $2^{t-1} = 2^{h-2s-1}$ choices of $(j_{s+1}, \ldots, j_{s+t})$, A_{h-2s} ways to choose $Z_{i_{s+1}}, \ldots, Z_{i_{s+t}}$, and $\binom{n-h+2s}{s}$ ways to choose $Z_{i_1}, \ldots, Z_{i_s}$; in this case, since $t \geq 1$, we have $s \leq (h-1)/2$ when h is odd, and $s \leq h/2 - 1$ when h is even;

(ii) if $t = 0$ ($s = h/2$), then $Z_{i_1}^+ Z_{i_1}^- \cdots Z_{i_s}^+ Z_{i_s}^-$ with even s are the only possible defining words; there are $\binom{n}{h/2}$ ways to choose such words when h is even and $s = h/2$. □

It follows from (11.13) that for designs of resolution III+,

$$A_3^* = 4A_3, \quad A_4^* = 8A_4 + \binom{n}{2}. \tag{11.14}$$

The result below follows from Theorem 11.22.

Corollary 11.23. *A regular design $\mathbf{X}_1$ has less aberration than another regular designs $\mathbf{X}_2$ if and only if $D(\mathbf{X}_1)$ has less aberration than $D(\mathbf{X}_2)$.*

Theorem 9.10 effectively shows that for $n = N/2$, maximal even designs are the only designs with $A_3 = 0$; therefore we have the following result.

Theorem 11.24. *Two-level maximal even designs are minimum aberration designs for $n = N/2$.*

The maximal designs with $n = 5N/16$ also have minimum aberration.

Theorem 11.25. *(Chen and Cheng, 2006; Butler, 2007) For $n = 5N/16$, $N = 16 \cdot 2^k$, $k \geq 0$, the maximal resolution IV+ design, which is the 2^{5-1} design defined by $I = ABCDE$ (for $k = 0$) or its repeated doubles (for $k \geq 1$), has minimum aberration.*

Proof. Since there is no maximal design with $5N/16 < n < N/2$, it suffices to show that the maximal design with $n = 5N/16$ has less aberration than any projection of the maximal even design. Let d be the maximal resolution IV+ design and d' be a projection of the maximal even design onto $5N/16$ factors. By (10.2), it suffices to show that $\sum_{i=1}^g [m_i(d)]^2 < \sum_{i=1}^g [m_i(d')]^2$. Since both designs are of resolution IV, this is the same as to show that $\sum_{i=1}^f [m_i(d)]^2 < \sum_{i=1}^f [m_i(d')]^2$.

By the observations in Example 11.3,

$$\sum_{i=1}^f [m_i(d)]^2 = 10 \cdot 2^k \cdot (2^k)^2 + (2^k - 1)(5 \cdot 2^{k-1})^2. \tag{11.15}$$

Since d' is a projection of the maximal even design, it has $8 \cdot 2^k - 1$ alias sets not containing main effects. Under d', all the two-factor interactions must appear in these alias sets. Therefore

$$\sum_{i=1}^{f} [m_i(d')]^2 \geq \left[\binom{5 \cdot 2^k}{2} / (8 \cdot 2^k - 1) \right]^2 (8 \cdot 2^k - 1). \tag{11.16}$$

Though tedious, it is straightforward to verify that the right-hand side of (11.16) is greater than that of (11.15). □

The maximal designs with $n = 9N/32$, however, do not have minimum aberration. We will return to this and present a stronger result than Theorem 11.25 in Section 11.10.

11.9 Minimum aberration even designs

The complementary design theory developed in Section 10.6 is not useful for finding two-level minimum aberration designs with $n < N/2$. This is because if $n < N/2$, then its complementary design has more than n factors. It follows from Theorem 11.14 that for $5N/16 < n < N/2$, $N = 2^k$, $k \geq 4$, regular designs of resolution IV must be projections of maximal even designs. Thus to construct a minimum aberration design in this case, instead of choosing n factors from the $N - 1$ factors of a saturated design, it is enough to choose from the $N/2$ factors of a maximal even design. A complementary design theory with the maximal even design as the universe, useful when n is close to $N/2$, was developed by Butler (2003a).

Throughout this section, for each even design d, we denote its complement in the maximal even design by $\overline{d}$.

Theorem 11.26. *(Butler, 2003a) For $5N/16 < n < N/2$, a regular 2^{n-p} design d has minimum aberration if and only if*

(i) *d is a projection of the maximal even design of size N, and*

(ii) *the complement of d in the maximal even design has minimum aberration among all the $(N/2 - n)$-factor projections of the maximal even design.*

Proof. We prove this result by using the equivalence of minimum aberration and minimum moment aberration. As in Section 10.10, for each design d, let $\mathbf{X}_d$ be the $N \times n$ matrix with each column corresponding to a factor and each row representing a treatment combination. We claim that for any pair $(d, \overline{d})$ of complementary subdesigns of the maximal even design, $H_t(d)$ and $H_t(\overline{d})$ differ by a constant that does not depend on d, where H_t is defined in (10.13). It follows that sequentially minimizing $H_t(d)$ over the n-factor projections of the maximal even design is equivalent to that of $H_t(\overline{d})$. Then the proof is finished since we have seen in Section 10.10 that minimum aberration is equivalent to minimum moment aberration.

To prove the claim, note that the maximal even design is of the form

$$\begin{bmatrix} -\mathbf{1}_{N/2} & \mathbf{Y} \\ \mathbf{1}_{N/2} & -\mathbf{Y} \end{bmatrix},$$

where $\mathbf{Y}$ is a saturated regular two-level design of size $N/2$. Let ϕ_{gh} and $\overline{\phi}_{gh}$ be the (g,h)th entry of $\mathbf{X}_d\mathbf{X}_d^T$ and $\mathbf{X}_{\overline{d}}\mathbf{X}_{\overline{d}}^T$, respectively.

Since d and $\overline{d}$ can be constructed by the foldover method, it is clear that

$$\sum_{g=1}^{N}\sum_{h=1}^{N}\left[\phi_{gh}\right]^t = \sum_{g=1}^{N}\sum_{h=1}^{N}\left[\overline{\phi}_{gh}\right]^t = 0 \text{ for all odd } t.$$

On the other hand, since $[-\mathbf{1}_{N/2}\ \ \mathbf{Y}]$ is a Hadamard matrix, all the pairwise inner products of its rows are equal to zero. Therefore

$$\phi_{gh}+\overline{\phi}_{gh} = 0 \text{ for all } g \neq h \text{ and } |g-h| \neq N/2. \tag{11.17}$$

Furthermore,

$$\phi_{gg} = n \text{ and } \overline{\phi}_{gg} = N/2 - n \text{ for all } g, \tag{11.18}$$

$$\phi_{gh} = -n \text{ and } \overline{\phi}_{gh} = -N/2 + n \text{ for } |g-h| = N/2. \tag{11.19}$$

It follows that for even t,

$$\sum_{g=1}^{N}\sum_{h=1}^{N}\left[\phi_{gh}\right]^t = \sum_{g=1}^{N}\left[\phi_{gg}\right]^t + \sum_{1\leq g,h\leq N,|h-g|=N/2}\left[\phi_{gh}\right]^t + \sum_{1\leq g,h\leq N, g\neq h, |g-h|\neq N/2}\left[\phi_{gh}\right]^t$$

$$= 2Nn^t + \sum_{1\leq g,h\leq N, g\neq h, |g-h|\neq N/2}\left[\overline{\phi}_{gh}\right]^t$$

$$= 2N\left[n^t - (N/2-n)^t\right] + \sum_{g=1}^{N}\sum_{h=1}^{N}\left[\overline{\phi}_{gh}\right]^t,$$

where the equalities follow from (11.17), (11.18), and (11.19). Thus $H_t(d)$ and $H_t(\overline{d})$ differ by a constant, as claimed. $\qquad\square$

Chen and Cheng (2009) derived explicit identities that relate the wordlength pattern of an even design to that of its complement in the maximal even design. The conclusion of Theorem 11.26 also follows from these identities. For instance, for any even design d,

$$A_4(d) = A_4(\overline{d}) + \left[\binom{n}{4} - \binom{2^{k-1}-n}{4}\right]\bigg/\left(2^{k-1}-3\right), \tag{11.20}$$

where $k = n - p$. Thus minimizing $A_4(d)$ among the even designs is equivalent to minimizing $A_4(\overline{d})$. It follows from (11.20) that

$$A_4(d) \geq \left[\binom{n}{4} - \binom{2^{k-1}-n}{4}\right]\bigg/\left(2^{k-1}-3\right)$$

for all even designs d. This gives a simple lower bound on A_4 for even designs. The lower bound is achieved if $A_4(\bar{d}) = 0$.

Proposition 11.27. *An even design of resolution IV whose complement in the maximal even design has resolution higher than IV has weak minimum aberration.*

For example, an even design of resolution VI can be obtained from a resolution V design by the method of foldover. The complement of such a design in the maximal even design minimizes the number of defining words of length four.

Chen and Cheng (2004) showed that the weak minimum aberration designs in Proposition 11.27 also have maximum estimation capacity among the resolution IV designs. Chen and Cheng (2009) used BCH codes (Bose and Ray-Chaudhuri, 1960; Hocquenghem, 1959) to construct such designs.

Using McWilliams identities and linear programming, Chen and Cheng (2009) derived a more general lower bound on the minimum number of defining words of length four that also applies to even designs whose complementary designs in the maximal even design have resolution IV.

It was mentioned in Section 10.6 that if a design minimizes $A_3(d)$, then its complement in the saturated regular design has minimum rank. Chen and Cheng (2009) showed that the complement of a minimum aberration even design in the maximal even design, however, has maximum rank. Therefore the structure of the complement of a minimum aberration even design in the maximal even design also depends on the run size. This makes the complementary design theory for maximal even designs less appealing since it is more difficult to compile a catalog of complementary designs of minimum aberration even designs. However, some results can still be derived.

The maximal even design with $N = 2^k$ corresponds to the $k \times N/2$ matrix $\mathbf{Y}$ that consists of the first k rows and columns $2^{k-1}, \ldots, 2^k - 1$ of the array in (9.22). These are the 2^{k-1}th column and its sums with the first $2^{k-1} - 1$ columns. Clearly $\mathbf{Y}$ has rank k. Let $\mathbf{b}_1, \ldots, \mathbf{b}_k$ be a set of k linearly independent columns of $\mathbf{Y}$. A convenient choice consists of the 2^{k-1}th column and its sums with columns $1, 2^1, 2^2, \ldots, 2^{k-2}$. For example, the maximal even design of size 32 is generated by the 5×16 matrix consisting of the first 5 rows and columns 16–31 of the array in (9.22), where columns 16, 17 $(= 16 + 1)$, 18 $(= 16 + 2)$, 20 $(= 16 + 4)$, 24 $(= 16 + 8)$ are linearly independent.

For $n < N/2$, let $\tilde{n} = N/2 - n$. If $\tilde{n} \leq k$, then by Theorem 11.26, the design obtained by deleting any $\tilde{n}$ linearly independent columns from the maximal even design has minimum aberration. For the case $\tilde{n} = k + 1$ and $\tilde{n}$ is even (respectively, odd), the complement of the $2^{\tilde{n}-1}$ design of resolution $\tilde{n}$ (respectively, $\tilde{n} - 1$) in the maximal even design is a minimum aberration design. Such a minimum aberration design can be obtained by deleting $\mathbf{b}_1, \ldots, \mathbf{b}_k$ and $\mathbf{b}_1 + \cdots + \mathbf{b}_k$ (respectively, $\mathbf{b}_1 + \cdots + \mathbf{b}_{k-1}$) from $\mathbf{Y}$ when k is even (respectively, odd).

For $\tilde{n} = k + 2$, write k as $k = 3m + r$, where $0 \leq r < 3$. Then a minimum aberration design can be constructed by deleting $k + 2$ columns of the form $\{\mathbf{b}_1, \ldots, \mathbf{b}_k, \mathbf{c}, \mathbf{d}\}$ from $\mathbf{Y}$, where for $r = 0$, $\mathbf{c} = \mathbf{b}_1 + \mathbf{b}_2 + \cdots + \mathbf{b}_{2m-1}, \mathbf{d} = \mathbf{b}_{m+1} + \mathbf{b}_{m+2} + \cdots + \mathbf{b}_{3m} + \mathbf{c}$;

for $r = 1$, $\mathbf{c} = \mathbf{b}_1 + \mathbf{b}_2 + \cdots + \mathbf{b}_{2m+1}, \mathbf{d} = \mathbf{b}_{m+1} + \mathbf{b}_{m+2} + \cdots + \mathbf{b}_{3m+1}$;

for $r = 2$, $\mathbf{c} = \mathbf{b}_1 + \mathbf{b}_2 + \cdots + \mathbf{b}_{2m+1}, \mathbf{d} = \mathbf{b}_{m+1} + \mathbf{b}_{m+2} + \cdots + \mathbf{b}_{3m+2} + \mathbf{c}$.

The proof of this is left as an exercise. We refer readers to Chen and Cheng (2009) for more results.

Example 11.5. Suppose $N = 64$ and $n = 24$. In this case $5N/16 = 20$; hence a minimum aberration design must be a 24-factor projection of the 32-factor maximal even design. The complementary design has $\tilde{n} = 8$ factors. Since $k = 6$ is a multiple of 3, a minimum aberration design can be constructed by deleting eight columns of the form $\mathbf{b}_1, \ldots, \mathbf{b}_6, \mathbf{c}, \mathbf{d}$ from the maximal even design, where $\mathbf{b}_1, \ldots, \mathbf{b}_6$ are linearly independent, $\mathbf{c} = \mathbf{b}_1 + \mathbf{b}_2 + \mathbf{b}_3$, and $\mathbf{d} = \mathbf{b}_1 + \mathbf{b}_2 + \mathbf{b}_4 + \mathbf{b}_5 + \mathbf{b}_6$. If we label the factors corresponding to $\mathbf{b}_1, \ldots, \mathbf{b}_6, \mathbf{c}, \mathbf{d}$ by $1, 2, 3, 4, 5, 6, 7, 8$, respectively, then these eight columns define a 2^{8-2} design with the defining relation $\mathbf{I} = \mathbf{1237} = \mathbf{124568} = \mathbf{345678}$.

11.10   Complementary design theory for doubling

The complementary design theory in Section 10.6 can be used to determine minimum aberration designs via their complements in the saturated design. When $5N/16 < n < N/2$, the theory developed in the previous section can be used to find minimum aberration designs via their complements in the maximal even design. For $9N/32 < n \leq 5N/16$, minimum aberration designs must be projections of the maximal even design or the maximal design with $5N/16$ factors. To determine the minimum aberration design, we need to compare the best projection of the maximal even design and the best projection of the maximal design with $5N/16$ factors. A question arises naturally: Can we develop a complementary design theory with the $5N/16$-factor maximal design as the universe to facilitate the search for its best projection? Is there such a complementary design theory for maximal designs with $9N/32$ factors, $17N/64$ factors, etc.? Since the designs in each of these families of maximal designs can be obtained from the smallest designs in the same family by doubling, is there a general complementary design theory for the maximal designs that can be constructed by doubling? Xu and Cheng (2008) provided a positive answer to these questions by deriving identities relating the wordlength patterns of pairs of complementary subdesigns of a doubled design.

Throughout this section, we make no distinction between a design (a collection of treatment combinations) and the matrix with the treatment combinations as its rows. Thus if $\mathbf{D}$ is the matrix associated with a design d, then the moments $M_k(d)$ defined in Section 10.10 and the word counts $A_k(d)$ are denoted by $M_k(\mathbf{D})$ and $A_k(\mathbf{D})$, respectively. We denote the two levels by 0 and 1 and use mod 2 arithmetic. For simplicity 1 and -1 are mapped to 0 and 1, respectively, to keep the algebraic structure. Then the double of a design $\mathbf{X}_0$ can be written as

$$\begin{bmatrix} \mathbf{X}_0 & \mathbf{X}_0 \\ \mathbf{X}_0 & \mathbf{X}_0 + \mathbf{J} \end{bmatrix},$$

where $\mathbf{J}$ is a matrix of all 1's. For each design $\mathbf{D}$, denote the Hamming weight of the

ith row of $\mathbf{D}$ by $w_i(\mathbf{D})$.

Theorem 11.28. *For an $N_0 \times n_0$ regular two-level design $\mathbf{X}_0$, let $\mathbf{X}$ be obtained from $\mathbf{X}_0$ by doubling it t times. Suppose $\mathbf{D}$ is a projection of $\mathbf{X}$ onto a subset of factors, and $\overline{\mathbf{D}}$ consists of all the columns of $\mathbf{X}$ that are not in $\mathbf{D}$. Then*

$$
\begin{aligned}
A_k(\mathbf{D}) \;=\; & (-1)^k A_k\left(\overline{\mathbf{D}}\right) + c_{k-1}A_{k-1}\left(\overline{\mathbf{D}}\right) + \cdots + c_1 A_1\left(\overline{\mathbf{D}}\right) + c_0 \\
& + d_k \Delta_k\left(\mathbf{D},\overline{\mathbf{D}}\right) + \cdots + d_1 \Delta_1\left(\mathbf{D},\overline{\mathbf{D}}\right),
\end{aligned}
\tag{11.21}
$$

where

$$
\Delta_j\left(\mathbf{D},\overline{\mathbf{D}}\right) = \sum_{i=1}^{N_0}\left[(w_i(\mathbf{D}))^j - \left(n_0 2^{t-1} - w_i\left(\overline{\mathbf{D}}\right)\right)^j \right],
\tag{11.22}
$$

and c_i, d_i are constants that do not depend on $\mathbf{D}$ or $\overline{\mathbf{D}}$.

The proof of Theorem 11.28 will be presented in Section 11.11.

To apply Theorem 11.28, we need to compute $w_i(\mathbf{D})$ and $w_i(\overline{\mathbf{D}})$. Note that $\mathbf{X}$ is $2^t N_0 \times 2^t n_0$, and each column of $\mathbf{X}_0$ generates 2^t columns of $\mathbf{X}$. Let $N = 2^t N_0$, and for each $j = 1, \ldots, n_0$, let f_j be the number of columns of $\overline{\mathbf{D}}$ that are generated by the jth column of $\mathbf{X}_0$. Then

$$
w_i\left(\overline{\mathbf{D}}\right) = \sum_{j=1}^{n_0} f_j x_{ij}, \text{ for } i = 1, \ldots, N_0,
\tag{11.23}
$$

$$
w_i(\mathbf{D}) = \sum_{j=1}^{n_0} (2^t - f_j) x_{ij} = 2^t w_i(\mathbf{X}_0) - w_i\left(\overline{\mathbf{D}}\right), \text{ for } i = 1, \ldots, N_0.
\tag{11.24}
$$

We show in two examples that Theorem 11.28 covers the complementary design theory for two-level saturated regular designs and that for maximal even designs as special cases. This is because the maximal even design can be constructed by repeatedly doubling the 2^1 design, and the two-level saturated regular design can be obtained by deleting the column of 0's from a repeated double.

Example 11.6. (Complementary design theory for two-level saturated designs) Let

$$
\mathbf{X}_0 = \begin{bmatrix} 0 & 0 \\ 0 & 1 \end{bmatrix}.
$$

Then $N_0 = n_0 = 2$, $N = 2^{t+1}$, and $\mathbf{X}$ is $2^{t+1} \times 2^{t+1}$, with the first column consisting of 0's. The saturated design of size N, denoted here by $\tilde{\mathbf{X}}$, is obtained by deleting the column of 0's from $\mathbf{X}$. An n-factor regular design $\mathbf{D}$ of size N consists of n columns of $\tilde{\mathbf{X}}$. Let $\tilde{\mathbf{D}}$ be the complement of $\mathbf{D}$ in $\tilde{\mathbf{X}}$; then $\overline{\mathbf{D}} = [\mathbf{0} \ \ \tilde{\mathbf{D}}]$, where $\mathbf{0}$ is a column of

0's. By (11.23) and (11.24), we have $w_1(\mathbf{D}) = w_1(\overline{\mathbf{D}}) = 0$, $w_2(\mathbf{D}) = 2^t - f_2$, $w_2(\overline{\mathbf{D}}) = f_2$. It follows from (11.22) that

$$
\begin{aligned}
\Delta_k(\mathbf{D}, \overline{\mathbf{D}}) &= \sum_{i=1}^{2} \left[(w_i(\mathbf{D}))^k - \left(2^t - w_i(\overline{\mathbf{D}}) \right)^k \right] \\
&= (2^t - f_2)^k - (2^t - f_2)^k - (2^t)^k \\
&= -2^{tk},
\end{aligned}
$$

which is a constant. Thus, by (11.21), sequentially minimizing $A_k(\mathbf{D})$ is equivalent to sequentially minimizing $(-1)^k A_k(\overline{\mathbf{D}})$. Since $\tilde{\mathbf{D}}$ is obtained by deleting a column of 0's from $\overline{\mathbf{D}}$, we have $A_1(\tilde{\mathbf{D}}) = A_2(\tilde{\mathbf{D}}) = 0$, and $A_k(\overline{\mathbf{D}}) = A_k(\tilde{\mathbf{D}}) + A_{k-1}(\tilde{\mathbf{D}})$ for $1 \leq k \leq 2^t - n$. It follows that sequentially minimizing $(-1)^k A_k(\tilde{\mathbf{D}})$ is also equivalent to sequentially minimizing $(-1)^k A_k(\overline{\mathbf{D}})$. Thus sequentially minimizing $A_k(\mathbf{D})$ is equivalent to sequentially minimizing $(-1)^k A_k(\tilde{\mathbf{D}})$. This is the complementary design theory for saturated designs obtained by Tang and Wu (1996) and Chen and Hedayat (1996).

Example 11.7. (Complementary design theory for maximal even designs) Let

$$
\mathbf{X}_0 = \begin{bmatrix} 0 \\ 1 \end{bmatrix}.
$$

Then $N = 2^{t+1}$ and the $2^{t+1} \times 2^t$ matrix $\mathbf{X}$ is a maximal even design. For a design $\mathbf{D}$ consisting of n columns of $\mathbf{X}$, it is easy to see that $w_1(\mathbf{D}) = w_1(\overline{\mathbf{D}}) = 0$, $w_2(\mathbf{D}) = n$, and $w_2(\overline{\mathbf{D}}) = 2^t - n$. Then

$$
\Delta_k(\mathbf{D}, \overline{\mathbf{D}}) = \sum_{i=1}^{2} \left[(w_i(\mathbf{D}))^k - \left(2^{t-1} - w_i(\overline{\mathbf{D}}) \right)^k \right] = -2^{(t-1)k} + n^k - (n - 2^{t-1})^k,
$$

which is a constant. It follows from (11.21) that sequentially minimizing $A_k(\mathbf{D})$ is equivalent to sequentially minimizing $(-1)^k A_k(\overline{\mathbf{D}})$. Now since both $\mathbf{D}$ and $\overline{\mathbf{D}}$ are even designs, we have $A_k(\mathbf{D}) = A_k(\overline{\mathbf{D}}) = 0$ for all odd k. Thus sequentially minimizing $A_4(\mathbf{D}), A_6(\mathbf{D}), \ldots$, etc. is equivalent to sequentially minimizing $A_4(\overline{\mathbf{D}}), A_6(\overline{\mathbf{D}}), \ldots$, etc. This, together with Theorem 11.14, implies Butler's (2003a) result stated in Theorem 11.26.

In general, the quantities $\Delta_k(\mathbf{D}, \overline{\mathbf{D}})$ that appear in the identities relating the wordlength pattern of $\mathbf{D}$ to that of $\overline{\mathbf{D}}$ depend on $\mathbf{X}_0$ and the frequencies $f_1, \ldots, f_{n_0}$. In the special cases treated in Examples 11.6 and 11.7, they are constants and do not depend on $f_1, \ldots, f_{n_0}$. This leads to simple complementary design theories for saturated and maximal even designs, but it is no longer the case for the other maximal designs. In general wordlength patterns of the complementary designs alone are not sufficient for characterizing minimum aberration projections. The result below is the counterpart of Examples 11.6 and 11.7 for maximal resolution IV designs with $5N/16$ factors. An additional condition on the frequencies $f_1, \ldots, f_{n_0}$ is needed.

Theorem 11.29. *Let* $\mathbf{X}_0$ *be the* 2_V^{5-1} *design defined by* $I = ABCDE$, *and* $\mathbf{X}$ *be the maximal design obtained from* $\mathbf{X}_0$ *by doubling it* t *times. Suppose* $\mathbf{D}^*$ *is an* n-*factor projection of* $\mathbf{X}$ *and* $\overline{\mathbf{D}}^*$ *is the complement of* $\mathbf{D}^*$ *in* $\mathbf{X}$. *For* $25N/128 \leq n \leq 5N/16$, $\mathbf{D}^*$ *has minimum aberration among all possible* n-*factor projections of* $\mathbf{X}$ *if*

 (i) $\overline{\mathbf{D}}^*$ *sequentially minimizes* $A_4(\overline{\mathbf{D}})$, $-A_5(\overline{\mathbf{D}})$, $A_6(\overline{\mathbf{D}})$, $-A_7(\overline{\mathbf{D}})$, $\ldots$, *etc.*

 (ii) $\left| f_i^* - f_j^* \right| \leq 1$ *for all* $1 \leq i < j \leq 5$, *where* f_1^*, f_2^*, f_3^*, f_4^*, *and* f_5^* *are the values of* f_1, f_2, f_3, f_4, *and* f_5 *for* $\overline{\mathbf{D}}^*$.

The proof of Theorem 11.29 is deferred to Section 11.11.

Xu and Cheng (2008) also established a result similar to Theorem 11.29 for complementary projections of the maximal resolution IV designs with $9N/32$ factors. Such a result is useful for comparing the best projections of maximal designs with $5N/16$ factors and those of maximal designs with $9N/32$ factors. It can be shown that for $17N/64 \leq n \leq 5N/16$, the best n-factor projection of the maximal design with $5N/16$ factors with respect to the minimum aberration criterion is better than those of the maximal even design and the maximal design with $9N/32$ factors, and that its best $(17N/64)$-factor projection is better than any of the five maximal designs with $17N/64$ factors. This leads to the following result.

Theorem 11.30. *For* $17N/64 \leq n \leq 5N/16$, *where* $N = 2^t$, $t \geq 5$, *a minimum aberration design with* N *runs and* n *factors must be a projection of the maximal design with* $5N/16$ *factors.*

Xu and Cheng (2008) proved the stronger result that for $N/4 + 1 \leq n \leq 5N/16$, the best n-factor projection of the maximal design with $5N/16$ factors is better than those of the maximal even design and the maximal design with $9N/32$ factors.

Combining Theorems 11.29 and 11.30, we have the following result.

Theorem 11.31. *For* $17N/64 \leq n \leq 5N/16$, *where* $N = 2^t$, $t \geq 5$, *suppose* $\mathbf{D}^*$ *is an* n-*factor projection of the resolution IV design* $\mathbf{X}$ *with* $5N/16$ *factors obtained by repeatedly doubling the* 2^{5-1} *design defined by* $I = ABCDE$. *If*

 (i) $\overline{\mathbf{D}}^*$ *sequentially minimizes* $A_4(\overline{\mathbf{D}})$, $-A_5(\overline{\mathbf{D}})$, $A_6(\overline{\mathbf{D}})$, $-A_7(\overline{\mathbf{D}})$, $\ldots$, *etc., among the* $(5N/16 - n)$-*factor projections of* $\mathbf{X}$, *and*

 (ii) $\left| f_i^* - f_j^* \right| \leq 1$ *for all* $1 \leq i < j \leq 5$, *where* f_1^*, f_2^*, f_3^*, f_4^*, *and* f_5^* *are the values of* f_1, f_2, f_3, f_4, *and* f_5 *for* $\overline{\mathbf{D}}^*$.

Then $\mathbf{D}^*$ *has minimum aberration among all the regular designs with* N *runs and* n *two-level factors.*

Complementary designs of n-factor projections with $1 \leq 5N/16 - n \leq 11$ that satisfy the two conditions in Theorem 11.29 can be found in Xu and Cheng (2008).

By deleting such designs from the 256-run maximal design with 80 factors, we obtain 256-run minimum aberration designs with $69 \leq n \leq 79$. Block (2003) attempted to construct minimum aberration designs from this maximal design by the simple approach of deleting one factor at a time. The result presented here confirms that, except for $n = 71$, the designs he obtained for $69 \leq n \leq 79$ indeed have minimum aberration.

11.11   Proofs of Theorems 11.28 and 11.29*

Proof of Theorem 11.28

It is easy to see that

$$w_i(\mathbf{D}) + w_i(\overline{\mathbf{D}}) = n_0 2^{t-1}, \text{ for } i = N_0 + 1, \ldots, N. \tag{11.25}$$

Since two-level regular designs can be considered as linear codes, the average pairwise distance of the runs is equal to their average Hamming weight; see Section 9.9. It follows that the quantity $M_k(\mathbf{D})$ defined in (10.12) can be written as

$$M_k(\mathbf{D}) = N^{-1} \sum_{i=1}^{N} [w_i(\mathbf{D})]^k. \tag{11.26}$$

Then

$$
\begin{aligned}
M_k(\mathbf{D}) &= \frac{1}{N} \sum_{i=1}^{N_0} [w_i(\mathbf{D})]^k + \frac{1}{N} \sum_{i=N_0+1}^{N} [w_i(\mathbf{D})]^k \\
&= \frac{1}{N} \sum_{i=1}^{N} \left[\left(n_0 2^{t-1} - w_i(\overline{\mathbf{D}}) \right)^k \right] + \frac{1}{N} \Delta_k(\mathbf{D}, \overline{\mathbf{D}}) \\
&= \frac{1}{N} \sum_{i=1}^{N} \sum_{h=0}^{k} \binom{k}{h} (n_0 2^{t-1})^{k-h} (-1)^h \left(w_i(\overline{\mathbf{D}}) \right)^h + \frac{1}{N} \Delta_k(\mathbf{D}, \overline{\mathbf{D}}) \\
&= \sum_{h=0}^{k} \binom{k}{h} (n_0 2^{t-1})^{k-h} (-1)^h M_h(\overline{\mathbf{D}}) + \frac{1}{N} \Delta_k(\mathbf{D}, \overline{\mathbf{D}}),
\end{aligned}
$$

where the first equality follows from (11.26), the second equality follows from (11.25) and (11.22), and the last equality follows from (11.26).

By applying Lemma 10.11 to $M_k(\mathbf{D})$ and $M_h(\overline{\mathbf{D}})$ on the two sides of the previous identity, we have

$$\sum_{i=0}^{k} Q_k(i;n) A_i(\mathbf{D}) = \sum_{i=0}^{k} \sum_{h=i}^{k} \left[\binom{k}{h} (n_0 2^{t-1})^{k-h} (-1)^h Q_h(i; n_0 2^t - n) \right] A_i(\overline{\mathbf{D}}) + \frac{\Delta_k(\mathbf{D}, \overline{\mathbf{D}})}{N}, \tag{11.27}$$

where $Q_k(i;n) = (-1)^i \left[\sum_{j=0}^{k} j! S(k,j) 2^{-j} \binom{n-i}{j-i} \right]$.

Now (11.21) follows from $Q_k(k;n) = (-1)^k k! / 2^k$. $\square$

Proof of Theorem 11.29

We have $N_0 = 16$, $n_0 = 5$, $N = 16 \cdot 2^t$, and $\mathbf{X}$ has $5N/16$ factors. Among the 16 rows of $\mathbf{X}_0$, one has all the five factors at level 0, ten have exactly two factors at level 1, and five have exactly four factors at level 1. Given an n-factor projection $\mathbf{D}$ of $\mathbf{X}$, for the row of $\mathbf{X}_0$ with all the factors at level 0, the corresponding values of $w_i(\overline{\mathbf{D}})$ and $w_i(\mathbf{D})$ are both equal to 0; for each row with exactly two factors (say factors j_1 and j_2) at level 1, the corresponding values of $w_i(\overline{\mathbf{D}})$ and $w_i(\mathbf{D})$ are $f_{j_1} + f_{j_2}$ and $2^{t+1} - f_{j_1} - f_{j_2}$, respectively; for each row with exactly four factors (say those other than the jth factor) at level 1, the corresponding values of $w_i(\overline{\mathbf{D}})$ and $w_i(\mathbf{D})$ are $5 \cdot 2^t - n - f_j$ and $4 \cdot 2^t - (5 \cdot 2^t - n - f_j) = n + f_j - 2^t$, respectively. Thus

$$\Delta_k(\mathbf{D}, \overline{\mathbf{D}}) = \sum_{j=1}^{5} \left[(n + f_j - 2^t)^k - (-5 \cdot 2^{t-1} + n + f_j)^k \right]$$
$$+ \sum_{1 \leq j_1 < j_2 \leq 5} \left[(2^{t+1} - f_{j_1} - f_{j_2})^k - (5 \cdot 2^{t-1} - f_{j_1} - f_{j_2})^k \right] - (5 \cdot 2^{t-1})^k.$$

$$(11.28)$$

Since $A_i(\mathbf{D}) = A_i(\overline{\mathbf{D}}) = 0$ for all $i \leq 3$, by (11.27), for $k = 4$, (11.21) reduces to

$$A_4(\mathbf{D}) = A_4(\overline{\mathbf{D}}) + d_4 \Delta_4(\mathbf{D}, \overline{\mathbf{D}}), \quad \text{where } d_4 > 0. \qquad (11.29)$$

We claim that in order to prove the theorem, it suffices to show that for $25 \cdot 2^{t-3} \leq n \leq 5 \cdot 2^t$, $\Delta_4(\mathbf{D}, \overline{\mathbf{D}})$ is minimized if and only if the frequencies f_1, f_2, f_3, f_4, and f_5 differ from one another by at most one. Suppose this claim is true and $\overline{\mathbf{D}}^*$ satisfies both (i) and (ii) of Theorem 11.29. Then it minimizes both $A_4(\overline{\mathbf{D}})$ and $\Delta_4(\mathbf{D}, \overline{\mathbf{D}})$, and hence, by (11.29), it also minimizes $A_4(\mathbf{D})$. On the other hand, any other design that minimizes $A_4(\mathbf{D})$ should also minimize both $A_4(\overline{\mathbf{D}})$ and $\Delta_4(\mathbf{D}, \overline{\mathbf{D}})$. By the claim, such designs must have f_1, f_2, f_3, f_4, and f_5 differ from one another by at most one. By (11.28), they all have the same values of $\Delta_k(\mathbf{D}, \overline{\mathbf{D}})$. Then by (11.21), sequentially minimizing $A_k(\mathbf{D})$ among such designs is equivalent to sequentially minimizing $(-1)^k A_k(\overline{\mathbf{D}})$. It follows that $\mathbf{D}^*$ has minimum aberration among the n-factor projections of $\mathbf{X}$.

To prove the claim, it suffices to show that if $f_1 \geq f_2 \geq f_3 \geq f_4 \geq f_5 \geq 0$ and $f_1 > f_5 + 1$, then $\Delta_4(\mathbf{D}, \overline{\mathbf{D}})$ is not minimized. By direct computation using (11.28), one can verify that the frequencies $f_1 - 1, f_2, f_3, f_4$, and $f_5 + 1$ produce a smaller value of $\Delta_4(\mathbf{D}, \overline{\mathbf{D}})$. $\square$

11.12 Coding and projective geometric connections*

Some of the results in this chapter originally appeared in the literature of projective geometry and coding theory.

When viewed as a set of n points in $\mathrm{PG}(n - p - 1, 2)$, a regular 2^{n-p} design $\mathbf{X}$ is of resolution IV+ if and only if the n points form a cap; see Section 9.11. A maximal resolution IV+ design is then equivalent to a maximal (or complete) cap: a set of points that do not contain a line, but a line is formed whenever a point is added to the set. Theorems 11.12 and 11.13 are crucial for the characterization of maximal caps. We will not reproduce the long proofs of these results here.

Let Σ_{n-p} be a $PG(n-p-1,2)$ embedded in Σ_{n-p+1}, a $PG(n-p,2)$, as in Section 9.11. Given a set H of n points in Σ_{n-p}, choose a point $\mathbf{v}$ in $\Sigma_{n-p+1} \setminus \Sigma_{n-p}$, and define the double of H as the set $D(H) = \{\mathbf{w} + \mathbf{v} : \mathbf{w} \in H\} \cup H$; this is also called the Plotkin construction. Then $D(H)$ contains $2n$ points. Without loss of generality, take $\mathbf{v}$ to be the point whose first $n - p$ components are zeros and the $(n - p + 1)$st component is equal to 1; cf. (9.22). Let $\mathbf{M}$ be the $(n - p + 1) \times 2n$ matrix with each point in $D(H)$ as a column. Then the set of all linear combinations of the $n - p + 1$ rows of $\mathbf{M}$ forms a $2N \times 2n$ matrix ($N = 2^{n-p}$) that defines a $(2N)$-run fractional factorial design for $2n$ two-level factors. It is easy to see that by switching the entries 0 and 1 of this matrix to 1 and -1, respectively, we obtain a doubled design as in (11.2).

The partial foldover construction in (11.4) is adapted from a geometric construction in Bruen and Wehlau (1999), and Theorem 11.17 is part (3) of Bruen and Wehlau's (1999) Theorem 4.1. We have provided a nongeometric proof.

When a regular fractional factorial design is viewed as a linear code, its resolution is the same as the minimum distance of the dual code. It can be shown that the estimation index is equal to the covering radius of the dual code. This is left as an exercise. Theorem 11.6 was stated in coding-theoretic language in Theorem 1.1 of Bruen, Haddad, and Wehlau (1998).

Exercises

11.1 Prove Proposition 11.1.

11.2 Prove Theorem 11.7.

11.3 Prove Proposition 11.18.

11.4 Verify (11.14).

11.5 Show that the weak minimum aberration designs in Proposition 11.27 have maximum estimation capacity over the resolution IV designs.
                                                    [Chen and Cheng (2004)]

11.6 Show that the maximal 2_{IV}^{10-5} design has maximum estimation capacity over all the regular 2^{10-5} designs.   [Chen and Cheng (2004)]

11.7 Show that, for $N/4 + 1 \leq n \leq 5N/16$, the best n-factor projection of the maximal design with $5N/16$ factors is better than that of the maximal even design with respect to the minimum aberration criterion. [Hint: Derive a lower bound on A_4 for projections of the maximal even design and an upper bound for projections of the maximal design with $5N/16$ factors.]
                                                    [Xu and Cheng (2008)]

11.8* Show that the estimation index of a regular fractional factorial design is equal to the covering radius of its dual code.

Chapter 12

Orthogonal Block Structures and Strata

In Chapter 5, the analyses of completely randomized designs, randomized complete block designs, and row-column designs were discussed. These designs have experimental units that are unstructured or have one of the two simplest structures obtained by crossing two factors or nesting one in the other. Nesting and crossing can be combined iteratively to form more complicated block structures. Such block structures and the more general orthogonal block structures, including their properties, Hasse diagrams, and appropriate statistical models, are discussed in this chapter. For the designs presented in Chapter 5, the eigenspaces of the patterned covariance matrix of the randomization model depend only on the block structure. There is one eigenspace, called a stratum, for each of the factors that define the block structure. These results are extended to more general block structures. Algorithms for determining the strata, including their degrees of freedom and orthogonal projections, are derived. The strata play an important role in the design and analysis of randomized experiments. We follow the mathematical framework presented in Bailey (1996, 2004, 2008) and include some results from these sources. Readers are advised to review the material in Chapters 3, 4, and 5.

12.1 Nesting and crossing operators

In a completely randomized experiment, the experimental units are unstructured. On the other hand, randomized block designs and row-column designs are based on the two simplest block structures that can be built from two sets of unstructured units by nesting and crossing, respectively.

For any two sets Ω_1 and Ω_2, let $\Omega_1 \times \Omega_2 = \{(w_1, w_2) : w_1 \in \Omega_1, w_2 \in \Omega_2\}$. Suppose $\mathcal{F}_1$ is a factor on Ω_1 and $\mathcal{F}_2$ is a factor on Ω_2. We define $\mathcal{F}_1 \times \mathcal{F}_2$ as the factor on $\Omega_1 \times \Omega_2$ such that (v_1, v_2) and (w_1, w_2) are in the same $(\mathcal{F}_1 \times \mathcal{F}_2)$-class if and only if v_1 and w_1 are in the same $\mathcal{F}_1$-class, and v_2 and w_2 are in the same $\mathcal{F}_2$-class. In other words, $\mathcal{F}_1 \times \mathcal{F}_2$ is a partition of $\Omega_1 \times \Omega_2$ according to the levels of $\mathcal{F}_1$ and $\mathcal{F}_2$ simultaneously.

For both a block design with n_1 blocks of size n_2 and a row-column design with n_1 rows and n_2 columns, each experimental unit can be indexed by a pair (w_1, w_2) with $w_1 \in \Omega_1$ and $w_2 \in \Omega_2$, where Ω_1 and Ω_2 are two sets of unstructured units of sizes n_1 and n_2, respectively. Then we can represent Ω as $\Omega_1 \times \Omega_2$.

We have seen in Section 4.2 that the block structure of a block design can be defined by three factors $\mathcal{U}$, $\mathcal{B}$, and $\mathcal{E}$. Let $\mathcal{U}_1$, $\mathcal{E}_1$, $\mathcal{U}_2$, and $\mathcal{E}_2$ be the universal and equality factors on Ω_1 and Ω_2, respectively. Then we have $\mathcal{U} = \mathcal{U}_1 \times \mathcal{U}_2$, $\mathcal{B} = \mathcal{E}_1 \times \mathcal{U}_2$, and $\mathcal{E} = \mathcal{E}_1 \times \mathcal{E}_2$. For instance, two units (v_1, v_2) and (w_1, w_2) have the same level of $\mathcal{E}_1 \times \mathcal{U}_2$ if and only if $v_1 = w_1$, that is, they are in the same block. On the other hand, the block structure of a row-column design can be defined by $\mathcal{U}$, $\mathcal{R}$, $\mathcal{C}$, and $\mathcal{E}$, with $\mathcal{U} = \mathcal{U}_1 \times \mathcal{U}_2$, $\mathcal{R} = \mathcal{E}_1 \times \mathcal{U}_2$, $\mathcal{C} = \mathcal{U}_1 \times \mathcal{E}_2$, and $\mathcal{E} = \mathcal{E}_1 \times \mathcal{E}_2$.

Thus the block structures of block and row-column designs can be built from $\mathfrak{B}_1 = \{\mathcal{U}_1, \mathcal{E}_1\}$ on Ω_1 and $\mathfrak{B}_2 = \{\mathcal{U}_2, \mathcal{E}_2\}$ on Ω_2. An important difference is that for block designs, among the four possible products of $\mathcal{U}$'s and $\mathcal{E}$'s, $\mathcal{U}_1 \times \mathcal{E}_2$ is missing. Two units (v_1, v_2) and (w_1, w_2) are in the same $(\mathcal{U}_1 \times \mathcal{E}_2)$-class if and only if $v_2 = w_2$. Under a block design one is *not interested in the level of $\mathcal{E}_2$ alone*. It is also necessary to specify the block to which each unit belongs. This is a special feature of nesting.

We state some simple properties before defining nesting and crossing operators.

Lemma 12.1. *Let $\mathcal{F}_1$ and $\mathcal{F}_2$ be two factors with $n_{\mathcal{F}_1}$ and $n_{\mathcal{F}_2}$ levels, respectively. Then*

(i) $\mathcal{F}_1 \times \mathcal{F}_2$ *has* $n_{\mathcal{F}_1} n_{\mathcal{F}_2}$ *levels;*

(ii) *If both $\mathcal{F}_1$ and $\mathcal{F}_2$ are uniform, then $\mathcal{F}_1 \times \mathcal{F}_2$ is uniform;*

(iii) $\mathcal{F}_1 \times \mathcal{F}_2 \preceq \mathcal{F}_1' \times \mathcal{F}_2'$ *if and only if $\mathcal{F}_1 \preceq \mathcal{F}_1'$ and $\mathcal{F}_2 \preceq \mathcal{F}_2'$;*

(iv) $\mathcal{U}_1 \times \mathcal{U}_2$ *is the universal factor on $\Omega_1 \times \Omega_2$;*

(v) $\mathcal{E}_1 \times \mathcal{E}_2$ *is the equality factor on $\Omega_1 \times \Omega_2$;*

(vi) $(\mathcal{F}_1 \times \mathcal{F}_2) \vee (\mathcal{G}_1 \times \mathcal{G}_2) = (\mathcal{F}_1 \vee \mathcal{G}_1) \times (\mathcal{F}_2 \vee \mathcal{G}_2)$;

(vii) $(\mathcal{F}_1 \times \mathcal{F}_2) \wedge (\mathcal{G}_1 \times \mathcal{G}_2) = (\mathcal{F}_1 \wedge \mathcal{G}_1) \times (\mathcal{F}_2 \wedge \mathcal{G}_2)$.

We adopt the definitions of crossing and nesting operators given in Bailey (1996). The crossing operator is considered first.

Definition 12.1. Given a block structure $\mathfrak{B}_1$ on Ω_1 and a block structure $\mathfrak{B}_2$ on Ω_2, $\mathfrak{B}_1 \times \mathfrak{B}_2$ is defined as the block structure on $\Omega_1 \times \Omega_2$ that consists of all the factors $\{\mathcal{F}_1 \times \mathcal{F}_2 : \mathcal{F}_1 \in \mathfrak{B}_1, \mathcal{F}_2 \in \mathfrak{B}_2\}$.

By parts (ii), (iv), and (v) of Lemma 12.1, $\mathfrak{B}_1 \times \mathfrak{B}_2$ contains both the universal and equality factors on $\Omega_1 \times \Omega_2$, and if all the factors in $\mathfrak{B}_1$ and $\mathfrak{B}_2$ are uniform, then the factors in $\mathfrak{B}_1 \times \mathfrak{B}_2$ are also uniform.

The two block structures $\mathfrak{B}_1$ and $\mathfrak{B}_2$ are embedded in $\mathfrak{B}_1 \times \mathfrak{B}_2$. Two units (v_1, v_2) and (w_1, w_2) in $\Omega_1 \times \Omega_2$ are in the same $(\mathcal{F}_1 \times \mathcal{U}_2)$-class if and only if v_1 and w_1 are in the same $\mathcal{F}_1$-class. Thus $\mathcal{F}_1$ induces the partition $\mathcal{F}_1 \times \mathcal{U}_2$ of $\Omega_1 \times \Omega_2$ according to the levels of $\mathcal{F}_1$. Likewise, $\mathcal{F}_2$ induces the partition $\mathcal{U}_1 \times \mathcal{F}_2$. Therefore we can identify $\mathcal{F}_1$ and $\mathcal{F}_2$ with $\mathcal{F}_1 \times \mathcal{U}_2$ and $\mathcal{U}_1 \times \mathcal{F}_2$, respectively. Note that $n_{\mathcal{F}_1 \times \mathcal{U}_2} = n_{\mathcal{F}_1}$ and $n_{\mathcal{U}_1 \times \mathcal{F}_2} = n_{\mathcal{F}_2}$. By part (vii) of Lemma 12.1, $\mathcal{F}_1 \times \mathcal{F}_2 = (\mathcal{F}_1 \times \mathcal{U}_2) \wedge (\mathcal{U}_1 \times \mathcal{F}_2)$; therefore if $\mathcal{F}_1$ and $\mathcal{F}_2$ are identified with $\mathcal{F}_1 \times \mathcal{U}_2$ and $\mathcal{U}_1 \times \mathcal{F}_2$, respectively, then we can write each factor $\mathcal{F}_1 \times \mathcal{F}_2$ in the crossed block structure $\mathfrak{B}_1 \times \mathfrak{B}_2$ as $\mathcal{F}_1 \wedge \mathcal{F}_2$.

Based on the partial order among the factors in $\mathfrak{B}_1 \times \mathfrak{B}_2$ given in part (iii) of Lemma 12.1 and the observations in the previous paragraph, the Hasse diagram for $\mathfrak{B}_1 \times \mathfrak{B}_2$ can be obtained from those of $\mathfrak{B}_1$ and $\mathfrak{B}_2$ as follows. We merge the top nodes $\mathcal{U}_1$ and $\mathcal{U}_2$ of the Hasse diagrams for $\mathfrak{B}_1$ and $\mathfrak{B}_2$ to form the top node $\mathcal{U}$ for $\mathfrak{B}_1 \times \mathfrak{B}_2$. The nodes of this merged diagram correspond to the factors $\mathcal{F}_1 \wedge \mathcal{F}_2$ with $\mathcal{F}_1 = \mathcal{U}_1$ or $\mathcal{F}_2 = \mathcal{U}_2$. We then add to the merged diagram $(|\mathfrak{B}_1| - 1)(|\mathfrak{B}_2| - 1)$ new nodes corresponding to $\mathcal{F}_1 \wedge \mathcal{F}_2$ for $\mathcal{F}_1 \in \mathfrak{B}_1$, $\mathcal{F}_2 \in \mathfrak{B}_2$, $\mathcal{F}_1 \neq \mathcal{U}_1$, and $\mathcal{F}_2 \neq \mathcal{U}_2$, where $|\mathfrak{B}_i|$ is the number of factors in $\mathfrak{B}_i$. Then attach the number of levels $n_{\mathcal{F}_1} n_{\mathcal{F}_2}$ to each such new node.

Example 12.1. To obtain the Hasse diagram for $n_1 \times n_2$, we start with the following Hasse diagrams for two sets of unstructured units of sizes n_1 and n_2, respectively:

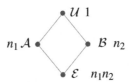

We merge the two $\mathcal{U}$ nodes, label the $\mathcal{E}_1$ and $\mathcal{E}_2$ nodes as $\mathcal{A}$ and $\mathcal{B}$, respectively, and add a new node corresponding to $\mathcal{E}_1 \wedge \mathcal{E}_2 = \mathcal{E}$. This results in the following Hasse diagram for the block structure $n_1 \times n_2$.

Figure 12.1 shows the Hasse diagram for the block structure $n_1 \times n_2 \times n_3$, which can be obtained from the Hasse diagram for $n_1 \times n_2$ and that for n_3 unstructured units by using the method described above. In general, the block structure $n_1 \times \cdots \times n_k$ is

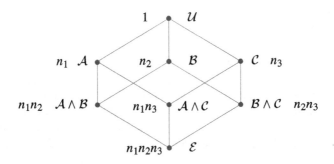

Figure 12.1 *Hasse diagram for the block structure $n_1 \times n_2 \times n_3$*

obtained by applying the crossing operator repeatedly to k sets of unstructured units $\Omega_1, \ldots, \Omega_k$ of sizes $n_1, \ldots, n_k$ (or, k factors $\mathcal{F}_1, \ldots, \mathcal{F}_k$ with $n_1, \ldots, n_k$ levels), respectively. It consists of all the 2^k factors of the form $\wedge_{i \in S} \mathcal{F}_i$, one for each $S \subseteq \{1, \ldots, k\}$. For $S = \{i_1, \ldots, i_h\}$, $\mathcal{F}_{i_1} \wedge \cdots \wedge \mathcal{F}_{i_h}$ is the $(n_{i_1} \cdots n_{i_h})$-level factor that partitions the $n_1 \cdots n_k$ units according to the levels of $\mathcal{F}_{i_1}, \ldots, \mathcal{F}_{i_h}$. If we apply the crossing operators repeatedly to n treatment factors $A_1, \ldots, A_n$ with $s_1, \ldots, s_n$ levels, respectively, then we obtain the factorial treatment structure $A_1 \times \cdots \times A_n$ discussed in Section 6.8. The Hasse diagram in Figure 12.1 is the same as that in Figure 6.1.

We now define the nesting operator.

Definition 12.2. Given a block structure $\mathfrak{B}_1$ on Ω_1 and a block structure $\mathfrak{B}_2$ on Ω_2, $\mathfrak{B}_1/\mathfrak{B}_2$ is the block structure on $\Omega_1 \times \Omega_2$ consisting of the factors in $\{\mathcal{F}_1 \times \mathcal{U}_2 : \mathcal{F}_1 \in \mathfrak{B}_1, \mathcal{F}_1 \neq \mathcal{E}_1\} \cup \{\mathcal{E}_1 \times \mathcal{F}_2 : \mathcal{F}_2 \in \mathfrak{B}_2\}$, where $\mathcal{E}_1$ is the equality factor on Ω_1 and $\mathcal{U}_2$ is the universal factor on Ω_2.

Example 12.2. Let Ω_1 and Ω_2 be two sets of unstructured units of sizes n_1 and n_2, respectively, and let $\mathcal{U}_1, \mathcal{U}_2, \mathcal{E}_1, \mathcal{E}_2$ be the associated universal and equality factors. We have seen that the block structure n_1/n_2 can be represented by $\mathfrak{B} = \{\mathcal{U}_1 \times \mathcal{U}_2, \mathcal{E}_1 \times \mathcal{U}_2, \mathcal{E}_1 \times \mathcal{E}_2\}$. It follows from Definition 12.2 that $\mathfrak{B} = \mathfrak{B}_1/\mathfrak{B}_2$, where $\mathfrak{B}_1 = \{\mathcal{U}_1, \mathcal{E}_1\}$ and $\mathfrak{B}_2 = \{\mathcal{U}_2, \mathcal{E}_2\}$.

In Definition 12.2, $\{\mathcal{F}_1 \times \mathcal{U}_2 : \mathcal{F}_1 \in \mathfrak{B}_1, \mathcal{F}_1 \neq \mathcal{E}_1\}$ preserves the block structure $\mathfrak{B}_1$. On the other hand, $\{\mathcal{E}_1 \times \mathcal{F}_2 : \mathcal{F}_2 \in \mathfrak{B}_2\}$ replaces each unit in Ω_1 with a copy of Ω_2 that has block structure $\mathfrak{B}_2$. Note that there is no factor of the form $\mathcal{F}_1 \times \mathcal{F}_2$ with $\mathcal{F}_1 \neq \mathcal{E}_1$ and $\mathcal{F}_2 \neq \mathcal{U}_2$. Thus for all $\mathcal{F}_1 \times \mathcal{F}_2 \in \mathfrak{B}_1/\mathfrak{B}_2$,

$$\text{if } \mathcal{F}_2 \neq \mathcal{U}_2, \text{ then } \mathcal{F}_1 = \mathcal{E}_1. \tag{12.1}$$

As in the case of block designs, when $\mathfrak{B}_2$ is nested in $\mathfrak{B}_1$, it is not enough to know the levels of the factors in $\mathfrak{B}_2$ alone. One also needs to know in which unit of Ω_1 they are nested.

Like $\mathfrak{B}_1 \times \mathfrak{B}_2$, $\mathfrak{B}_1/\mathfrak{B}_2$ contains both the universal and equality factors on $\Omega_1 \times \Omega_2$, and if all the factors in $\mathfrak{B}_1$ and $\mathfrak{B}_2$ are uniform, then the factors in $\mathfrak{B}_1/\mathfrak{B}_2$ are also uniform.

Under $\mathfrak{B}_1/\mathfrak{B}_2$, each factor $\mathcal{F}_1$ in $\mathfrak{B}_1$ induces the partition $\mathcal{F}_1 \times \mathcal{U}_2$ of $\Omega_1 \times \Omega_2$ according to the levels of $\mathcal{F}_1$, and each factor $\mathcal{F}_2$ in $\mathfrak{B}_2$ also induces the partition $\mathcal{E}_1 \times \mathcal{F}_2$. However, although $\mathcal{F}_1 \times \mathcal{U}_2$ has $n_{\mathcal{F}_1}$ levels, $\mathcal{E}_1 \times \mathcal{F}_2$ has $n_{\mathcal{F}_2} |\Omega_1|$ levels. This is because we have one copy of $\mathfrak{B}_2$ nested in each of $|\Omega_1|$ units: two units (v_1, v_2) and (w_1, w_2) are in the same $(\mathcal{E}_1 \times \mathcal{F}_2)$-class if and only if v_2 and w_2 are in the same $\mathcal{F}_2$-class *and* $v_1 = w_1$. The partial order among the factors in $\mathfrak{B}_1/\mathfrak{B}_2$ is that $\mathcal{F}_1 \times \mathcal{U}_2 \preceq \mathcal{F}_1' \times \mathcal{U}_2 \Leftrightarrow \mathcal{F}_1 \preceq \mathcal{F}_1'$, $\mathcal{E}_1 \times \mathcal{F}_2 \preceq \mathcal{E}_1 \times \mathcal{F}_2' \Leftrightarrow \mathcal{F}_2 \preceq \mathcal{F}_2'$, and $\mathcal{E}_1 \times \mathcal{F}_2 \preceq \mathcal{F}_1 \times \mathcal{U}_2$ for all $\mathcal{F}_1$ and $\mathcal{F}_2$. Based on this, the Hasse diagram for $\mathfrak{B}_1/\mathfrak{B}_2$ can be obtained from those of $\mathfrak{B}_1$ and $\mathfrak{B}_2$ by placing the Hasse diagram for $\mathfrak{B}_1$ on top of that for $\mathfrak{B}_2$, merging the bottom node of the former (corresponding to $\mathcal{E}_1$) with the top node of the latter (corresponding to $\mathcal{U}_2$), and then replacing the number of levels $n_{\mathcal{F}_2}$ for each factor $\mathcal{F}_2$ in the Hasse diagram for $\mathfrak{B}_2$ with $n_{\mathcal{F}_2} |\Omega_1|$.

Example 12.3. Suppose $\mathfrak{B}_1$ is the bock structure for n_1 unstructured units and $\mathfrak{B}_2$ is defined by the formula $n_2 \times n_3$. Then $\mathfrak{B}_1/\mathfrak{B}_2$ produces a block structure with the formula $n_1/(n_2 \times n_3)$. Write $\mathfrak{B}_1$ and $\mathfrak{B}_2$ as $\mathfrak{B}_1 = \{\mathcal{U}_1, \mathcal{E}_1\}$ and $\mathfrak{B}_2 = \{\mathcal{U}_2, \mathcal{R}, \mathcal{C}, \mathcal{E}_2\}$, where $\mathcal{R}$ has n_2 levels and $\mathcal{C}$ has n_3 levels. Then $\mathfrak{B}_1/\mathfrak{B}_2 = \{\mathcal{E}_1 \times \mathcal{E}_2, \mathcal{E}_1 \times \mathcal{R}, \mathcal{E}_1 \times \mathcal{C}, \mathcal{E}_1 \times \mathcal{U}_2, \mathcal{U}_1 \times \mathcal{U}_2\}$. These give the five factors $\mathcal{E}$ $(= \mathcal{E}_1 \times \mathcal{E}_2)$, $\mathcal{R}'$ $(= \mathcal{E}_1 \times \mathcal{R})$, $\mathcal{C}'$ $(= \mathcal{E}_1 \times \mathcal{C})$, $\mathcal{B}$ $(= \mathcal{E}_1 \times \mathcal{U}_2)$, and $\mathcal{U}$ $(= \mathcal{U}_1 \times \mathcal{U}_2)$ that define the block structure for nested row-column designs discussed in Section 4.2; also see Example 1.3. The Hasse diagram for the block structure $n_1/(n_2 \times n_3)$ given in Figure 4.3(b) can be obtained by placing the Hasse diagram for $n_2 \times n_3$, such as that in Example 12.1, underneath the Hasse diagram for n_1 unstructured units and merging the top node of the former with the bottom node of the latter. After the merge, the numbers of levels for all the factors in the Hasse diagram for $n_2 \times n_3$ are multiplied by n_1.

Example 12.4. Example 1.1 involves a block structure that can be described by the formula $n_1/[(n_2/n_3/n_4) \times n_5]$. The $n_1 n_2 n_3 n_4 n_5$ units are partitioned into n_1 blocks of size $n_2 n_3 n_4 n_5$, each block is partitioned into n_2 plots of size $n_3 n_4 n_5$, each plot is partitioned into n_3 subplots of size $n_4 n_5$, each subplot is partitioned into n_4 sub-subplots of size n_5, and each block is also partitioned into n_5 horizontal strips. Denote these factors by $\mathcal{B}$ (blocks; n_1 levels), $\mathcal{P}$ (plots; $n_1 n_2$ levels), $\mathcal{S}$ (subplots; $n_1 n_2 n_3$ levels), $\mathcal{SS}$ (sub-subplots; $n_1 n_2 n_3 n_4$ levels), and $\mathcal{ST}$ (strips; $n_1 n_5$ levels). Then we have $\mathcal{SS} \prec \mathcal{S} \prec \mathcal{P} \prec \mathcal{B}$ and $\mathcal{ST} \prec \mathcal{B}$. To construct the Hasse diagram for $n_1/[(n_2/n_3/n_4) \times n_5]$, we first cross

$$\bullet \mathcal{U} \quad 1$$
$$\bullet \mathcal{ST} \quad n_5$$

with

$$\bullet \mathcal{U} \quad 1$$
$$\bullet \mathcal{P} \quad n_2$$
$$\bullet \mathcal{S} \quad n_2 n_3$$
$$\bullet \mathcal{SS} \quad n_2 n_3 n_4$$

Then we place the resulting Hasse diagram for the block structure $(n_2/n_3/n_4) \times n_5$ underneath that for n_1 unstructured units to form the Hasse diagram for $n_1/[(n_2/n_3/n_4) \times n_5]$ in Figure 12.2.

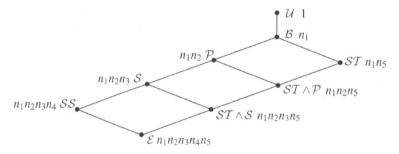

Figure 12.2 *Hasse diagram for the block structure $n_1/[(n_2/n_3/n_4) \times n_5]$*

12.2   Simple block structures

Simple block structures can be defined inductively as follows.

Definition 12.3. The block structure $\mathfrak{B} = \{\mathcal{U}, \mathcal{E}\}$ of unstructured units is a simple block structure. If $\mathfrak{B}_1$ and $\mathfrak{B}_2$ are simple block structures, then $\mathfrak{B}_1 \times \mathfrak{B}_2$ and $\mathfrak{B}_1/\mathfrak{B}_2$ are also simple block structures.

We will need the following result later.

Theorem 12.2. *Let $\mathfrak{B}$ be a simple block structure. Then*

(i) *All the factors in $\mathfrak{B}$ are uniform;*

(ii) $\mathcal{E} \in \mathfrak{B}$;

(iii) $\mathcal{U} \in \mathfrak{B}$;

(iv) *if $\mathcal{F}, \mathcal{G} \in \mathfrak{B}$, then $\mathcal{F} \vee \mathcal{G} \in \mathfrak{B}$;*

(v) *if $\mathcal{F}, \mathcal{G} \in \mathfrak{B}$, then $\mathcal{F} \wedge \mathcal{G} \in \mathfrak{B}$.*

Proof. Properties (i)–(iii) have been mentioned earlier. For (iv) and (v), we prove the more general result that if two block structures $\mathfrak{B}_1$ and $\mathfrak{B}_2$ are both closed under $\vee$ and $\wedge$, then $\mathfrak{B}_1/\mathfrak{B}_2$ and $\mathfrak{B}_1 \times \mathfrak{B}_2$ are also closed under $\vee$ and $\wedge$. By induction, it then follows that all simple block structures are closed under $\vee$ and $\wedge$.

The factors in $\mathfrak{B}_1 \times \mathfrak{B}_2$ are of the form $\mathcal{F}_1 \times \mathcal{F}_2$, and those in $\mathfrak{B}_1/\mathfrak{B}_2$ are of the form $\mathcal{F}_1 \times \mathcal{U}_2$ or $\mathcal{E}_1 \times \mathcal{F}_2$, where $\mathcal{F}_1 \in \mathfrak{B}_1$ and $\mathcal{F}_2 \in \mathfrak{B}_2$. It follows immediately from parts (vi) and (vii) of Lemma 12.1 that if $\mathfrak{B}_1$ and $\mathfrak{B}_2$ are closed under $\vee$ and $\wedge$, then $\mathfrak{B}_1 \times \mathfrak{B}_2$ and $\mathfrak{B}_1/\mathfrak{B}_2$ are also closed under $\vee$ and $\wedge$. $\square$

In the rest of this section, we give an explicit description of the factors that make up a simple block structure. Suppose a simple block structure $\mathfrak{B}$ on $\Omega_1 \times \cdots \times \Omega_m$ is obtained by iterating the nesting and crossing operators, with $\mathfrak{B}_i = \{\mathcal{U}_i, \mathcal{E}_i\}$, $i = 1, \ldots, m$, as the building blocks, where $\mathcal{U}_i$ and $\mathcal{E}_i$ are the universal and equality factors, respectively, on a set Ω_i of n_i unstructured units. We say that $\mathcal{E}_j$ dominates $\mathcal{E}_i$ (or $\mathcal{E}_j$

is an *ancestor* of $\mathcal{E}_i$) if at a certain stage in building $\mathfrak{B}$, a structure of the form $\mathfrak{F}_1/\mathfrak{F}_2$ is formed, where $\mathfrak{B}_j$ is one of the building blocks of $\mathfrak{F}_1$ and $\mathfrak{B}_i$ is a building block of $\mathfrak{F}_2$. For example, suppose $\mathfrak{B} = (\mathfrak{B}_1 \times \mathfrak{B}_2)/(\mathfrak{B}_3/\mathfrak{B}_4)$; then $\mathcal{E}_1$, $\mathcal{E}_2$, and $\mathcal{E}_3$ are ancestors of $\mathcal{E}_4$ and $\mathcal{E}_1$, $\mathcal{E}_2$ are ancestors of $\mathcal{E}_3$, but $\mathcal{E}_1$ and $\mathcal{E}_2$ are not ancestors of each other.

Each factor in $\mathfrak{B}$ is of the form $\mathcal{F}_1 \times \cdots \times \mathcal{F}_m$, where $\mathcal{F}_i = \mathcal{U}_i$ or $\mathcal{E}_i$. When no $\mathcal{E}_i$ is an ancestor of any other $\mathcal{E}_j$, we obtain the *largest* block structure $\mathfrak{B}_1 \times \cdots \times \mathfrak{B}_m$, which consists of *all* the 2^m factors of the form $\mathcal{F}_1 \times \cdots \times \mathcal{F}_m$. Otherwise, the relation of ancestry among the $\mathcal{E}_i$'s imposes some constraints on the factors in $\mathfrak{B}$, resulting in a proper subset of $\mathfrak{B}_1 \times \cdots \times \mathfrak{B}_m$. Specifically, by (12.1), $\mathfrak{B}$ consists of all factors of the form $\mathcal{F}_1 \times \cdots \times \mathcal{F}_m$, where $\mathcal{F}_i = \mathcal{U}_i$ or $\mathcal{E}_i$, such that

$$\text{If } \mathcal{F}_i = \mathcal{E}_i, \text{ then } \mathcal{F}_j = \mathcal{E}_j \text{ for all } j \text{ such that } \mathcal{E}_j \text{ is an ancestor of } \mathcal{E}_i. \tag{12.2}$$

As in the discussion following (12.1), knowing the level of $\mathcal{E}_i$ is not enough without knowing the levels of all its ancestors.

We can label each factor in $\mathfrak{B}$ by the set of $\mathcal{E}_i$'s with $\mathcal{F}_i = \mathcal{E}_i$. If $\mathcal{E}_i$ appears in such a label, then all of its ancestors must also be present. Therefore each simple block structure can be represented by a collection of subsets of $\{\mathcal{E}_1, \ldots, \mathcal{E}_m\}$. Each factor in a simple block structure is a partition of $\Omega_1 \times \cdots \times \Omega_m$ according to the levels of the $\mathcal{E}_i$'s that appear in its label. If the label of a factor is a subset of that of another factor, then the latter is nested in the former. The universal and equality factors correspond to $\emptyset$ and $\{\mathcal{E}_1, \ldots, \mathcal{E}_m\}$, respectively.

Example 12.5. Consider the simple block structure $\mathfrak{B} = \mathfrak{B}_1/(\mathfrak{B}_2 \times \mathfrak{B}_3)$ with $\mathfrak{B}_1 = \{\mathcal{U}_1, \mathcal{E}_1\}$, $\mathfrak{B}_2 = \{\mathcal{U}_2, \mathcal{E}_2\}$, and $\mathfrak{B}_3 = \{\mathcal{U}_3, \mathcal{E}_3\}$. Suppose we write $\mathcal{E}_1$, $\mathcal{E}_2$, and $\mathcal{E}_3$ as B (block), R (row), and C (column), respectively. Then B is an ancestor of both R and C. By using (12.2), we can write down the five factors in $\mathfrak{B}$: $\mathcal{U}_1 \times \mathcal{U}_2 \times \mathcal{U}_3$ $(= \mathcal{U})$, $B \times \mathcal{U}_2 \times \mathcal{U}_3$, $B \times R \times \mathcal{U}_3$, $B \times \mathcal{U}_2 \times C$, and $B \times R \times C$ $(= \mathcal{E})$. For example, since B is an ancestor of both R and C, B must appear whenever at least one of R and C appears. (We need to know which block each row or column is in.) The five factors in $\mathfrak{B}$ can be labeled by $\emptyset$, $\{B\}$, $\{B, R\}$, $\{B, C\}$, and $\{B, R, C\}$, respectively.

Let $\mathcal{E}_i^*$ be the factor $\mathcal{F}_1 \times \cdots \times \mathcal{F}_m$ in $\mathfrak{B}$ with

$$\mathcal{F}_j = \begin{cases} \mathcal{E}_i, & \text{if } j = i; \\ \mathcal{E}_j, & \text{if } \mathcal{E}_j \text{ is an ancestor of } \mathcal{E}_i; \\ \mathcal{U}_j, & \text{otherwise.} \end{cases} \tag{12.3}$$

Then $\mathcal{E}_i^*$ is the coarsest factor in $\mathfrak{B}$ with $\mathcal{F}_i = \mathcal{E}_i$. Note that $\mathcal{E}_i^*$ partitions the units according to the levels of $\mathcal{E}_i$ and all its ancestors and is labeled by these factors. While $\mathcal{E}_i$ has n_i levels, the number of levels for $\mathcal{E}_i^*$ is

$$n_{\mathcal{E}_i^*} = n_i \cdot \prod_{j:\mathcal{E}_j \text{ is an ancestor of } \mathcal{E}_i} n_j. \tag{12.4}$$

Furthermore, $\mathcal{E}_j$ is an ancestor of $\mathcal{E}_i$ if and only if $\mathcal{E}_i^* \prec \mathcal{E}_j^*$. Therefore (12.4) can also be written as

$$n_{\mathcal{E}_i^*} = \prod_{j:\mathcal{E}_i^* \preceq \mathcal{E}_j^*} n_j. \tag{12.5}$$

Example 12.6. (Example 12.5 continued) We have $\mathcal{B}^* = \mathcal{B} \times \mathcal{U}_2 \times \mathcal{U}_3$, $\mathcal{R}^* = \mathcal{B} \times \mathcal{R} \times \mathcal{U}_3$, and $\mathcal{C}^* = \mathcal{B} \times \mathcal{U}_2 \times \mathcal{C}$. Note that these are the factors that partition the units into n_1 blocks, $n_1 n_2$ rows, and $n_1 n_3$ columns, respectively, and are what we denoted by $\mathcal{B}$, $\mathcal{R}'$, and $\mathcal{C}'$ in Example 12.3 and Section 4.2.

12.3   Statistical models

We will generalize the statistical models (3.9)–(3.10) for block designs, (3.16)–(3.17) for row-column designs, (3.20)–(3.21) for nested row-column designs, and (3.22)–(3.23) for blocked split-plot designs to experiments with more general block structures.

For each of the simple block structures studied in Chapter 3, the covariance matrix $\mathbf{V}$ is assumed to be a patterned matrix with constant covariance for the observations on each pair of units that satisfy one of several relations. We first show that these relations, and hence $\mathbf{V}$, can be expressed succinctly in terms of the factors that define the block structure.

For each factor $\mathcal{F}$ in a block structure $\mathfrak{B}$ (not necessarily a simple block structure), define an $N \times N$ matrix $\mathbf{A}_{\mathcal{F}}$ by

$$[\mathbf{A}_{\mathcal{F}}]_{vw} = \begin{cases} 1, & \text{if } \mathcal{F} \text{ is the finest factor in } \mathfrak{B} \text{ such that units } v \text{ and } w \\ & \text{are in the same } \mathcal{F}\text{-class,} \\ 0, & \text{otherwise,} \end{cases} \tag{12.6}$$

where $[\mathbf{A}_{\mathcal{F}}]_{vw}$ is the (v,w)th entry of $\mathbf{A}_{\mathcal{F}}$. In other words, $[\mathbf{A}_{\mathcal{F}}]_{vw} = 1$ if and only if *units v and w are in the same $\mathcal{F}$-class but are in different classes of any factor that is immediately below $\mathcal{F}$ in the Hasse diagram.* It is clear that for any block structure, $[\mathbf{A}_{\mathcal{E}}]_{vw} = 1$ if and only if $v = w$. Therefore we have $\mathbf{A}_{\mathcal{E}} = \mathbf{I}_N$.

As an example, consider the block structure row $\times$ column. It is easy to see (for example, from the Hasse diagram) that $[\mathbf{A}_{\mathcal{R}}]_{vw} = 1$ if and only if units v and w are in the same row, but $v \neq w$; that is, they are in the same row and different columns. Similarly, $[\mathbf{A}_{\mathcal{C}}]_{vw} = 1$ if and only if units v and w are in the same column but different rows, and $[\mathbf{A}_{\mathcal{U}}]_{vw} = 1$ if and only if units v and w are in different rows and different columns. Each of the four $(0,1)$-matrices $\mathbf{A}_{\mathcal{E}}$, $\mathbf{A}_{\mathcal{R}}$, $\mathbf{A}_{\mathcal{C}}$, and $\mathbf{A}_{\mathcal{U}}$ can be considered as an adjacency matrix describing one of the four aforementioned relations between each pair of experimental units. Since each of the N^2 pairs of units satisfies *exactly one* of the four relations, we have

$$\mathbf{J} = \mathbf{A}_{\mathcal{E}} + \mathbf{A}_{\mathcal{R}} + \mathbf{A}_{\mathcal{C}} + \mathbf{A}_{\mathcal{U}}. \tag{12.7}$$

Then the covariance matrix $\mathbf{V}$ for a randomized row-column design in (3.17) can be expressed as

$$\mathbf{V} = \sigma^2 (\mathbf{A}_{\mathcal{E}} + \rho_1 \mathbf{A}_{\mathcal{R}} + \rho_2 \mathbf{A}_{\mathcal{C}} + \rho_3 \mathbf{A}_{\mathcal{U}}). \tag{12.8}$$

Similarly for unstructured units,

$$\mathbf{J} = \mathbf{A}_{\mathcal{E}} + \mathbf{A}_{\mathcal{U}}, \tag{12.9}$$

and the covariance matrix $\mathbf{V}$ for a completely randomized experiment in (3.6) can be expressed as

$$\mathbf{V} = \sigma^2(\mathbf{A}_{\mathcal{E}} + \rho\mathbf{A}_{\mathcal{U}}); \tag{12.10}$$

for block designs,

$$\mathbf{J} = \mathbf{A}_{\mathcal{E}} + \mathbf{A}_{\mathcal{B}} + \mathbf{A}_{\mathcal{U}}, \tag{12.11}$$

and the $\mathbf{V}$ in (3.10) can be expressed as

$$\mathbf{V} = \sigma^2(\mathbf{A}_{\mathcal{E}} + \rho_1\mathbf{A}_{\mathcal{B}} + \rho_2\mathbf{A}_{\mathcal{U}}); \tag{12.12}$$

for blocked split-plot designs,

$$\mathbf{J} = \mathbf{A}_{\mathcal{E}} + \mathbf{A}_{\mathcal{P}} + \mathbf{A}_{\mathcal{B}} + \mathbf{A}_{\mathcal{U}}, \tag{12.13}$$

and the $\mathbf{V}$ in (3.23) can be expressed as

$$\mathbf{V} = \sigma^2(\mathbf{A}_{\mathcal{E}} + \rho_1\mathbf{A}_{\mathcal{P}} + \rho_2\mathbf{A}_{\mathcal{B}} + \rho_3\mathbf{A}_{\mathcal{U}}); \tag{12.14}$$

finally, for nested row-column designs,

$$\mathbf{J} = \mathbf{A}_{\mathcal{E}} + \mathbf{A}_{\mathcal{R}'} + \mathbf{A}_{\mathcal{C}'} + \mathbf{A}_{\mathcal{B}} + \mathbf{A}_{\mathcal{U}}, \tag{12.15}$$

and the $\mathbf{V}$ in (3.21) can be expressed as

$$\mathbf{V} = \sigma^2(\mathbf{A}_{\mathcal{E}} + \rho_1\mathbf{A}_{\mathcal{R}'} + \rho_2\mathbf{A}_{\mathcal{C}'} + \rho_3\mathbf{A}_{\mathcal{B}} + \rho_4\mathbf{A}_{\mathcal{U}}). \tag{12.16}$$

In general, the following result holds.

Proposition 12.3. *If $\mathfrak{B}$ is a block structure that is closed under $\wedge$, then for any two units v and w, there is a unique $\mathcal{F} \in \mathfrak{B}$ that is the finest factor such that v and w are in the same $\mathcal{F}$-class. In this case the matrix $\mathbf{A}_{\mathcal{F}}$ in (12.6) is well-defined for each factor $\mathcal{F} \in \mathfrak{B}$, and*

$$\mathbf{J} = \sum_{\mathcal{F}:\mathcal{F} \in \mathfrak{B}} \mathbf{A}_{\mathcal{F}}. \tag{12.17}$$

Proof. Since all the units are in the same $\mathcal{U}$-class and $\mathcal{U} \in \mathfrak{B}$, for any pair of units (v, w), there is at least one $\mathcal{G} \in \mathfrak{B}$ such that v and w are in the same $\mathcal{G}$-class. Furthermore, if v and w are in the same $\mathcal{G}_1$-class and also in the same $\mathcal{G}_2$-class, then they must be in the same $(\mathcal{G}_1 \wedge \mathcal{G}_2)$-class. Let $\mathcal{F} = \wedge\{\mathcal{G} \in \mathfrak{B} : v \text{ and } w \text{ are in the same } \mathcal{G}\text{-class}\}$. Then v and w are in the same $\mathcal{F}$-class. Since $\mathfrak{B}$ is closed under $\wedge$, we have $\mathcal{F} \in \mathfrak{B}$. Clearly $\mathcal{F}$ is the unique finest factor in $\mathfrak{B}$ such that v and w are in the same $\mathcal{F}$-class. Then (12.17) follows. $\square$

Equation (12.17) is a generalization of (12.7), (12.9), (12.11), (12.13), and (12.15). It says that the $\mathbf{A}_{\mathcal{F}}$'s induce a partition of $\Omega \times \Omega$ into disjoint classes, one for each factor in $\mathfrak{B}$: (v, w) and (v', w') are in the same class if and only if there is an $\mathcal{F} \in \mathfrak{B}$ such that $[\mathbf{A}_{\mathcal{F}}]_{vw} = [\mathbf{A}_{\mathcal{F}}]_{v'w'} = 1$. Each of these classes corresponds to one kind of relation between pairs of units.

For experiments with block structures that are *closed under* $\wedge$ (so that the $\mathbf{A}_{\mathcal{F}}$'s are well-defined), we assume that

$$E(\mathbf{y}) = \mu \mathbf{1}_N + \mathbf{X}_T \boldsymbol{\alpha}, \tag{12.18}$$

and

$$\mathbf{V} = \text{cov}(\mathbf{y}) = \sigma^2 \sum_{\mathcal{F}:\mathcal{F} \in \mathfrak{B}} \rho_{\mathcal{F}} \mathbf{A}_{\mathcal{F}}, \text{ where } \rho_{\mathcal{E}} = 1. \tag{12.19}$$

This applies to simple block structures which, by Theorem 12.2(v), are closed under $\wedge$. The covariance matrix in (12.19) assumes a constant correlation between the observations on any pair of units that are in the same class of the partition of $\Omega \times \Omega$ induced by the $\mathbf{A}_{\mathcal{F}}$'s described above. This generalizes (12.8), (12.10), (12.12), (12.14), and (12.16).

In Section 3.6, we showed that the covariance matrices for the randomization models under completely randomized designs, block designs, row-column designs, blocked split-plot designs, and nested row-column designs are patterned matrices that can be expressed in the form of (12.19). We will show in Section 12.14 that the same result holds for general simple block structures. This was claimed by Nelder (1965a), and was proved by Bailey, Praeger, Rowley, and Speed (1983) for *poset block structures*, which include all the simple block structures. Such a result, however, is not true for all the block structures.

12.4 Poset block structures

We have seen in Section 12.2 that a simple block structure $\mathfrak{B}$ on $n_1 \cdots n_m$ units is constructed from m sets of unstructured units $\Omega_1, \ldots, \Omega_m$ of sizes $n_1, \ldots, n_m$, respectively. Let $\mathfrak{B}_i = \{\mathcal{U}_i, \mathcal{E}_i\}$. Recall the notion of domination defined in Section 12.2. Write $\mathcal{E}_i < \mathcal{E}_j$ if $\mathcal{E}_j$ dominates $\mathcal{E}_i$. Then $\mathfrak{B}$ consists of all the factors $\mathcal{F}_1 \times \cdots \times \mathcal{F}_m$ with $\mathcal{F}_i = \mathcal{U}_i$ or $\mathcal{E}_i$ that satisfy (12.2):

$$\text{If } \mathcal{F}_i = \mathcal{E}_i, \text{ then } \mathcal{F}_j = \mathcal{E}_j \text{ for all } j \text{ such that } \mathcal{E}_i < \mathcal{E}_j. \tag{12.20}$$

We write $\mathcal{E}_i \leq \mathcal{E}_j$ if $\mathcal{E}_i < \mathcal{E}_j$ or $\mathcal{E}_i = \mathcal{E}_j$. The relation $\leq$ so defined is a partial order on $\{\mathcal{E}_1, \ldots, \mathcal{E}_m\}$.

The construction described above can be applied to an *arbitrary* partially ordered set, also called a *poset*. Given any partial order $\leq$ on $\{\mathcal{E}_1, \ldots, \mathcal{E}_m\}$, the block structure on $\Omega_1 \times \cdots \times \Omega_m$ consisting of all the factors $\mathcal{F}_1 \times \cdots \times \mathcal{F}_m$ with $\mathcal{F}_i = \mathcal{U}_i$ or $\mathcal{E}_i$ that satisfy (12.20) is called a *poset block structure*. All simple block structures are poset block structures. We refer readers to Example 10 of Bailey (1991) for an experiment with a poset block structure that cannot be obtained by iterations of nesting and crossing.

The definition of poset block structures was formally given in Speed and Bailey (1982), though it goes back to some earlier work such as Throckmorton (1961) and Zyskind (1962).

It is easy to see that all poset block structures are closed under $\wedge$. Therefore all the results in Section 12.3 also apply to poset block structures.

Randomization of a poset block structure can be performed by, for each $i, 1 \leq i \leq m$, randomly permuting the levels of $\mathcal{E}_i$ among each set of units with fixed levels of all the ancestors of $\mathcal{E}_i$. Bailey et al. (1983) proved that the covariance matrix of the resulting randomization model is of the form (12.19).

12.5   Orthogonal block structures

Not all the examples in Chapter 1 have simple or poset block structures. Example 1.4 is about an experiment with three or four processing stages, where at each stage the experimental units are divided into four groups. A simple or poset block structure would require $4^3 = 64$ units for three stages and $4^4 = 256$ units for four stages. When there are only 16 experimental runs, the block structure can be regarded as an *incomplete* crossing of some four-level unit factors. In this case, the experimental units constitute a *fraction* of those with the block structure $4 \times 4 \times 4$ or $4 \times 4 \times 4 \times 4$.

Example 1.5 in Chapter 1 also involves incomplete crossing of unit factors. In an experiment with two processing stages where at each of the two stages the experimental units are to be divided into eight groups, complete crossing would require 64 runs. One is forced to use a fraction when only 32 runs are conducted. Thus this is also not a simple or poset block structure.

Speed and Bailey (1982) introduced a more general class of block structures, named *orthogonal block structures* in Bailey (1985).

Definition 12.4. A block structure $\mathfrak{B}$ consisting of nonequivalent factors is called an *orthogonal block structure* if the following conditions hold:

(i) All the factors in $\mathfrak{B}$ are uniform;

(ii) $\mathcal{E} \in \mathfrak{B}$;

(iii) $\mathcal{U} \in \mathfrak{B}$;

(iv) if $\mathcal{F}, \mathcal{G} \in \mathfrak{B}$, then $\mathcal{F} \wedge \mathcal{G} \in \mathfrak{B}$;

(v) if $\mathcal{F}, \mathcal{G} \in \mathfrak{B}$, then $\mathcal{F} \vee \mathcal{G} \in \mathfrak{B}$;

(vi) the factors in $\mathfrak{B}$ are pairwise orthogonal.

Theorem 12.4. *Poset block structures (and hence simple block structures) are orthogonal block structures.*

Proof. That poset block structures satisfy (i)–(v) in Definition 12.4 can easily be established. It remains to show that the factors in any poset block structure are pairwise orthogonal. Since all the factors in a poset block structure are of the form $\mathcal{F}_1 \times \cdots \times \mathcal{F}_m$ with $\mathcal{F}_i = \mathcal{U}_i$ or $\mathcal{E}_i$, and $\mathcal{U}_i$ and $\mathcal{E}_i$ are orthogonal to each other, it suffices to show that if $\mathcal{F}_1$ and $\mathcal{G}_1$ are orthogonal factors on Ω_1, and $\mathcal{F}_2$ and $\mathcal{G}_2$ are orthogonal factors on Ω_2, then $\mathcal{F}_1 \times \mathcal{F}_2$ and $\mathcal{G}_1 \times \mathcal{G}_2$ are orthogonal. We note that

$$\mathbf{P}_{V_{\mathcal{F}_1 \times \mathcal{F}_2}} = \mathbf{P}_{V_{\mathcal{F}_1}} \otimes \mathbf{P}_{V_{\mathcal{F}_2}}.$$

By Theorem 4.6, $\mathbf{P}_{V_{\mathcal{F}_1}} \mathbf{P}_{V_{\mathcal{G}_1}} = \mathbf{P}_{V_{\mathcal{G}_1}} \mathbf{P}_{V_{\mathcal{F}_1}}$ and $\mathbf{P}_{V_{\mathcal{F}_2}} \mathbf{P}_{V_{\mathcal{G}_2}} = \mathbf{P}_{V_{\mathcal{G}_2}} \mathbf{P}_{V_{\mathcal{F}_2}}$. It is enough to show that $\mathbf{P}_{V_{\mathcal{F}_1 \times \mathcal{F}_2}} \mathbf{P}_{V_{\mathcal{G}_1 \times \mathcal{G}_2}} = \mathbf{P}_{V_{\mathcal{G}_1 \times \mathcal{G}_2}} \mathbf{P}_{V_{\mathcal{F}_1 \times \mathcal{F}_2}}$. This holds since $\mathbf{P}_{V_{\mathcal{F}_1 \times \mathcal{F}_2}} \mathbf{P}_{V_{\mathcal{G}_1 \times \mathcal{G}_2}} =$
$(\mathbf{P}_{V_{\mathcal{F}_1}} \otimes \mathbf{P}_{V_{\mathcal{F}_2}})(\mathbf{P}_{V_{\mathcal{G}_1}} \otimes \mathbf{P}_{V_{\mathcal{G}_2}}) = (\mathbf{P}_{V_{\mathcal{F}_1}} \mathbf{P}_{V_{\mathcal{G}_1}}) \otimes (\mathbf{P}_{V_{\mathcal{F}_2}} \mathbf{P}_{V_{\mathcal{G}_2}}) = (\mathbf{P}_{V_{\mathcal{G}_1}} \mathbf{P}_{V_{\mathcal{F}_1}}) \otimes (\mathbf{P}_{V_{\mathcal{G}_2}} \mathbf{P}_{V_{\mathcal{F}_2}}) =$
$(\mathbf{P}_{V_{\mathcal{G}_1}} \otimes \mathbf{P}_{V_{\mathcal{G}_2}})(\mathbf{P}_{V_{\mathcal{F}_1}} \otimes \mathbf{P}_{V_{\mathcal{F}_2}}) = \mathbf{P}_{V_{\mathcal{G}_1 \times \mathcal{G}_2}} \mathbf{P}_{V_{\mathcal{F}_1 \times \mathcal{F}_2}}$. $\qquad\square$

Since orthogonal block structures are closed under $\wedge$, by Proposition 12.3, the patterns-of-covariance form (12.19) of $\mathbf{V}$ is well defined for orthogonal block structures.

Example 12.7. (Example 1.4 revisited) This involves an orthogonal block structure that is not a simple block structure. In an experiment with three processing stages, let $\mathcal{R}$ and $\mathcal{C}$ be the factors defined by row and column partitions. Suppose a Latin square is used for the third-stage partitioning, and let the resulting factor be denoted by $\mathcal{L}$. Then all the factors are uniform. They are also pairwise orthogonal since the condition of proportional frequencies holds for all three pairs of $\mathcal{R}$, $\mathcal{C}$, and $\mathcal{L}$. It is easy to see that $\mathcal{R} \vee \mathcal{C} = \mathcal{R} \vee \mathcal{L} = \mathcal{C} \vee \mathcal{L} = \mathcal{U}$, and $\mathcal{R} \wedge \mathcal{C} = \mathcal{R} \wedge \mathcal{L} = \mathcal{C} \wedge \mathcal{L} = \mathcal{E}$. Let $\mathfrak{B} = \{\mathcal{E}, \mathcal{R}, \mathcal{C}, \mathcal{L}, \mathcal{U}\}$. Then $\mathfrak{B}$ is an orthogonal block structure with the partial order

$$\mathcal{E} \prec \mathcal{R} \prec \mathcal{U}, \mathcal{E} \prec \mathcal{C} \prec \mathcal{U}, \mathcal{E} \prec \mathcal{L} \prec \mathcal{U}.$$

Suppose there is a fourth stage, and a pair of orthogonal Latin squares is used for the third- and fourth-stage partitions. Let the factors defined by these two Latin squares be $\mathcal{L}_1$ and $\mathcal{L}_2$, and let $\mathfrak{B}$ be $\{\mathcal{E}, \mathcal{R}, \mathcal{C}, \mathcal{L}_1, \mathcal{L}_2, \mathcal{U}\}$. Then $\mathfrak{B}$ is also an orthogonal block structure. Its Hasse diagram is shown in Figure 12.3.

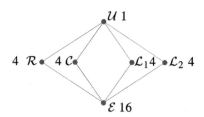

Figure 12.3 *Hasse diagram for the block structure in Example 1.4*

12.6 Models with random effects

A model of the form (12.18)–(12.19) can also be obtained by assuming a set of uncorrelated random effects for the levels of each factor $\mathcal{F} \in \mathfrak{B}$. Specifically, for each $v = 1, \ldots, N$, let y_v be the observation on the vth unit and $\phi(v) \in \{1, \ldots, t\}$ be the treatment assigned to unit v, and for each $\mathcal{F} \in \mathfrak{B}$, let $\mathcal{F}(v) \in \{1, \ldots, n_{\mathcal{F}}\}$ be the $\mathcal{F}$-class to which unit v belongs (or the level of $\mathcal{F}$ at unit v). Suppose

$$y_v = \mu + \alpha_{\phi(v)} + \sum_{\mathcal{F} \in \mathfrak{B}} \beta^{\mathcal{F}}_{\mathcal{F}(v)}, \tag{12.21}$$

where μ and the treatment effects $\alpha_1, \ldots, \alpha_t$ are unknown constants, and $\{\beta^{\mathcal{F}}_1, \ldots, \beta^{\mathcal{F}}_{n_{\mathcal{F}}}\}_{\mathcal{F} \in \mathfrak{B}}$ are uncorrelated random variables with $E(\beta^{\mathcal{F}}_k) = 0$ and $\text{var}(\beta^{\mathcal{F}}_k) = \sigma^2_{\mathcal{F}}$. Then we can write (12.21) in the matrix form

$$\mathbf{y} = \mu \mathbf{1}_N + \mathbf{X}_T \boldsymbol{\alpha} + \sum_{\mathcal{F} \in \mathfrak{B}} \mathbf{X}_{\mathcal{F}} \boldsymbol{\beta}^{\mathcal{F}},$$

where $\boldsymbol{\alpha} = (\alpha_1, \ldots, \alpha_t)^T$, $\boldsymbol{\beta}^{\mathcal{F}} = (\beta^{\mathcal{F}}_1, \ldots, \beta^{\mathcal{F}}_{n_{\mathcal{F}}})^T$, and $\mathbf{X}_{\mathcal{F}}$ is the incidence matrix between the units and factor $\mathcal{F}$. Then

$$E(\mathbf{y}) = \mu \mathbf{1}_N + \mathbf{X}_T \boldsymbol{\alpha},$$

and

$$\mathbf{V} = \sum_{\mathcal{F} \in \mathfrak{B}} \sigma^2_{\mathcal{F}} \mathbf{X}_{\mathcal{F}} \mathbf{X}^T_{\mathcal{F}} = \sum_{\mathcal{F} \in \mathfrak{B}} \sigma^2_{\mathcal{F}} \mathbf{R}_{\mathcal{F}}, \tag{12.22}$$

where the second equality follows from (4.1). In this form of the covariance matrix, we have one variance component for each of the factors that define the block structure. The second equality in (12.22) shows that it can be expressed as a linear combination of the relation matrices of these factors.

Theorem 12.5. *If a block structure $\mathfrak{B}$ is closed under $\wedge$, then the covariance matrix $\mathbf{V}$ in (12.22) can be expressed as in (12.19).*

Proof. It follows from (12.22) that for any two units v and w,

$$\text{cov}(y_v, y_w) = \sum_{\mathcal{G} \in \mathfrak{B}: v \text{ and } w \text{ are in the same } \mathcal{G}\text{-class}} \sigma^2_{\mathcal{G}}. \tag{12.23}$$

Let $\mathcal{F}$ be the finest factor in $\mathfrak{B}$ such that units v and w are in the same $\mathcal{F}$-class. If units v and w are in the same $\mathcal{G}$-class, then by the definition of $\mathcal{F}$, we must have $\mathcal{F} \preceq \mathcal{G}$. On the other hand, if $\mathcal{F} \preceq \mathcal{G}$, then v and w are also in the same $\mathcal{G}$-class. This and (12.23) imply that

$$\text{cov}(y_v, y_w) = \sum_{\mathcal{G} \in \mathfrak{B}: \mathcal{F} \preceq \mathcal{G}} \sigma^2_{\mathcal{G}}.$$

Then $\mathbf{V}$ is as in (12.19) with

$$\sigma^2 = \sum_{\mathcal{G}: \mathcal{G} \in \mathfrak{B}} \sigma^2_{\mathcal{G}}, \tag{12.24}$$

and

$$\rho_{\mathcal{F}} = \frac{\sum_{\mathcal{G} \in \mathfrak{B}: \mathcal{F} \preceq \mathcal{G}} \sigma^2_{\mathcal{G}}}{\sum_{\mathcal{G}: \mathcal{G} \in \mathfrak{B}} \sigma^2_{\mathcal{G}}}. \tag{12.25}$$

$\square$

In this case, suppose

$$\sigma_{\mathcal{F}}^2 > 0 \text{ for all } \mathcal{F} \neq \mathcal{U}. \tag{12.26}$$

Then it follows from (12.25) that

$$\mathcal{F}_1 \prec \mathcal{F}_2 \Rightarrow \rho_{\mathcal{F}_1} > \rho_{\mathcal{F}_2}. \tag{12.27}$$

We have seen special cases of (12.24) and (12.25) in (3.13) and (3.19), and special cases of (12.27) for block and row-column designs in Sections 3.3 and 3.4, respectively.

In recent literature on experiments with multiple processing stages, typically a mixed-effect model such as (12.21) is assumed, with one set of random effects for the partition at each stage. For the block structure $\mathfrak{B} = \{\mathcal{E}, \mathcal{R}, \mathcal{C}, \mathcal{L}, \mathcal{U}\}$ with the partitions defined by the rows, columns, and letters of a Latin square mentioned in Example 12.7, Preece, Bailey, and Patterson (1978) showed that the largest group of permutations that preserve the partitions by rows, columns, and letters of the Latin square may not be transitive, and for some Latin squares, transitivity holds but the resulting covariance matrix may not be in the form of (12.19).

We refer readers to Brien and Bailey (2006, 2009, 2010) and the references cited there for the analysis of experiments that involve multiple randomizations. Bailey and Brien (2010) derived the randomization model and its analysis for experiments with a chain of randomizations.

12.7 Strata

The main result of this section is the spectral decomposition of the matrix $\mathbf{V}$ in (12.19) for orthogonal block structures. Theorem 12.6 shows that for a block structure $\mathfrak{B}$ that is closed under $\vee$ and has pairwise orthogonal factors, the N-dimensional space $V_{\mathcal{E}}$ can be decomposed as an orthogonal sum of subspaces, with one subspace associated with each factor in $\mathfrak{B}$ (or each node of its Hasse diagram). Theorem 12.7 shows that for orthogonal block structures, each of these subspaces is an eigenspace of $\mathbf{V}$ unless it is the null space.

Theorem 12.6. *Let $\mathfrak{B}$ be a set of nonequivalent factors satisfying the following conditions:*

(i) the factors in $\mathfrak{B}$ are pairwise orthogonal;

(ii) if $\mathcal{F}, \mathcal{G} \in \mathfrak{B}$, then $\mathcal{F} \vee \mathcal{G} \in \mathfrak{B}$.

For each factor $\mathcal{F} \in \mathfrak{B}$, let

$$W_{\mathcal{F}} = V_{\mathcal{F}} \ominus \left(\sum_{\mathcal{G} \in \mathfrak{B}:\mathcal{F} \prec \mathcal{G}} V_{\mathcal{G}} \right). \tag{12.28}$$

Then the spaces $W_{\mathcal{F}}$, $\mathcal{F} \in \mathfrak{B}$, are mutually orthogonal. Furthermore, for each $\mathcal{F} \in \mathfrak{B}$,

$$V_{\mathcal{F}} = \bigoplus_{\mathcal{G} \in \mathfrak{B}:\mathcal{F} \preceq \mathcal{G}} W_{\mathcal{G}}. \tag{12.29}$$

In particular, if $\mathcal{E} \in \mathfrak{B}$, then

$$V_{\mathcal{E}} = \mathbb{R}^N = \underset{\mathcal{F}:\mathcal{F}\in\mathfrak{B}}{\oplus} W_{\mathcal{F}}. \tag{12.30}$$

Theorem 12.7. *Let $\mathfrak{B}$ be an orthogonal block structure and $\mathbf{V}$ be as in (12.19). Then $\mathbf{V}$ can be expressed as*

$$\mathbf{V} = \sum_{\mathcal{F}:\mathcal{F}\in\mathfrak{B}} \xi_{\mathcal{F}} \mathbf{P}_{W_{\mathcal{F}}}, \tag{12.31}$$

where $\mathbf{P}_{W_{\mathcal{F}}}$ is the orthogonal projection matrix onto $W_{\mathcal{F}}$.

It follows from (12.30) that

$$\mathbf{P}_{W_{\mathcal{F}}}\mathbf{P}_{W_{\mathcal{G}}} = \mathbf{0} \text{ for all } \mathcal{F} \neq \mathcal{G}, \text{ and } \sum_{\mathcal{F}:\mathcal{F}\in\mathfrak{B}} \mathbf{P}_{W_{\mathcal{F}}} = \mathbf{I}_N. \tag{12.32}$$

The spaces $W_{\mathcal{F}}$ depend only on the block structure and do not depend on the entries of $\mathbf{V}$. The expression in (12.31) gives the spectral decomposition of $\mathbf{V}$. Note that some of the $W_{\mathcal{F}}$'s may be $\{\mathbf{0}\}$ (see Remark 12.2 in Section 12.10). If $W_{\mathcal{F}} \neq \{\mathbf{0}\}$, then $\xi_{\mathcal{F}}$ is an eigenvalue of $\mathbf{V}$ and $W_{\mathcal{F}}$ is the associated eigenspace; in this connection we refer readers to Remark 5.1. Each nonnull $W_{\mathcal{F}}$, $\mathcal{F} \in \mathfrak{B}$, is called a *stratum*, and $\dim(W_{\mathcal{F}})$ is called its degrees of freedom. If $\mathfrak{B}$ is a simple block structure, then $W_{\mathcal{F}} \neq \{\mathbf{0}\}$ for all $\mathcal{F} \in \mathfrak{B}$.

Since $\mathcal{U}$ is the coarsest factor, by (12.28),

$$W_{\mathcal{U}} = V_{\mathcal{U}}, \dim(W_{\mathcal{U}}) = 1,$$

and

$$\mathbf{P}_{W_{\mathcal{U}}} = \mathbf{P}_{V_{\mathcal{U}}} = \frac{1}{N}\mathbf{J}_N.$$

We call $V_{\mathcal{U}}$ the mean stratum. Then

$$\mathbb{R}^N \ominus V_{\mathcal{U}} = \underset{\mathcal{F}:\mathcal{F}\in\mathfrak{B},\mathcal{F}\neq\mathcal{U}}{\oplus} W_{\mathcal{F}}.$$

Thus the strata other than the mean stratum give an orthogonal decomposition of the space of all *unit contrasts*.

We have

$$\mathrm{cov}\left(\mathbf{P}_{W_{\mathcal{F}}}\mathbf{y}\right) = \mathbf{P}_{W_{\mathcal{F}}}\mathbf{V}\mathbf{P}_{W_{\mathcal{F}}} = \mathbf{P}_{W_{\mathcal{F}}}\left[\sum_{\mathcal{G}:\mathcal{G}\in\mathfrak{B}} \xi_{\mathcal{G}}\mathbf{P}_{W_{\mathcal{G}}}\right]\mathbf{P}_{W_{\mathcal{F}}} = \xi_{\mathcal{F}}\mathbf{P}_{W_{\mathcal{F}}}$$

where the last equality follows from (12.32). In particular, for any $\mathbf{c} \in W_{\mathcal{F}}$,

$$\mathrm{var}\left(\mathbf{c}^T\mathbf{y}\right) = \mathbf{c}^T\left[\sum_{\mathcal{G}:\mathcal{G}\in\mathfrak{B}} \xi_{\mathcal{G}}\mathbf{P}_{W_{\mathcal{G}}}\right]\mathbf{c} = \xi_{\mathcal{F}}\mathbf{c}^T\mathbf{P}_{W_{\mathcal{F}}}\mathbf{c} = \xi_{\mathcal{F}}\|\mathbf{c}\|^2.$$

For two different strata $\mathcal{F}_1$ and $\mathcal{F}_2$,

$$\text{cov}\left(\mathbf{P}_{W_{\mathcal{F}_1}}\mathbf{y}, \mathbf{P}_{W_{\mathcal{F}_2}}\mathbf{y}\right) = \mathbf{P}_{W_{\mathcal{F}_1}}\mathbf{V}\mathbf{P}_{W_{\mathcal{F}_2}} = \mathbf{P}_{W_{\mathcal{F}_1}}\left[\sum_{\mathcal{G}:\mathcal{G}\in\mathfrak{B}} \xi_{\mathcal{G}}\mathbf{P}_{W_{\mathcal{G}}}\right]\mathbf{P}_{W_{\mathcal{F}_2}} = \mathbf{0}.$$

Nelder (1965a) proved Theorem 12.7 for simple block structures. He also showed how to obtain the spectral decomposition of $\mathbf{V}$ in (12.31) from the block structure formula and gave simple rules for determining the degrees of freedom and projections onto the eigenspaces. We first discuss implications of Theorem 12.7 in the next section. Then we present Nelder's rules with illustrative examples in Section 12.9, and an alternative method based on Hasse diagrams in Section 12.10. The proofs, including those of Theorems 12.6 and 12.7, are deferred to later sections.

Remark 12.1. Theorem 12.6 also applies when $\mathfrak{B}$ is a structure on the treatments. Theorem 6.2 is a special case of Theorem 12.6 when we apply the latter to factorial treatment structures. We have used Theorem 6.2 to decompose the space of treatment contrasts into orthogonal components representing various factorial effects.

12.8 Null ANOVA

The observation vector $\mathbf{y}$ can be projected onto different strata, resulting in the decomposition

$$\mathbf{y} - \mathbf{P}_{V_{\mathcal{U}}}\mathbf{y} = \sum_{\mathcal{F}\in\mathfrak{B}:\mathcal{F}\neq\mathcal{U}} \mathbf{P}_{W_{\mathcal{F}}}\mathbf{y}.$$

Then we have

$$\left\|\mathbf{y} - \mathbf{P}_{V_{\mathcal{U}}}\mathbf{y}\right\|^2 = \sum_{\mathcal{F}\in\mathfrak{B}:\mathcal{F}\neq\mathcal{U}} \left\|\mathbf{P}_{W_{\mathcal{F}}}\mathbf{y}\right\|^2 \qquad (12.33)$$

with the breakdown of degrees of freedom

$$N - 1 = \sum_{\mathcal{F}\in\mathfrak{B}:\mathcal{F}\neq\mathcal{U}} \dim(W_{\mathcal{F}}).$$

Suppose $\alpha_1 = \cdots = \alpha_t$. Then for $\mathcal{F}\neq\mathcal{U}$,

$$\begin{aligned}
\text{E}\left(\left\|\mathbf{P}_{W_{\mathcal{F}}}\mathbf{y}\right\|^2\right) &= [\text{E}(\mathbf{y})]^T\mathbf{P}_{W_{\mathcal{F}}}\text{E}(\mathbf{y}) + \text{tr}\left(\mathbf{P}_{W_{\mathcal{F}}}\mathbf{V}\right) \\
&= \text{tr}\left(\xi_{\mathcal{F}}\mathbf{P}_{W_{\mathcal{F}}}\right) \\
&= \xi_{\mathcal{F}}\dim(W_{\mathcal{F}}).
\end{aligned}$$

The second equality holds since, when $\alpha_1 = \cdots = \alpha_t$, $\text{E}\mathbf{y} \in V_{\mathcal{U}}$, which is orthogonal to $W_{\mathcal{F}}$ for all $\mathcal{F}\neq\mathcal{U}$. Then

$$\text{E}\left(\frac{1}{\dim(W_{\mathcal{F}})}\left\|\mathbf{P}_{W_{\mathcal{F}}}\mathbf{y}\right\|^2\right) = \xi_{\mathcal{F}} \text{ for } \mathcal{F}\neq\mathcal{U}. \qquad (12.34)$$

The eigenvalues $\xi_{\mathcal{F}}$ are called *stratum variances*. The decomposition of the total sum of squares in (12.33) is the null ANOVA. We have seen special cases of this result for the three classes of designs studied in Chapter 5.

Suppose there are s factors $\mathcal{F}_1, \ldots, \mathcal{F}_s$ other than $\mathcal{U}$ with $W_{\mathcal{F}_i} \neq \{\mathbf{0}\}$. Then (12.33) is summarized in the following null ANOVA table.

Source of variation	Sum of squares	Degrees of freedom	Mean square	E(MS)
$\mathcal{F}_1$	$\left\|\mathbf{P}_{W_{\mathcal{F}_1}}\mathbf{y}\right\|^2$	$\dim\left(W_{\mathcal{F}_1}\right)$	$\frac{1}{\dim(W_{\mathcal{F}_1})}\left\|\mathbf{P}_{W_{\mathcal{F}_1}}\mathbf{y}\right\|^2$	ξ_1
$\vdots$	$\vdots$	$\vdots$	$\vdots$	$\vdots$
$\mathcal{F}_s$	$\left\|\mathbf{P}_{W_{\mathcal{F}_s}}\mathbf{y}\right\|^2$	$\dim\left(W_{\mathcal{F}_s}\right)$	$\frac{1}{\dim(W_{\mathcal{F}_s})}\left\|\mathbf{P}_{W_{\mathcal{F}_s}}\mathbf{y}\right\|^2$	ξ_s
Total	$\left\|\mathbf{y} - \mathbf{P}_{V_\mathcal{U}}\mathbf{y}\right\|^2$	$N-1$		

12.9 Nelder's rules

We describe Nelder's (1965a) rules for determining the degrees of freedom and projections onto the strata, which are needed for completing the null ANOVA. The first step is to determine the number of strata and how the degrees of freedom are split up. For the block structure n_1/n_2, we have the *degree-of-freedom identity*

$$n_1 n_2 = 1 + v_1 + n_1 v_2,$$

where $v_i = n_i - 1$. The three terms on the right-hand side specify the degrees of freedom of the mean, interblock, and intrablock strata, respectively. The degree-of-freedom identity for the block structure $n_1 \times n_2$ is

$$n_1 n_2 = 1 + v_1 + v_2 + v_1 v_2.$$

We define the nesting and crossing functions as

$$\mathcal{N}(n_1, n_2) = 1 + v_1 + n_1 v_2 \tag{12.35}$$

and

$$\mathcal{C}(n_1, n_2) = 1 + v_1 + v_2 + v_1 v_2,$$

respectively.

The arguments in the $\mathcal{N}$ and $\mathcal{C}$ functions can be substituted by other $\mathcal{N}$ and $\mathcal{C}$ functions. The degree-of-freedom identities for more complex block structures can be obtained from their block structure formulas by expanding the corresponding $\mathcal{N}$ and $\mathcal{C}$ functions. Then the number of terms in the expansion is the number of strata, and each term in the expansion is the dimension (degrees of freedom) of the corresponding stratum.

For example, the degree-of-freedom identity for $n_1/(n_2 \times n_3)$ can be obtained by expanding $\mathcal{N}(n_1, \mathcal{C}(n_2, n_3))$. There are a few rules that must be followed. Generally, we cannot destroy terms by algebraic manipulation (e.g., $1+v$ cannot be replaced by n) except that we may remove identical terms when they are subtracted, and replace $1 \cdot x$ with x. Another important rule is that the n_1 term that appears on the right-hand side of (12.35) must be the algebraic sum of all the terms in the expansion of n_1.

Nelder's rules will be justified in Section 12.15.

Example 12.8. To determine the degree-of-freedom identity for the block structure $n_1/(n_2/n_3)$, we expand

$$
\begin{aligned}
\mathcal{N}(n_1, \mathcal{N}(n_2, n_3)) &= 1 + v_1 + n_1[\mathcal{N}(n_2, n_3) - 1] \\
&= 1 + v_1 + n_1(v_2 + n_2 v_3) \\
&= 1 + v_1 + n_1 v_2 + n_1 n_2 v_3. \quad\quad (12.36)
\end{aligned}
$$

This shows that there are four strata with 1, $n_1 - 1$, $n_1(n_2 - 1)$, and $n_1 n_2(n_3 - 1)$ degrees of freedom, respectively. Note that (12.36) can also be obtained by expanding $\mathcal{N}(\mathcal{N}(n_1, n_2), n_3)$. To illustrate one of the rules mentioned earlier, we have $\mathcal{N}(\mathcal{N}(n_1, n_2), n_3) = \mathcal{N}(1 + v_1 + n_1 v_2, n_3)$, which must be expanded as $1 + [(1 + v_1 + n_1 v_2) - 1] + \mathbf{n_1 n_2 v_3}$. It then yields the correct result as in (12.36).

Example 12.9. The degree-of-freedom identity for $n_1/(n_2 \times n_3)$ follows from the expansion of $\mathcal{N}(n_1, \mathcal{C}(n_2, n_3))$:

$$
\begin{aligned}
\mathcal{N}(n_1, \mathcal{C}(n_2, n_3)) &= 1 + v_1 + n_1(\mathcal{C}(n_2, n_3) - 1) \\
&= 1 + v_1 + n_1(v_2 + v_3 + v_2 v_3) \\
&= 1 + v_1 + n_1 v_2 + n_1 v_3 + n_1 v_2 v_3. \quad\quad (12.37)
\end{aligned}
$$

From the degree-of-freedom identity, we can write down a *yield identity*, which gives orthogonal projections of $\mathbf{y}$ onto all the strata. For convenience, each unit is indexed by multi-subscripts, and as before, dot notation is used for averaging. We first expand each term in the degree-of-freedom identity as a function of the n_i's; then we replace each term in the expansion with a mean of the y's that is averaged over the subscripts for which the corresponding n_i's are absent. The linear combination of the y means obtained from the same term in the degree-of-freedom identity gives an entry of the orthogonal projection of $\mathbf{y}$ onto the corresponding stratum.

Example 12.10. (Example 12.8 revisited) From (12.36), we obtain

$$
n_1 n_2 n_3 = 1 + (n_1 - 1) + (n_1 n_2 - n_1) + (n_1 n_2 n_3 - n_1 n_2).
$$

This gives the yield identity

$$
y_{ijk} = y_{\cdots} + (y_{i\cdot\cdot} - y_{\cdots}) + (y_{ij\cdot} - y_{i\cdot\cdot}) + (y_{ijk} - y_{ij\cdot}). \quad\quad (12.38)
$$

For example, the first term $n_1 n_2 n_3$ in the degree-of-freedom identity is replaced by y_{ijk} since all of n_1, n_2, n_3 are present, and $n_1 - 1$ is replaced by $y_{i\cdot\cdot} - y_{\cdots}$ since n_2 and n_3 are not present in n_1, and all three are absent from 1. For the block structure $n_1/(n_2/n_3)$, the last three terms on the right side of (12.38) are the entries of the orthogonal projections of $\mathbf{y}$ onto the three strata other than $V_{\mathcal{U}}$. It follows that the corresponding sums of squares in the null ANOVA are $\sum_{i=1}^{n_1} n_2 n_3(y_{i\cdot\cdot} - y_{\cdots})^2$, $\sum_{i=1}^{n_1} \sum_{j=1}^{n_2} n_3(y_{ij\cdot} - y_{i\cdot\cdot})^2$, and $\sum_{i=1}^{n_1} \sum_{j=1}^{n_2} \sum_{k=1}^{n_3}(y_{ijk} - y_{ij\cdot})^2$.

We explained in Section 4.2 that the block structure $(n_1/n_2)/n_3$ can be defined by the four factors $\mathcal{E}$, $\mathcal{P}$, $\mathcal{B}$, and $\mathcal{U}$. By an argument similar to those used to derive (5.21) and (5.28), it is a tedious but routine exercise to show that for the covariance structure given in (3.23), $\mathbf{V}$ can be written in the form

$$\mathbf{V} = \xi_{\mathcal{U}}\mathbf{P}_{V_{\mathcal{U}}} + \xi_{\mathcal{B}}\left(\mathbf{P}_{V_{\mathcal{B}}} - \mathbf{P}_{V_{\mathcal{U}}}\right) + \xi_{\mathcal{P}}\left(\mathbf{P}_{V_{\mathcal{P}}} - \mathbf{P}_{V_{\mathcal{B}}}\right) + \xi_{\mathcal{E}}\left(\mathbf{I} - \mathbf{P}_{V_{\mathcal{P}}}\right)$$

for some $\xi_{\mathcal{U}}$, $\xi_{\mathcal{B}}$, $\xi_{\mathcal{P}}$, and $\xi_{\mathcal{E}}$. Then we see that $\mathbf{V}$ is as in (12.31):

$$\mathbf{V} = \xi_{\mathcal{U}}\mathbf{P}_{W_{\mathcal{U}}} + \xi_{\mathcal{B}}\mathbf{P}_{W_{\mathcal{B}}} + \xi_{\mathcal{P}}\mathbf{P}_{W_{\mathcal{P}}} + \xi_{\mathcal{E}}\mathbf{P}_{W_{\mathcal{E}}},$$

where

$$\mathbf{P}_{W_{\mathcal{U}}} = \mathbf{P}_{V_{\mathcal{U}}},$$

$$\mathbf{P}_{W_{\mathcal{B}}} = \mathbf{P}_{V_{\mathcal{B}}} - \mathbf{P}_{V_{\mathcal{U}}} = \mathbf{P}_{V_{\mathcal{B}} \ominus V_{\mathcal{U}}},$$

$$\mathbf{P}_{W_{\mathcal{P}}} = \mathbf{P}_{V_{\mathcal{P}}} - \mathbf{P}_{V_{\mathcal{B}}} = \mathbf{P}_{V_{\mathcal{P}} \ominus V_{\mathcal{B}}},$$

$$\mathbf{P}_{W_{\mathcal{E}}} = \mathbf{I} - \mathbf{P}_{V_{\mathcal{P}}} = \mathbf{P}_{V_{\mathcal{P}}^{\perp}}.$$

The yield identity in (12.38) indeed shows the projections of $\mathbf{y}$ onto $V_{\mathcal{U}}$, $V_{\mathcal{B}} \ominus V_{\mathcal{U}}$, $V_{\mathcal{P}} \ominus V_{\mathcal{B}}$, and $V_{\mathcal{P}}^{\perp}$, respectively. By (12.34), if $\alpha_1 = \cdots = \alpha_t$, then

$$E\left[\frac{1}{n_1 - 1}\sum_{i=1}^{n_1} n_2 n_3 (y_{i\cdot\cdot} - y_{\cdot\cdot\cdot})^2\right] = \xi_{\mathcal{B}},$$

$$E\left[\frac{1}{n_1(n_2 - 1)}\sum_{i=1}^{n_1}\sum_{j=1}^{n_2} n_3 (y_{ij\cdot} - y_{i\cdot\cdot})^2\right] = \xi_{\mathcal{P}},$$

$$E\left[\frac{1}{n_1 n_2(n_3 - 1)}\sum_{i=1}^{n_1}\sum_{j=1}^{n_2}\sum_{k=1}^{n_3} (y_{ijk} - y_{ij\cdot})^2\right] = \xi_{\mathcal{E}}.$$

This shows that $\xi_{\mathcal{B}}$, $\xi_{\mathcal{P}}$, and $\xi_{\mathcal{E}}$ are the between-block variance, between-whole-plot variance within blocks, and between-subplot variance within whole-plots, respectively. We call $W_{\mathcal{B}}$, $W_{\mathcal{P}}$, and $W_{\mathcal{E}}$ the block, whole-plot, and subplot strata, respectively. When the mean stratum is included, there is one stratum for each of the four factors $\mathcal{E}$, $\mathcal{P}$, $\mathcal{B}$, and $\mathcal{U}$. In Example 1.2, $n_1 = n_2 = 4$, and $n_3 = 2$; hence $W_{\mathcal{B}}, W_{\mathcal{P}}$, and $W_{\mathcal{E}}$ have, respectively, 3, 12, and 16 degrees of freedom.

Example 12.11. (Example 12.9 revisited) From (12.37),

$$n_1 n_2 n_3 = 1 + (n_1 - 1) + (n_1 n_2 - n_1) + (n_1 n_3 - n_1) + (n_1 n_2 n_3 - n_1 n_2 - n_1 n_3 + n_1).$$

This gives the yield identity

$$y_{ijk} = y_{\cdots} + (y_{i\cdot\cdot} - y_{\cdots}) + (y_{ij\cdot} - y_{i\cdot\cdot}) + (y_{i\cdot k} - y_{i\cdot\cdot}) + (y_{ijk} - y_{ij\cdot} - y_{i\cdot k} + y_{i\cdot\cdot}). \quad (12.39)$$

Therefore for the block structure $n_1/(n_2 \times n_3)$, the strata other than $V_\mathcal{U}$ have degrees of freedom equal to $n_1 - 1$, $n_1(n_2 - 1)$, $n_1(n_3 - 1)$, and $n_1(n_2 - 1)(n_3 - 1)$, and the corresponding sums of squares in the null ANOVA are $\sum_{i=1}^{n_1} n_2 n_3 (y_{i..} - y_{...})^2$, $\sum_{i=1}^{n_1} \sum_{j=1}^{n_2} n_3 (y_{ij.} - y_{i..})^2$, $\sum_{i=1}^{n_1} \sum_{k=1}^{n_3} n_2 (y_{i.k} - y_{i..})^2$, and $\sum_{i=1}^{n_1} \sum_{j=1}^{n_2} \sum_{k=1}^{n_3} (y_{ijk} - y_{ij.} - y_{i.k} + y_{i..})^2$. The yield identity in (12.39) gives the projections of $\mathbf{y}$ onto $V_\mathcal{U}$, $V_\mathcal{B} \ominus V_\mathcal{U}$, $V_{\mathcal{R}'} \ominus V_\mathcal{B}$, $V_{\mathcal{C}'} \ominus V_\mathcal{B}$, and $[V_\mathcal{U} \oplus (V_\mathcal{B} \ominus V_\mathcal{U}) \oplus (V_{\mathcal{R}'} \ominus V_\mathcal{B}) \oplus (V_{\mathcal{C}'} \ominus V_\mathcal{B})]^\perp = (V_{\mathcal{R}'} + V_{\mathcal{C}'})^\perp$, where $\mathcal{E}$, $\mathcal{R}'$, $\mathcal{C}'$, $\mathcal{B}$, and $\mathcal{U}$ are the five factors defining the block structure, as discussed in Section 4.2. This gives the five strata $W_\mathcal{U}$, $W_\mathcal{B}$, $W_{\mathcal{R}'}$, $W_{\mathcal{C}'}$, and $W_\mathcal{E}$, respectively. Again if $\alpha_1 = \cdots = \alpha_t$, then

$$
E \left[\frac{1}{n_1 - 1} \sum_{i=1}^{n_1} n_2 n_3 (y_{i..} - y_{...})^2 \right] = \xi_\mathcal{B},
$$

$$
E \left[\frac{1}{n_1(n_2 - 1)} \sum_{i=1}^{n_1} \sum_{j=1}^{n_2} n_3 (y_{ij.} - y_{i..})^2 \right] = \xi_{\mathcal{R}'},
$$

$$
E \left[\frac{1}{n_1(n_3 - 1)} \sum_{i=1}^{n_1} \sum_{k=1}^{n_3} n_2 (y_{i.k} - y_{i..})^2 \right] = \xi_{\mathcal{C}'},
$$

$$
E \left[\frac{1}{n_1(n_2 - 1)(n_3 - 1)} \sum_{i=1}^{n_1} \sum_{j=1}^{n_2} \sum_{k=1}^{n_3} (y_{ijk} - y_{ij.} - y_{i.k} + y_{i..})^2 \right] = \xi_\mathcal{E}.
$$

We may call $W_\mathcal{B}$, $W_{\mathcal{R}'}$, $W_{\mathcal{C}'}$, and $W_\mathcal{E}$ block, row, column, and unit strata, respectively. Again there is one stratum for each of the five factors $\mathcal{E}$, $\mathcal{R}'$, $\mathcal{C}'$, $\mathcal{B}$, and $\mathcal{U}$ that define the block structure. In Example 1.3, $n_1 = 2$, $n_2 = n_3 = 4$; hence $W_\mathcal{B}$, $W_{\mathcal{R}'}$, $W_{\mathcal{C}'}$, and $W_\mathcal{E}$ have, respectively, 1, 6, 6, and 18 degrees of freedom.

Example 12.12. (Example 1.1 revisited) We obtain the degree-of-freedom identity for $n_1/[(n_2/n_3/n_4) \times n_5]$ by expanding $\mathcal{N}(n_1, \mathcal{C}(\mathcal{N}(\mathcal{N}(n_2, n_3), n_4), n_5))$:

$$
\mathcal{N}(n_1, \mathcal{C}(\mathcal{N}(\mathcal{N}(n_2, n_3), n_4), n_5))
$$
$$
= 1 + v_1 + n_1 v_2 + n_1 n_2 v_3 + n_1 n_2 n_3 v_4 + n_1 v_5 + n_1 v_2 v_5 + n_1 n_2 v_3 v_5 + n_1 n_2 n_3 v_4 v_5.
$$

In this case, there are nine strata including the mean stratum. After obtaining the yield identity, one can see that the projections onto the nine strata have components $y_{.....}$, $y_{i....} - y_{.....}$, $y_{ij...} - y_{i....}$, $y_{ijk..} - y_{ij...}$, $y_{ijkl.} - y_{ijk..}$, $y_{i...m} - y_{i....}$, $y_{ij..m} - y_{ij...} - y_{i...m} + y_{i....}$, $y_{ijk.m} - y_{ijk..} - y_{ij..m} + y_{ij...}$, $y_{ijklm} - y_{ijkl.} - y_{ijk.m} + y_{ijk..}$. The null ANOVA can be obtained as usual. For $n_1 = 4$, $n_2 = 3$, $n_3 = 2$, $n_4 = 3$, $n_5 = 7$ as in Example 1.1, the degrees of freedom for the eight strata other than the mean stratum are 3, 8, 12, 48, 24, 48, 72, and 288, respectively.

12.10 Determining strata from Hasse diagrams

Nelder's rules only apply to simple block structures. For general orthogonal block structures, the strata, their degrees of freedom, and orthogonal projections can be determined by using Hasse diagrams. There is one W space for each node of the Hasse diagram.

It follows from (12.29) that

$$W_{\mathcal{F}} = V_{\mathcal{F}} \ominus \left(\bigoplus_{\mathcal{G} \in \mathfrak{B}: \mathcal{F} \prec \mathcal{G}} W_{\mathcal{G}} \right). \tag{12.40}$$

Two immediate consequences of (12.40) are

$$\dim(W_{\mathcal{F}}) = n_{\mathcal{F}} - \sum_{\mathcal{G} \in \mathfrak{B}: \mathcal{F} \prec \mathcal{G}} \dim(W_{\mathcal{G}}) \tag{12.41}$$

and

$$\mathbf{P}_{W_{\mathcal{F}}} = \mathbf{P}_{V_{\mathcal{F}}} - \sum_{\mathcal{G} \in \mathfrak{B}: \mathcal{F} \prec \mathcal{G}} \mathbf{P}_{W_{\mathcal{G}}}. \tag{12.42}$$

Since $\left[\sum_{\mathcal{G} \in \mathfrak{B}: \mathcal{F} \prec \mathcal{G}} V_{\mathcal{G}} \right]^{\perp} = \bigcap_{\mathcal{G} \in \mathfrak{B}: \mathcal{F} \prec \mathcal{G}} V_{\mathcal{G}}^{\perp}$, we can also write (12.28) as

$$W_{\mathcal{F}} = V_{\mathcal{F}} \cap \left(\bigcap_{\mathcal{G} \in \mathfrak{B}: \mathcal{F} \prec \mathcal{G}} V_{\mathcal{G}}^{\perp} \right). \tag{12.43}$$

We have noted in Section 12.7 that $W_{\mathcal{U}} = V_{\mathcal{U}}$, $\dim(W_{\mathcal{U}}) = 1$, and $\mathbf{P}_{V_{\mathcal{U}}} = \frac{1}{N}\mathbf{J}$. One can start from the top node $\mathcal{U}$ and work down the paths of the Hasse diagram to determine all the strata, their degrees of freedom, and orthogonal projections by repeatedly using (12.40), (12.41), and (12.42). In this process, (12.29) and (12.43) are also useful. The degrees of freedom of each $W_{\mathcal{F}}$ can be computed by subtracting from $n_{\mathcal{F}}$ the already computed degrees of freedom of the strata associated with all the nodes above the node corresponding to $\mathcal{F}$. Likewise, the orthogonal projection of $\mathbf{y}$ onto $W_{\mathcal{F}}$ is obtained by subtracting from $\mathbf{P}_{V_{\mathcal{F}}}\mathbf{y}$ the orthogonal projections of $\mathbf{y}$ onto the strata associated with all the nodes above the node corresponding to $\mathcal{F}$. A typical entry of $\mathbf{P}_{V_{\mathcal{F}}}\mathbf{y}$ is the average of the y values within the same level of $\mathcal{F}$; see (4.4). We give some examples below to illustrate how to use these results.

We use the Hasse diagram in Figure 4.3(b) for the block structure $n_1/(n_2 \times n_3)$ to determine the strata, their dimensions, and orthogonal projections. Let y_{ijk} be the observation at the intersection of the jth row and kth column of the ith block. We start from the top: $W_{\mathcal{U}} = V_{\mathcal{U}}$, $\dim(W_{\mathcal{U}}) = 1$, $\mathbf{P}_{W_{\mathcal{U}}} = \frac{1}{n_1 n_2 n_3}\mathbf{J}$, and $\mathbf{P}_{W_{\mathcal{U}}}\mathbf{y} = [y_{\ldots}]_{n_1 n_2 n_3 \times 1}$. Then since $\mathcal{U}$ is the only node that is above $\mathcal{B}$, $W_{\mathcal{B}} = V_{\mathcal{B}} \ominus W_{\mathcal{U}} = V_{\mathcal{B}} \ominus V_{\mathcal{U}}$, $\dim(W_{\mathcal{B}}) = n_1 - \dim(W_{\mathcal{U}}) = n_1 - 1$, $\mathbf{P}_{W_{\mathcal{B}}} = \mathbf{P}_{V_{\mathcal{B}}} - \mathbf{P}_{V_{\mathcal{U}}}$, and $\mathbf{P}_{W_{\mathcal{B}}}\mathbf{y} = [y_{i\cdot\cdot} - y_{\ldots}]_{n_1 n_2 n_3 \times 1}$; by the fact that there are two nodes $\mathcal{B}$ and $\mathcal{U}$ above $\mathcal{R}'$, $W_{\mathcal{R}'} = V_{\mathcal{R}'} \ominus (W_{\mathcal{B}} \oplus W_{\mathcal{U}}) = V_{\mathcal{R}'} \ominus V_{\mathcal{B}}$, $\dim(W_{\mathcal{R}'}) = \dim(V_{\mathcal{R}'}) - \dim(W_{\mathcal{B}}) - \dim(W_{\mathcal{U}}) = n_1 n_2 - (n_1 - 1) - 1 = n_1(n_2 - 1)$, $\mathbf{P}_{W_{\mathcal{R}'}} = \mathbf{P}_{V_{\mathcal{R}'}} - \mathbf{P}_{V_{\mathcal{B}}}$, and $\mathbf{P}_{W_{\mathcal{R}'}}\mathbf{y} = [y_{ij\cdot} - y_{i\cdot\cdot}]_{n_1 n_2 n_3 \times 1}$; likewise, $W_{\mathcal{C}'} = V_{\mathcal{C}'} \ominus V_{\mathcal{B}}$, $\dim(W_{\mathcal{C}'}) = n_1(n_3 - 1)$, $\mathbf{P}_{W_{\mathcal{C}'}} = \mathbf{P}_{V_{\mathcal{C}'}} - \mathbf{P}_{V_{\mathcal{B}}}$, $\mathbf{P}_{W_{\mathcal{C}'}}\mathbf{y} = [y_{i\cdot k} - y_{i\cdot\cdot}]_{n_1 n_2 n_3 \times 1}$; $W_{\mathcal{E}} = V_{\mathcal{E}} \ominus (W_{\mathcal{R}'} \oplus W_{\mathcal{C}'} \oplus W_{\mathcal{B}} \oplus W_{\mathcal{U}}) = V_{\mathcal{E}} \ominus (V_{\mathcal{R}'} + V_{\mathcal{C}'}) = (V_{\mathcal{R}'} + V_{\mathcal{C}'})^{\perp}$, $\dim(W_{\mathcal{E}}) = \dim(V_{\mathcal{E}}) - \dim(W_{\mathcal{R}'}) - \dim(W_{\mathcal{C}'}) - \dim(W_{\mathcal{B}}) - \dim(W_{\mathcal{U}}) = n_1 n_2 n_3 - n_1(n_2 - 1) - n_1(n_3 - 1) - (n_1 - 1) - 1 = n_1(n_2 - 1)(n_3 - 1)$, $\mathbf{P}_{W_{\mathcal{E}}} = \mathbf{P}_{V_{\mathcal{E}}} - \mathbf{P}_{W_{\mathcal{R}'}} - \mathbf{P}_{W_{\mathcal{C}'}} - \mathbf{P}_{W_{\mathcal{B}}} - \mathbf{P}_{W_{\mathcal{U}}} = \mathbf{I} - (\mathbf{P}_{V_{\mathcal{R}'}} - \mathbf{P}_{V_{\mathcal{B}}}) - (\mathbf{P}_{V_{\mathcal{C}'}} - \mathbf{P}_{V_{\mathcal{B}}}) - (\mathbf{P}_{V_{\mathcal{B}}} - \mathbf{P}_{V_{\mathcal{U}}}) - \mathbf{P}_{V_{\mathcal{U}}} = \mathbf{I} - \mathbf{P}_{V_{\mathcal{R}'}} - \mathbf{P}_{V_{\mathcal{C}'}} + \mathbf{P}_{V_{\mathcal{B}}}$, and $\mathbf{P}_{W_{\mathcal{E}}}\mathbf{y} = [y_{ijk} - y_{ij\cdot} - y_{i\cdot k} + y_{i\cdot\cdot}]_{n_1 n_2 n_3 \times 1}$.

Example 12.13. (**Example 1.1 revisited**) The Hasse diagram for $n_1/[(n_2/n_3/n_4) \times n_5]$ in Figure 12.2 has nine nodes corresponding to $\mathcal{U}$, $\mathcal{B}$, $\mathcal{P}$, $\mathcal{S}$, $\mathcal{SS}$, $\mathcal{ST}$, $\mathcal{ST} \wedge \mathcal{P}$, $\mathcal{ST} \wedge \mathcal{S}$, and $\mathcal{ST} \wedge \mathcal{SS} = \mathcal{E}$. Therefore there are nine strata (including $V_{\mathcal{U}}$), which can be obtained by applying (12.40), (12.41), (12.42), (12.43), and (12.29). The eight strata other than $W_{\mathcal{U}}$ are $W_{\mathcal{B}} = V_{\mathcal{B}} \ominus V_{\mathcal{U}}$, $W_{\mathcal{P}} = V_{\mathcal{P}} \ominus V_{\mathcal{B}}$, $W_{\mathcal{S}} = V_{\mathcal{S}} \ominus V_{\mathcal{P}}$, $W_{\mathcal{SS}} = V_{\mathcal{SS}} \ominus V_{\mathcal{S}}$, $W_{\mathcal{ST}} = V_{\mathcal{ST}} \ominus V_{\mathcal{B}}$, $W_{\mathcal{ST} \wedge \mathcal{P}} = V_{\mathcal{ST} \wedge \mathcal{P}} \ominus (V_{\mathcal{ST}} + V_{\mathcal{P}})$, $W_{\mathcal{ST} \wedge \mathcal{S}} = V_{\mathcal{ST} \wedge \mathcal{S}} \ominus (V_{\mathcal{ST} \wedge \mathcal{P}} + V_{\mathcal{S}})$, and $W_{\mathcal{E}} = (V_{\mathcal{ST} \wedge \mathcal{S}} + V_{\mathcal{SS}})^{\perp}$.

Example 12.14. (**Example 1.4 revisited**) As discussed in Section 12.5, the block structure is $\mathfrak{B} = \{\mathcal{E}, \mathcal{R}, \mathcal{C}, \mathcal{L}_1, \mathcal{L}_2, \mathcal{U}\}$, where $\mathcal{R}$, $\mathcal{C}$, $\mathcal{L}_1$, and $\mathcal{L}_2$ are partitions of the 16 units by rows, columns, and symbols of two orthogonal 4×4 Latin squares. Since $\mathfrak{B}$ is an orthogonal block structure, Theorem 12.7 can be applied to models with $\mathbf{V}$ as in (12.19), e.g., the mixed-effect model presented in Section 12.6. We conclude that there are six strata including the mean stratum: $W_{\mathcal{U}} = V_{\mathcal{U}}$, $W_{\mathcal{R}} = V_{\mathcal{R}} \ominus V_{\mathcal{U}}$, $W_{\mathcal{C}} = V_{\mathcal{C}} \ominus V_{\mathcal{U}}$, $W_{\mathcal{L}_1} = V_{\mathcal{L}_1} \ominus V_{\mathcal{U}}$, $W_{\mathcal{L}_2} = V_{\mathcal{L}_2} \ominus V_{\mathcal{U}}$, and $W_{\mathcal{E}} = V_{\mathcal{E}} \ominus [(V_{\mathcal{R}} \ominus V_{\mathcal{U}}) \oplus (V_{\mathcal{C}} \ominus V_{\mathcal{U}}) \oplus (V_{\mathcal{L}_1} \ominus V_{\mathcal{U}}) \oplus (V_{\mathcal{L}_2} \ominus V_{\mathcal{U}}) \oplus V_{\mathcal{U}}] = V_{\mathcal{E}} \ominus (V_{\mathcal{R}} + V_{\mathcal{C}} + V_{\mathcal{L}_1} + V_{\mathcal{L}_2})$. Except for the mean stratum, each of the other five strata has three degrees of freedom.

Remark 12.2. In Example 12.14, suppose $\mathfrak{B} = \{\mathcal{E}, \mathcal{R}, \mathcal{C}, \mathcal{L}_1, \mathcal{L}_2, \mathcal{L}_3, \mathcal{U}\}$, where $\mathcal{L}_1$, $\mathcal{L}_2$, and $\mathcal{L}_3$ are partitions of the 16 units by the symbols of three mutually orthogonal 4×4 Latin squares. Then $W_{\mathcal{R}}$, $W_{\mathcal{C}}$, $W_{\mathcal{L}_1}$, and $W_{\mathcal{L}_2}$ are as in Example 12.14. We have $W_{\mathcal{L}_3} = V_{\mathcal{L}_3} \ominus V_{\mathcal{U}}$ and $W_{\mathcal{E}} = \{\mathbf{0}\}$. In this case, $W_{\mathcal{E}}$ is not a stratum.

12.11 Proofs of Theorems 12.6 and 12.7

Proof of Theorem 12.6

We first show that for any two different factors $\mathcal{F}$ and $\mathcal{G}$, $W_{\mathcal{F}}$ and $W_{\mathcal{G}}$ are orthogonal. If $\mathcal{F} \prec \mathcal{G}$ (that is, $\mathcal{F} \vee \mathcal{G} = \mathcal{G}$), then by (12.28), clearly $W_{\mathcal{F}}$ and $W_{\mathcal{G}}$ are orthogonal. If $\mathcal{G} \neq \mathcal{F} \vee \mathcal{G}$, then $\mathcal{G} \prec \mathcal{F} \vee \mathcal{G}$; again by (12.28), we have

$$W_{\mathcal{G}} \subseteq V_{\mathcal{G}} \ominus V_{\mathcal{F} \vee \mathcal{G}} \tag{12.44}$$

and

$$W_{\mathcal{F}} \subseteq V_{\mathcal{F}}. \tag{12.45}$$

Since $\mathcal{F}$ and $\mathcal{G}$ are orthogonal factors,

$$V_{\mathcal{G}} \ominus V_{\mathcal{F} \vee \mathcal{G}} \text{ and } V_{\mathcal{F}} \text{ are orthogonal.} \tag{12.46}$$

It follows from (12.44), (12.45), and (12.46) that $W_{\mathcal{F}}$ and $W_{\mathcal{G}}$ are orthogonal.

We now prove (12.29). If there is no factor $\mathcal{G}$ such that $\mathcal{F} \prec \mathcal{G}$, then $W_{\mathcal{F}} = V_{\mathcal{F}}$, and (12.29) holds trivially. The general case can be proved by induction. Suppose (12.29) holds for all the factors that are coarser than $\mathcal{F}$. Then

$$\sum_{\mathcal{G} \in \mathfrak{B}: \mathcal{F} \prec \mathcal{G}} V_{\mathcal{G}} = \sum_{\mathcal{G} \in \mathfrak{B}: \mathcal{F} \prec \mathcal{G}} \left[\bigoplus_{\mathcal{H} \in \mathfrak{B}: \mathcal{G} \preceq \mathcal{H}} W_{\mathcal{H}} \right] = \bigoplus_{\mathcal{G} \in \mathfrak{B}: \mathcal{F} \prec \mathcal{G}} W_{\mathcal{G}}.$$

By (12.28), $V_{\mathcal{F}} = W_{\mathcal{F}} \oplus \left[\sum_{\mathcal{G}\in\mathfrak{B}:\mathcal{F}\prec\mathcal{G}} V_{\mathcal{G}}\right] = W_{\mathcal{F}} \oplus \left[\bigoplus_{\mathcal{G}\in\mathfrak{B}:\mathcal{F}\prec\mathcal{G}} W_{\mathcal{G}}\right] = \bigoplus_{\mathcal{G}\in\mathfrak{B}:\mathcal{F}\preceq\mathcal{G}} W_{\mathcal{G}}.$ □

Proof of Theorem 12.7

We first establish a relation between the $\mathbf{A}_{\mathcal{F}}$'s and the $\mathbf{R}_{\mathcal{F}}$'s defined in Section 4.3. Recall that $\mathbf{R}_{\mathcal{F}}$ is the $(0,1)$-matrix such that its (v,w)th entry is 1 if and only if units v and w are in the same $\mathcal{F}$-class. Suppose units v and w are in the same $\mathcal{F}$-class and let $\mathcal{G}$ be the finest factor such that they are in the same $\mathcal{G}$-class. Then we must have $\mathcal{G} \preceq \mathcal{F}$. Therefore

$$\mathbf{R}_{\mathcal{F}} = \sum_{\mathcal{G}\in\mathfrak{B}:\mathcal{G}\preceq\mathcal{F}} \mathbf{A}_{\mathcal{G}}$$

for all $\mathcal{F} \in \mathfrak{B}$. It follows that

$$\mathbf{A}_{\mathcal{F}} = \mathbf{R}_{\mathcal{F}} - \sum_{\mathcal{G}\in\mathfrak{B}:\mathcal{G}\prec\mathcal{F}} \mathbf{A}_{\mathcal{G}}. \tag{12.47}$$

At the bottom of the Hasse diagram, we have $\mathbf{A}_{\mathcal{E}} = \mathbf{R}_{\mathcal{E}} = \mathbf{I}_N$. Starting from $\mathcal{E}$ and applying (12.47) repeatedly, we can work up the Hasse diagram to show that each $\mathbf{A}_{\mathcal{F}}$ is a linear combination of the $\mathbf{R}_{\mathcal{G}}$'s with $\mathcal{G} \preceq \mathcal{F}$. Therefore $\mathbf{V}$ is of the form

$$\mathbf{V} = \sum_{\mathcal{F}:\mathcal{F}\in\mathfrak{B}} \delta_{\mathcal{F}}\mathbf{R}_{\mathcal{F}} \tag{12.48}$$

for some constants $\delta_{\mathcal{F}}$. Since all the factors $\mathcal{F}$ in $\mathfrak{B}$ are uniform, by (4.5), $\mathbf{R}_{\mathcal{F}} = \frac{N}{n_{\mathcal{F}}}\mathbf{P}_{V_{\mathcal{F}}}$. Thus for each $\mathbf{y}$, if $\mathbf{y} \in V_{\mathcal{F}}$, then $\mathbf{R}_{\mathcal{F}}\mathbf{y} = \frac{N}{n_{\mathcal{F}}}\mathbf{y}$, and if $\mathbf{y} \in V_{\mathcal{F}}^{\perp}$, then $\mathbf{R}_{\mathcal{F}}\mathbf{y} = \mathbf{0}$. By Theorem 12.6, if $\mathbf{y} \in W_{\mathcal{F}}$, then $\mathbf{y} \in V_{\mathcal{G}}$ for all $\mathcal{G} \preceq \mathcal{F}$, and $\mathbf{y} \in V_{\mathcal{G}}^{\perp}$ for other $\mathcal{G}$'s. Therefore it follows from (12.48) that for each $\mathbf{y} \in W_{\mathcal{F}}$

$$\mathbf{V}\mathbf{y} = \left[\sum_{\mathcal{G}\in\mathfrak{B}:\mathcal{G}\preceq\mathcal{F}} \frac{N}{n_{\mathcal{G}}}\delta_{\mathcal{G}}\right] \mathbf{y}.$$

This shows that if $W_{\mathcal{F}} \neq \{\mathbf{0}\}$, then the vectors in $W_{\mathcal{F}}$ are eigenvectors with the same eigenvalue.   □

12.12   Models with random effects revisited

It follows from Theorem 12.5 that Theorem 12.7 can be applied to the mixed-effect model (12.21) for orthogonal block structures. The following theorem, due to Tjur (1984), shows that for block structures that satisfy conditions (i)–(iii) and (v), (vi) in the definition of orthogonal block structures, the conclusion of Theorem 12.7 also holds for the mixed-effect model (12.21). In this case, the condition that the block structure is closed under ∧ can be dropped.

Theorem 12.8. *Let $\mathfrak{B}$ be a set of nonequivalent factors satisfying the following conditions:*

(i) *All the factors in $\mathfrak{B}$ are uniform;*

(ii) *$\mathcal{E} \in \mathfrak{B}$;*

(iii) *$\mathcal{U} \in \mathfrak{B}$;*

(iv) *if $\mathcal{F}, \mathcal{G} \in \mathfrak{B}$, then $\mathcal{F} \vee \mathcal{G} \in \mathfrak{B}$;*

(v) *the factors in $\mathfrak{B}$ are pairwise orthogonal.*

Then under the mixed-effect model (12.21), the subspaces $W_{\mathcal{F}}$, $\mathcal{F} \in \mathfrak{B}$, defined in Theorem 12.6 are eigenspaces of $\mathbf{V}$, provided that $W_{\mathcal{F}} \neq \{\mathbf{0}\}$.

Proof. By (12.22),

$$
\begin{aligned}
\mathbf{V} &= \sum_{\mathcal{G}:\mathcal{G}\in\mathfrak{B}} \sigma_{\mathcal{G}}^2 \mathbf{R}_{\mathcal{G}} \\
&= \sum_{\mathcal{G}:\mathcal{G}\in\mathfrak{B}} \sigma_{\mathcal{G}}^2 \frac{N}{n_{\mathcal{G}}} \mathbf{P}_{V_{\mathcal{G}}} \\
&= \sum_{\mathcal{G}:\mathcal{G}\in\mathfrak{B}} \sigma_{\mathcal{G}}^2 \frac{N}{n_{\mathcal{G}}} \left[\sum_{\mathcal{F}\in\mathfrak{B}:\mathcal{G}\preceq\mathcal{F}} \mathbf{P}_{W_{\mathcal{F}}} \right] \\
&= \sum_{\mathcal{F}:\mathcal{F}\in\mathfrak{B}} \left[\sum_{\mathcal{G}\in\mathfrak{B}:\mathcal{G}\preceq\mathcal{F}} \sigma_{\mathcal{G}}^2 \frac{N}{n_{\mathcal{G}}} \right] \mathbf{P}_{W_{\mathcal{F}}},
\end{aligned}
$$

where the third equality follows from (12.42). Therefore each nonnull $W_{\mathcal{F}}, \mathcal{F} \in \mathfrak{B}$, is an eigenspace of $\mathbf{V}$ with eigenvalue

$$
\xi_{\mathcal{F}} = \sum_{\mathcal{G}\in\mathfrak{B}:\mathcal{G}\preceq\mathcal{F}} \frac{N}{n_{\mathcal{G}}} \sigma_{\mathcal{G}}^2. \tag{12.49}
$$

$\square$

It follows from (12.49) that under the assumption in (12.26),

$$
\mathcal{F}_1 \prec \mathcal{F}_2 \Rightarrow \xi_{\mathcal{F}_1} < \xi_{\mathcal{F}_2}. \tag{12.50}
$$

In particular, the stratum variance $\xi_{\mathcal{E}}$ of the bottom stratum is the smallest among all the stratum variances. Condition (12.50) (for example, the interblock variance is larger than the intrablock variance) is guaranteed under the mixed-effect model (12.21). In general, it would have to be assumed.

Example 12.15. Consider the block structure $\mathfrak{B}_1 \times \mathfrak{B}_2 \times \mathfrak{B}_3$, where $\mathfrak{B}_i = \{\mathcal{E}_i, \mathcal{U}_i\}$, with $\mathcal{E}_i$ and $\mathcal{U}_i$ being the equality and universal factors, respectively, on Ω_i. Under the randomization model, there are eight strata corresponding to the eight factors $\mathcal{U}$, $\mathcal{E}_1$, $\mathcal{E}_2$, $\mathcal{E}_3$, $\mathcal{E}_1 \wedge \mathcal{E}_2$, $\mathcal{E}_1 \wedge \mathcal{E}_3$, $\mathcal{E}_2 \wedge \mathcal{E}_3$, and $\mathcal{E}_1 \wedge \mathcal{E}_2 \wedge \mathcal{E}_3 = \mathcal{E}$. We also obtain the same strata under model (12.21) with one set of random effects for each of the eight factors. In both cases, the covariance matrix $\mathbf{V}$ is of the form in (12.19) and has eight different entries. Suppose we have model (12.21) with $\mathfrak{B} = \{\mathcal{U}, \mathcal{E}_1, \mathcal{E}_2, \mathcal{E}_3, \mathcal{E}\}$ instead. Then Theorem 12.7 cannot be applied to $\mathfrak{B}$ directly since it is not closed under $\wedge$.

However, the conditions in Theorem 12.8 are satisfied, and it can be applied to show that there are five strata corresponding to $\mathcal{U}, \mathcal{E}_1, \mathcal{E}_2, \mathcal{E}_3$, and $\mathcal{E}$, even though $\mathbf{V}$ also has eight different entries. Alternatively, since in this case the model is the same as that for the simple block structure $\mathcal{B}_1 \times \mathcal{B}_2 \times \mathcal{B}_3$ with the constraint $\sigma^2_{\mathcal{E}_1 \wedge \mathcal{E}_2} = \sigma^2_{\mathcal{E}_1 \wedge \mathcal{E}_3} = \sigma^2_{\mathcal{E}_2 \wedge \mathcal{E}_3} = 0$, we can apply Theorem 12.7 to $\mathcal{B}_1 \times \mathcal{B}_2 \times \mathcal{B}_3$. It follows from the above constraint that $\xi_{\mathcal{E}_1 \wedge \mathcal{E}_2} = \xi_{\mathcal{E}_1 \wedge \mathcal{E}_3} = \xi_{\mathcal{E}_2 \wedge \mathcal{E}_3} = \xi_{\mathcal{E}}$. Then $W_{\mathcal{E}_1 \wedge \mathcal{E}_2}$, $W_{\mathcal{E}_1 \wedge \mathcal{E}_3}$, $W_{\mathcal{E}_2 \wedge \mathcal{E}_3}$, and $W_{\mathcal{E}}$ ($= W_{\mathcal{E}_1 \wedge \mathcal{E}_2 \wedge \mathcal{E}_3}$) collapse into one single stratum, yielding the same result as obtained by using Theorem 12.8 for the block structure $\{\mathcal{U}, \mathcal{E}_1, \mathcal{E}_2, \mathcal{E}_3, \mathcal{E}\}$.

12.13 Experiments with multiple processing stages

In factorial experiments with multiple processing stages, at each stage the experimental units are partitioned into disjoint classes, and for each of the treatment factors processed at that stage, its levels are randomly assigned to various classes of the partition, with the same level assigned to all the units in the same class. In this section, we derive strata for this kind of experiment under the commonly used mixed-effect model. Design construction will be discussed in the next chapter.

The partition of the units at the ith stage defines a factor $\mathcal{F}_i$ on Ω. Suppose there are h stages. Let $\mathcal{B} = \{\mathcal{U}, \mathcal{E}, \mathcal{F}_1, \ldots, \mathcal{F}_h\}$ if $\mathcal{F}_i \neq \mathcal{E}$ for all i; otherwise let $\mathcal{B} = \{\mathcal{U}, \mathcal{F}_1, \ldots, \mathcal{F}_h\}$. Consider model (12.21) with one set of random effects for each factor in $\mathcal{B}$. This is the model adopted by Mee and Bates (1998), Bingham et al. (2008), and Ranjan, Bingham, and Dean (2009), except that they do not include a random term for $\mathcal{U}$. The absence of a random term for $\mathcal{U}$ is equivalent to that $\sigma^2_{\mathcal{U}} = 0$.

Example 12.16. In an experiment with two processing stages, suppose the levels of certain treatment factors are assigned to some batches of material at the first stage, and each batch is divided into several samples to receive the levels of the other factors at the second stage. Then we have $\mathcal{F}_2 = \mathcal{E}$. In this case $\mathcal{F}_2$ (the subplots) is nested in $\mathcal{F}_1$ (the whole-plots). In a strip-plot design, neither $\mathcal{F}_1$ nor $\mathcal{F}_2$ (say rows and columns) is nested in the other, and both are different from $\mathcal{E}$. In this case, $\mathcal{E}$ is added to the block structure.

Throughout this section, we assume that $\mathcal{F}_1, \ldots, \mathcal{F}_h$ are uniform and mutually orthogonal. If, in addition, $\mathcal{B}$ is closed under $\vee$, then the conditions in Theorem 12.8 are satisfied. In this case the strata can be determined by the rule given in Theorem 12.6. For example, if $\mathcal{F}_i \neq \mathcal{E}$ for all i, then

$$W_{\mathcal{F}_i} = V_{\mathcal{F}_i} \ominus \left(\sum_{j:\mathcal{F}_i \prec \mathcal{F}_j} V_{\mathcal{F}_j} \right), \quad \xi_{\mathcal{F}_i} = \sum_{j:\mathcal{F}_j \preceq \mathcal{F}_i} \frac{N}{n_{\mathcal{F}_j}} \sigma^2_{\mathcal{F}_j},$$

$$\dim \left(W_{\mathcal{F}_i} \right) = n_{\mathcal{F}_i} - 1 - \sum_{j:\mathcal{F}_i \prec \mathcal{F}_j} \dim \left(W_{\mathcal{F}_j} \right),$$

$$W_{\mathcal{E}} = \left[\sum_{i=1}^{h} V_{\mathcal{F}_i} \right]^{\perp}, \quad \xi_{\mathcal{E}} = \sigma^2_{\mathcal{E}}, \quad \dim(W_{\mathcal{E}}) = N - 1 - \sum_{i=1}^{h} \dim \left(W_{\mathcal{F}_i} \right).$$

The strata are all the nonnull W spaces. In the special case where the factors

$\mathcal{F}_1,\ldots,\mathcal{F}_h$ are not nested in one another, we have

$$W_{\mathcal{F}_i} = V_{\mathcal{F}_i} \ominus V_{\mathcal{U}} = C_{\mathcal{F}_i}, \; \xi_{\mathcal{F}_i} = \frac{N}{n_{\mathcal{F}_i}}\sigma^2_{\mathcal{F}_i} + \sigma^2_{\mathcal{E}}, \; \dim(W_{\mathcal{F}_i}) = n_{\mathcal{F}_i} - 1, \; i = 1,\ldots,h,$$

$$W_{\mathcal{E}} = \left[\textstyle\sum_{i=1}^h V_{\mathcal{F}_i}\right]^{\perp}, \; \xi_{\mathcal{E}} = \sigma^2_{\mathcal{E}}, \; \dim(W_{\mathcal{E}}) = N - 1 - \textstyle\sum_{i=1}^h (n_{\mathcal{F}_i} - 1).$$

$$(12.51)$$

In this case, each $C_{\mathcal{F}_i}$, the space of all the contrasts of the levels of $\mathcal{F}_i$, is a stratum, and $W_{\mathcal{E}}$ is a stratum if $\sum_{i=1}^h (n_{\mathcal{F}_i} - 1) < N - 1$.

The factors $\mathcal{F}_1,\ldots,\mathcal{F}_h$ define an $N \times h$ array with each row corresponding to an experimental unit and the ith column corresponding to $\mathcal{F}_i$ such that the entry on the wth row and ith column is the level of $\mathcal{F}_i$ at unit w. We show that when $\mathcal{F}_1,\ldots,\mathcal{F}_h$ are uniform, mutually orthogonal, and are not nested in one another, $\{\mathcal{U},\mathcal{E},\mathcal{F}_1,\ldots,\mathcal{F}_h\}$ is closed under $\vee$ if and only if the array defined above is an *orthogonal array of strength two*.

Proposition 12.9. *(Cheng and Tsai, 2011) Let $\mathcal{F}_1,\ldots,\mathcal{F}_h$ be uniform and mutually orthogonal factors on Ω. Then the following conditions are equivalent:*

(i) *$\mathcal{F}_1,\ldots,\mathcal{F}_h$ define an orthogonal array of strength two;*

(ii) *$\mathcal{F}_1,\ldots,\mathcal{F}_h$ are not nested in one another, and $\mathfrak{B} = \{\mathcal{U},\mathcal{E},\mathcal{F}_1,\ldots,\mathcal{F}_h\}$ is closed under $\vee$;*

(iii) *$C_{\mathcal{F}_1},\ldots,C_{\mathcal{F}_h}$ are mutually orthogonal.*

In this case, the conditions in Theorem 12.8 are satisfied, and under the mixed-effect model (12.21) there are h or $h+1$ strata in addition to the mean stratum, depending on whether $\sum_{i=1}^h (n_{\mathcal{F}_i} - 1) = N - 1$ or $\sum_{i=1}^h (n_{\mathcal{F}_i} - 1) < N - 1$.

Proof.
(i) $\Rightarrow$ (ii) Suppose $\mathcal{F}_1,\ldots,\mathcal{F}_h$ define an orthogonal array of strength two. Then they have pairwise proportional frequencies. By Proposition 4.1, for all $i \neq j$, $\mathcal{F}_i \vee \mathcal{F}_j = \mathcal{U} \in \mathfrak{B}$. Therefore $\mathfrak{B}$ is closed under $\vee$. It is also clear that if two factors have proportional frequencies, then neither can be nested in the other.

(ii) $\Rightarrow$ (iii) Suppose $\mathcal{F}_1,\ldots,\mathcal{F}_h$ are not nested in one another, and $\mathfrak{B} = \{\mathcal{U}, \mathcal{E}, \mathcal{F}_1,\ldots,\mathcal{F}_h\}$ is closed under $\vee$. Then, for any $i \neq j$, $\mathcal{F}_i \vee \mathcal{F}_j \neq \mathcal{F}_l$ for all $1 \leq l \leq h$. Since $\mathcal{F}_i \vee \mathcal{F}_j \in \mathfrak{B}$, we must have $\mathcal{F}_i \vee \mathcal{F}_j = \mathcal{U}$. It follows that $V_{\mathcal{F}_i} \cap V_{\mathcal{F}_j} = V_{\mathcal{U}}$. This, together with the orthogonality of $\mathcal{F}_i$ and $\mathcal{F}_j$, implies that $C_{\mathcal{F}_i} = V_{\mathcal{F}_i} \ominus V_{\mathcal{U}}$ is orthogonal to $C_{\mathcal{F}_j} = V_{\mathcal{F}_j} \ominus V_{\mathcal{U}}$.

(iii) $\Rightarrow$ (i) Suppose $C_{\mathcal{F}_1},\ldots,C_{\mathcal{F}_h}$ are mutually orthogonal. Then, by Theorem 2.5, for all $i \neq j$, $\mathcal{F}_i$ and $\mathcal{F}_j$ have proportional frequencies. Since $\mathcal{F}_1,\ldots,\mathcal{F}_h$ are uniform, they define an orthogonal array of strength two.

The strata are $C_{\mathcal{F}_1},\ldots,C_{\mathcal{F}_h}$, and $\left[\sum_{i=1}^h V_{\mathcal{F}_i}\right]^{\perp}$ (if $\sum_{i=1}^h (n_{\mathcal{F}_i} - 1) < N - 1$) as given in (12.51). $\qquad\square$

Remark 12.3. The orthogonal arrays in Proposition 12.9 are arrays of units, instead of treatments. We may call them *unit arrays*. Example 1.4 involves four processing stages. The partitions of units are based on a pair of orthogonal 4×4 Latin squares which, by Theorem 8.10, is equivalent to an $OA(16, 4^4, 2)$.

In a multi-stratum experiment, when normal or half-normal plots are used to judge the significance of factorial effects, those estimated in different strata should be evaluated separately, with one plot for each stratum. Schoen (1999) recommended a minimum of seven points on each plot to separate active effects from inactive ones. This means that there should be at least eight classes at each stage to produce at least seven degrees of freedom for each $W_{\mathcal{F}_i}$. Larger $n_{\mathcal{F}_i}$ may render pairwise proportional frequencies not possible. We have seen in Example 1.5 that if we have to partition 32 experimental units into eight classes of size four at both stages, then since $8^2 > 32$, not all $\mathcal{F}_1$-classes can meet with every $\mathcal{F}_2$-class. It follows that $\mathcal{F}_1$ and $\mathcal{F}_2$ cannot have proportional frequencies, so $C_{\mathcal{F}_1}$ and $C_{\mathcal{F}_2}$ are not orthogonal. In this case if we want $\mathcal{F}_1$ and $\mathcal{F}_2$ to be orthogonal to each other, then $C_{\mathcal{F}_1} \cap C_{\mathcal{F}_2} \neq \{\mathbf{0}\}$. Overlapping of such contrast spaces was observed in Bingham et al. (2008) and further studied in Ranjan, Bingham, and Dean (2009). Cheng and Tsai (2011) elucidated this phenomenon using the general theory of block structures.

Let $\mathfrak{B} = \{\mathcal{E}, \mathcal{U}, \mathcal{F}_1, \mathcal{F}_2\}$. Since $C_{\mathcal{F}_1} \cap C_{\mathcal{F}_2} = C_{\mathcal{F}_1 \vee \mathcal{F}_2}$, if $C_{\mathcal{F}_1} \cap C_{\mathcal{F}_2} \neq \{\mathbf{0}\}$, then $\mathcal{F}_1 \vee \mathcal{F}_2$ is strictly nested in $\mathcal{U}$. It follows that $\mathcal{F}_1 \vee \mathcal{F}_2 \notin \mathfrak{B}$. Therefore $\mathfrak{B}$ is not closed under $\vee$ and is not an orthogonal block structure. For example, consider the partitions of 32 units in Figure 12.4, where the units are those marked with x, with the eight rows and eight columns of size four forming partitions of the 32 units at the two stages.

X	X	X	X				
X	X	X	X				
X	X	X	X				
X	X	X	X				
				X	X	X	X
				X	X	X	X
				X	X	X	X
				X	X	X	X

Figure 12.4 *Layout of the units in Example 1.5*

Here $\mathcal{B} = \mathcal{F}_1 \vee \mathcal{F}_2$ splits the 32 units into two blocks of size 16. This creates a two-level *pseudo factor* in which both $\mathcal{F}_1$ and $\mathcal{F}_2$ are nested. Since $\mathfrak{B}$ is not an orthogonal block structure, Theorem 12.6 cannot be applied. Even Theorem 12.8 is not applicable since it also requires the closure of $\vee$. However, if we add $\mathcal{B}$ to $\{\mathcal{E}, \mathcal{F}_1, \mathcal{F}_2, \mathcal{U}\}$, then $\mathfrak{B}' = \{\mathcal{E}, \mathcal{F}_1, \mathcal{F}_2, \mathcal{B}, \mathcal{U}\}$ is closed under $\vee$ and the conditions of Theorem 12.8 are satisfied. Cheng and Tsai (2011) observed that the model

$$\mathbf{y} = \mu \mathbf{1}_N + \mathbf{X}_T \boldsymbol{\alpha} + \mathbf{X}_{\mathcal{E}} \boldsymbol{\beta}^{\mathcal{E}} + \mathbf{X}_{\mathcal{F}_1} \boldsymbol{\beta}^{\mathcal{F}_1} + \mathbf{X}_{\mathcal{F}_2} \boldsymbol{\beta}^{\mathcal{F}_2} + \mathbf{X}_{\mathcal{U}} \boldsymbol{\beta}^{\mathcal{U}} \tag{12.52}$$

is the same as

$$y = \mu \mathbf{1}_N + \mathbf{X}_T \boldsymbol{\alpha} + \mathbf{X}_\mathcal{E} \boldsymbol{\beta}^\mathcal{E} + \mathbf{X}_{\mathcal{F}_1} \boldsymbol{\beta}^{\mathcal{F}_1} + \mathbf{X}_{\mathcal{F}_2} \boldsymbol{\beta}^{\mathcal{F}_2} + \mathbf{X}_\mathcal{B} \boldsymbol{\beta}^\mathcal{B} + \mathbf{X}_\mathcal{U} \boldsymbol{\beta}^\mathcal{U}$$

as long as we assume $\sigma_\mathcal{B}^2 = 0$. Then we can apply Theorem 12.8 to $\mathfrak{B}'$ to conclude that there are *five*, not four, strata under (12.52): there is an extra stratum corresponding to the pseudo factor $\mathcal{B}$. If $\sigma_{\mathcal{F}_1}^2 > 0$ and $\sigma_{\mathcal{F}_2}^2 > 0$, then by (12.50), $\xi_\mathcal{B} > \xi_{\mathcal{F}_1}$ and $\xi_\mathcal{B} > \xi_{\mathcal{F}_2}$. Thus we end up with the same number of strata and degrees of freedom as in Example 1.3. With the two-level pseudo factor included, the block structure essentially becomes $2/(4 \times 4)$.

We can also explain the creation of the extra stratum by observing that a comparison of the two levels of the pseudo factor $\mathcal{B}$ is both a row contrast and a column contrast. Thus it picks up both row and column random effects, resulting in a larger stratum variance. The extra stratum $W_\mathcal{B} = C_\mathcal{B} = C_{\mathcal{F}_1} \cap C_{\mathcal{F}_2}$ results from the overlapping of $C_{\mathcal{F}_1}$ and $C_{\mathcal{F}_2}$.

In both Examples 1.4 and 1.5, the experimental units are fractions of the level combinations of some completely crossed unit factors. We have an orthogonal array of strength two in Example 1.4, but not in Example 1.5.

In general, suppose $\mathcal{F}_1, \ldots, \mathcal{F}_h$ are uniform and pairwise orthogonal, but $\mathfrak{B} = \{\mathcal{U}, \mathcal{E}, \mathcal{F}_1, \ldots, \mathcal{F}_h\}$ is not closed under $\vee$. One can add $\vee_{j=1}^k \mathcal{F}_{i_j}$ to $\mathfrak{B}$ for each subset of factors $\{\mathcal{F}_{i_1}, \ldots, \mathcal{F}_{i_k}\}$ such that $\vee_{j=1}^k \mathcal{F}_{i_j} \notin \mathfrak{B}$. Let $\mathfrak{B}'$ be the expanded block structure. Then $\mathfrak{B}'$ is the smallest block structure containing $\mathfrak{B}$ that is closed under $\vee$. By Theorem 4.9, the factors in $\mathfrak{B}'$ are mutually orthogonal.

Note that, however, the supremum of two uniform factors may not be uniform. Therefore, even though the expanded block structure $\mathfrak{B}'$ satisfies conditions (ii), (iii), (iv), and (v) of Theorem 12.8, it may not satisfy condition (i), which needs to be checked separately. If all the factors added to $\mathfrak{B}$ are uniform, then the conditions in Theorem 12.8 are satisfied by $\mathfrak{B}'$. Since model (12.21) for $\mathfrak{B}'$ is the same as the original model for $\mathfrak{B}$ if we make $\sigma_\mathcal{F}^2$ equal to zero for all the added pseudo factors, Theorem 12.8 can be used to determine the strata.

Suppose the uniform and pairwise orthogonal factors $\mathcal{F}_1, \ldots, \mathcal{F}_h$ are not nested in one another. Then, as shown in Proposition 12.9, $\mathfrak{B}$ is closed under $\vee$ if and only if $C_{\mathcal{F}_1}, \ldots, C_{\mathcal{F}_h}$ are mutually orthogonal. We call the case where $C_{\mathcal{F}_1}, \ldots, C_{\mathcal{F}_h}$ are mutually orthogonal the *nonoverlapping case*. This is when $\mathcal{F}_1, \ldots, \mathcal{F}_h$ define an orthogonal array of strength two. Otherwise, not all of $C_{\mathcal{F}_1}, \ldots, C_{\mathcal{F}_h}$ are mutually orthogonal. Complications arise in this *overlapping case*. If for all the subsets $\{i_1, \ldots, i_k\}$ of $\{1, \ldots, h\}$ such that $\cap_{j=1}^k C_{\mathcal{F}_{i_j}} \neq \{\mathbf{0}\}$, the pseudo factors $\vee_{j=1}^k \mathcal{F}_{i_j}$ are uniform, then there is an extra stratum associated with each of these pseudo factors, with $\xi_{\vee_{j=1}^k \mathcal{F}_{i_j}} = \sigma_\mathcal{E}^2 + \sum_{j=1}^k (N/n_{\mathcal{F}_{i_j}}) \sigma_{\mathcal{F}_{i_j}}^2$, while $\xi_{\mathcal{F}_i} = \sigma_\mathcal{E}^2 + (N/n_{\mathcal{F}_i}) \sigma_{\mathcal{F}_i}^2$ for all $1 \leq i \leq h$. The extra strata have larger variances than those associated with the factors from which they are generated.

In general, the strata, degrees of freedom, and projections onto the strata can be determined by applying (12.40), (12.41), and (12.42) to the expanded Hasse diagram with the nodes for the pseudo factors included.

12.14 Randomization justification of the models for simple block structures*

For a block structure $\mathfrak{B}$ defined on Ω, we say that a permutation π of the units of Ω is $\mathfrak{B}$-allowable if, for any $\mathcal{F} \in \mathfrak{B}$, any two units v and w of Ω are in the same $\mathcal{F}$-class if and only if $\pi(v)$ and $\pi(w)$ are in the same $\mathcal{F}$-class. Randomization can be carried out by choosing a permutation of Ω randomly from the set of allowable permutations.

We first present two propositions that characterize allowable permutations for block structures obtained by nesting and crossing operators. For any permutation π_1 of Ω_1 and any permutation π_2 of Ω_2, let $\pi_1 \times \pi_2$ denote the permutation of $\Omega_1 \times \Omega_2$ such that $(\pi_1 \times \pi_2)((v_1, v_2)) = (\pi_1(v_1), \pi_2(v_2))$ for all $(v_1, v_2) \in \Omega_1 \times \Omega_2$.

Proposition 12.10. *Given two block structures $\mathfrak{B}_1$ and $\mathfrak{B}_2$ defined on Ω_1 and Ω_2, respectively, a permutation π of $\Omega_1 \times \Omega_2$ is $(\mathfrak{B}_1 \times \mathfrak{B}_2)$-allowable if and only if there exist a $\mathfrak{B}_1$-allowable permutation π_1 of Ω_1 and a $\mathfrak{B}_2$-allowable permutation π_2 of Ω_2 such that $\pi = \pi_1 \times \pi_2$.*

Proof. Suppose π is $(\mathfrak{B}_1 \times \mathfrak{B}_2)$-allowable, $\pi((v_1, v_2)) = (w_1, w_2)$, and $\pi((v_1', v_2')) = (w_1', w_2')$. If $v_1 = v_1'$, then since (v_1, v_2) and (v_1', v_2') are in the same $(\mathcal{E}_1 \times \mathcal{U}_2)$-class, (w_1, w_2) and (w_1', w_2') must also be in the same $(\mathcal{E}_1 \times \mathcal{U}_2)$-class. Then $w_1 = w_1'$. Similarly, if $v_2 = v_2'$, then $w_2 = w_2'$. This shows that π must be of the form $\pi_1 \times \pi_2$. Clearly π is $(\mathfrak{B}_1 \times \mathfrak{B}_2)$-allowable if and only if π_1 is $\mathfrak{B}_1$-allowable and π_2 is $\mathfrak{B}_2$-allowable. □

Proposition 12.11. *Given two block structures $\mathfrak{B}_1$ and $\mathfrak{B}_2$ defined on Ω_1 and Ω_2, respectively, a permutation π of $\Omega_1 \times \Omega_2$ is $(\mathfrak{B}_1/\mathfrak{B}_2)$-allowable if and only if there exist a $\mathfrak{B}_1$-allowable permutation π_1 of Ω_1 and, for each $w \in \Omega_1$, a $\mathfrak{B}_2$-allowable permutation π^w of Ω_2 such that $\pi((v_1, v_2)) = (\pi_1(v_1), \pi^{\pi_1(v_1)}(v_2))$ for all $(v_1, v_2) \in \Omega_1 \times \Omega_2$.*

Proof. Suppose π is $(\mathfrak{B}_1/\mathfrak{B}_2)$-allowable. We have $\mathcal{E}_1 \times \mathcal{U}_2 \in \mathfrak{B}_1/\mathfrak{B}_2$. Similar to the proof of Proposition 12.10, if $\pi((v_1, v_2)) = (w_1, w_2)$, $\pi((v_1', v_2')) = (w_1', w_2')$, and $v_1 = v_1'$, then $w_1 = w_1'$. This implies the existence of a permutation π_1 of Ω_1 such that for all $v_2 \in \Omega_2$, the first components of $\pi((v_1, v_2))$ are equal to $\pi_1(v_1)$. Their second components, for fixed v_1, must cover each element of Ω_2 once. This implies the existence of π^w for each $w \in \Omega_1$ as stated. Clearly, π is $(\mathfrak{B}_1/\mathfrak{B}_2)$-allowable if and only if π_1 is $\mathfrak{B}_1$-allowable and each π^w is $\mathfrak{B}_2$-allowable. □

By Proposition 12.10, under each $(\mathfrak{B}_1 \times \mathfrak{B}_2)$-allowable permutation of $\Omega_1 \times \Omega_2$, a $\mathfrak{B}_1$-allowable permutation and a $\mathfrak{B}_2$-allowable permutation are applied to the units of Ω_1 and Ω_2 separately. Suppose there are ϕ_1 permutations of Ω_1 that are $\mathfrak{B}_1$-allowable and ϕ_2 permutations of Ω_2 that are $\mathfrak{B}_2$-allowable. Then there are a total of $\phi_1\phi_2$ permutations of $\Omega_1 \times \Omega_2$ that are $(\mathfrak{B}_1 \times \mathfrak{B}_2)$-allowable. Picking one $(\mathfrak{B}_1 \times \mathfrak{B}_2)$-allowable permutation randomly is equivalent to drawing a $\mathfrak{B}_1$-allowable permutation of Ω_1 randomly and independently drawing a $\mathfrak{B}_2$-allowable permutation of Ω_2 randomly.

By Proposition 12.11, under each $(\mathfrak{B}_1/\mathfrak{B}_2)$-allowable permutation of $\Omega_1 \times \Omega_2$, a $\mathfrak{B}_1$-allowable permutation is first applied to Ω_1, and then a separate $\mathfrak{B}_2$-allowable permutation is applied to each of the $|\Omega_1|$ sets $\{(w_1, w_2) : w_2 \in \Omega_2\}$ of $|\Omega_2|$ units that share the same $w_1 \in \Omega_1$. Suppose there are ϕ_1 permutations of Ω_1 that are $\mathfrak{B}_1$-allowable and ϕ_2 permutations of Ω_2 that are $\mathfrak{B}_2$-allowable. Then there are $\phi_1 \phi_2^{|\Omega_1|}$ $(\mathfrak{B}_1/\mathfrak{B}_2)$-allowable permutations of $\Omega_1 \times \Omega_2$. Picking one $(\mathfrak{B}_1/\mathfrak{B}_2)$-allowable permutation randomly is equivalent to drawing a $\mathfrak{B}_1$-allowable permutation of Ω_1 randomly and, for each $w \in \Omega_1$, independently drawing a $\mathfrak{B}_2$-allowable permutation of Ω_2 randomly.

In view of these discussions, randomization of an experiment with a simple block structure is performed by following, at each stage of the block structure formula, the recipes for crossing and nesting described in the previous two paragraphs.

Proposition 12.12. *For any simple block structure, the set of all the allowable permutations of the experimental units is a transitive group.*

Proof. The set of allowable permutations for a simple block structure is clearly a group. It was shown in the paragraph following the statement of Theorem 3.2 that the allowable permutations for b/k and $r \times c$ are transitive. Using the same argument, it can be shown that if the $\mathfrak{B}_1$-allowable permutations and $\mathfrak{B}_2$-allowable permutations are transitive, then the $(\mathfrak{B}_1 \times \mathfrak{B}_2)$-allowable and $(\mathfrak{B}_1/\mathfrak{B}_2)$-allowable permutations are also transitive. The transitivity of allowable permutations for any simple block structure then follows by mathematical induction. The details are left as an exercise. $\square$

In algebraic terms, the allowable permutations are automorphisms of the block structure, and the group of allowable permutations is the automorphism group.

We have shown in the proof of Theorem 3.2 that if the set of allowable permutations is a transitive group, then the N^2 pairs of units in $\Omega \times \Omega$ are partitioned into disjoint orbits, and, under the randomization model, $\text{cov}(y_v, y_w)$ is a constant for all (v, w) in the same orbit. By Proposition 12.12, this holds for simple block structures. On the other hand, we have seen in Section 12.3 that for each simple block structure $\mathfrak{B}$, the $\mathbf{A}_{\mathcal{F}}$'s, $\mathcal{F} \in \mathfrak{B}$, also induce a partition of $\Omega \times \Omega$, with (v, w) and (v', w') in the same class if and only if there is an $\mathcal{F} \in \mathfrak{B}$ such that $[\mathbf{A}_{\mathcal{F}}]_{vw} = [\mathbf{A}_{\mathcal{F}}]_{v'w'} = 1$, and that under a covariance matrix of the form (12.19), $\text{cov}(y_v, y_w)$ is a constant for all (v, w) that belong to the same class in this partition. Therefore, to show that the covariance matrices under the randomization models for simple block structures are of the form (12.19), it is enough to show that the two aforementioned partitions of $\Omega \times \Omega$ coincide. We prove this by mathematical induction. Assume that it holds for both $\mathfrak{B}_1$ and $\mathfrak{B}_2$. We need to show that it is also true for $\mathfrak{B}_1/\mathfrak{B}_2$ and $\mathfrak{B}_1 \times \mathfrak{B}_2$.

We first consider $\mathfrak{B}_1 \times \mathfrak{B}_2$. Suppose $(v_1, v_2), (w_1, w_2), (v'_1, v'_2), (w'_1, w'_2) \in \Omega_1 \times \Omega_2$, $\mathcal{F}_1 \times \mathcal{F}_2$ is the finest factor in $\mathfrak{B}_1 \times \mathfrak{B}_2$ that has (v_1, v_2) and (w_1, w_2) in the same class and is also the finest factor that has (v'_1, v'_2) and (w'_1, w'_2) in the same class. We need to show that there is a $(\mathfrak{B}_1 \times \mathfrak{B}_2)$-allowable permutation π such that $\pi((v_1, v_2)) = (v'_1, v'_2)$ and $\pi((w_1, w_2)) = (w'_1, w'_2)$. It follows from the definition of the crossing operator that $\mathcal{F}_1$ is the finest factor in $\mathfrak{B}_1$ that has v_1 and w_1 (and also v'_1 and w'_1) in the same class, and $\mathcal{F}_2$ is the finest factor in $\mathfrak{B}_2$ that has v_2 and w_2 (and also v_2' and w_2') in the same class. Then, by the induction hypothesis, there is a $\mathfrak{B}_1$-allowable permutation π_1 and a $\mathfrak{B}_2$-allowable permutation π_2 such that $\pi_1(v_1) = v_1'$, $\pi_1(w_1) = w_1'$, and $\pi_2(v_2) = v_2', \pi_2(w_2) = w_2'$. Then $\pi_1 \times \pi_2$ is a $(\mathfrak{B}_1 \times \mathfrak{B}_2)$-allowable permutation such that $(\pi_1 \times \pi_2)((v_1, v_2)) = (v_1', v_2')$, and $(\pi_1 \times \pi_2)((w_1, w_2)) = (w_1', w_2')$.

We have shown that if $((v_1, v_2), (w_1, w_2))$ and $((v_1', v_2'), (w_1', w_2'))$ are in the same class of the partition of $(\Omega_1 \times \Omega_2) \times (\Omega_1 \times \Omega_2)$ induced by the $\mathbf{A}_{\mathcal{F}}$'s for the factors in $\mathfrak{B}_1 \times \mathfrak{B}_2$, then there is a $(\mathfrak{B}_1 \times \mathfrak{B}_2)$-allowable permutation $\pi_1 \times \pi_2$ such that $(\pi_1 \times \pi_2)((v_1, v_2)) = (v_1', v_2')$, and $(\pi_1 \times \pi_2)((w_1, w_2)) = (w_1', w_2')$. We can reverse the arguments to show that the converse is also true.

Now consider $\mathfrak{B}_1/\mathfrak{B}_2$. The factors in $\mathfrak{B}_1/\mathfrak{B}_2$ are of the form $\mathcal{F}_1 \times \mathcal{U}_2$ with $\mathcal{F}_1 \in \mathfrak{B}_1$, $\mathcal{F}_1 \neq \mathcal{E}_1$, or $\mathcal{E}_1 \times \mathcal{F}_2$ with $\mathcal{F}_2 \in \mathfrak{B}_2$. Suppose $\mathcal{E}_1 \times \mathcal{F}_2$ is the finest factor in $\mathfrak{B}_1/\mathfrak{B}_2$ that has (v_1, v_2) and (w_1, w_2) in the same class and is also the finest factor that has (v_1', v_2') and (w_1', w_2') in the same class. Then $v_1 = w_1$, $v_1' = w_1'$, and $\mathcal{F}_2$ is the finest factor in $\mathfrak{B}_2$ that has v_2 and w_2 (and also v_2' and w_2') in the same class. By the induction hypothesis, the latter implies that there is a $\mathfrak{B}_2$-allowable permutation π_2 such that $\pi_2(v_2) = v_2'$ and $\pi_2(w_2) = w_2'$. By the transitivity of allowable permutations, there is a $\mathfrak{B}_1$-allowable permutation π_1 such that $\pi_1(v_1) = \pi_1(w_1) = v_1' = w_1'$. Let $\pi = \pi_1 \times \pi_2$. Then π is a $(\mathfrak{B}_1/\mathfrak{B}_2)$-allowable permutation such that $\pi((v_1, v_2)) = (v_1', v_2')$ and $\pi((w_1, w_2)) = (w_1', w_2')$. Note that in terms of the notation in Proposition 12.11, in this case $\pi^w = \pi_2$ for all $w \in \Omega_1$.

On the other hand, suppose $\mathcal{F}_1 \times \mathcal{U}_2$, $\mathcal{F}_1 \neq \mathcal{E}_1$, is the finest factor in $\mathfrak{B}_1/\mathfrak{B}_2$ that has (v_1, v_2) and (w_1, w_2) in the same class and is also the finest factor that has (v_1', v_2') and (w_1', w_2') in the same class. Then v_1 and w_1 (and also v_1' and w_1') are different units in Ω_1, and $\mathcal{F}_1$ is the finest factor in $\mathfrak{B}_1$ that has v_1 and w_1 (and also v_1' and w_1') in the same class. Again by the induction hypothesis, there is a $\mathfrak{B}_1$-allowable permutation π_1 such that $\pi_1(v_1) = v_1'$ and $\pi_1(w_1) = w_1'$. Since $\mathfrak{B}_2$-allowable permutations are transitive, there are $\mathfrak{B}_2$-allowable permutations π_2 and π_2' such that $\pi_2(v_2) = v_2'$ and $\pi_2'(w_2) = w_2'$. Using the notation in Proposition 12.11, for $w \in \Omega_1$, let

$$\pi^w = \begin{cases} \pi_2, & \text{if } w = v_1'; \\ \pi_2', & \text{if } w = w_1'; \\ \text{any } \mathfrak{B}_2\text{-allowable permutation}, & \text{otherwise.} \end{cases}$$

Then the permutation π of $\Omega_1 \times \Omega_2$ with $\pi((u_1, u_2)) = (\pi_1(u_1), \pi^{\pi_1(u_1)}(u_2))$, as defined in Proposition 12.11, is $(\mathfrak{B}_1/\mathfrak{B}_2)$-allowable, $\pi((v_1, v_2)) = (v_1', v_2')$, and $\pi((w_1, w_2)) = (w_1', w_2')$.

We have shown that if $((v_1, v_2), (w_1, w_2))$ and $((v_1', v_2'), (w_1', w_2'))$ are in the same class of the partition of $(\Omega_1 \times \Omega_2) \times (\Omega_1 \times \Omega_2)$ induced by the $\mathbf{A}_{\mathcal{F}}$'s for the factors in $\mathfrak{B}_1/\mathfrak{B}_2$, then there is a $(\mathfrak{B}_1/\mathfrak{B}_2)$-allowable permutation π such that $\pi((v_1, v_2)) = (v_1', v_2')$, and $\pi((w_1, w_2)) = (w_1', w_2')$. We can reverse the arguments to show that the converse is also true. $\qquad \square$

12.15 Justification of Nelder's rules[*]

We justify here Nelder's rule for determining the strata in the case of simple block structures. Given a factor $\mathcal{F}_1$ on Ω_1 and a factor $\mathcal{F}_2$ on Ω_2, by the definition of

$\mathcal{F}_1 \times \mathcal{F}_2$, we have

$$V_{\mathcal{F}_1 \times \mathcal{F}_2} = V_{\mathcal{F}_1} \otimes V_{\mathcal{F}_2} = \{\mathbf{v}_1 \otimes \mathbf{v}_2 : \mathbf{v}_1 \in V_{\mathcal{F}_1}, \mathbf{v}_2 \in V_{\mathcal{F}_2}\}. \tag{12.53}$$

Given a block structure $\mathcal{B}_1$ on Ω_1 and a block structure $\mathcal{B}_2$ on Ω_2, $\mathcal{B}_1 \times \mathcal{B}_2$ consists of all the factors of the form $\mathcal{F}_1 \times \mathcal{F}_2$, where $\mathcal{F}_1 \in \mathcal{B}_1$ and $\mathcal{F}_2 \in \mathcal{B}_2$. By (12.53) and (12.28),

$$
\begin{aligned}
W_{\mathcal{F}_1 \times \mathcal{F}_2} &= V_{\mathcal{F}_1 \times \mathcal{F}_2} \ominus \left(\sum_{(\mathcal{G}_1, \mathcal{G}_2) \in \mathcal{B}_1 \times \mathcal{B}_2 : \mathcal{F}_1 \times \mathcal{F}_2 \prec \mathcal{G}_1 \times \mathcal{G}_2} V_{\mathcal{G}_1 \times \mathcal{G}_2} \right) \\
&= V_{\mathcal{F}_1} \otimes V_{\mathcal{F}_2} \ominus \left[\sum_{(\mathcal{G}_1, \mathcal{G}_2) \in \mathcal{B}_1 \times \mathcal{B}_2 : \mathcal{F}_1 \times \mathcal{F}_2 \prec \mathcal{G}_1 \times \mathcal{G}_2} V_{\mathcal{G}_1} \otimes V_{\mathcal{G}_2} \right] \\
&= \left[V_{\mathcal{F}_1} \ominus \left(\sum_{\mathcal{G}_1 \in \mathcal{B}_1 : \mathcal{F}_1 \prec \mathcal{G}_1} V_{\mathcal{G}_1} \right) \right] \otimes \left[V_{\mathcal{F}_2} \ominus \left(\sum_{\mathcal{G}_2 \in \mathcal{B}_2 : \mathcal{F}_2 \prec \mathcal{G}_2} V_{\mathcal{G}_2} \right) \right] \\
&= W_{\mathcal{F}_1} \otimes W_{\mathcal{F}_2}.
\end{aligned}
$$

This shows that the strata for $\mathcal{B}_1 \times \mathcal{B}_2$ are all the pairwise products $W_{\mathcal{F}_1} \otimes W_{\mathcal{F}_2}$ of strata for $\mathcal{B}_1$ and $\mathcal{B}_2$, with

$$\dim(W_{\mathcal{F}_1 \times \mathcal{F}_2}) = \dim(W_{\mathcal{F}_1})\dim(W_{\mathcal{F}_2}) \tag{12.54}$$

and

$$\mathbf{P}_{W_{\mathcal{F}_1 \times \mathcal{F}_2}} = \mathbf{P}_{W_{\mathcal{F}_1}} \otimes \mathbf{P}_{W_{\mathcal{F}_2}}. \tag{12.55}$$

On the other hand, $\mathcal{B}_1/\mathcal{B}_2$ consists of all the $\mathcal{F}_1 \times \mathcal{U}_2$, where $\mathcal{F}_1 \in \mathcal{B}_1, \mathcal{F}_1 \neq \mathcal{E}_1$, and all the $\mathcal{E}_1 \times \mathcal{F}_2$, where $\mathcal{F}_2 \in \mathcal{B}_2$. We have

$$
\begin{aligned}
W_{\mathcal{F}_1 \times \mathcal{U}_2} &= V_{\mathcal{F}_1 \times \mathcal{U}_2} \ominus \left(\sum_{\mathcal{G}_1 \in \mathcal{B}_1 : \mathcal{F}_1 \prec \mathcal{G}_1} V_{\mathcal{G}_1 \times \mathcal{U}_2} \right) \\
&= \left[V_{\mathcal{F}_1} \otimes V_{\mathcal{U}_2} \right] \ominus \left(\sum_{\mathcal{G}_1 \in \mathcal{B}_1 : \mathcal{F}_1 \prec \mathcal{G}_1} \left[V_{\mathcal{G}_1} \otimes V_{\mathcal{U}_2} \right] \right) \\
&= W_{\mathcal{F}_1} \otimes V_{\mathcal{U}_2}. \tag{12.56}
\end{aligned}
$$

For $W_{\mathcal{E}_1 \times \mathcal{F}_2}$, we split into two cases depending on whether $\mathcal{F}_2 = \mathcal{U}_2$ or not. If $\mathcal{F}_2 = \mathcal{U}_2$, then by the same calculation that leads to (12.56),

$$W_{\mathcal{E}_1 \times \mathcal{U}_2} = W_{\mathcal{E}_1} \otimes V_{\mathcal{U}_2}.$$

For $\mathcal{F}_2 \neq \mathcal{U}_2$,

$$
\begin{aligned}
W_{\mathcal{E}_1 \times \mathcal{F}_2} &= V_{\mathcal{E}_1 \times \mathcal{F}_2} \ominus \left(\sum_{\mathcal{G}_1 \times \mathcal{G}_2 \in \mathcal{B}_1/\mathcal{B}_2 : \mathcal{E}_1 \times \mathcal{F}_2 \prec \mathcal{G}_1 \times \mathcal{G}_2} V_{\mathcal{G}_1 \times \mathcal{G}_2} \right) \\
&= V_{\mathcal{E}_1 \times \mathcal{F}_2} \ominus \left(\sum_{\mathcal{G}_2 \in \mathcal{B}_2 : \mathcal{F}_2 \prec \mathcal{G}_2} V_{\mathcal{E}_1 \times \mathcal{G}_2} + \sum_{\mathcal{G}_1 \in \mathcal{B}_1 : \mathcal{G}_1 \neq \mathcal{E}_1} V_{\mathcal{G}_1 \times \mathcal{U}_2} \right),
\end{aligned}
$$

where the second equality holds since for any $\mathcal{G}_1 \times \mathcal{G}_2 \in \mathfrak{B}_1/\mathfrak{B}_2$, by the definition of the nesting operator, if $\mathcal{G}_1 \neq \mathcal{E}_1$, then $\mathcal{G}_2 = \mathcal{U}_2$. Since $\mathcal{E}_1 \times \mathcal{G}_2 \prec \mathcal{G}_1 \times \mathcal{U}_2$, we have $V_{\mathcal{G}_1 \times \mathcal{U}_2} \subseteq V_{\mathcal{E}_1 \times \mathcal{G}_2}$. It follows that

$$
\begin{aligned}
W_{\mathcal{E}_1 \times \mathcal{F}_2} &= V_{\mathcal{E}_1 \times \mathcal{F}_2} \ominus \left(\sum_{\mathcal{G}_2 \in \mathfrak{B}_2 : \mathcal{F}_2 \prec \mathcal{G}_2} V_{\mathcal{E}_1 \times \mathcal{G}_2} \right) \\
&= \left[V_{\mathcal{E}_1} \otimes V_{\mathcal{F}_2} \right] \ominus \left(\sum_{\mathcal{G}_2 \in \mathfrak{B}_2 : \mathcal{F}_2 \prec \mathcal{G}_2} V_{\mathcal{E}_1} \otimes V_{\mathcal{G}_2} \right) \\
&= V_{\mathcal{E}_1} \otimes W_{\mathcal{F}_2}.
\end{aligned}
$$

Therefore the strata for $\mathfrak{B}_1/\mathfrak{B}_2$ are all the products $W_{\mathcal{F}_1} \otimes V_{\mathcal{U}_2}$, where $W_{\mathcal{F}_1}$ is a stratum for $\mathfrak{B}_1$, and all the products $V_{\mathcal{E}_1} \otimes W_{\mathcal{F}_2}$, where $W_{\mathcal{F}_2}$ is a stratum for $\mathfrak{B}_2$ with $\mathcal{F}_2 \neq \mathcal{U}_2$. We have

$$\dim\left(W_{\mathcal{F}_1 \times \mathcal{U}_2}\right) = \dim\left(W_{\mathcal{F}_1}\right) \dim\left(V_{\mathcal{U}_2}\right) = \dim\left(W_{\mathcal{F}_1}\right), \tag{12.57}$$

$$\mathbf{P}_{W_{\mathcal{F}_1 \times \mathcal{U}_2}} = \mathbf{P}_{W_{\mathcal{F}_1}} \otimes \mathbf{P}_{V_{\mathcal{U}_2}}, \tag{12.58}$$

$$\dim\left(W_{\mathcal{E}_1 \times \mathcal{F}_2}\right) = \dim\left(V_{\mathcal{E}_1}\right) \dim\left(W_{\mathcal{F}_2}\right) = |\Omega_1| \dim\left(W_{\mathcal{F}_2}\right), \tag{12.59}$$

and

$$\mathbf{P}_{W_{\mathcal{E}_1 \times \mathcal{F}_2}} = \mathbf{P}_{W_{\mathcal{E}_1}} \otimes \mathbf{P}_{W_{\mathcal{F}_2}} = \mathbf{I}_{|\Omega_1|} \otimes \mathbf{P}_{W_{\mathcal{F}_2}}. \tag{12.60}$$

Let $\mathfrak{B}$ be a simple block structure on $n_1 \cdots n_m$ units obtained by applying the nesting and crossing operators to m sets of unstructured units of sizes $n_1, \ldots, n_m$. Each set of n_i unstructured units has two strata $V_{\mathcal{U}_i}$ and $W_{\mathcal{E}_i}$ with orthogonal projection matrices $\frac{1}{n_i}\mathbf{J}_{n_i}$ and $\mathbf{I}_{n_i} - \frac{1}{n_i}\mathbf{J}_{n_i}$, respectively. Nelder's rules of expanding the crossing and nesting functions to obtain the degree-of-freedom identity, thereby determining the number of strata and their degrees of freedom, follow from (12.54), (12.57), and (12.59). The special rule that the n_1 term that appears on the right-hand side of (12.35) must be the algebraic sum of all the terms in the expansion of n_1 is a consequence of (12.59). It follows from (12.55), (12.58), and (12.60) that the orthogonal projection matrix onto each stratum of $\mathfrak{B}$ is of the form $\mathbf{P}_1 \otimes \cdots \otimes \mathbf{P}_m$, where each $\mathbf{P}_i$ is $\mathbf{I}_{n_i}$, $\frac{1}{n_i}\mathbf{J}_{n_i}$, or $\mathbf{I}_{n_i} - \frac{1}{n_i}\mathbf{J}_{n_i}$. Expanding the product, we see that it must be of the form $\sum \pm (\mathbf{Q}_1 \otimes \cdots \otimes \mathbf{Q}_m)$, where each $\mathbf{Q}_i$ is either $\mathbf{I}_{n_i}$ or $\frac{1}{n_i}\mathbf{J}_{n_i}$. Nelder's rule of using the yield identity to determine orthogonal projections onto the strata is a consequence of these observations.

Exercises

12.1   Prove parts (vi) and (vii) of Lemma 12.1.

12.2   Show that if $\mathfrak{B}_1$ and $\mathfrak{B}_2$ are orthogonal block structures on Ω_1 and Ω_2, respectively, then $\mathfrak{B}_1 \times \mathfrak{B}_2$ and $\mathfrak{B}_1/\mathfrak{B}_2$ are orthogonal block structures on $\Omega_1 \times \Omega_2$, and if $\mathfrak{B}_1$ and $\mathfrak{B}_2$ are poset block structures, then $\mathfrak{B}_1 \times \mathfrak{B}_2$ and $\mathfrak{B}_1/\mathfrak{B}_2$ are poset block structures

12.3 Verify that the strata for the block structure $n_1/[(n_2/n_3/n_4) \times n_5]$ are as given in Example 12.13.

12.4 Determine the null ANOVA, including strata, degrees of freedom, orthogonal projections onto the strata, and the corresponding sums of squares for the simple block structure $n_1/(n_2 \times n_3)/n_4$.

12.5 Use Theorem 12.7 to show the following result stated in the paragraph preceding Example 10.5: in a two-level factorial experiment, if the prior distribution of the treatment effects $\boldsymbol{\alpha}$ is stationary, then the covariance matrix of the factorial effects is diagonal, and their variances are the eigenvalues of cov($\boldsymbol{\alpha}$) multiplied by a constant.

12.6 Complete the proof of Proposition 12.12 and the statement at the end of Section 12.14.

12.7 Let G be a finite Abelian group under an operation +. Each subgroup H and its cosets constitute a uniform partition $\mathcal{F}_H$ of G. Show that for any two subgroups H and K of G, $\mathcal{F}_H \wedge \mathcal{F}_K = \mathcal{F}_{H \cap K}$ and $\mathcal{F}_H \vee \mathcal{F}_K = \mathcal{F}_{H+K}$, where $H+K = \{h+k : h \in H, k \in K\}$. Furthermore, $\mathcal{F}_H$ and $\mathcal{F}_K$ are orthogonal. Thus, any set of subgroups, containing G and $\{0\}$, that is closed under intersection and sum defines an orthogonal block structure on G. This is a specialized version of a result on *group block structures* in Section 8.6 of Bailey (2004).

12.8 In the third paragraph of Section 7.8, for each subgroup H of $\mathbb{Z}_{s_1} \oplus \cdots \oplus \mathbb{Z}_{s_n}$, The annihilator H^0 of H and its cosets give a partition of $\mathbb{Z}_{s_1} \oplus \cdots \oplus \mathbb{Z}_{s_n}$. Denote this partition by $\mathcal{F}_{H^0}$. Use the previous exercise to show that for any two subgroups H_1 and H_2 of $\mathbb{Z}_{s_1} \oplus \cdots \oplus \mathbb{Z}_{s_n}$, $\mathcal{F}_{H_1^0} \vee \mathcal{F}_{H_2^0} = \mathcal{F}_{(H_1 \cap H_2)^0}$. Similarly, each set V of k linearly independent vectors in EG(n,s) defines a pencil of $(n-k)$-flats, where s is a prime or a prime power. Denote this partition of EG(n,s) by $\mathcal{F}_{V^0}$. Show that if V_1 and V_2 are two sets of linearly independent vectors in EG(n,s) such that the vectors in $V_1 \cup V_2$ are linearly independent, then $\mathcal{F}_{V_1^0} \vee \mathcal{F}_{V_2^0} = \mathcal{F}_{(V_1 \cap V_2)^0}$. For an application of this, see Section 13.11, p. 284.

Chapter 13

Complete Factorial Designs with Orthogonal Block Structures

Design and analysis of complete factorial experiments with incomplete blocks and split-plots were discussed in Chapter 7. More complicated block structures are considered in this chapter. We first discuss blocked split-plot and blocked strip-plot experiments before presenting a general procedure for constructing complete factorial designs with simple block structures. The method of design key and the construction of designs for experiments with multiple processing stages are also discussed. We present a general notion of orthogonal designs and derive a result for checking design orthogonality, which can also be used to characterize the unit contrasts in each stratum and to determine the strata in which various treatment factorial effects are estimated under an orthogonal design. Such a result is useful for the design and analysis of experiments with multiple strata. Construction and selection of fractional factorial designs for multi-stratum experiments will be presented in the next chapter.

13.1 Orthogonal designs

Under (12.18) and (12.31), if $V_T \ominus V_U \subset W_F$ for some F, then (5.9) is satisfied. By Theorem 5.2, the design can be analyzed in a straightforward manner. Each treatment contrast $\sum_{j=1}^{t} c_j \alpha_j$ has a simple estimator $\sum_{j=1}^{t} c_j \overline{T}_j$, with $\mathrm{var}(\sum_{j=1}^{t} c_j \overline{T}_j) = \zeta_F \sum_{j=1}^{t} c_j^2 / r_j$, where $\overline{T}_j$ is the jth treatment mean, and r_j is the number of replications of the jth treatment. Let the space W in (5.11) be W_F. Then $\|\mathbf{P}_{W_F}\mathbf{y}\|^2$ in the null ANOVA identity (12.33) can be decomposed as in (5.12):

$$\|\mathbf{P}_{W_F}\mathbf{y}\|^2 = \|\mathbf{P}_{V_T \ominus V_U}\mathbf{y}\|^2 + \|\mathbf{P}_{W_F \ominus (V_T \ominus V_U)}\mathbf{y}\|^2 \qquad (13.1)$$

with the breakdown of degrees of freedom

$$\dim(W_F) = (t-1) + [\dim(W_F) - (t-1)].$$

The treatment sum of squares $\|\mathbf{P}_{V_T \ominus V_U}\mathbf{y}\|^2$ is easily computed by using (5.13), and the computation of the stratum sum of squares $\|\mathbf{P}_{W_F}\mathbf{y}\|^2$ was discussed in Sections 12.9 and 12.10. Then the residual sum of squares $\|\mathbf{P}_{W_F \ominus (V_T \ominus V_U)}\mathbf{y}\|^2$ can be obtained by subtraction, and $s^2 = \left\| \mathbf{P}_{W_{\mathcal{F}} \ominus (V_T \ominus V_{\mathcal{U}})} \mathbf{y} \right\|^2 / [\dim(W_{\mathcal{F}}) - (t - 1)]$ is an unbiased estimator of $\xi_{\mathcal{F}}$, provided that $\dim(W_{\mathcal{F}}) > t - 1$. Substituting $\left\| \mathbf{P}_{W_{\mathcal{F}}} \mathbf{y} \right\|^2$ in the null ANOVA at the end of Section 12.8 by the two terms on the right side of (13.1), we obtain the following full ANOVA table:

Source	SS	d.f.	MS	E(MS)
$\mathcal{G}$	$\left\| \mathbf{P}_{W_{\mathcal{G}}} \mathbf{y} \right\|^2$	$\dim(W_{\mathcal{G}})$	$\cdots$	$\xi_{\mathcal{G}}$
$\vdots$	$\vdots$	$\vdots$	$\vdots$	$\vdots$
$\mathcal{F}$	$\left\| \mathbf{P}_{W_{\mathcal{F}}} \mathbf{y} \right\|^2$	$\dim(W_{\mathcal{F}})$		
Treatments	$\sum_{j=1}^{t} r_j \left(\overline{T}_j - y_. \right)^2$	$t - 1$	$\cdots$	$\xi_{\mathcal{F}} + \frac{1}{t-1} \left[\sum_{j=1}^{t} r_j (\alpha_j - \overline{\alpha})^2 \right]$
Residual	By subtraction	$\dim(W_{\mathcal{F}}) - (t - 1)$	$\cdots$	$\xi_{\mathcal{F}}$
$\vdots$	$\vdots$	$\vdots$	$\vdots$	$\vdots$
Total	$\left\| \mathbf{y} - \mathbf{P}_{V_{\mathcal{U}}} \mathbf{y} \right\|^2$	$N - 1$		

This covers the three designs discussed in Chapter 5: completely randomized designs, randomized complete block designs, and Latin square designs.

Under a complete factorial design with incomplete blocks, $V_T \ominus V_{\mathcal{U}}$ cannot lie entirely in one single stratum. We have seen in Chapter 7 how to construct such designs with some factorial effects estimated in the interblock stratum and the other effects estimated in the intrablock stratum. We say that a design is *orthogonal* if $V_T \ominus V_{\mathcal{U}}$, which corresponds to all the treatment contrasts, can be decomposed orthogonally as

$$V_T \ominus V_{\mathcal{U}} = \oplus_l Z_l, \tag{13.2}$$

such that for each l,

$$Z_l \subset W_{\mathcal{F}} \text{ for some } \mathcal{F}, \tag{13.3}$$

and the Z_l's together cover the treatment contrasts of interest.

Similar to Theorem 5.2, we have the result below for such orthogonal designs.

Theorem 13.1. *Under (12.18), (12.31), (13.2), and (13.3), the best linear unbiased estimator of any treatment contrast $\mathbf{c}^T \alpha$ with $\mathbf{c}^* \in Z_l$, where $\mathbf{c}^*$ is defined in (5.2), is $\mathbf{c}^T \widehat{\alpha} = \sum_{j=1}^{t} c_j \overline{T}_j$, with $\mathrm{var}(\sum_{j=1}^{t} c_j \overline{T}_j) = \xi_{\mathcal{F}} \sum_{j=1}^{t} c_j^2 / r_j$.*

We say that the contrasts $\mathbf{c}^T \alpha$ with $\mathbf{c}^* \in Z_l$ are estimated in $W_{\mathcal{F}}$.

Let

$$Z_{\mathcal{F}} = \oplus_{l: Z_l \subset W_{\mathcal{F}}} Z_l.$$

If $Z_{\mathcal{F}} \neq \{\mathbf{0}\}$, then the stratum sum of squares $\left\| \mathbf{P}_{W_{\mathcal{F}}} \mathbf{y} \right\|^2$ can be decomposed as

$$\left\| \mathbf{P}_{W_{\mathcal{F}}} \mathbf{y} \right\|^2 = \left\| \mathbf{P}_{Z_{\mathcal{F}}} \mathbf{y} \right\|^2 + \left\| \mathbf{P}_{W_{\mathcal{F}} \ominus Z_{\mathcal{F}}} \mathbf{y} \right\|^2, \tag{13.4}$$

giving the ANOVA in stratum $W_\mathcal{F}$, where $\left\|\mathbf{P}_{Z_\mathcal{F}}\mathbf{y}\right\|^2$ is called the treatment sum of squares in $W_\mathcal{F}$, and $\left\|\mathbf{P}_{W_\mathcal{F}\ominus Z_\mathcal{F}}\mathbf{y}\right\|^2$ is the residual sum of squares in that stratum. It is easy to see that $W_\mathcal{F}\ominus Z_\mathcal{F}$ is orthogonal to $V_\mathcal{T}\ominus V_\mathcal{U}$. By the same argument used to prove Proposition 5.3, if $\dim(W_\mathcal{F}) > \dim(Z_\mathcal{F})$, then $\left\|\mathbf{P}_{W_\mathcal{F}\ominus Z_\mathcal{F}}\mathbf{y}\right\|^2 / [\dim(W_\mathcal{F})-\dim(Z_\mathcal{F})]$ is an unbiased estimator of $\xi_\mathcal{F}$.

By (13.2), we have

$$\left\|\mathbf{P}_{V_\mathcal{T}\ominus V_\mathcal{U}}\mathbf{y}\right\|^2 = \sum_\mathcal{F} \left\|\mathbf{P}_{Z_\mathcal{F}}\mathbf{y}\right\|^2 .$$

Under designs satisfying (13.2) and (13.3), again we have simple estimates of the treatment contrasts of interest and simple ANOVA, but not all these contrasts are estimated in the same stratum, resulting in different precisions for treatment contrasts estimated in different strata.

The notion of orthogonal designs defined in (13.2) and (13.3) is the most useful if the decomposition in (13.2) is determined by the treatment structure. For example, when the treatments have a factorial structure, the Z_l's should correspond to various main effects and interactions. To apply Theorem 13.1 or to perform the ANOVA in (13.4), one needs to determine the stratum that contains each Z_l. We discuss how to do this in Sections 13.4 and 13.10.

13.2 Blocked complete factorial split-plot designs

Construction of complete factorial split-plot designs was considered in Section 7.10. Sometimes the experimental runs cannot be performed under homogeneous conditions. This leads to blocking of split-plot designs, where the whole-plots are divided into blocks (e.g., days or laboratories).

Suppose there are m blocks of a whole-plots, with each whole-plot split into b subplots. The mab experimental units have the structure (m blocks)/(a whole-plots)/(b subplots). This block structure consists of four unit factors $\mathcal{E} \prec \mathcal{P} \prec \mathcal{B} \prec \mathcal{U}$, where $\mathcal{P}$ (whole-plots) has ma levels and $\mathcal{B}$ (blocks) has m levels. Then there are three strata other than the mean stratum: $W_\mathcal{B} = V_\mathcal{B} \ominus V_\mathcal{U}$ (block stratum), $W_\mathcal{P} = V_\mathcal{P} \ominus V_\mathcal{B}$ (whole-plot stratum), and $W_\mathcal{E} = V_\mathcal{P}^\perp$ (subplot stratum), with $m-1$, $m(a-1)$, and $ma(b-1)$ degrees of freedom, respectively; see Example 12.10.

We first consider the simple case of two treatment factors A and B, where A has a levels and B has b levels. Each level of factor A is assigned to one whole-plot in each block, and each level of B is assigned to one subplot in each whole-plot. Thus A is a whole-plot treatment factor and B is a subplot treatment factor.

Since each treatment combination appears once in each block, by proportional frequencies, $\mathbf{c}^* \perp W_\mathcal{B}$ for all the treatment contrasts $\mathbf{c}^T\boldsymbol{\alpha}$. If $\mathbf{c}^T\boldsymbol{\alpha}$ is a main-effect contrast of A, then we have $\mathbf{c}^* \in V_\mathcal{P} \ominus V_\mathcal{U}$ since A has the same level on all the subplots in each whole-plot. Therefore $\mathbf{c}^* \in W_\mathcal{P}$. By using the orthogonality between the main effect of A and the other factorial effects, one can show that if $\mathbf{c}^T\boldsymbol{\alpha}$ is a main-effect contrast of B or an interaction contrast of A and B, then $\mathbf{c}^* \perp V_\mathcal{P}$, and hence $\mathbf{c}^* \in W_{\mathcal{E}}$. Therefore we have an orthogonal design. The main-effect contrasts of A are estimated in $W_{\mathcal{P}}$, and the main-effect contrasts of B as well as the interaction contrasts of A and B are estimated in $W_{\mathcal{E}}$. Estimates of these contrasts and the associated sums of squares can be computed in the same simple manner as under complete randomization. The following is a skeleton of the ANOVA.

Source of variation	d.f.
Block stratum	$m-1$
Whole-plot stratum	
A	$a-1$
Residual	$(m-1)(a-1)$
Subplot stratum	
B	$b-1$
AB	$(a-1)(b-1)$
Residual	$a(m-1)(b-1)$
Total	$mab-1$

When it is not possible to have a complete replication within each block, under an orthogonal design some factorial effects have to be confounded with blocks. Suppose the block structure is (s^g blocks)/(s^{q-g} whole-plots)/(s^{n-q} subplots), where s is a prime number or a prime power, and there are n treatment factors each with s levels, n_1 of which are whole-plot factors, and $n_2 = n - n_1$ are subplot treatment factors. We first use the method presented in Section 7.10 to construct a complete factorial design for the block structure (s^q whole-plots)/(s^{n-q} subplots) and then apply the usual method of constructing blocked factorial designs to divide the s^q whole-plots into s^g disjoint groups of size s^{q-g}. For this purpose, we need g independent blocking effects (words) that are either factorial effects of whole-plot treatment factors or splitting effects (when $n_1 < q$). Under the resulting design, these g independent effects and their generalized interactions are confounded with blocks; factorial effects of whole-plot treatment factors and splitting effects that are not confounded with blocks are estimated in the whole-plot stratum, and the remaining factorial effects are estimated in the subplot stratum. One should avoid confounding important effects with blocks or whole-plots if possible. For example, it is generally desirable to avoid confounding main effects of whole-plot treatment factors with blocks.

In order to have a complete factorial design, we must have $n_1 \leq q$. We consider four cases:

Case (i) $n_1 = q$.

In this case, the number of whole-plot treatment factor level combinations is equal to the number of whole-plots, and the number of subplot treatment factor level combinations is equal to the number of subplots per whole-plot. Each level combination of the whole-plot treatment factors is assigned to one whole-plot, and each level combination of the subplot treatment factors is assigned to one subplot in each whole-plot. Blocking is accomplished by choosing g independent words that involve whole-plot treatment factors only.

When $n_1 < q$, $q - n_1$ independent splitting words $\mathbf{a}_{n_1+1}, \ldots, \mathbf{a}_q$ are used to partition each set of $s^{n_2} = s^{n-n_1}$ treatment combinations $\mathbf{x}$ with all the whole-plot treatment factors at constant levels into s^{q-n_1} disjoint groups of size s^{n-q} according to the values of $\mathbf{a}_{n_1+1}^T \mathbf{x}, \ldots, \mathbf{a}_q^T \mathbf{x}$, and the treatment combinations in the same group are assigned to one whole-plot. Recall that the whole-plot variance $\xi_{\mathcal{P}}$ is expected to be smaller than the block variance ξ_B, and this holds under the mixed-effect model (12.21) with positive variance components; see (12.50). In this case, to obtain a more efficient design, one should avoid replicating the same whole-plot treatment factor level combination in the same block if possible and, if such within-block replication cannot be avoided (for example, when $n_1 < q - g$), then it should be kept at the minimum. That is, if $n_1 \leq q - g$, then, in each block it is better to have s^{q-g-n_1} replicates of a complete factorial than a replicated fractional factorial of the whole-plot treatment factors. On the other hand, if $n_1 > q - g$, then, in each block, we require an *unreplicated* $s^{n_1-(n_1-q+g)}$ design of the whole-plot treatment factors, instead of replications of a smaller fraction. The blocking words must be splitting words or words that involve whole-plot treatment factors only. It is easy to see that under the aforementioned requirement, as many of the independent blocking words as possible should be splitting words. This is because if more words that involve whole-plot treatment factors only than necessary are used to do blocking, then there will be more within-block replications of the whole-plot treatment factor level combinations than minimally possible.

Case (iia) $n_1 < q$ and $g = q - n_1$.
In this case, all $g = q - n_1$ independent blocking words must be splitting words. The $s^g = s^{q-n_1}$ blocks of whole-plots can be obtained by putting the treatment combinations with the same values of $\mathbf{a}_{n_1+1}^T \mathbf{x}, \ldots, \mathbf{a}_q^T \mathbf{x}$ in the same block. In other words, the splitting effects play the double roles of splitting and blocking the whole plots, and are confounded with blocks.

Case (iib) $n_1 < q$ and $g > q - n_1$.
In this case, besides the independent splitting words $\mathbf{a}_{n_1+1}, \ldots, \mathbf{a}_q$, we need $g - (q - n_1)$ additional independent words that involve the whole-plot treatment factors only to further divide each of the s^{q-n_1} sets in the partition obtained by $\mathbf{a}_{n_1+1}, \ldots,$ $\mathbf{a}_q$ into $s^{g-(q-n_1)}$ smaller sets, producing a total of s^g blocks as desired.

Case (iic) $n_1 < q$ and $g < q - n_1$.
In this case, we can choose g of the $q - n_1$ independent splitting words to double as blocking words.

McLeod and Brewster (2004) proposed three methods, called pure whole-plot blocking, separation, and mixed blocking, for constructing blocked split-plot fractional factorial designs. For constructing complete factorial designs, these three methods are equivalent to what we describe here for cases (i), (iia), and (iib), respectively.

Remark 13.1. Technically, the construction described above is equivalent to the following procedure that applies in general to block structures that involve two layers of nesting. We use a set of q linearly independent vectors $\mathbf{a}_1, \ldots, \mathbf{a}_q$ in $EG(n, s)$ to partition the s^n treatment combinations into s^q sets of size s^{n-q}, with those in the same set assigned to the same whole-plot, and use a *subset* of g of these words, say $\mathbf{a}_1, \ldots, \mathbf{a}_g$, to divide the treatment combinations into s^g sets, with those in the same set assigned to the same block. The treatment combinations in each whole-plot are solutions $\mathbf{x} = (x_1, \ldots, x_n)^T$ to the equations $\mathbf{a}_i^T \mathbf{x} = b_i$, $i = 1, \ldots, q$, where $b_i \in GF(s)$, and each block is the solution set of the equations $\mathbf{a}_i^T \mathbf{x} = b_i$, $i = 1, \ldots, g$. The factorial effects defined by nonzero linear combinations of $\mathbf{a}_1, \ldots, \mathbf{a}_g$ are confounded with blocks, and those which are defined by nonzero linear combinations of $\mathbf{a}_1, \ldots, \mathbf{a}_q$ but are not confounded with blocks are estimated in the whole-plot stratum. The remaining factorial effects are estimated in the subplot stratum. For blocked split-plot designs, since main effects of whole-plot treatment factors must be confounded with whole-plots, without loss of generality, n_1 of the q words $\mathbf{a}_1, \ldots, \mathbf{a}_q$ can be an arbitrary set of independent words that involve whole-plot treatment factors only. Therefore effectively we only need to choose $q - n_1$ independent splitting words (if $n_1 < q$) and g independent blocking words.

Example 13.1. (Example 1.2 revisited) A design is needed for the block structure (four blocks)/(four whole-plots)/(two subplots), three two-level whole-plot treatment factors, and three two-level subplot treatment factors. We consider here the simpler case of three whole-plot treatment factors (A, B, C) and two subplot treatment factors (S, T) so that one complete replicate can be accommodated; a full discussion of the original example will be given in Sections 14.4 and 14.13. Then $n = 5$, $n_1 = 3$, $g = 2$, and $q = 4$. This falls in case (iib). An unblocked design was already constructed in Example 7.11 by using ST as a splitting word. To divide the 16 whole-plots into four blocks, we need two independent blocking words, one of which is a splitting word. Suppose we use BST as a blocking word; note that BST is also a splitting word since it is the generalized interaction of B and ST. Then we need another word that involves the whole-plot treatment factors only, say ABC. Then the first (principal) block consists of the four whole-plots containing the treatment combinations that have even numbers of letters in common with both ABC and BST:

(1)	ac	bct	abt
st	acst	bcs	abs

The other three blocks can be generated from the principal block in the usual way.

c	a	bt	$abct$
cst	ast	bs	$abcs$

t	act	bc	ab
s	acs	$bcst$	$abst$

ct	at	b	abc
cs	as	bst	$abcst$

Bypassing the construction in Example 7.11 (and following the construction in Remark 13.1), we can use A, B, ABC, and BST to divide the 32 treatment combinations into 16 whole-plots and then use ABC and BST to divide the 16 whole-plots into four blocks of size 4. We first identify the treatment combinations that have even numbers of letters in common with each of A, B, ABC, and BST: (1) and st. These two treatment combinations constitute the first whole-plot in the first block. Multiplying (1) and st by a treatment combination that has even numbers of letters in common with both ABC and BST, say ac, we obtain the two treatment combinations in the second whole-plot of the first block: ac and $acst$. Multiplying the four treatment combinations that have been generated by another treatment combination that has even numbers of letters in common with both ABC and BST, say bct, we complete the first block. Iterating twice the process of multiplying the treatment combinations that have been generated by one that has not appeared (for example, c and t), we generate the same design (before randomization). The following shows the strata in which various factorial effects are estimated.

Strata	Effects
Blocks	BST, ABC, $ACST$
Whole-plots	A, B, C, AB, AC, BC, ST, AST, CST, $ABST$, $BCST$, $ABCST$
Subplots	S, T, AS, AT, BS, BT, CS, CT, ABS, ABT, ACS, ACT, BCS, BCT $ABCS$, $ABCT$

The design in the above example can be constructed by the method of design key. This will be discussed in Section 13.7.

13.3 Blocked complete factorial strip-plot designs

Strip-plot designs were briefly discussed in Section 7.11. Suppose the block structure is $(s^g \text{ blocks})/[(s^p \text{ rows}) \times (s^q \text{ columns})]$, defined by factors $\mathcal{E}$, $\mathcal{R}'$, $\mathcal{C}'$, $\mathcal{B}$, and $\mathcal{U}$, where s is a prime number or power of a prime number. There are four strata other than the mean stratum: $W_{\mathcal{B}} = V_{\mathcal{B}} \ominus V_{\mathcal{U}}$ (block stratum), $W_{\mathcal{R}'} = V_{\mathcal{R}'} \ominus V_{\mathcal{B}}$ (row stratum), $W_{\mathcal{C}'} = V_{\mathcal{C}'} \ominus V_{\mathcal{B}}$ (column stratum), and $W_{\mathcal{E}} = (V_{\mathcal{R}'} + V_{\mathcal{C}'})^{\perp}$ (unit stratum), with $s^g - 1$, $s^g(s^p - 1)$, $s^g(s^q - 1)$, and $s^g(s^p - 1)(s^q - 1)$ degrees of freedom, respectively.

Consider the case where n_1 of the n treatment factors have constant levels on all the units in the same row, and the other $n_2 = n - n_1$ treatment factors have constant levels on all the units in the same column. We call these factors row and column treatment factors, respectively. For a single-replicate complete factorial $(n_1 + n_2 = g + p + q)$, we must have $n_1 \leq g + p$ and $n_2 \leq g + q$; thus $n_1 \geq p$ and $n_2 \geq q$. In this case, a blocked strip-plot design can be obtained as the *product* of a blocked complete factorial of the row treatment factors in $s^{n_1 - p}$ blocks (called the *row design*) and a blocked complete factorial of the column treatment factors in $s^{n_2 - q}$ blocks (called the *column design*). Each block of the row design is coupled with each block of the column design to form a block of s^p rows and s^q columns, with the same level combination of the row (respectively, column) treatment factors appearing on all the units in the same row (respectively, column). This produces a total of $s^{n_1 - p} \times s^{n_2 - q} = s^g$ blocks for the strip-plot design. The construction requires the selection of $n_1 - p$ independent blocking words involving row treatment factors only and $n_2 - q$ independent blocking words involving column treatment factors only. Under the constructed design, all the treatment factorial effects generated by independent blocking effects are confounded with blocks, all the factorial effects of row treatment factors that are not confounded with blocks and their generalized interactions with the blocking effects are estimated in the row stratum, all the factorial effects of column treatment factors that are not confounded with blocks and their generalized interactions with the blocking effects are estimated in the column stratum, and the remaining treatment factorial effects are estimated in the bottom stratum.

Example 13.2. (revisit of Example 1.3) This example called for a design with ten two-level treatment factors, six of which were configurations of washers and four were configurations of dryers. Each of these factors must be kept at the same level in each washing or drying cycle, respectively. Here we only consider the case of three washer factors and two dryer factors so that one complete replicate can be accommodated. A full discussion of the original example is deferred to Sections 14.5 and 14.14. Let the washer factors be A, B, and C, and the dryer factors be S and T. Identify each washer with a row and each dryer with a column. Then the block structure is (two blocks)/[(four rows)×(four columns)]. The four combinations of S and T are assigned to different columns in each block, but since there are eight combinations of A, B, and C, one of their factorial effects must be confounded with blocks. Suppose we choose to confound the two-factor interaction AC with blocks. Then the four level combinations of A, B, C with neither or both of A and C at the high level are assigned to different rows in block 1, and the other four combinations are assigned to different rows in block 2. This results in the following design (before randomization):

(1)	s	t	st
b	bs	bt	bst
ac	acs	act	acst
abc	abcs	abct	abcst

a	as	at	ast
ab	abs	abt	abst
c	cs	ct	cst
bc	bcs	bct	bcst

The following shows the strata in which various factorial effects are estimated.

Strata	Effects
Blocks	AC
Rows	A, B, C, AB, BC, ABC
Columns	$S, T, ST, ACS, ACT, ACST$
Units	$AS, AT, BS, BT, CS, CT, AST, BST, CST, ABS$
	$ABT, BCS, BCT, ABST, BCST, ABCS, ABCT, ABCST$

Remark 13.2. In general, for constructing a single-replicate complete factorial for an experiment with the block structure $(s^g$ blocks$)/[(s^p$ rows$) \times (s^q$ columns$)]$, where s is a prime number or power of a prime number, we choose a set of n linearly independent vectors $\mathbf{a}_1, \ldots, \mathbf{a}_n$ in $\mathrm{EG}(n, s)$ to partition the treatment combinations into blocks, rows, and columns. Specifically, each block is the solution set of the equations $\mathbf{a}_i^T \mathbf{x} = b_i$, $i = 1, \ldots, g$, each row consists of solutions to the equations $\mathbf{a}_i^T \mathbf{x} = b_i$, $i = 1, \ldots, p + g$, and each column consists of solutions to the equations $\mathbf{a}_i^T \mathbf{x} = b_i$, $i = 1, \ldots, g, p + g + 1, \ldots, n$, where $b_i \in \mathrm{GF}(s)$. Then the factorial effects defined by nonzero linear combinations of $\mathbf{a}_1, \ldots, \mathbf{a}_g$ are confounded with blocks. The factorial effects defined by nonzero linear combinations of $\mathbf{a}_1, \ldots, \mathbf{a}_{p+g}$, except those that are confounded with blocks, are confounded with rows, and the factorial effects defined by nonzero linear combinations of $\mathbf{a}_1, \ldots, \mathbf{a}_g, \mathbf{a}_{p+g+1}, \ldots, \mathbf{a}_n$, except those that are confounded with blocks, are confounded with columns. The remaining factorial effects are estimated in the bottom stratum. When choosing the vectors $\mathbf{a}_1, \ldots, \mathbf{a}_n$, one must observe the constraints imposed on the experiment. In the case of blocked strip-plot designs, since the main effects of the row and column treatment factors must be confounded with rows and columns, respectively, without loss of generality, we can choose $\mathbf{a}_{g+1}, \ldots, \mathbf{a}_{p+g}$ to be those defining the main effects of p row treatment factors, and $\mathbf{a}_{p+g+1}, \ldots, \mathbf{a}_n$ to be those defining the main effects of q column treatment factors. Therefore effectively we only need to choose g independent blocking words, $n_1 - p$ of which involve row treatment factors only and $n_2 - q$ involve column treatment factors only, as we have seen in the procedure described in the paragraph preceding Example 13.2.

13.4 Contrasts in the strata of simple block structures

The construction of orthogonal designs for single-replicate factorial experiments with incomplete blocks, split-plots, and strip-plots discussed in Sections 7.2, 7.10, 7.11, 13.2, and 13.3 can be extended to arbitrary simple block structures. Two different but equivalent methods of construction will be presented in the next two sections. In the first method, we use independent factorial effects to partition the treatment combinations into blocks, rows, columns, and, more generally, the classes of various factors in the given block structure. We need to be able to identify the strata in which the treatment factorial effects are estimated. In the second method based on design keys, we choose unit contrasts to be aliases of treatment factorial effects so that these effects are estimated in designated strata. In order to implement either method, we need to explicitly characterize the unit contrasts in each stratum.

Each factor in a simple block structure $\mathfrak{B}$ is of the form $\mathcal{F}_1 \times \cdots \times \mathcal{F}_m$ with $\mathcal{F}_i = \mathcal{U}_i$ or $\mathcal{E}_i$, where $\mathcal{U}_i$ and $\mathcal{E}_i$ are the universal and equality factors on a set Ω_i of n_i unstructured units. Let $\mathfrak{B}_i = \{\mathcal{U}_i, \mathcal{E}_i\}$, $i = 1, \ldots, m$. Then the block structure $\mathfrak{F} = \mathfrak{B}_1 \times \cdots \times \mathfrak{B}_m$, obtained by crossing the m sets of unstructured units, consists of all the 2^m factors $\mathcal{F}_1 \times \cdots \times \mathcal{F}_m$ with $\mathcal{F}_i = \mathcal{U}_i$ or $\mathcal{E}_i$. This is the largest simple block structure in the sense that

$$\mathfrak{B} \subseteq \mathfrak{F} \tag{13.5}$$

for all the simple block structures $\mathfrak{B}$ constructed from $\mathfrak{B}_1, \ldots, \mathfrak{B}_m$. Note that $\mathfrak{F}$ is analogous to the factorial treatment structure. It has $2^m - 1$ strata besides the mean stratum, with one stratum corresponding to each nonempty subset of $\{\mathcal{E}_1, \ldots, \mathcal{E}_m\}$. We abuse the language by calling the contrasts in each of these strata factorial effects (main effects and interactions) of the unit factors $\mathcal{E}_1, \ldots, \mathcal{E}_m$.

For any $\mathcal{F} = \mathcal{F}_1 \times \cdots \times \mathcal{F}_m \in \mathfrak{F}$ and any simple block structure $\mathfrak{B}$, $\mathcal{F}$ may not be a factor in $\mathfrak{B}$. Let $\mathcal{F}^*$ be the coarsest factor $\mathcal{G}$ in $\mathfrak{B}$ such that $\mathcal{G} \preceq \mathcal{F}$. Then $\mathcal{F}^* = \mathcal{G}_1 \times \cdots \times \mathcal{G}_m$, where

$$\mathcal{G}_i = \begin{cases} \mathcal{E}_i, & \text{if there is a } j \text{ such that } \mathcal{E}_i \text{ is an ancestor of } \mathcal{E}_j \text{ and } \mathcal{F}_j = \mathcal{E}_j; \\ \mathcal{F}_i & \text{otherwise.} \end{cases}$$

This extends the definition of $\mathcal{E}_i^*$ in (12.3): if $\mathcal{F}$ is the factor in $\mathfrak{F}$ that corresponds to $\mathcal{E}_i$ ($\mathcal{F}_i = \mathcal{E}_i$ and $\mathcal{F}_j = \mathcal{U}_j$ for all $j \neq i$), then $\mathcal{F}^* = \mathcal{E}_i^*$.

The fact that $\mathfrak{B} \subseteq \mathfrak{F}$ for any simple block structure $\mathfrak{B}$ leads to the following useful result for identifying the unit contrasts in each stratum of $\mathfrak{B}$.

Theorem 13.2. *Let $\mathfrak{B}$ be a simple block structure and $\mathfrak{F}$ be as described above. For $\mathcal{F} \in \mathfrak{F}$ (respectively, $\mathfrak{B}$), let $W_{\mathcal{F}}^{\mathfrak{F}}$ (respectively, $W_{\mathcal{F}}^{\mathfrak{B}}$) be the stratum of $\mathfrak{F}$ (respectively, $\mathfrak{B}$) associated with $\mathcal{F}$. Then $W_{\mathcal{F}}^{\mathfrak{F}} \subseteq W_{\mathcal{F}^*}^{\mathfrak{B}}$ for all $\mathcal{F} \in \mathfrak{F}$. In particular, if $\mathcal{F} = \mathcal{F}_1 \times \cdots \times \mathcal{F}_m$, where $\mathcal{F}_i = \mathcal{E}_i$, and $\mathcal{F}_j = \mathcal{U}_j$ for all $j \neq i$, then $W_{\mathcal{F}}^{\mathfrak{F}} \subseteq W_{\mathcal{E}_i^*}^{\mathfrak{B}}$. That is, the main-effect contrasts of the unit factor $\mathcal{E}_i$ fall in the stratum $W_{\mathcal{E}_i^*}^{\mathfrak{B}}$.*

Theorem 13.2 will be proved in Section 13.8 as a corollary of a more general result.

It follows from Theorem 13.2 that each stratum of a simple block structure $\mathfrak{B}$ is the direct sum of some strata of $\mathfrak{F}$. Specifically, for any $\mathcal{G} \in \mathfrak{B}$,

$$W_{\mathcal{G}}^{\mathfrak{B}} = \bigoplus_{\mathcal{F} \in \mathfrak{F}: \mathcal{F}^* = \mathcal{G}} W_{\mathcal{F}}^{\mathfrak{F}}. \tag{13.6}$$

In Section 7.6, we have already used the fact that under the block structure $\mathfrak{B}_B/\mathfrak{B}_P$ for a block design, with $\mathfrak{B}_B = \{\mathcal{U}_B, \mathcal{E}_B\}$ and $\mathfrak{B}_P = \{\mathcal{U}_P, \mathcal{E}_P\}$, where $\mathcal{E}_B$ defines the blocks and $\mathcal{E}_P$ defines the plots (units), respectively, the interblock stratum consists of all the main-effect contrasts of $\mathcal{E}_B$, and the intrablock stratum is generated by the main-effect contrasts of $\mathcal{E}_P$ and interaction contrasts of $\mathcal{E}_B$ and $\mathcal{E}_P$. This is a simple consequence of (13.6).

We pointed out in Section 12.2 that each factor $\mathcal{F}_1 \times \cdots \times \mathcal{F}_m$ in $\mathfrak{B}$ (and $\mathfrak{F}$) and the associated stratum can be labeled by the $\mathcal{E}_i$'s with $\mathcal{F}_i = \mathcal{E}_i$. Then the factors in $\mathfrak{F}$ and the associated strata are labeled by the set of all the 2^m subsets of $\{\mathcal{E}_1, \ldots, \mathcal{E}_m\}$. The result in (13.6) provides a simple way of determining all the strata of a simple block structure and the unit contrasts in each stratum.

Rule for determining the strata of a simple block structure

For any $\mathcal{F} \in \mathfrak{F}$, determine the label of $\mathcal{F}^*$, and hence $\mathcal{F}^*$ itself, by adding to the label of $\mathcal{F}$ all the $\mathcal{E}_j$'s that are ancestors of at least one $\mathcal{E}_i$ in the label of $\mathcal{F}$. Each distinct $\mathcal{F}^*$ corresponds to one stratum of $\mathfrak{B}$. All the $\mathcal{F}$'s that produce the same $\mathcal{F}^*$ appear on the right-hand side of (13.6) for $\mathcal{G} = \mathcal{F}^*$.

Example 13.3. (Example 12.5 revisited) Consider the simple block structure $\mathfrak{B} = \mathfrak{B}_B/(\mathfrak{B}_R \times \mathfrak{B}_C)$ with $\mathfrak{B}_B = \{\mathcal{U}_B, \mathcal{E}_B\}$, $\mathfrak{B}_R = \{\mathcal{U}_R, \mathcal{E}_R\}$, and $\mathfrak{B}_C = \{\mathcal{U}_C, \mathcal{E}_C\}$, where $\mathcal{E}_B$, $\mathcal{E}_R$, and $\mathcal{E}_C$ correspond to blocks, rows and columns, respectively. We have that $\mathcal{E}_B$ is an ancestor of both $\mathcal{E}_R$ and $\mathcal{E}_C$. By the rule given above, for the $\mathcal{F}$ in $\mathfrak{F}$ that is labeled by $\{\mathcal{E}_R, \mathcal{E}_C\}$, we need to add $\mathcal{E}_B$ to $\{\mathcal{E}_R, \mathcal{E}_C\}$ to obtain the label of $\mathcal{F}^*$. Applying this to all the eight factors in $\mathfrak{F}$, we obtain the following correspondence between the factors in $\mathfrak{F}$ and those in $\mathfrak{B}$ as given in (13.6).

Label of $\mathcal{F}$	Label of $\mathcal{F}^*$
$\emptyset$	$\emptyset$
$\{\mathcal{E}_B\}$	$\{\mathcal{E}_B\}$
$\{\mathcal{E}_R\}, \{\mathcal{E}_B, \mathcal{E}_R\}$	$\{\mathcal{E}_B, \mathcal{E}_R\}$
$\{\mathcal{E}_C\}, \{\mathcal{E}_B, \mathcal{E}_C\}$	$\{\mathcal{E}_B, \mathcal{E}_C\}$
$\{\mathcal{E}_R, \mathcal{E}_C\}, \{\mathcal{E}_B, \mathcal{E}_R, \mathcal{E}_C\}$	$\{\mathcal{E}_B, \mathcal{E}_R, \mathcal{E}_C\}$

This shows that besides $\mathcal{U}$, the block structure $\mathfrak{B}_B/(\mathfrak{B}_R \times \mathfrak{B}_C)$ has four factors labeled by $\{\mathcal{E}_B\}$, $\{\mathcal{E}_B, \mathcal{E}_R\}$, $\{\mathcal{E}_B, \mathcal{E}_C\}$, and $\{\mathcal{E}_B, \mathcal{E}_R, \mathcal{E}_C\}$. These are the factors denoted by B, $\mathcal{R}'$, C', and $\mathcal{E}$, respectively, in Section 4.2. Denote the associated strata by W_B, $W_{\mathcal{R}'}$, $W_{C'}$, and $W_{\mathcal{E}}$, respectively. Then we have

$$W_B = W_{\mathcal{E}_B}^{\mathfrak{F}},$$

$$W_{\mathcal{R}'} = W_{\mathcal{E}_R}^{\mathfrak{F}} \oplus W_{\mathcal{E}_B \mathcal{E}_R}^{\mathfrak{F}},$$

$$W_{C'} = W_{\mathcal{E}_C}^{\mathfrak{F}} \oplus W_{\mathcal{E}_B \mathcal{E}_C}^{\mathfrak{F}},$$

$$W_{\mathcal{E}} = W_{\mathcal{E}_R \mathcal{E}_C}^{\mathfrak{F}} \oplus W_{\mathcal{E}_B \mathcal{E}_R \mathcal{E}_C}^{\mathfrak{F}}.$$

In other words, the block stratum $W_{\mathcal{E}_B}$ consists of all the main-effect contrasts of $\mathcal{E}_B$, the row stratum $W_{\mathcal{R}'}$ is generated by the main-effect contrasts of $\mathcal{E}_R$ and interaction contrasts of $\mathcal{E}_B$ and $\mathcal{E}_R$, the column stratum $W_{C'}$ is generated by the main-effect contrasts of $\mathcal{E}_C$ and interaction contrasts of $\mathcal{E}_B$ and $\mathcal{E}_C$, and the bottom stratum $W_{\mathcal{E}}$ is generated by all the two-factor interaction contrasts of $\mathcal{E}_R$ and $\mathcal{E}_C$ and three-factor interaction contrasts of $\mathcal{E}_B$, $\mathcal{E}_R$, and $\mathcal{E}_C$. Each of the seven factorial components of $\mathcal{E}_B$, $\mathcal{E}_R$, and $\mathcal{E}_C$ is a subspace of a certain stratum of $\mathfrak{B}$.

Example 13.4. Consider the block structure $\mathfrak{B} = \mathfrak{B}_B/\mathfrak{B}_P/\mathfrak{B}_S$ with $\mathfrak{B}_B = \{\mathcal{U}_B, \mathcal{E}_B\}$, $\mathfrak{B}_P = \{\mathcal{U}_P, \mathcal{E}_P\}$, and $\mathfrak{B}_S = \{\mathcal{U}_S, \mathcal{E}_S\}$, where $\mathcal{E}_B$, $\mathcal{E}_P$, and $\mathcal{E}_S$ correspond to blocks, whole-plots, and subplots, respectively. We have $\mathcal{E}_S < \mathcal{E}_P < \mathcal{E}_B$. Thus the factors in $\mathfrak{B}$ besides $\mathcal{U}$ are labeled by $\{\mathcal{E}_B\}$, $\{\mathcal{E}_B, \mathcal{E}_P\}$, and $\{\mathcal{E}_B, \mathcal{E}_P, \mathcal{E}_S\}$, respectively. These are what we denoted by $\mathcal{B}$, $\mathcal{P}$, and $\mathcal{E}$, respectively, in Section 13.2. The three strata other than the mean stratum are

$$W_{\mathcal{B}} = W_{\mathcal{E}_B}^{\mathfrak{F}},$$

$$W_{\mathcal{P}} = W_{\mathcal{E}_P}^{\mathfrak{F}} \oplus W_{\mathcal{E}_B\mathcal{E}_P}^{\mathfrak{F}},$$

$$W_{\mathcal{E}} = W_{\mathcal{E}_B\mathcal{E}_P\mathcal{E}_S}^{\mathfrak{F}} \oplus W_{\mathcal{E}_P\mathcal{E}_S}^{\mathfrak{F}} \oplus W_{\mathcal{E}_B\mathcal{E}_S}^{\mathfrak{F}} \oplus W_{\mathcal{E}_S}^{\mathfrak{F}}.$$

That is, the contrasts in the block stratum $W_{\mathcal{B}}$ are the same as the main effect contrasts of $\mathcal{E}_B$, the whole-plot stratum $W_{\mathcal{P}}$ is generated by the main effect contrasts of $\mathcal{E}_P$ and interaction contrasts of $\mathcal{E}_B$ and $\mathcal{E}_P$, and the subplot stratum $W_{\mathcal{E}}$ is generated by the main effect contrasts of $\mathcal{E}_S$, interaction contrasts of $\mathcal{E}_P$ and $\mathcal{E}_S$, interaction contrasts of $\mathcal{E}_B$ and $\mathcal{E}_S$, and those of $\mathcal{E}_B$, $\mathcal{E}_P$, and $\mathcal{E}_S$.

Example 13.5. (Example 1.1 revisited) The block structure is given by $\mathfrak{B} = \mathfrak{B}_B/[(\mathfrak{B}_P/\mathfrak{B}_S/\mathfrak{B}_{SS}) \times \mathfrak{B}_{ST}]$, with $\mathfrak{B}_X = \{\mathcal{U}_X, \mathcal{E}_X\}$, where $\mathcal{E}_B$, $\mathcal{E}_P$, $\mathcal{E}_S$, $\mathcal{E}_{SS}$, and $\mathcal{E}_{ST}$ correspond to blocks, plots, subplots, sub-subplots, and strips, respectively. We have $\mathcal{E}_{SS} < \mathcal{E}_S < \mathcal{E}_P < \mathcal{E}_B$, and $\mathcal{E}_{ST} < \mathcal{E}_B$. By using the rule given earlier, we can determine that there are eight strata other than the mean stratum:

$$W_{\mathcal{B}} = W_{\mathcal{E}_B}^{\mathfrak{F}},$$

$$W_{\mathcal{P}} = W_{\mathcal{E}_B\mathcal{E}_P}^{\mathfrak{F}} \oplus W_{\mathcal{E}_P}^{\mathfrak{F}},$$

$$W_{\mathcal{S}} = W_{\mathcal{E}_B\mathcal{E}_P\mathcal{E}_S}^{\mathfrak{F}} \oplus W_{\mathcal{E}_P\mathcal{E}_S}^{\mathfrak{F}} \oplus W_{\mathcal{E}_B\mathcal{E}_S}^{\mathfrak{F}} \oplus W_{\mathcal{E}_S}^{\mathfrak{F}},$$

$$W_{\mathcal{SS}} = W_{\mathcal{E}_B\mathcal{E}_P\mathcal{E}_S\mathcal{E}_{SS}}^{\mathfrak{F}} \oplus W_{\mathcal{E}_B\mathcal{E}_P\mathcal{E}_{SS}}^{\mathfrak{F}} \oplus W_{\mathcal{E}_B\mathcal{E}_S\mathcal{E}_{SS}}^{\mathfrak{F}} \oplus W_{\mathcal{E}_P\mathcal{E}_S\mathcal{E}_{SS}}^{\mathfrak{F}}$$
$$\oplus W_{\mathcal{E}_B\mathcal{E}_{SS}}^{\mathfrak{F}} \oplus W_{\mathcal{E}_P\mathcal{E}_{SS}}^{\mathfrak{F}} \oplus W_{\mathcal{E}_S\mathcal{E}_{SS}}^{\mathfrak{F}} \oplus W_{\mathcal{E}_{SS}}^{\mathfrak{F}},$$

$$W_{\mathcal{ST}} = W_{\mathcal{E}_B\mathcal{E}_{ST}}^{\mathfrak{F}} \oplus W_{\mathcal{E}_{ST}}^{\mathfrak{F}},$$

$$W_{\mathcal{P} \wedge \mathcal{ST}} = W_{\mathcal{E}_B\mathcal{E}_P\mathcal{E}_{ST}}^{\mathfrak{F}} \oplus W_{\mathcal{E}_P\mathcal{E}_{ST}}^{\mathfrak{F}},$$

$$W_{\mathcal{S} \wedge \mathcal{ST}} = W_{\mathcal{E}_B\mathcal{E}_P\mathcal{E}_S\mathcal{E}_{ST}}^{\mathfrak{F}} \oplus W_{\mathcal{E}_P\mathcal{E}_S\mathcal{E}_{ST}}^{\mathfrak{F}} \oplus W_{\mathcal{E}_B\mathcal{E}_S\mathcal{E}_{ST}}^{\mathfrak{F}} \oplus W_{\mathcal{E}_S\mathcal{E}_{ST}}^{\mathfrak{F}},$$

$$W_{\mathcal{E}} = W_{\bar{\mathcal{E}}_B \mathcal{E}_P \mathcal{E}_S \mathcal{E}_{SS} \mathcal{E}_{ST}}^{\mathfrak{F}} \oplus W_{\bar{\mathcal{E}}_B \mathcal{E}_P \mathcal{E}_{SS} \mathcal{E}_{ST}}^{\mathfrak{F}} \oplus W_{\bar{\mathcal{E}}_B \mathcal{E}_S \mathcal{E}_{SS} \mathcal{E}_{ST}}^{\mathfrak{F}} \oplus W_{\bar{\mathcal{E}}_P \mathcal{E}_S \mathcal{E}_{SS} \mathcal{E}_{ST}}^{\mathfrak{F}}$$

$$\oplus W_{\bar{\mathcal{E}}_B \mathcal{E}_{SS} \mathcal{E}_{ST}}^{\mathfrak{F}} \oplus W_{\bar{\mathcal{E}}_P \mathcal{E}_{SS} \mathcal{E}_{ST}}^{\mathfrak{F}} \oplus W_{\bar{\mathcal{E}}_S \mathcal{E}_{SS} \mathcal{E}_{ST}}^{\mathfrak{F}} \oplus W_{\bar{\mathcal{E}}_{SS} \mathcal{E}_{ST}}^{\mathfrak{F}}.$$

We point out that all the results in this section are also applicable to poset block structures.

13.5   Construction of complete factorial designs with simple block structures

As in the previous section, consider a simple block structure $\mathfrak{B}$ consisting of factors of the form $\mathcal{F}_1 \times \cdots \times \mathcal{F}_m$ where $\mathcal{F}_i = \mathcal{U}_i$ or $\mathcal{E}_i$, and $\mathcal{E}_i$ has n_i levels. Then $N = n_1 \cdots n_m$. Suppose there are n treatment factors each with s levels, where s is a prime number or power of a prime number, and each n_i is a power of s, say $n_i = s^{l_i}$ with $\sum_{i=1}^{m} l_i = n$. Then $N = s^n$, and we have a single-replicate complete factorial experiment. For each i, $1 \leq i \leq m$, recall that $\mathcal{E}_i^*$ has $\tilde{n}_i$ levels, where

$$\tilde{n}_i = \prod_{j: \mathcal{E}_i^* \preceq \mathcal{E}_j^*} n_j;$$

see (12.4) and (12.5). Let $\tilde{l}_i = \sum_{j: \mathcal{E}_i^* \preceq \mathcal{E}_j^*} l_j$. Then $\tilde{n}_i = s^{\tilde{l}_i}$. To construct a complete factorial design with block structure $\mathfrak{B}$, for each i, one needs to choose a set of $\tilde{l}_i$ linearly independent vectors from EG(n, s) to partition the s^n treatment combinations into $\tilde{n}_i$ disjoint classes of equal size, with all the treatment combinations in the same class assigned to the same level of $\mathcal{E}_i^*$. To maintain the nesting structure, if $\mathcal{E}_i^*$ is nested in $\mathcal{E}_j^*$ (that is, if $\mathcal{E}_j$ is an ancestor of $\mathcal{E}_i$), then the $\tilde{l}_j$ linearly independent vectors chosen for $\mathcal{E}_j^*$ must be a subset of those for $\mathcal{E}_i^*$. This can be implemented as follows. For each $i = 1, \ldots, m$, choose a set Δ_i of l_i linearly independent vectors from EG(n, s) such that the vectors in $\cup_{i=1}^m \Delta_i$ are linearly independent, so $\cup_{i=1}^m \Delta_i$ is a basis of EG(n, s). Then for each i, use the $\tilde{l}_i$ linearly independent vectors in $\cup_{j: \mathcal{E}_i^* \preceq \mathcal{E}_j^*} \Delta_j$ to partition the treatment combinations into $\tilde{n}_i = s^{\tilde{l}_i}$ disjoint sets of equal size: two treatment combinations $\mathbf{x}_1$ and $\mathbf{x}_2$ are in the same set if and only if $\mathbf{a}^T \mathbf{x}_1 = \mathbf{a}^T \mathbf{x}_2$ for all $\mathbf{a} \in \cup_{j: \mathcal{E}_i^* \preceq \mathcal{E}_j^*} \Delta_j$.

Note that if one of $\mathcal{E}_1^*, \ldots, \mathcal{E}_m^*$, say $\mathcal{E}_m^*$, is nested in all the others, then in the construction described in the previous paragraph, there is no need to choose Δ_m. In this case, $\mathcal{E}_m^*$ is the equality factor $\mathcal{E}$, and the vectors in $\cup_{i=1}^{m-1} \Delta_i$ would have already partitioned the s^n treatment combinations into s^n / n_m sets of size n_m as desired. For example, for the block structure s^q / s^{n-q} considered in Section 7.2, we only need one set of linearly independent vectors $\mathbf{a}_1, \ldots, \mathbf{a}_q$ to partition the s^n treatment combinations into s^q blocks of size s^{n-q}, whereas for the block structure $s^p \times s^{n-p}$ discussed in Section 7.9, two sets of linearly independent vectors are needed for the row and column partitions. Likewise we need two sets of vectors for $s^g / s^{q-g} / s^{n-q}$ (Section 13.2) and three sets for $s^g / (s^p \times s^q)$ (Section 13.3).

Under this construction, the stratum in which each treatment factorial effect is estimated can be determined as follows. The treatment factorial effects defined by the vectors in Δ_i and their nonzero linear combinations are contrasts in $W_{\mathcal{E}_i}^{\mathfrak{F}}$, the "main-effect" contrasts of $\mathcal{E}_i$. By Theorem 13.2, these treatment factorial effects are estimated in the stratum of $\mathfrak{B}$ associated with $\mathcal{E}_i^*$. For any nonzero vector $\mathbf{a} \in EG(n, s)$, if $\mathbf{a} = \sum_{j=1}^{h} \mathbf{a}_{i_j}$, where $\mathbf{a}_{i_j}$ is a nonzero linear combination of the l_{i_j} vectors in Δ_{i_j}, $1 \leq i_1 < \cdots < i_h \leq m$, then the treatment factorial effects defined by $\mathbf{a}$ are contrasts in $W_{\mathcal{E}_{i_1} \cdots \mathcal{E}_{i_h}}^{\mathfrak{F}}$, which are "interaction" contrasts of $\mathcal{E}_{i_1}, \ldots,$ and $\mathcal{E}_{i_h}$. These treatment factorial effects are estimated in $W_{\mathcal{G}}^{\mathfrak{B}}$, where $\mathcal{G}$ is determined by the rule given in Section 13.4.

When one of $\mathcal{E}_1^*, \ldots, \mathcal{E}_m^*$, say $\mathcal{E}_m^*$, is nested in all the others, the rules given in the previous paragraph can be used to determine the strata where the treatment factorial effects defined by nonzero linear combinations of the vectors in $\cup_{i=1}^{m-1} \Delta_i$ are estimated. All the other factorial effects are estimated in the bottom stratum.

Sometimes there are requirements that certain treatment factorial effects must be estimated in specific strata; for example, in split-plot experiments main effects of whole-plot and subplot treatment factors must be estimated in the whole-plot and subplot strata, respectively. Such requirements impose some constraints on choosing the vectors (words) in the Δ_i's. Either these conditions are spelled out and enforced in the construction/search of eligible designs, or one needs to check that, after the vectors in the Δ_i's have been chosen, all the requirements are satisfied. For this purpose, the rule given above for determining the stratum in which each treatment factorial effect is estimated is useful. Franklin and Bailey's (1977) algorithm, discussed in Section 9.7, in particular the version for selecting blocking schemes, can be extended to search for appropriate vectors in each Δ_i.

Example 13.6. (Example 13.2 revisited) We have $s = 2$, $n = 5$, $m = 3$, $l_1 = 1$, $l_2 = l_3 = 2$. For the design constructed in Example 13.2, $\Delta_1 = \{(1, 0, 1, 0, 0)^T\}$, $\Delta_2 = \{(1, 0, 0, 0, 0)^T, (0, 1, 0, 0, 0)^T\}$, and $\Delta_3 = \{(0, 0, 0, 1, 0)^T, (0, 0, 0, 0, 1)^T\}$. The single vector $(1, 0, 1, 0, 0)^T$ in Δ_1 is used to partition the 32 treatment combinations into two blocks according to the values of $x_1 + x_3$ (mod 2). This causes the two-factor interaction AC to be confounded with blocks. Since the rows are nested in the blocks, the two vectors in Δ_2 together with $(1, 0, 1, 0, 0)^T$ are used to partition the treatment combinations into eight rows of size four, with four rows in each block. Then the main effects and interaction of A and B are confounded with rows. Since $C = (AC)(A)$, $BC = (AC)(AB)$, and $ABC = (AC)(B)$, these effects are defined by vectors in $W_{\mathcal{E}_B \mathcal{E}_R}^{\mathfrak{F}}$. It follows from $W_{\mathcal{E}_B \mathcal{E}_R}^{\mathfrak{F}} \subseteq W_{\mathcal{E}_B \mathcal{E}_R}^{\mathfrak{B}} = W_{\mathcal{R}'}$ that the treatment factorial effects C, BC, and ABC are also estimated in the row stratum. Likewise, S, T, and ST are estimated in the column stratum. Since $ACS = (AC)(S)$, $ACT = (AC)(T)$, and $ACST = (AC)(ST)$, ACS, ACT, and $ACST$ are defined by vectors in $W_{\mathcal{E}_B \mathcal{E}_C}^{\mathfrak{F}} \subseteq W_{\mathcal{E}_B \mathcal{E}_C}^{\mathfrak{B}} = W_{C'}$; therefore they are estimated in the column stratum. All the other treatment factorial effects are defined by vectors in $W_{\mathcal{E}_R \mathcal{E}_C}^{\mathfrak{F}}$ or $W_{\mathcal{E}_B \mathcal{E}_R \mathcal{E}_C}^{\mathfrak{F}}$, both of which are subspaces of $W_{\mathcal{E}_B \mathcal{E}_R \mathcal{E}_C}^{\mathfrak{B}} = W_{\mathcal{E}}$. Hence all these effects are estimated in the bottom stratum.

Example 13.7. (Example 13.1 revisited) In this example we use A, B, ABC, and BST to divide the 32 treatments into 16 groups of size 2, with those in the same group allocated to the same whole-plot. We also use ABC and BST to divide the 16 whole-plots into four blocks of size 4. In the notations of this section, we have $\Delta_1 = \{(1,1,1,0,0)^T, (0,1,0,1,1)^T\}$, and $\Delta_2 = \{(1,0,0,0,0)^T,(0,1,0,0,0)^T\}$.

13.6 Design keys

Consider the construction of single-replicate complete factorial designs with simple block structures as in the previous section. We follow the same notations and call $\mathcal{E}_1,\ldots,\mathcal{E}_m$ the unit factors. It is convenient to represent each level of $\mathcal{E}_i$ by a combination of s-level "pseudo" factors $\mathcal{E}_i^1,\ldots,\mathcal{E}_i^{l_i}$ if $l_i > 1$; then the experimental units can be thought of as all the s^n combinations of the n pseudo factors $\{\mathcal{E}_i^j\}_{1 \le i \le m, 1 \le j \le l_i}$.

As in Section 7.6, we can describe the relation between the experimental units and the pseudo unit factors by an $n \times s^n$ matrix $\mathbf{Y}$ such that for each j, $1 \le j \le s^n$, the jth column of $\mathbf{Y}$ is the level combination of the pseudo unit factors corresponding to the jth unit. Let

$$\mathbf{X} = \mathbf{KY},$$

where the design key matrix $\mathbf{K}$ is an $n \times n$ matrix with entries from GF(s). Then a design is obtained by assigning the treatment combination represented by each column of $\mathbf{X}$ to the corresponding unit. The same argument as in Section 7.6 shows that under the constructed design, the main effect of the ith treatment factor coincides with the factorial effect of pseudo unit factors defined by the ith row of $\mathbf{K}$. Again we call the latter the unit alias of the former. We achieve the objective of estimating the main effect of a treatment factor in a designated stratum by choosing its unit alias to be an effect of the pseudo unit factors in that stratum. As discussed in Section 13.4, all the contrasts in each stratum of $\mathfrak{B}$ can be regarded as "factorial" effects of $\mathcal{E}_1,\ldots,\mathcal{E}_m$. Each of these effects is in turn a factorial effect of the n pseudo unit factors. The rule given in Section 13.4 can be used to identify the factorial effects of $\mathcal{E}_1,\ldots,\mathcal{E}_m$, and therefore the factorial effects of the n pseudo unit factors, that fall in each stratum of $\mathfrak{B}$.

In order to generate a complete factorial design, $\mathbf{K}$ must be nonsingular. The unit aliases of the treatment main effects can be used to determine unit aliases of the treatment interactions and therefore the strata in which they are estimated. Furthermore, the columns of $\mathbf{K}$ give a set of independent generators of the treatment combinations that can be used to generate the design.

Example 13.8. (Examples 13.2 and 13.3 revisited) For simplicity, we write $\mathcal{E}_B$, $\mathcal{E}_R$, and $\mathcal{E}_C$ as $\mathcal{B}$, $\mathcal{R}$, and $\mathcal{C}$, respectively, and represent the levels of $\mathcal{R}$ (respectively, $\mathcal{C}$) by the level combinations of pseudo factors $\mathcal{R}_1$ and $\mathcal{R}_2$ (respectively, $\mathcal{C}_1$ and $\mathcal{C}_2$). By the observations in Example 13.3, the four strata consist of effects of the following forms:

Block stratum:	$\mathcal{B}$;
Row stratum:	$\mathcal{R}_1, \mathcal{R}_2, \mathcal{R}_1\mathcal{R}_2, \mathcal{B}\mathcal{R}_1, \mathcal{B}\mathcal{R}_2, \mathcal{B}\mathcal{R}_1\mathcal{R}_2$;
Column stratum:	$\mathcal{C}_1, \mathcal{C}_2, \mathcal{C}_1\mathcal{C}_2, \mathcal{B}\mathcal{C}_1, \mathcal{B}\mathcal{C}_2, \mathcal{B}\mathcal{C}_1\mathcal{C}_2$;

Unit stratum: $\mathcal{R}_1\mathcal{C}_1, \mathcal{R}_1\mathcal{C}_2, \mathcal{R}_1\mathcal{C}_1\mathcal{C}_2, \mathcal{R}_2\mathcal{C}_1, \mathcal{R}_2\mathcal{C}_2, \mathcal{R}_2\mathcal{C}_1\mathcal{C}_2, \mathcal{R}_1\mathcal{R}_2\mathcal{C}_1,$
$\mathcal{R}_1\mathcal{R}_2\mathcal{C}_2, \mathcal{R}_1\mathcal{R}_2\mathcal{C}_1\mathcal{C}_2, \mathcal{B}\mathcal{R}_1\mathcal{C}_1, \mathcal{B}\mathcal{R}_1\mathcal{C}_2, \mathcal{B}\mathcal{R}_1\mathcal{C}_1\mathcal{C}_2, \mathcal{B}\mathcal{R}_2\mathcal{C}_1,$
$\mathcal{B}\mathcal{R}_2\mathcal{C}_2, \mathcal{B}\mathcal{R}_2\mathcal{C}_1\mathcal{C}_2, \mathcal{B}\mathcal{R}_1\mathcal{R}_2\mathcal{C}_1, \mathcal{B}\mathcal{R}_1\mathcal{R}_2\mathcal{C}_2, \mathcal{B}\mathcal{R}_1\mathcal{R}_2\mathcal{C}_1\mathcal{C}_2.$

We need to choose unit aliases of the main effects of treatment factors A, B, and C (respectively, S and T) from the row (respectively, column) stratum. A factorial effect of the unit factors is in the block (respectively, row, column, or unit) stratum if it involves only the $\mathcal{B}$'s (respectively, at least one $\mathcal{R}$ but no $\mathcal{C}$'s, at least one $\mathcal{C}$ but no $\mathcal{R}$'s, or at least one $\mathcal{R}$ and at least one $\mathcal{C}$). Suppose the unit aliases of A, B, C, S, and T are $\mathcal{R}_1$, $\mathcal{R}_2$, $\mathcal{B}\mathcal{R}_1$, $\mathcal{C}_1$, and $\mathcal{C}_2$, respectively. Then the design key matrix can be written as

$$
\mathbf{K} = \begin{array}{ccccc}
\mathcal{R}_1 & \mathcal{R}_2 & \mathcal{C}_1 & \mathcal{C}_2 & \mathcal{B} \\
\begin{bmatrix}
1 & 0 & 0 & 0 & 0 \\
0 & 1 & 0 & 0 & 0 \\
0 & 0 & 1 & 0 & 0 \\
0 & 0 & 0 & 1 & 0 \\
1 & 0 & 0 & 0 & 1
\end{bmatrix} & & & & \begin{array}{c} A \\ B \\ S \\ T \\ C \end{array}
\end{array}
\tag{13.7}
$$

The two-factor interaction AC, with unit alias $\mathcal{R}_1(\mathcal{B}\mathcal{R}_1) = \mathcal{B}$, is confounded with blocks. Five independent generators ac, b, s, t, and c, obtained from the columns of $\mathbf{K}$, are used to generate the 32 treatment combinations in the Yates order. The first two generators produce the four treatment combinations in the first column of the first block. The next two generators can be used to complete the first block, and the last generator produces the second block. This gives the same design as constructed in Example 13.2 subject to row and column permutations within the same block.

When the numbers of levels of the treatment factors are different powers of the same s, one can also represent the levels of each treatment factor by the level combinations of some s-level pseudo treatment factors.

The following extension of the design key templates (7.9) and (7.11) for block and row-column designs derived in Remark 9.6 and Exercise 9.5, respectively, can be proved by using similar arguments.

Theorem 13.3. *(Cheng and Tsai, 2013) Suppose $\mathfrak{B}_1$ and $\mathfrak{B}_2$ are simple block structures on s^{m_1} and s^{m_2} units, respectively, where s is a prime number or power of a prime number, and $n = m_1 + m_2$. Then, subject to factor relabeling, a complete s^n factorial design with block structure $\mathfrak{B}_1/\mathfrak{B}_2$ or $\mathfrak{B}_1 \times \mathfrak{B}_2$ can be constructed by using a design key of the form*

$$
\mathbf{K} = \begin{bmatrix} \mathbf{K}_2 & \mathbf{A}_2 \\ \mathbf{A}_1 & \mathbf{K}_1 \end{bmatrix},
$$

where $\mathbf{K}_1$ and $\mathbf{K}_2$ are design keys for complete s^{m_1} and s^{m_2} factorial designs with block structures $\mathfrak{B}_1$ and $\mathfrak{B}_2$, respectively, and $\mathbf{A}_1$ and $\mathbf{A}_2$ are some matrices, with the first m_2 columns of $\mathbf{K}$ corresponding to the unit factors in $\mathfrak{B}_2$ and the last m_1

columns corresponding to the unit factors in $\mathfrak{B}_1$. In the case where the s^{m_1} units in $\mathfrak{B}_1$ are unstructured ($\mathfrak{B}_1$ consists of the universal and equality factors on s^{m_1} units), a complete s^n factorial design with the block structure $\mathfrak{B}_1/\mathfrak{B}_2$ can be constructed by using a design key of the form

$$\mathbf{K} = \begin{bmatrix} \mathbf{K}_2 & \mathbf{0} \\ \mathbf{A}_1 & \mathbf{I}_{m_1} \end{bmatrix}. \tag{13.8}$$

The proof of Theorem 13.3 is left as an exercise. Applications to blocked split-plot and blocked strip-plot designs are given in the next section.

13.7 Design key templates for blocked split-plot and strip-plot designs

We first consider the construction of a complete s^n blocked split-plot design with the block structure $s^g/s^{q-g}/s^{n-q}$, where s is a prime number or power of a prime number, and n_1 of the n treatment factors are whole-plot factors. As in Section 13.2, we require that within-block replications of the same whole-plot treatment factor level combination be kept at the minimum. Each experimental unit can be represented by a combination of s-level pseudo unit factors $\mathcal{S}_1, \ldots, \mathcal{S}_{n-q}, \mathcal{P}_1, \ldots, \mathcal{P}_{q-g}, \mathcal{B}_1, \ldots, \mathcal{B}_g$. By Example 13.4, a factorial effect of $\mathcal{S}_1, \ldots, \mathcal{S}_{n-q}, \mathcal{P}_1, \ldots, \mathcal{P}_{q-g}, \mathcal{B}_1, \ldots, \mathcal{B}_g$ is in the block stratum if and only if it involves the $\mathcal{B}$'s only, is in the whole-plot stratum if and only if it involves at least one $\mathcal{P}$ but does not involve the $\mathcal{S}$'s, and is in the subplot stratum if and only if it involves at least one $\mathcal{S}$.

Theorem 13.4. *(Cheng and Tsai, 2013) Suppose it is required that no treatment main effect be confounded with blocks. In addition, the requirement of minimizing within-block replications of the same whole-plot treatment factor level combination as described in Section 13.2 is imposed. Then, subject to factor relabeling and the constraints described below, a blocked complete factorial split-plot design can be constructed by using a design key matrix of the form*

$$\mathbf{K} = \begin{bmatrix} \mathbf{I}_{n-q} & \mathbf{0} & \mathbf{0} \\ \mathbf{C}_1 & \mathbf{I}_{q-g} & \mathbf{0} \\ \mathbf{C}_2 & \mathbf{C}_3 & \mathbf{I}_g \end{bmatrix}, \tag{13.9}$$

where the first $n-q$ columns of $\mathbf{K}$ correspond to $\mathcal{S}_1, \ldots, \mathcal{S}_{n-q}$, the next $q-g$ columns correspond to $\mathcal{P}_1, \ldots, \mathcal{P}_{q-g}$, the last g columns correspond to $\mathcal{B}_1, \ldots, \mathcal{B}_g$, the first $n-q$ rows of $\mathbf{K}$ correspond to subplot treatment factors, the next n_1 rows correspond to whole-plot treatment factors, and all the remaining rows correspond to subplot treatment factors if $n_1 < q$. Furthermore, all the first $n_1 - (q-g)$ rows of $\mathbf{C}_3$ are nonzero if $n_1 > q-g$, the first n_1 rows of $[\mathbf{C}_1^T \ \mathbf{C}_2^T]^T$ are zero, and all its last $q-n_1$ rows are nonzero if $n_1 < q$.

Proof. By Theorem 13.3, a design key can be derived by applying (13.8) with $m_1 = g$. The matrix $\mathbf{K}_2$ in (13.8) is a design key for a complete s^{n-g} factorial in s^{q-g} whole-plots each containing s^{n-q} subplots and can be obtained from (7.13), with q replaced by $q - g$. Thus, without loss of generality, we have a design key of the form in (13.9). That the first $n - q$ rows of $\mathbf{K}$ must correspond to subplot treatment factors follows from the fact that all the rows of $\mathbf{I}_{n-q}$ are nonzero.

The $q - g$ columns of $\mathbf{K}$ corresponding to $\mathcal{P}_1, \ldots, \mathcal{P}_{q-g}$ are used to generate the whole-plot treatment factor level combinations in the first block. When $n_1 \geq q - g$, it is required that these treatment combinations be distinct. They can be generated by using a certain set of $q - g$ whole-plot treatment factors as basic factors. Without loss of generality, we may associate these factors with the rows of $\mathbf{K}$ that correspond to those of $\mathbf{I}_{q-g}$. Then all the entries of $\mathbf{C}_1$ must be zero. In this case, if $n_1 > q - g$, then $n_1 - (q - g)$ of the last g rows of $\mathbf{K}$ must also correspond to whole-plot treatment factors. Let these be the first $n_1 - (q - g)$ of the last g rows of $\mathbf{K}$. Then the corresponding rows of $\mathbf{C}_2$ are zero. Furthermore, the corresponding rows of $\mathbf{C}_3$ cannot be zero; otherwise main effects of the corresponding whole-plot treatment factors would be confounded with blocks. The last $q - n_1$ rows of $\mathbf{C}_2$ are also nonzero if $q > n_1$.

On the other hand, if $n_1 < q - g$, then all the whole-plot treatment factor level combinations must be replicated the same number of times among those generated by the columns of $\mathbf{K}$ that correspond to $\mathcal{P}_1, \ldots, \mathcal{P}_{q-g}$. These s^{q-g} treatment combinations can be generated by using the n_1 whole-plot treatment factors and $q - g - n_1$ subplot treatment factors as basic factors. Without loss of generality, let the first n_1 of the last q rows of $\mathbf{K}$ correspond to whole-plot treatment factors, and let the remaining $q - n_1$ rows correspond to subplot treatment factors. Then the first n_1 rows of $\mathbf{C}_1$ must be zero, and its remaining rows as well as all the rows of $\mathbf{C}_2$ must be nonzero. □

For $n_1 \geq q - g$, $\mathbf{K}$ is of the form

$$\mathbf{K} = \begin{bmatrix} \mathbf{I}_{n-q} & \mathbf{0} & \mathbf{0} \\ \mathbf{0} & \mathbf{I}_{q-g} & \mathbf{0} \\ \mathbf{C}_2 & \mathbf{C}_3 & \mathbf{I}_g \end{bmatrix}. \tag{13.10}$$

Then

$$\mathbf{K}^{-1} = \begin{bmatrix} \mathbf{I}_{n-q} & \mathbf{0} & \mathbf{0} \\ \mathbf{0} & \mathbf{I}_{q-g} & \mathbf{0} \\ -\mathbf{C}_2 & -\mathbf{C}_3 & \mathbf{I}_g \end{bmatrix}.$$

In particular

$$\mathbf{K}^{-1} = \mathbf{K} \text{ for } s = 2.$$

Therefore, as in the discussion following (7.10) in Section 7.7, in this case, a set of g independent blocking words can be identified from the last g rows of $\mathbf{K}$. Each of these rows with the first $n - g$ components replaced by their additive inverses (unchanged when $s = 2$) defines a blocking word.

One single design key template can be used to generate all possible designs in all cases and, as shown below, it reduces to what McLeod and Brewster (2004) called pure whole-plot blocking, separation, and mixed blocking, respectively, in cases (i), (iia), and (iib) discussed before Remark 13.1 in Section 13.2.

(i) $n_1 = q$.

In this case (13.9) reduces to

$$\mathbf{K} = \begin{bmatrix} \mathbf{I}_{n-q} & \mathbf{0} & \mathbf{0} \\ \mathbf{0} & \mathbf{I}_{q-g} & \mathbf{0} \\ \mathbf{0} & \mathbf{C} & \mathbf{I}_g \end{bmatrix},$$

where all the g rows of $\mathbf{C}$ are nonzero. The construction requires choosing the g rows of $\mathbf{C}$. This is equivalent to choosing g independent whole-plot treatment interactions for dividing the whole-plots into s^g blocks.

(iia) $n_1 < q$ and $g = q - n_1$.

In this case, the matrix $\mathbf{C}_2$ in (13.10) has nonzero rows. It is easy to see that the g blocking words identified from the last g rows of $\mathbf{K}$ are also splitting words, since each of them involves at least one subplot treatment factor, and its unit alias does not involve any subplot unit factor. Therefore the construction is equivalent to using $g = q - n_1$ independent words that double as splitting and blocking words.

(iib) $n_1 < q$ and $g > q - n_1$.

In this case, among the g independent blocking words identified from $\mathbf{K}$, those identified from the last $q - n_1$ rows are also splitting words, and the other $g - (q - n_1)$ involve whole-plot treatment factors only. This is because the first $g - (q - n_1)$ rows of $\mathbf{C}_2$ in (13.10) are zero vectors. Thus the construction is equivalent to using a set of $q - n_1$ independent splitting words, which double as blocking words, and $g - (q - n_1)$ additional independent blocking words involving whole-plot-treatment factors only.

(iic) $n_1 < q$ and $g < q - n_1$

In this case $\mathbf{K}^{-1}$ cannot be obtained from the design key matrix $\mathbf{K}$ in (13.9) by changing the signs of $\mathbf{C}_1$, $\mathbf{C}_2$, and $\mathbf{C}_3$. Instead,

$$\mathbf{K}^{-1} = \begin{bmatrix} \mathbf{I}_{n-q} & \mathbf{0} & \mathbf{0} \\ -\mathbf{C}_1 & \mathbf{I}_{q-g} & \mathbf{0} \\ -\mathbf{C}_2 + \mathbf{C}_3\mathbf{C}_1 & -\mathbf{C}_3 & \mathbf{I}_g \end{bmatrix}. \tag{13.11}$$

A set of $q - n_1$ independent splitting words can be identified from the last $q - n_1$ rows of $\mathbf{K}^{-1}$, out of which the last g are independent blocking words. Therefore the construction is equivalent to using a set of $q - n_1$ independent splitting words, with g of which doubling as blocking words. We point out that the $q - n_1 - g$ independent splitting words that are not blocking words can be identified directly from the rows of $\mathbf{K}$ that correspond to the last $q - n_1 - g$ rows of $\mathbf{I}_{q-g}$ in (13.9) in the same way as described before.

Example 13.9. (Example 13.1 revisited) For $n = 5$, $n_1 = 3$, $q = 4$, $g = 2$, (13.9) reduces to

$$
\begin{array}{ccccc}
\mathcal{S} & \mathcal{P}_1 & \mathcal{P}_2 & \mathcal{B}_1 & \mathcal{B}_2 \\
\begin{bmatrix}
1 & 0 & 0 & 0 & 0 \\
0 & 1 & 0 & 0 & 0 \\
0 & 0 & 1 & 0 & 0 \\
0 & * & * & 1 & 0 \\
1 & * & * & 0 & 1
\end{bmatrix} &
\begin{array}{c}
S \\ A \\ B \\ C \\ T
\end{array}
\end{array}
$$

where S and T are subplot treatment factors and A, B, C are whole-plot treatment factors. At least one of the two blank entries in the fourth row must be nonzero. Fill in the blank cells as follows to complete a design key:

$$
\mathbf{K} =
\begin{array}{ccccc}
\mathcal{S} & \mathcal{P}_1 & \mathcal{P}_2 & \mathcal{B}_1 & \mathcal{B}_2 \\
\begin{bmatrix}
1 & 0 & 0 & 0 & 0 \\
0 & 1 & 0 & 0 & 0 \\
0 & 0 & 1 & 0 & 0 \\
0 & 1 & 1 & 1 & 0 \\
1 & 0 & 1 & 0 & 1
\end{bmatrix} &
\begin{array}{c}
S \\ A \\ B \\ C \\ T
\end{array}
\end{array}
$$

This example falls in case (iib). We can identify the splitting word BST from the last row of $\mathbf{K}$. Two independent blocking words can be identified from the last two rows: the splitting word BST and an additional blocking word ABC that involves whole-plot treatment factors only. The five columns of the design key matrix yield independent generators st, ac, bct, c, and t for writing down the 32 treatment combinations in the Yates order. Then the first two treatment combinations in the resulting sequence are in the first whole-plot of the first block, and each subsequent set of two treatment combinations is in the same whole-plot. Furthermore, the first four whole-plots are in the same block, and each subsequent set of four whole-plots is also in the same block. This reproduces the design constructed in Example 13.1.

Example 13.10. Consider $s = 2$, $n = 5$, $n_1 = 2$, $q = 4$, $g = 1$ and the design key

$$
\mathbf{K} =
\begin{array}{cccccc}
\mathcal{S} & \mathcal{P}_1 & \mathcal{P}_2 & \mathcal{P}_3 & \mathcal{B} \\
\begin{bmatrix}
1 & 0 & 0 & 0 & 0 \\
0 & 1 & 0 & 0 & 0 \\
0 & 0 & 1 & 0 & 0 \\
1 & 0 & 0 & 1 & 0 \\
1 & 1 & 1 & 1 & 1
\end{bmatrix} &
\begin{array}{c}
S \\ A \\ B \\ T \\ U
\end{array}
\end{array}
$$

There are three unfilled entries in the design key template: the second, third, and fourth entries of the last row. We fill all of them with 1 in the design key displayed above. This example falls in case (iic). By (13.11), we have the inverse design key

$$\mathbf{K}^{-1} = \begin{array}{c} \begin{array}{ccccc} S & A & B & T & U \end{array} \\ \begin{bmatrix} 1 & 0 & 0 & 0 & 0 \\ 0 & 1 & 0 & 0 & 0 \\ 0 & 0 & 1 & 0 & 0 \\ 1 & 0 & 0 & 1 & 0 \\ 0 & 1 & 1 & 1 & 1 \end{bmatrix} \begin{array}{c} \mathcal{S} \\ \mathcal{P}_1 \\ \mathcal{P}_2 \\ \mathcal{P}_3 \\ \mathcal{B} \end{array} \end{array}$$

The first four rows of $\mathbf{K}$ are equal to those of $\mathbf{K}^{-1}$. From the last two rows of $\mathbf{K}^{-1}$, we can identify two independent splitting words ST and $ABTU$, where $ABTU$ is also a blocking word. Note that ST can also be identified directly from $\mathbf{K}$, but not $ABTU$.

Now we apply Theorem 13.3 to obtain a design key template for blocked strip-plot designs. Consider the construction of a complete s^n factorial design with the block structure $(s^g$ blocks$)/(s^p$ rows $\times$ s^q columns$)$. As in Example 13.8, represent each experimental unit by a combination of n unit factors $\mathcal{R}_1, \dots, \mathcal{R}_p, \mathcal{C}_1, \dots, \mathcal{C}_q, \mathcal{B}_1, \dots, \mathcal{B}_g$. In (13.8), let $m_1 = g$ and $\mathbf{K}_2$ be given by (7.15). Then we have the following design key template given in Cheng and Tsai (2013):

$$\mathbf{K} = \begin{bmatrix} \mathbf{I}_p & \mathbf{0} & \mathbf{0} \\ \mathbf{0} & \mathbf{I}_q & \mathbf{0} \\ \mathbf{A} & \mathbf{B} & \mathbf{I}_g \end{bmatrix}, \tag{13.12}$$

where the first p columns correspond to $\mathcal{R}_1, \dots, \mathcal{R}_p$, the next q columns correspond to $\mathcal{C}_1, \dots, \mathcal{C}_q$, the last g columns correspond to $\mathcal{B}_1, \dots, \mathcal{B}_g$, the first p rows correspond to row treatment factors, the next q rows correspond to column treatment factors, and the last g rows correspond to the remaining row or column treatment factors. For each of the last g rows associated with a column (respectively, row) treatment factor, the corresponding row of $\mathbf{A}$ (respectively, $\mathbf{B}$) must be zero. The same rule as given before can be used to identify a set of g independent blocking words from the last g rows of $\mathbf{K}$.

Example 13.11. (Example 13.8 revisited) By (13.12), a design key template for this example is as follows:

$$\begin{array}{c} \begin{array}{ccccc} \mathcal{R}_1 & \mathcal{R}_2 & \mathcal{C}_1 & \mathcal{C}_2 & \mathcal{B} \end{array} \\ \begin{bmatrix} 1 & 0 & 0 & 0 & 0 \\ 0 & 1 & 0 & 0 & 0 \\ 0 & 0 & 1 & 0 & 0 \\ 0 & 0 & 0 & 1 & 0 \\ * & * & 0 & 0 & 1 \end{bmatrix} \begin{array}{c} A \\ B \\ S \\ T \\ C \end{array} \end{array}$$

There are only two entries to be filled, at least one of which must be nonzero. The design key (13.7) was obtained by choosing the first entry to be 1 and the second entry to be 0. This amounts to confounding AC with blocks.

13.8 Proof of Theorem 13.2

We prove Theorem 13.2 as a corollary of the following more general result, which is essentially Theorem 10.11 of Bailey (2008).

Theorem 13.5. *Let $\mathfrak{F}$ and $\mathfrak{B}$ be two block structures on Ω. Suppose the following conditions hold:*

(i) the factors in $\mathfrak{F}$ are pairwise orthogonal;

(ii) $\mathfrak{F}$ is closed under $\vee$;

(iii) the factors in $\mathfrak{B}$ are pairwise orthogonal;

(iv) $\mathfrak{B}$ is closed under $\vee$;

(v) $\mathcal{E} \in \mathfrak{B}$;

(vi) the factors in $\mathfrak{B}$ are orthogonal to those in $\mathfrak{F}$;

(vii) for any $\mathcal{F} \in \mathfrak{F}$ and $\mathcal{G} \in \mathfrak{B}$, $\mathcal{F} \vee \mathcal{G} \in \mathfrak{F}$.

For $\mathcal{F} \in \mathfrak{F}$ and $\mathcal{G} \in \mathfrak{G}$, let $W_{\mathcal{F}}$ and $W_{\mathcal{G}}$ be the strata associated with $\mathcal{F}$ and $\mathcal{G}$ under $\mathfrak{F}$ and $\mathfrak{B}$, respectively. Furthermore, for any $\mathcal{F} \in \mathfrak{F}$, let $\mathcal{G}$ be the coarsest factor in $\mathfrak{B}$ such that $\mathcal{G} \preceq \mathcal{F}$. Then such a factor $\mathcal{G}$ exists and is unique, and $W_{\mathcal{F}} \subseteq W_{\mathcal{G}}$.

Proof. By Theorem 12.6, conditions (i)–(iv) assure the existence of the strata. For any factor $\mathcal{F} \in \mathfrak{F}$, the existence of the coarsest factor $\mathcal{G} \in \mathfrak{B}$ such that $\mathcal{G} \preceq \mathcal{F}$ follows from the following facts: (a) $\mathcal{E} \preceq \mathcal{F}$, (b) if $\mathcal{G}_1 \preceq \mathcal{F}$ and $\mathcal{G}_2 \preceq \mathcal{F}$, then $\mathcal{G}_1 \vee \mathcal{G}_2 \preceq \mathcal{F}$, and (c) $\mathfrak{B}$ is closed under $\vee$. One can take $\mathcal{G}$ to be $\mathcal{G}_1 \vee \cdots \vee \mathcal{G}_h$, where $\mathcal{G}_1, \ldots, \mathcal{G}_h$ are all the factors in $\mathfrak{B}$ such that $\mathcal{G}_i \preceq \mathcal{F}$.

Then $V_{\mathcal{F}} \subseteq V_{\mathcal{G}}$; therefore $W_{\mathcal{F}} \subseteq V_{\mathcal{G}}$. We claim that $W_{\mathcal{F}} \subseteq V_{\mathcal{H}}^{\perp}$ for all $\mathcal{H} \in \mathfrak{B}$ such that $\mathcal{G} \prec \mathcal{H}$. If this is true, then

$$W_{\mathcal{F}} \subseteq V_{\mathcal{G}} \cap \left[\bigcap_{\mathcal{H}:\mathcal{H} \in \mathfrak{B}, \mathcal{G} \prec \mathcal{H}} V_{\mathcal{H}}^{\perp} \right] = W_{\mathcal{G}}.$$

Let $\mathcal{H}$ be a factor in $\mathfrak{B}$ such that $\mathcal{G} \prec \mathcal{H}$. To show $W_{\mathcal{F}} \subseteq V_{\mathcal{H}}^{\perp}$, we note that by (vii), $\mathcal{F} \vee \mathcal{H} \in \mathfrak{F}$. We have $\mathcal{F} \preceq \mathcal{F} \vee \mathcal{H}$. However, we cannot have $\mathcal{F} = \mathcal{F} \vee \mathcal{H}$. This is because if $\mathcal{F} = \mathcal{F} \vee \mathcal{H}$, then $\mathcal{H} \preceq \mathcal{F}$, which together with $\mathcal{G} \prec \mathcal{H}$ would contradict the fact that $\mathcal{G}$ is the coarsest factor in $\mathfrak{B}$ such that $\mathcal{G} \preceq \mathcal{F}$. Therefore $\mathcal{F} \prec \mathcal{F} \vee \mathcal{H}$. Then

$$W_{\mathcal{F}} = V_{\mathcal{F}} \cap \left[\bigcap_{\mathcal{F}':\mathcal{F}' \in \mathfrak{F}, \mathcal{F} \prec \mathcal{F}'} V_{\mathcal{F}'}^{\perp} \right] \subseteq V_{\mathcal{F}} \cap V_{\mathcal{F} \vee \mathcal{H}}^{\perp} = V_{\mathcal{F}} \ominus (V_{\mathcal{F}} \cap V_{\mathcal{H}}), \qquad (13.13)$$

where the last equality holds since $V_{\mathcal{F} \vee \mathcal{H}} = V_{\mathcal{F}} \cap V_{\mathcal{H}}$. On the other hand, $\mathcal{F}$ and $\mathcal{H}$ are orthogonal (by (vi)); so $V_{\mathcal{F}} \ominus (V_{\mathcal{F}} \cap V_{\mathcal{H}})$ is orthogonal to $V_{\mathcal{H}}$. This and (13.13) imply that $W_{\mathcal{F}}$ is orthogonal to $V_{\mathcal{H}}$. □

Proof of Theorem 13.2

Theorem 13.2 is an immediate consequence of Theorem 13.5. Conditions (i)–(v) hold since both $\mathfrak{F}$ and $\mathfrak{B}$ are orthogonal block structures. Furthermore, conditions

(vi) and (vii) follow from (13.5).   □

In addition to providing a simple rule for determining the unit contrasts in each stratum of a simple (and poset) block structure, Theorem 13.5 also gives a sufficient condition for design orthogonality. We present this application in Section 13.10 after a discussion of treatment structures in the next section.

13.9   Treatment structures

We first examine a decomposition such as (13.2) for general treatment structures. A comparative experiment has three components: a set Ξ of t treatments, a set Ω of N experimental units, and a function $\phi : \Omega \to \Xi$ that assigns a treatment $\phi(w)$ to each $w \in \Omega$. Like block structures, each set of factors on Ξ defines a treatment structure, and for each treatment structure $\mathfrak{T}$ that consists of pairwise orthogonal factors and is closed under $\vee$, Theorem 12.6 can also be applied to construct mutually orthogonal spaces W_F, one for each $F \in \mathfrak{T}$:

$$W_F = V_F \ominus \left(\sum_{G: G \in \mathfrak{T}, F \prec G} V_G \right). \tag{13.14}$$

We always assume that $\mathfrak{T}$ contains the universal factor U on the treatments. Then each W_F is orthogonal to V_U. If $\mathfrak{T}$ also contains the equality factor E, then (13.14) gives an orthogonal decomposition of the space of treatment contrasts as

$$V_E \ominus V_U = \bigoplus_{F: F \in \mathfrak{T}, F \neq U} W_F. \tag{13.15}$$

We have seen an application of this to the factorial treatment structure in Theorem 6.2.

For each factor $F \in \mathfrak{T}$, the function $\phi : \Omega \to \Xi$ induces a factor F^ϕ on Ω: two units v and w are in the same F^ϕ-class if and only if the treatments assigned to them, $\phi(v)$ and $\phi(w)$, are in the same F-class. The treatment structure $\mathfrak{T}$ then induces a "block" structure $\mathfrak{T}^\phi = \{F^\phi : F \in \mathfrak{T}\}$ on Ω.

For equireplicate designs (all the treatments are assigned to the same number of units), $\mathfrak{T}^\phi$ and $\mathfrak{T}$ are the same structurewise. We say that they are isomorphic. For example, for any two treatment factors $F, G \in \mathfrak{T}$, $F^\phi \vee G^\phi = (F \vee G)^\phi$, $F^\phi \wedge G^\phi = (F \wedge G)^\phi$, F and G are orthogonal if and only if F^ϕ and G^ϕ are orthogonal, etc.

For each $t \times 1$ vector $\mathbf{v} \in V_F$, define $\mathbf{v}^\phi$ as the $N \times 1$ vector in V_Ξ such that the component of $\mathbf{v}^\phi$ corresponding to unit $w \in \Omega$ is equal to the component of $\mathbf{v}$ corresponding to the treatment assigned to unit w. Then $V_{F^\phi} = \{\mathbf{v}^\phi : \mathbf{v} \in V_F\}$. In particular, $V_{E^\phi} = V_\mathcal{T}$ and $V_{U^\phi} = V_\mathcal{U}$. For equireplicate designs, if we define $\mathbf{v}^*$ as in (5.2),

$$v_w^* = v_{\phi(w)}/q, \ 1 \leq w \leq N,$$

where q is the number of replications of each treatment, then

$$\mathbf{v}^\phi = q\mathbf{v}^*. \tag{13.16}$$

If the assumptions in Theorem 12.6 are satisfied by $\mathfrak{T}$, then they are also satisfied by $\mathfrak{T}^\phi$. When $E \in \mathfrak{T}$, the decomposition in (13.15) carries over to Ω, and structurewise it is the same as

$$V_T \ominus V_U = \bigoplus_{F \in \mathfrak{T}, F \neq U} W_{F\phi}, \qquad (13.17)$$

where for each $F \in \mathfrak{T}$,

$$W_{F\phi} = \{\mathbf{v}^\phi : \mathbf{v} \in W_F\}. \qquad (13.18)$$

Therefore for equireplicate designs, essentially there is no difference between a treatment structure and the structure it induces on Ω. Note that we can always add E to $\mathfrak{T}$, so (13.17) always holds.

In the rest of this chapter, we only consider equireplicate designs. In this case, for any factor $F \in \mathfrak{T}$, if $W_{F\phi} \subseteq W_{\mathcal{G}}$ for some $\mathcal{G} \in \mathfrak{B}$, then by (13.16), (13.18), and Theorem 13.1, the best linear unbiased estimator of any treatment contrast $\mathbf{c}^T \alpha$ with $\mathbf{c} \in W_F$ is $\mathbf{c}^T \widehat{\alpha} = \sum_{j=1}^t c_j \overline{T}_j$, with $\text{var}(\sum_{j=1}^t c_j \overline{T}_j) = \xi_{\mathcal{G}} \sum_{j=1}^t c_j^2 / q$. If for each $F \in \mathfrak{T}$, $W_{F\phi} \subseteq W_{\mathcal{G}}$ for some $\mathcal{G} \in \mathfrak{B}$, then by (13.17), we have an orthogonal design in the sense of (13.2) and (13.3). For simplicity sometimes we drop the superscript ϕ in F^ϕ and $\mathfrak{T}^\phi$ when there is no danger of confusion.

Example 13.12. For the design in (7.2), the decomposition $V_E \ominus V_U = W_A \oplus W_B \oplus W_{A \wedge B}$ resulting from the treatment factorial structure induces a corresponding decomposition $V_T \ominus V_U = W_{A\phi} \oplus W_{B\phi} \oplus W_{(A \wedge B)\phi}$, with $W_{A\phi}$, $W_{B\phi} \subseteq W_{\mathcal{E}}$, and $W_{(A \wedge B)\phi} \subseteq W_{\mathcal{B}}$. So it is an orthogonal design in the sense of (13.2) and (13.3). As we have seen in Chapter 7, the main-effect contrasts of A and B are estimated in the intrablock stratum, and their interaction contrasts are estimated in the interblock stratum.

13.10 Checking design orthogonality

By applying Theorem 13.5 to a block structure $\mathfrak{B}$ and the structure $\mathfrak{T}^\phi$ induced by a treatment structure $\mathfrak{T}$, we have the following result (Theorem 10.11 of Bailey (2008)) that gives a sufficient condition for design orthogonality.

Theorem 13.6. *Let $\mathfrak{B}$ be a block structure on Ω and $\mathfrak{T}$ be a treatment structure on Ξ. Suppose*

(i) the factors in $\mathfrak{B}$ are pairwise orthogonal;

(ii) $\mathfrak{B}$ is closed under $\vee$;

(iii) the factors in $\mathfrak{T}$ are pairwise orthogonal;

(iv) $\mathfrak{T}$ is closed under $\vee$;

(v) $\mathcal{E} \in \mathfrak{B}$;

(vi) the factors in $\mathfrak{T}^\phi$ are orthogonal to those in $\mathfrak{B}$;

(vii) for any $F \in \mathfrak{T}$ and $\mathcal{G} \in \mathfrak{B}$, $F^\phi \vee \mathcal{G} \in \mathfrak{T}^\phi$.

For any $F \in \mathfrak{T}$, let $\mathcal{G}$ be the coarsest factor in $\mathfrak{B}$ such that $\mathcal{G} \preceq F^\phi$. Such a factor $\mathcal{G}$ exists and is unique, and $W_{F\phi} \subseteq W_{\mathcal{G}}$.

Note that conditions (i), (ii), and (v) are satisfied by any simple (and orthogonal) block structure; (iii) and (iv) are satisfied by the factorial treatment structure discussed in Section 6.8 if the treatment combinations are equireplicate. The only conditions that need to be checked in these applications are (vi) and (vii).

Example 13.13. Consider the design in (7.2). We have $\mathcal{B} = \{\mathcal{E}, \mathcal{B}, \mathcal{U}\}$. Instead of taking $\{A \wedge B, A, B, U\}$ to be the treatment structure as in Section 6.8, we consider the partitions given by the main-effect and interaction contrasts. Each contrast defines a partition of the treatment combinations into two disjoint sets and therefore defines a two-level factor. We denote the three factors so defined by A, B, and AB, and let $\mathfrak{T} = \{U, A, B, AB\}$. Then it is easy to see that conditions (i)–(vii) in Theorem 13.6 are satisfied. For example, (iii) follows from the orthogonality of the main-effect and interaction contrasts (the partitions defined by these contrasts have proportional frequencies); for (iv), we have $A \vee B = A \vee AB = B \vee AB = U$; (vi) follows from design construction; for (vii), we have $(AB)^\phi \vee \mathcal{B} = (AB)^\phi \in \mathfrak{T}^\phi$ (since $\mathcal{B} \preceq (AB)^\phi$) and $A^\phi \vee \mathcal{B} = B^\phi \vee \mathcal{B} = \mathcal{U}$. The treatment and block structures can be represented by the following Hasse diagrams:

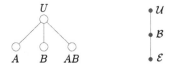

The merged Hasse diagram

shows that the coarsest factors $\mathcal{G}$ in $\mathcal{B}$ such that $\mathcal{G} \preceq F^\phi$ for $F = A$, B, and AB are $\mathcal{E}$, $\mathcal{E}$, and $\mathcal{B}$, respectively. Therefore by Theorem 13.6 the main-effect contrasts of both treatment factors are estimated in the intrablock stratum and the interaction contrasts are estimated in the interblock stratum.

Example 13.14. (Example 1.1 revisited) The Hasse diagram for the block structure was given in Figure 12.2, and the strata were derived in Sections 12.9 and 12.10. Let the four treatment factors variety/herbicide combinations, dates and rates of herbicide application, and weed species be denoted by V, D, R, and W, respectively. The treatment structure $\mathfrak{T}$ consists of U, V, D, R, W, and all their infimums. By the way the design is constructed, ignoring the superscript ϕ, we have $V \succ \mathcal{P}$, $D \succ \mathcal{S}$, $R \succ \mathcal{SS}$, and $W \succ \mathcal{ST}$. It follows that

$$V \wedge D \succ P \wedge S = S, \quad V \wedge R \succ P \wedge SS = SS,$$

$$V \wedge W \succ P \wedge ST, \quad D \wedge R \succ S \wedge SS = SS,$$

$$D \wedge W \succ S \wedge ST, \quad R \wedge W \succ SS \wedge ST,$$

$$V \wedge D \wedge R \succ P \wedge S \wedge SS = SS,$$

$$V \wedge D \wedge W \succ P \wedge S \wedge ST = S \wedge ST,$$

$$V \wedge R \wedge W \succ P \wedge SS \wedge ST = SS \wedge ST,$$

$$D \wedge R \wedge W \succ S \wedge SS \wedge ST = SS \wedge ST,$$

$$V \wedge D \wedge R \wedge W \succ P \wedge S \wedge SS \wedge ST = SS \wedge ST.$$

The conditions in Theorem 13.6 are clearly satisfied. Therefore this is an orthogonal design. We have $W_V \subseteq W_P$; W_D, $W_{V \wedge D} \subseteq W_S$; W_R, $W_{V \wedge R}$, $W_{D \wedge R}$, $W_{V \wedge D \wedge R} \subseteq W_{SS}$; $W_{V \wedge W} \subseteq W_{P \wedge ST}$; $W_W \subseteq W_{ST}$; $W_{D \wedge W}$, $W_{V \wedge D \wedge W} \subseteq W_{S \wedge ST}$; $W_{R \wedge W}$, $W_{V \wedge R \wedge W}$, $W_{D \wedge R \wedge W}$, $W_{V \wedge D \wedge R \wedge W} \subseteq W_{SS \wedge ST}$. Except for W_B (the block stratum), each of the other seven strata contains some treatment information, with each main-effect and interaction component estimated in one single stratum. A skeleton of the ANOVA table is shown in Table 13.1.

Großmann (2014) developed a computer package "AutomaticAnova" for automating the ANOVA of orthogonal designs for multi-stratum experiments.

13.11 Experiments with multiple processing stages: the nonoverlapping case

Suppose there are h processing stages and n treatment factors with s levels each. At the ith stage the s^n units are partitioned into s_i disjoint classes and the same level of each ith-stage treatment factor is assigned to all the units in the same class. Let the partition at the ith stage be $\mathcal{F}_i$, and $\mathfrak{B} = \{\mathcal{E}\} \cup \{\mathcal{U}, \mathcal{F}_1, \ldots, \mathcal{F}_h\}$.

A single-replicate complete factorial design in this setting can be constructed by partitioning the s^n treatment combinations into s_i disjoint classes of size s^n/s_i for each $1 \leq i \leq h$ in such a way that each ith-stage treatment factor has the same level in all the treatment combinations in the same class. Under this construction, each $\mathcal{F}_i$ coincides with a partition of the treatment combinations.

Suppose s is a prime or a prime power, and each s_i is a power of s, say $s_i = s^{r_i}$. We can choose r_i linearly independent vectors (words) $\mathbf{a}_1^i, \ldots, \mathbf{a}_{r_i}^i$ from EG(n, s) to perform the partition at the ith stage: the treatment combinations $\mathbf{x}$ with the same value of $(\mathbf{a}_l^i)^T \mathbf{x}$ for all $l = 1, \ldots, r_i$ are placed in the same class. The words must be chosen so that each ith-stage treatment factor has a constant level in the same $\mathcal{F}_i$-class. If the partition at stage i is to be nested in that of stage j (hereafter for brevity we say that stage i is nested in stage j), then $\{\mathbf{a}_1^j, \ldots, \mathbf{a}_{r_j}^j\}$ should be a subset of $\{\mathbf{a}_1^i, \ldots, \mathbf{a}_{r_i}^i\}$. However, if a certain stage, say the hth stage, is to produce classes of size one (that is, the equality factor $\mathcal{E}$), then $\{\mathbf{a}_1^h, \ldots, \mathbf{a}_{r_h}^h\}$ is not needed; see the discussion in the second paragraph of Section 13.5.

Table 13.1 *ANOVA for Example 13.14*

Source of variation	degrees of freedom
$W_{\mathcal{B}}$	3
$W_{\mathcal{P}}$	
Variety	2
residual	6
$W_{\mathcal{ST}}$	
Weed	6
residual	18
$W_{\mathcal{S}}$	
Day	1
Variety $\wedge$ Day	2
residual	9
$W_{\mathcal{P} \wedge \mathcal{ST}}$	
Variety $\wedge$ Weed	12
residual	36
$W_{\mathcal{SS}}$	
Rate	2
Variety $\wedge$ Rate	4
Day $\wedge$ Rate	2
Variety $\wedge$ Day $\wedge$ Rate	4
residual	36
$W_{\mathcal{S} \wedge \mathcal{ST}}$	
Day $\wedge$ Weed	6
Variety $\wedge$ Day $\wedge$ Weed	12
residual	54
$W_{\mathcal{SS} \wedge \mathcal{ST}}$	
Rate $\wedge$ Weed	12
Variety $\wedge$ Rate $\wedge$ Weed	24
Day $\wedge$ Rate $\wedge$ Weed	12
Variety $\wedge$ Day $\wedge$ Rate $\wedge$ Weed	24
residual	216
Total	503

Due to the geometric construction, each class of $\mathcal{F}_1,\ldots,\mathcal{F}_h$ is a flat in $EG(n,s)$. It follows that $\mathcal{F}_1,\ldots,\mathcal{F}_h$ are uniform and mutually orthogonal. Under model (12.21), if $\mathfrak{B}$ is also closed under $\vee$, then by Theorem 12.8, the strata are as given in Section 12.13. That is, they are the nonnull spaces among

$$W_{\mathcal{F}_i} = C_{\mathcal{F}_i} \ominus (\sum_{j:\mathcal{F}_i \prec \mathcal{F}_j} C_{\mathcal{F}_j}), \ i = 1,\ldots,h,$$

and

$$W_{\mathcal{E}} = \left[\sum_{i=1}^{h} V_{\mathcal{F}_i}\right]^{\perp} \text{ if } \mathcal{F}_i \neq \mathcal{E} \text{ for all } i. \tag{13.19}$$

Under the given construction, $C_{\mathcal{F}_i}$ is the $(s^{r_i} - 1)$-dimensional space of treatment contrasts defined by the pencil $P(\mathbf{a}_1^i,\ldots,\mathbf{a}_{r_i}^i)$. The linear space consisting of $\mathbf{a}_1^i,\ldots,\mathbf{a}_{r_i}^i$ and their linear combinations is called a randomization defining contrast subgroup (RDCS) by Bingham et al. (2008). We refer to it as the RDCS at stage i. A treatment factorial effect is estimated in $W_{\mathcal{F}_i}$, $1 \leq i \leq h$, if and only if it is defined by a vector in the RDCS at stage i, but not in the RDCS at stage j for any j such that $\mathcal{F}_i \prec \mathcal{F}_j$. All the other treatment factorial effects, if there are any, are estimated in $W_{\mathcal{E}}$.

If the factors $\mathcal{F}_1,\ldots,\mathcal{F}_h$ are not nested in one another, then we have

$$W_{\mathcal{F}_i} = C_{\mathcal{F}_i} \text{ for } 1 \leq i \leq h, \text{ and } W_{\mathcal{E}} = \left[\sum_{i=1}^{h} V_{\mathcal{F}_i}\right]^{\perp}.$$

In this case, a treatment factorial effect is estimated in $W_{\mathcal{F}_i}$, $1 \leq i \leq h$, if and only if it is defined by a vector that is in the RDCS at stage i. All the other treatment factorial effects, if there are any, are estimated in $W_{\mathcal{E}}$.

The results above require the assumption that $\mathfrak{B}$ is closed with respect to $\vee$. Now we derive conditions under which this holds. We split the discussion into two cases depending on whether or not there is nesting among the factors in $\mathfrak{F}$.

Case 1. Some stages are nested in others

For any $i \neq j$, if $\mathcal{F}_i \prec \mathcal{F}_j$, then $\mathcal{F}_i \vee \mathcal{F}_j = \mathcal{F}_j$, and we have $\mathcal{F}_i \vee \mathcal{F}_j \in \mathfrak{B}$. If neither $\mathcal{F}_i \prec \mathcal{F}_j$ nor $\mathcal{F}_j \prec \mathcal{F}_i$, then in order that $\mathcal{F}_i \vee \mathcal{F}_j \in \mathfrak{B}$, we must have (1) $\mathcal{F}_i \vee \mathcal{F}_j = \mathcal{F}_k$ for some $k \neq i$, j, or (2) $\mathcal{F}_i \vee \mathcal{F}_j = \mathcal{U}$.

In case (1), $\mathcal{F}_k$ is the finest partition in $\mathfrak{B}$ that nests both $\mathcal{F}_i$ and $\mathcal{F}_j$, and by Corollary 4.3, we must have $C_{\mathcal{F}_k} = C_{\mathcal{F}_i} \cap C_{\mathcal{F}_j}$. This can be achieved by making $\{\mathbf{a}_1^k,\cdots,\mathbf{a}_{r_k}^k\} = \{\mathbf{a}_1^i,\cdots,\mathbf{a}_{r_i}^i\} \cap \{\mathbf{a}_1^j,\cdots,\mathbf{a}_{r_j}^j\}$ and that the $r_i + r_j - r_k$ vectors in $\{\mathbf{a}_1^i,\cdots,\mathbf{a}_{r_i}^i\} \cup \{\mathbf{a}_1^j,\cdots,\mathbf{a}_{r_j}^j\}$ are linearly independent; also see Exercise 12.8.

In case (2), since $C_{\mathcal{F}_i} \cap C_{\mathcal{F}_j} = \{0\}$, $C_{\mathcal{F}_i}$ and $C_{\mathcal{F}_j}$ are orthogonal to each other (or equivalently, $\{\mathbf{a}_1^i,\cdots,\mathbf{a}_{r_i}^i\}$ and $\{\mathbf{a}_1^j,\cdots,\mathbf{a}_{r_j}^j\}$ are linearly independent).

To summarize, if the vectors $\mathbf{a}_1^i,\cdots,\mathbf{a}_{r_i}^i$, $1 \leq i \leq h$, can be chosen to satisfy the following conditions, then the strata in which the treatment factorial effects are estimated are as described in the paragraph following (13.19):

(i) $\mathbf{a}_1^i, \cdots, \mathbf{a}_{r_i}^i$ are linearly independent for all $1 \le i \le h$.

(ii) If stage i is nested in stage j, then $\{\mathbf{a}_1^j, \cdots, \mathbf{a}_{r_j}^j\} \subset \{\mathbf{a}_1^i, \cdots, \mathbf{a}_{r_i}^i\}$.

(iii) If stages i and j are not nested in each other and there is no other stage that nests both of them, then $\mathbf{a}_1^i, \cdots, \mathbf{a}_{r_i}^i, \mathbf{a}_1^j, \cdots, \mathbf{a}_{r_j}^j$ are linearly independent.

(iv) Suppose stages i and j are not nested in each other and both are nested in another stage. Let k be the coarsest such stage. Then $\{\mathbf{a}_1^k, \cdots, \mathbf{a}_{r_k}^k\} = \{\mathbf{a}_1^i, \cdots, \mathbf{a}_{r_i}^i\} \cap \{\mathbf{a}_1^j, \cdots, \mathbf{a}_{r_j}^j\}$, and the vectors in $\{\mathbf{a}_1^i, \cdots, \mathbf{a}_{r_i}^i\} \cup \{\mathbf{a}_1^j, \cdots, \mathbf{a}_{r_j}^j\}$ are linearly independent.

One also needs to impose the following condition to assure that the main effect contrasts of all the ith-stage treatment factors are estimated in $W_{\mathcal{F}_i}$.

(v) The vector that defines the main effect of any ith-stage treatment factor is a linear combination of $\mathbf{a}_1^i, \ldots, \mathbf{a}_{r_i}^i$, but is not a linear combination of $\mathbf{a}_1^j, \ldots, \mathbf{a}_{r_j}^j$ for any j such that stage i is nested in stage j.

Case 2. No nesting among the stages

In this case (i)–(v) reduce to the following:

(a) For any $1 \le i \ne j \le h$, $\mathbf{a}_1^i, \ldots, \mathbf{a}_{r_i}^i, \mathbf{a}_1^j, \ldots, \mathbf{a}_{r_j}^j$ are linearly independent.

(b) The vector that defines the main effect of any ith-stage treatment factor is a linear combination of $\mathbf{a}_1^i, \ldots, \mathbf{a}_{r_i}^i$.

A Franklin and Bailey (1977) type of algorithm can be used to search for designs that satisfy conditions (i)–(v) when there is some nesting among the stages and those that satisfy (a) and (b) when there is no nesting. To save the efforts in searching for eligible designs, as many of the vectors that define main effects of the ith stage treatment factors as possible should be included in the words that are used to perform the partition for $\mathcal{F}_i$, and by default for all $\mathcal{F}_j$ with $\mathcal{F}_j \prec \mathcal{F}_i$, but not for any other $\mathcal{F}_j$.

No design satisfying the stated conditions can be found if it is not possible for $\mathfrak{B}$ to be closed with respect to $\vee$. Then we have an overlapping case as discussed in Section 12.13. This will be considered in the next section.

Although computer searches can be used to see whether designs satisfying (i)–(v) (or (a) and (b)) exist, it is useful to have theoretical results on the existence of such designs. We present below some results on the case of no nesting among the partitions at different stages.

Let G_i be the RDCS at stage i. Recall that G_i consists of $\mathbf{a}_1^i, \ldots, \mathbf{a}_{r_i}^i$ and all their linear combinations, and that $C_{\mathcal{F}_i}$ is the $(s^{r_i} - 1)$-dimensional space of treatment contrasts defined by the nonzero vectors in G_i. Condition (a) is equivalent to the mutual orthogonality of $C_{\mathcal{F}_1}, \ldots, C_{\mathcal{F}_h}$. As shown in Proposition 12.9, this is equivalent to that partitions at the h stages define an $OA(N, s_1 \times \cdots \times s_h, 2)$, where $s_i = s^{r_i}$. In Section 9.8 we have seen how one can use h sets of vectors $\mathbf{a}_1^i, \ldots, \mathbf{a}_{r_i}^i$, $1 \le i \le h$, to form h disjoint groups of factorial effects and use the method of replacement to construct an $OA(N, s_1 \times \cdots \times s_h, 2)$. We use an example to illustrate how such an orthogonal array provides a template for performing the partitions at various stages.

Example 13.15. Suppose $s = 2$, $n = 4$, $h = 4$, and $s_1 = s_2 = s_3 = s_4 = 4$. Example 9.11 shows how the 15 factorial effects in a 2^4 design can be grouped into five disjoint sets of the form (**a, b, ab**). We can pick four of the five groups, say (**1, 2, 12**), (**3, 4, 34**), (**13, 24, 1234**), and (**23, 124, 134**), to construct an OA($16, 4^4, 2$). Figure 13.1 shows the 16 level combinations of a complete 2^4 factorial in the standard order, followed by an OA($16, 4^4, 2$) constructed from the four chosen disjoint groups of factorial effects, with one four-level factor obtained from each group by the method of replacement. Each column of the OA($16, 4^4, 2$) can be used to partition the 16 treatment combinations into four disjoint classes of size four: those corresponding to the same level in each column of the orthogonal array are placed in the same class. Suppose the ith column is used to perform the partition at the ith stage, $1 \leq i \leq 4$. Then the factorial effects estimated in various strata are as follows:

$$
\begin{aligned}
W_{\mathcal{F}_1} &: \quad 1, 2, 12 \\
W_{\mathcal{F}_2} &: \quad 3, 4, 34 \\
W_{\mathcal{F}_3} &: \quad 13, 24, 1234 \\
W_{\mathcal{F}_4} &: \quad 23, 124, 134 \\
W_{\mathcal{E}} &: \quad 14, 123, 234.
\end{aligned}
$$

The effects that are estimated in $\mathcal{F}_i$ are those that are in the ith group, and the other effects are estimated in the bottom stratum. If the levels of factors 1, 2, 3, and 4 are to be assigned at stages 1, 2, 3, and 4, respectively, then we must have their main effects estimated in $W_{\mathcal{F}_1}$, $W_{\mathcal{F}_2}$, $W_{\mathcal{F}_3}$, and $W_{\mathcal{F}_4}$, respectively. This requirement is not satisfied here. Some transformation (relabeling of treatment factorial effects) is needed to produce a design that satisfies the requirement. We continue in Example 13.16.

Orthogonal arrays constructed by the method discussed in Section 9.8 or grouping schemes that are available in the literature (such as those surveyed in Ranjan, Bingham, and Dean (2009)), as demonstrated in Example 13.15, may not have the treatment main effects estimated in the correct strata. We have seen how the method of design key can be used to construct factorial designs with simple block structures such that the treatment main effects are estimated in designated strata. It is also useful in the present setting. Suppose, as in Example 13.15, we already have a scheme for partitioning the treatment combinations that are listed in the standard order, but the main effects are not estimated in desired strata. For each $1 \leq i \leq h$, we can choose factorial effects from the ith stage RDCS, one for each of the ith stage treatment factors. The idea is to re-order the treatment combinations so that the contrasts that define the chosen factorial effects in the original order become main-effect contrasts of the ith-stage treatment factors in the new order. We call these factorial effects "aliases" of the corresponding treatment main effects. This construction can be done by using a design key. In this case both the rows and columns of the design key are

1	2	3	4	1,2,12	3,4,34	13,24,1234	23,124,134
0	0	0	0	0	0	0	0
1	0	0	0	2	0	2	1
0	1	0	0	1	0	1	3
1	1	0	0	3	0	3	2
0	0	1	0	0	2	2	2
1	0	1	0	2	2	0	3
0	1	1	0	1	2	3	1
1	1	1	0	3	2	1	0
0	0	0	1	0	1	1	1
1	0	0	1	2	1	3	0
0	1	0	1	1	1	0	2
1	1	0	1	3	1	2	3
0	0	1	1	0	3	3	3
1	0	1	1	2	3	1	2
0	1	1	1	1	3	2	0
1	1	1	1	3	3	0	1

Figure 13.1 An $OA(16, 4^4, 2)$ constructed by replacement

labeled by treatment factors. The chosen "aliases" of the main effects, which must be linearly independent, are used to write down the rows of the design key matrix. One can then re-order the treatment combinations by arranging them in the Yates order with respect to the generators identified from the columns of the design key matrix. Using the grouping scheme on hand to partition the re-ordered treatment combinations produces a design that has levels of the treatment factors set at correct stages.

Example 13.16. (Examples 1.4 and 13.15 revisited) In Example 13.15, suppose the levels of factors 1, 2, 3, and 4 are to be assigned at stages 1, 2, 3, and 4, respectively. Suppose we pick **1**, **3**, **24**, and **134**, one from each RDCS, to serve as "aliases" of the main effects **1**, **2**, **3**, and **4**, respectively. Then we have the design key matrix

$$K = \begin{matrix} & 1 & 2 & 3 & 4 & \\ & \begin{bmatrix} 1 & 0 & 0 & 0 \\ 0 & 0 & 1 & 0 \\ 0 & 1 & 0 & 1 \\ 1 & 0 & 1 & 1 \end{bmatrix} & \begin{matrix} 1 \\ 2 \\ 3 \\ 4 \end{matrix} \end{matrix}$$

For example, since **24** is the alias of **3**, the third row of **K** has entries 1 corresponding to factors 2 and 4. Write down the 16 treatment combinations in the Yates order with respect to the generators identified from the columns of **K**, while keeping the $OA(16, 4^4, 2)$ in Figure 13.1 intact. The re-ordered treatment combinations are displayed in Figure 13.2, with the four generators shown in boldface. The words $\{1, 2, 12\}$, $\{3, 4, 34\}$, $\{13, 24, 1234\}$, and $\{23, 124, 134\}$ in the original RDCS's are transformed to $\{1, 1234, 234\}$, $\{2, 124, 14\}$, $\{12, 3, 123\}$, and $\{134, 13, 4\}$, respec-

1	2	3	4	1, 1234, 234	2, 124, 14	12, 3, 123	134, 13, 4
0	0	0	0	0	0	0	0
1	0	0	1	2	0	2	1
0	0	1	0	1	0	1	3
1	0	1	1	3	0	3	2
0	1	0	1	0	2	2	2
1	1	0	0	2	2	0	3
0	1	1	1	1	2	3	1
1	1	1	0	3	2	1	0
0	0	1	1	0	1	1	1
1	0	1	0	2	1	3	0
0	0	0	1	1	1	0	2
1	0	0	0	3	1	2	3
0	1	1	0	0	3	3	3
1	1	1	1	2	3	1	2
0	1	0	0	1	3	2	0
1	1	0	1	3	3	0	1

Figure 13.2 *The design in Figure 13.1 reordered*

tively. For example, the alias of **1234**, which is in the new first-stage RDCS, is **(1)(3)(24)(134) = 2**, an effect in the original first-stage RDCS. The aliases of the other factorial effects can be similarly determined. Such information can also be obtained from $\mathbf{K}^{-1}$. Now there is one main effect in each RDCS as desired. One can see that in Figure 13.2, for each $1 \leq i \leq 4$, treatment factor i has a constant level in all the treatment combinations that correspond to the same level in the ith column of the orthogonal array. (Here the levels in the orthogonal array represent the four classes in each partition, and are labeled by 0, 1, 2, and 3.) This produces the design shown in Example 1.4 when the OA($16, 4^4, 2$) is represented by two mutually orthogonal Latin squares. Note that other choices of four disjoint groups of factorial effects and different choices of effects from these groups as aliases of treatment main effects produce other possibly nonisomorphic designs. A search is needed if one wants to find good designs with respect to some criteria.

The construction of orthogonal arrays of the form OA($s^n, s^u \times (s^{r_1})^{k_1} \times \cdots \times (s^{r_h})^{k_h}, 2$) in Wu, Zhang, and Wang (1992) and Hedayat, Pu, and Stufken (1992) is useful for constructing designs where the partitions at different stages have different numbers of classes. Some results on the construction of such designs can also be found in Ranjan, Bingham, and Dean (2009).

13.12 Experiments with multiple processing stages: the overlapping case

When $\mathfrak{B} = \{\mathcal{E}\} \cup \{\mathcal{U}, \mathcal{F}_1, \ldots, \mathcal{F}_h\}$ is not closed under $\vee$, for any $\mathcal{F}_{i_1}, \ldots, \mathcal{F}_{i_k}$ such that $\vee_{j=1}^k \mathcal{F}_{i_j} \notin \mathfrak{B}$, $\vee_{j=1}^k \mathcal{F}_{i_j}$ needs to be added to $\mathfrak{B}$. Under the geometric construction described in the third paragraph of the previous section, it is easy to see that all such pseudo factors are uniform. Therefore by the discussions in Section 12.13, Theorem 12.8 is applicable to the expanded block structure. In this case there is one extra stratum associated with each pseudo factor $\vee_{j=1}^{k} \mathcal{F}_{i_j}$, and we have $\xi_{\mathcal{F}_{i_1} \vee \cdots \vee \mathcal{F}_{i_k}} > \xi_{\mathcal{F}_{i_j}}$ for each j. No extra stratum will be introduced if a search finds a design that satisfies conditions (i)–(v) (or (a) and (b) in the case of no nesting) in the previous section.

Since $C_{\vee_{j=1}^{k} \mathcal{F}_{i_j}} = \cap_{j=1}^{k} C_{\mathcal{F}_{i_j}}$, in the case of no nesting, if a design is constructed without enforcing condition (a) in the previous section, then the extra strata, if there are any, are precisely those $\cap_{j=1}^{k} C_{\mathcal{F}_{i_j}}$ that are not equal to $\{\mathbf{0}\}$. These strata result from the overlap of the $C_{\mathcal{F}_i}$'s. In this case, unlike experiments with simple block structures, the number of strata and their degrees of freedom are design dependent. Since the extra strata have larger variances, it is desirable to avoid them. If this is not possible, then we may want to minimize the number and dimensions of such overlaps. It is desirable to have just one common overlap. However, Ranjan, Bingham, and Mukerjee (2010) pointed out that sometimes it may be unavoidable to have some low-order effects in an overlap. In this case if the overlap is small, then we would not be able to assess the effects therein since the associated normal probability plot is not informative. They suggested that it may be advantageous to have a large enough overlap so that more effects can be assessed. See Ranjan, Bingham, and Mukerjee (2010) for more discussions of overlapping strategies and further results.

Example 13.17. (Example 1.5 revisited) Consider a complete 2^5 experiment with two processing stages. The treatment combinations are partitioned into eight classes at both stages, where neither stage is nested in the other. We have seen in Section 12.13 that the overlap of $C_{\mathcal{F}_1}$ and $C_{\mathcal{F}_2}$ cannot be avoided. In this case an overlap of just one degree of freedom is the best we can do. Then $C_{\mathcal{F}_1} \cap C_{\mathcal{F}_2} = C_{\mathcal{F}_1 \vee \mathcal{F}_2} = C_{\mathcal{B}}$, where $\mathcal{B}$ is a two-level pseudo factor that nests both $\mathcal{F}_1$ and $\mathcal{F}_2$. Formally the block structure is the same as $2/(4*4)$. Therefore we can proceed as in the construction of blocked strip-plot designs discussed in Section 13.3. Suppose three of the treatment factors have their levels set at the first stage, and those of the other two factors are set at the second stage. Then we need to choose five independent factorial effects X_1, X_2, X_3, X_4, and X_5 of the treatment factors, use X_1, X_2, X_3 to divide the 32 treatment combinations into eight classes of size four, and then use X_1, X_4, and X_5 to do the second-stage partitioning. Then X_1 is estimated in the stratum corresponding to the pseudo block, with the largest variance. The effects X_2, X_3 and all the four generalized interactions of X_1, X_2, and X_3 are estimated in $W_{\mathcal{F}_1}$; likewise the effects X_4, X_5 and all the four generalized interactions of X_1, X_4, and X_5 are estimated in $W_{\mathcal{F}_2}$. The remaining 18 factorial effects are estimated in the bottom stratum. We note that $W_{\mathcal{F}_i} = C_{\mathcal{F}_i} \ominus C_{\mathcal{B}}$ and is no longer equal to $C_{\mathcal{F}_i}$. Suppose the first-stage treatment factors are A, B, and C, and the second-stage treatment factors are S and T. Then one of the factorial effects of A, B, and C must be estimated in the stratum corresponding to the pseudo block. Let X_1 be AC. Then X_2 and X_3 can be any two of the three main effects of A, B, and C, and X_4 and X_5 are the main effects of S and T. The contrasts X_1, X_2, X_3, X_4, and X_5 chosen above are the same as those used to construct the design in Example 13.2, revisited in Sections 13.5 and 13.6. The design constructed in Example 13.2 can be applied directly to the current setting, resulting in the design shown below.

(1)	s	t	st				
b	bs	bt	bst				
ac	acs	act	$acst$				
abc	$abcs$	$abct$	$abcst$				
				a	as	at	ast
				ab	abs	abt	$abst$
				c	cs	ct	cst
				bc	bcs	bct	$bcst$

As determined in Example 13.6, A, B, C, AB, BC, and ABC are estimated in $W_{\mathcal{F}_1}$, S, T, ST, ACS, ACT, and $ACST$ are estimated in $W_{\mathcal{F}_2}$, AC is estimated in $W_{\mathcal{F}_1 \vee \mathcal{F}_2}$, and the rest are estimated in the bottom stratum.

Exercises

13.1 Verify the effect-confounding stated at the end of the paragraph preceding Example 13.2.

13.2 Verify the strata stated in Example 13.5 and the entries of Table 13.1.

13.3 Use the rule on p. 267 to determine the strata and the decomposition in (13.6) for all the strata of the simple block structure $n_1/(n_2 \times n_3)/n_4$.

13.4 Show that under the construction described in the third paragraph of Section 13.11, for any $1 \le i_1 < \cdots < i_k \le h, \vee_{j=1}^{k} \mathcal{F}_{i_j}$ is a uniform factor.

13.5 Prove Theorem 13.3.

13.6 Verify that the design in Figure 13.2 is the same as the one displayed at the end of Example 1.4 on p.11. Note that in Figure 13.2, the four classes of each partition are represented by 0, 1, 2, and 3.

13.7 According to Exercises 6.3 and 6.4, if the treatment combinations in a complete factorial experiment are observed in the Yates order in an equally spaced time sequence, then the estimates of k-factor interaction contrasts are orthogonal to a $(k-1)$-th degree polynomial time trend. Show how the method of design key can be used to reorder the treatment combinations so that all the main-effect contrasts are orthogonal to higher-order time trends. Construct a run order of a complete 2^4 experiment such that all the main-effect contrasts are orthogonal to linear and quadratic time trends.

Chapter 14

Multi-Stratum Fractional Factorial Designs

In this chapter we extend the design and analysis of multi-stratum complete factorial experiments discussed in Chapter 13 to regular fractional factorial designs. Independent treatment factorial effects are chosen to define regular fractions and to partition the treatment combinations into disjoint classes for different levels of various unit factors in a given block structure. This general procedure is applied to several specific cases, including fractional factorial designs for blocked, split-plot, and strip-plot experiments. The issue of design selection is also addressed. For unstructured experimental units, minimum aberration is a popular criterion for selecting fractional factorial designs under the hierarchical assumption. This criterion is extended to fractional factorial experiments with multiple strata. In addition to effect aliasing, stratum variances are also taken into account in the design selection. The examples in Chapter 1 are revisited.

14.1 A general procedure

The construction of designs for single-replicate factorial experiments with simple block structures presented in Section 13.5 can easily be extended to fractional factorial designs. Suppose there are n treatment factors each with s levels, and the number of units is s^{n-p}, where s is a prime number or power of a prime number. Furthermore, the s^{n-p} units have a simple block structure $\mathcal{B}$ consisting of factors of the form $\mathcal{F}_1 \times \cdots \times \mathcal{F}_m$ where $\mathcal{F}_i = \mathcal{U}_i$ or $\mathcal{E}_i$, and $\mathcal{E}_i$ has n_i levels with $n_i = s^{l_i}$. Then $\sum_{i=1}^{m} l_i = n - p$. One can construct a complete factorial design for a set of $n - p$ basic factors by using the method of Section 13.5, and use interactions of the basic factors to define p additional factors. The choices of Δ_i, $1 \leq i \leq m$, and their roles in the construction are the same as in Section 13.5, except that we also need to choose a set Δ_0 of p linearly independent vectors from $\mathrm{EG}(n,s)$ as independent defining words to define a regular s^{n-p} fraction. We need $\cup_{i=0}^{m} \Delta_i$ to be a basis of $\mathrm{EG}(n,s)$. Again, as noted in Section 13.5, there is no need to choose Δ_m if the corresponding unit factor in the block structure is nested in all the others.

The strata in which factorial effects of the basic factors (and hence their aliases) are estimated can be determined as in Section 13.5. Separate half-normal probability plots should be used to examine the significance of the effects estimated in different strata.

Design keys can also be used to construct fractional factorial designs with simple block structures. As in Section 7.6, one unit alias is selected for the main effect of each treatment factor, resulting in an $n \times (n - p)$ design key matrix $\mathbf{K}$. In order to generate s^{n-p} distinct treatment combinations, the $n - p$ columns of $\mathbf{K}$ must be linearly independent. One can first write down a design key for the complete factorial of basic factors and then add one row for each added factor. The condition that the columns of the design key are linearly independent can be built into a template for the complete factorial of basic factors. Unit aliases of the treatment interactions can be determined from those of the main effects and can be used to identify the stratum in which each interaction is estimated. These unit aliases also provide information about effect aliasing: two treatment factorial effects are aliased if and only if their unit aliases are aliased.

14.2 Construction of blocked regular fractional factorial designs

To construct a regular s^{n-p} fractional factorial design and divide the treatment combinations into s^q blocks of size s^{n-p-q}, where $1 \leq q < n - p$, we need $p + q$ linearly independent vectors $\mathbf{a}_1, \ldots, \mathbf{a}_p, \mathbf{b}_1, \ldots, \mathbf{b}_q$ from EG(n, s). The p vectors $\mathbf{a}_1, \ldots, \mathbf{a}_p$ are used to construct a regular fraction that consists of the s^{n-p} treatment combinations $\mathbf{x}$ satisfying

$$\mathbf{a}_i^T \mathbf{x} = 0, \ i = 1, \ldots, p. \tag{14.1}$$

The q additional vectors $\mathbf{b}_1, \ldots, \mathbf{b}_q$ are used to divide these s^{n-p} treatment combinations into s^q blocks. All the treatment combinations $\mathbf{x}$ that have the same values of $\mathbf{b}_i^T \mathbf{x}$, $i = 1, \ldots, q$, are placed in the same block. Thus each block consists of all the treatment combinations $\mathbf{x}$ that satisfy (14.1) and

$$\mathbf{b}_i^T \mathbf{x} = c_i, \ i = 1, \ldots, q,$$

where $c_1, \ldots, c_q \in \text{GF}(s)$. We call $\mathbf{a}_1, \ldots, \mathbf{a}_p$ independent treatment defining words and $\mathbf{b}_1, \ldots, \mathbf{b}_q$ independent *blocking words*.

Each $\mathbf{b}_j$, $j = 1, \ldots, q$, defines a factor on the set of treatment combinations by placing those treatment combinations $\mathbf{x}$ with the same value of $\mathbf{b}_j^T \mathbf{x}$ in the same class. We denote such factors by $\mathcal{B}_1, \ldots, \mathcal{B}_q$, and call them *blocking factors*.

Example 14.1. Consider the design with $n = 6$ and $p = q = 2$ displayed in (14.2). The first six columns of the array constitute a 2^{6-2}_{IV} design defined by $I = ACDE = ABCF = BDEF$ and with the factor levels coded by 1 and -1. The levels of the blocking factors $\mathcal{B}_1$ and $\mathcal{B}_2$, defined by $\mathcal{B}_1 = AC$ and $\mathcal{B}_2 = ABD$, are shown in the last two columns. The 16 treatment combinations are partitioned into four blocks of size four according to the levels of $\mathcal{B}_1$ and $\mathcal{B}_2$. Underneath the array, the four blocks are shown with the treatment combinations written in shorthand notation.

$$
\begin{array}{cccccccc}
A & B & C & D & E & F & B_1 & B_2 \\
-1 & -1 & -1 & -1 & -1 & -1 & 1 & -1 \\
1 & -1 & -1 & -1 & 1 & 1 & -1 & 1 \\
-1 & 1 & -1 & -1 & -1 & 1 & 1 & 1 \\
1 & 1 & -1 & -1 & 1 & -1 & -1 & -1 \\
-1 & -1 & 1 & -1 & 1 & 1 & -1 & -1 \\
1 & -1 & 1 & -1 & -1 & -1 & 1 & 1 \\
-1 & 1 & 1 & -1 & 1 & -1 & -1 & 1 \\
1 & 1 & 1 & -1 & -1 & 1 & 1 & -1 \\
-1 & -1 & -1 & 1 & 1 & -1 & 1 & 1 \\
1 & -1 & -1 & 1 & -1 & 1 & -1 & -1 \\
-1 & 1 & -1 & 1 & 1 & 1 & 1 & -1 \\
1 & 1 & -1 & 1 & -1 & -1 & -1 & 1 \\
-1 & -1 & 1 & 1 & -1 & 1 & -1 & 1 \\
1 & -1 & 1 & 1 & 1 & -1 & 1 & -1 \\
-1 & 1 & 1 & 1 & -1 & -1 & -1 & -1 \\
1 & 1 & 1 & 1 & 1 & 1 & 1 & 1
\end{array}
\tag{14.2}
$$

(1)	aef	bf	abe
$abcf$	bce	ac	cef
$bdef$	abd	de	adf
$acde$	cdf	$abcdef$	bcd

This design can also be constructed by first identifying the treatment combinations that have even numbers of letters in common with all of $ACDE$, $ABCF$, AC, and ABD: (1), $abcf$, $bdef$, $acde$. These treatment combinations form the first block. The remaining blocks can be obtained by successively multiplying all the treatment combinations in the blocks that have been generated by treatment combinations that have even numbers of letters in common with $ACDE$ and $ABCF$, but have not appeared.

Suppose **b** is a nonzero linear combination of $\mathbf{b}_1, \ldots, \mathbf{b}_q$. Then, as in Section 7.2, the treatment contrasts defined by $P(\mathbf{b})$ are confounded with blocks. An additional complication here is that their aliases are also confounded with blocks. Hereafter we only consider designs under which all the main-effect contrasts are neither aliased with other main-effect contrasts nor confounded with blocks.

Example 14.2. Under the blocked design constructed in Example 14.1, AC, ABD, their generalized interaction $(AC)(ABD) = BCD$, and all the aliases of these three effects are confounded with blocks. The 15 alias sets of factorial effects are shown on the next page. The effects in the three underlined alias sets are confounded with blocks.

$$A = BCF = CDE = ABDEF$$
$$B = ACF = ABCDE = DEF$$
$$C = ABF = ADE = BCDEF$$
$$D = ABCDF = ACE = BEF$$
$$E = ABCEF = ACD = BDF$$
$$F = ABC = ACDEF = BDE$$
$$AB = CF = BCDE = ADEF$$
$$\underline{AC = BF = DE = ABCDEF}$$
$$AD = BCDF = CE = ABEF$$
$$AE = BCEF = CD = ABDF$$
$$AF = BC = CDEF = ABDE$$
$$BD = ACDF = ABCE = EF$$
$$BE = ACEF = ABCD = DF$$
$$\underline{ABD = CDF = BCE = AEF}$$
$$\underline{ABE = CEF = BCD = ADF}$$

The array in (14.2) can be viewed as a 2^{8-4} design with the defining relation

$$I = ABCF = ACDE = AC\mathcal{B}_1 = ABD\mathcal{B}_2 = BDEF = BF\mathcal{B}_1 = CDF\mathcal{B}_2 = DE\mathcal{B}_1 = BCE\mathcal{B}_2$$
$$= BCD\mathcal{B}_1\mathcal{B}_2 = ABCDEF\mathcal{B}_1 = AEF\mathcal{B}_2 = ADF\mathcal{B}_1\mathcal{B}_2 = ABE\mathcal{B}_1\mathcal{B}_2 = CEF\mathcal{B}_1\mathcal{B}_2,$$

which can be obtained from the four independent defining words $ABCF$, $ACDE$, $AC\mathcal{B}_1$, and $ABD\mathcal{B}_2$. The 12 treatment factorial effects that are confounded with blocks are those that are aliased with $\mathcal{B}_1$, $\mathcal{B}_2$, or $\mathcal{B}_1\mathcal{B}_2$ and can be identified from the defining words that contain at least one of $\mathcal{B}_1$ and $\mathcal{B}_2$. We will show in Section 14.9 that this design is optimal with respect to a certain criterion.

In general, a regular s^{n-p} fractional factorial design in s^q blocks of size s^{n-p-q} can be viewed as an $s^{(n+q)-(p+q)}$ design with $p+q$ independent defining words, where n of the $n+q$ factors are genuine treatment factors, and the remaining q factors are blocking factors. Then there are two kinds of defining words. The defining words that do not contain blocking factors are called *treatment defining words*, and those that contain at least one blocking factor are called *block defining words*. The treatment defining words can be used to determine aliasing among the treatment factorial effects, while the treatment factorial effects that are confounded with blocks can be identified from the block defining words.

The design in Example 14.1 can be constructed by using the design key matrix

$$
\begin{array}{cccc}
\mathcal{P}_1 & \mathcal{P}_2 & \mathcal{B}_1 & \mathcal{B}_2 \\
\end{array}
$$
$$
\begin{bmatrix}
1 & 0 & 0 & 0 \\
0 & 1 & 0 & 0 \\
1 & 0 & 1 & 0 \\
1 & 1 & 0 & 1 \\
1 & 1 & 1 & 1 \\
0 & 1 & 1 & 0 \\
\end{bmatrix}
\begin{array}{c}
A \\ B \\ C \\ D \\ E \\ F
\end{array}
$$

The first four rows constitute a design key for a blocked 2^4 complete factorial for A, B, C, D and can be obtained by using the template in (7.9). Two independent blocking words AC and ABD can be identified from the third and fourth rows of $\mathbf{K}$. There are two extra rows for the added factors E and F. These rows are the sums of the rows corresponding to the basic factors whose interactions (ACD and ABC, respectively) are used to define E and F. Using the four independent generators $acde$, $bdef$, cef, and de obtained from the columns of $\mathbf{K}$ to generate 16 treatment combinations in the Yates order, and putting each set of four consecutive treatment combinations in the same block, we produce the same design as in Example 14.1.

14.3   Fractional factorial split-plot designs

Suppose there are n treatment factors each with s levels and s^q whole-plots each containing s^{m-q} subplots, where s is a prime number or power of a prime number. Among the n treatment factors, n_1 are whole-plot factors and the other $n_2 = n - n_1$ factors are subplot factors. The number of subplots is s^m, and fractional factorial split-plot designs are those with $s^n > s^m$. As in Section 7.10, we allow the possibility that $q > n_1$. For $q \geq n_1$, it is assumed that each whole-plot treatment factor level combination is assigned to s^{q-n_1} whole-plots; for $q < n_1$, we require that the design on the whole-plots be an $s^{n_1-(n_1-q)}$ fractional factorial design. Let $p_1 = \max(0, n_1 - q)$, and $p_2 = n_1 + n_2 - m - p_1$. Then $m = (n_1 + n_2) - (p_1 + p_2)$, and we have an s^{n-p} fractional factorial design for the treatment factors, where $p = p_1 + p_2$.

One can follow the same construction procedure described in Section 14.2, subject to the constraint that main effects of the whole-plot treatment factors are confounded with whole-plot contrasts, and main effects of the subplot treatment factors are confounded with subplot contrasts.

Denote each treatment combination by $\mathbf{x} = (x_1, \ldots, x_{n_1}, x_{n_1+1}, \ldots, x_{n_1+n_2})^T$, where $x_1, \ldots, x_{n_1}$ are levels of the whole-plot treatment factors, and $x_{n_1+1}, \ldots, x_{n_1+n_2}$ are levels of the subplot treatment factors. Then, as in Section 14.2, we need $p_1 + p_2 + q$ linearly independent vectors $\mathbf{a}_1, \ldots, \mathbf{a}_{p_1+p_2+q}$ from EG$(n_1 + n_2, s)$, where $\mathbf{a}_1, \ldots, \mathbf{a}_{p_1+p_2}$ are used to define an $s^{(n_1+n_2)-(p_1+p_2)}$ regular fraction of the $n_1 + n_2$ treatment factors, and the other q independent words $\mathbf{a}_{p_1+p_2+1}, \ldots, \mathbf{a}_{p_1 \mid p_2 \mid q}$ are used to partition the $s^{(n_1+n_2)-(p_1+p_2)}$ treatment combinations into s^q disjoint sets, each of which is assigned to one whole-plot. If $p_1 > 0$, then the first p_1 defining words $\mathbf{a}_1, \ldots, \mathbf{a}_{p_1}$ are used to construct a regular $s^{n_1-p_1}$ fraction of the whole-plot treatment factors. These words involve whole-plot treatment factors only, with all their last n_2 components equal to zero. The factorial effects defined by nonzero linear combinations of $\mathbf{a}_1, \ldots, \mathbf{a}_{p_1}$ are called *whole-plot defining effects*.

The p_2 defining words $\mathbf{a}_{p_1+1}, \ldots, \mathbf{a}_{p_1+p_2}$ involve at least some subplot treatment factors and may or may not involve whole-plot treatment factors. The factorial effects defined by nonzero vectors that are linear combinations of $\mathbf{a}_1, \ldots, \mathbf{a}_{p_1+p_2}$, but are not linear combinations of $\mathbf{a}_1, \ldots, \mathbf{a}_{p_1}$, are called *subplot defining effects*.

We divide into two cases according to whether $n_1 \geq q$ or $n_1 < q$.

Case 1: $n_1 \geq q$

In this case, $q = n_1 - p_1$. Since the whole-plot treatment main effects must be confounded with whole plots, without loss of generality, we may let the q words $\mathbf{a}_{p_1+p_2+1}, \ldots, \mathbf{a}_{p_1+p_2+q}$ be those that define the main effects of the n_1 whole-plot treatment factors (when $p_1 = 0$; that is, $n_1 = q$), or the main effects of basic factors for an $s^{n_1-p_1}$ fraction of whole-plot treatment factors (when $p_1 > 0$). Thus it suffices to select $p_1 + p_2$ independent defining words, and the construction of a regular fractional factorial split-plot design is equivalent to that of an $s^{n_1+n_2-(p_1+p_2)}$ design with the following constraints on the defining words:

(S1) p_1 of the independent defining words involve whole-plot treatment factors only.
(S2) No generalized interaction of the other p_2 independent defining words involves whole-plot treatment factors only.
(S3) There is no defining word in which exactly one factor is a subplot treatment factor and all the other factors are whole-plot treatment factors.

Condition (S2) is to assure that $\mathbf{a}_1, \ldots, \mathbf{a}_{p_1+p_2+q}$ are linearly independent. Condition (S3) prevents any subplot treatment factor from becoming a whole-plot factor. The $s^{n_1+n_2-p_1-p_2}$ treatment combinations in the fraction are divided into $s^{n_1-p_1}$ groups of size $s^{n_2-p_2}$, with those having the same level combination of the whole-plot treatment factors in the same group and assigned to the same whole-plot. Under the constructed design, all the factorial effects of whole-plot treatment factors and their aliases are estimated in the whole-plot stratum, and all the other treatment factorial effects are estimated in the subplot stratum.

Case 2: $n_1 < q$

In this case, $p_1 = 0$. We need $p_2 + q$ independent words $\mathbf{a}_1, \ldots, \mathbf{a}_{p_2}, \mathbf{a}_{p_2+1}, \ldots, \mathbf{a}_{p_2+q}$, where $\mathbf{a}_1, \ldots, \mathbf{a}_{p_2}$ are subplot defining words. As in Case 1, we may let the n_1 words $\mathbf{a}_{p_2+1}, \ldots, \mathbf{a}_{p_2+n_1}$ be those that define the main effects of the n_1 whole-plot treatment factors. The last $q - n_1$ words $\mathbf{a}_{p_2+n_1+1}, \ldots, \mathbf{a}_{p_2+q}$, all of which involve at least some subplot treatment factors, are independent splitting words. Recall that the treatment factorial effects defined by nonzero vectors that are linear combinations of $\mathbf{a}_{p_2+1}, \ldots, \mathbf{a}_{p_2+q}$, but are not linear combinations of $\mathbf{a}_{p_2+1}, \ldots, \mathbf{a}_{p_2+n_1}$, are called *splitting effects*.

Therefore we only need to choose $p_2 + q - n_1$ independent words, including p_2 subplot defining words and $q - n_1$ splitting words. These independent words must satisfy the following conditions.

(S4) No linear combination of $\mathbf{a}_1, \ldots, \mathbf{a}_{p_2}, \mathbf{a}_{p_2+n_1+1}, \ldots, \mathbf{a}_{p_2+q}$ involves whole-plot treatment factors only, that is, no word involving whole-plot treatment factors only can be generated by the independent subplot defining words and splitting words.
(S5) No linear combination of $\mathbf{a}_{p_2+1}, \ldots, \mathbf{a}_{p_2+q}$ defines the main effect of a subplot treatment factor, that is, no subplot treatment main effect can be generated by the

splitting effects and whole-plot treatment factorial effects.

Condition (S4) is to make sure that $\mathbf{a}_1, \ldots, \mathbf{a}_{p_2+q}$ are linearly independent. Condition (S5) is to prevent any subplot treatment main effect from being confounded with whole-plot contrasts. Then all the factorial effects of whole-plot treatment factors, all the splitting effects, and all their aliases are estimated in the whole-plot stratum. All the other treatment factorial effects are estimated in the subplot stratum.

For design key construction of fractional factorial split-plot designs, the template in (7.13) with n replaced by $m = n - p$ can be used for constructing split-plot complete factorials of basic factors. (For $n_1 \geq q$, this template reduces to (7.14), with $n - q$ and q replaced by $n_2 - p_2$ and $n_1 - p_1$, respectively.) Then there is an additional row for each added factor subject to the constraint that the unit aliases of subplot treatment main effects must contain at least one $\mathcal{S}$, and those of whole-plot treatment main effects must not contain any $\mathcal{S}$.

Example 14.3. Suppose there are eight whole-plots, each containing two subplots, and an experiment with five two-level whole-plot treatment factors (A, B, C, D, E) and three two-level subplot treatment factors (U, V, W) is to be performed. This calls for a 2^{5-2} design on the whole-plots. Then we have $q = 3$, $n_1 = 5$, $n_2 = 3$, $p_1 = 2$, and $p_2 = 2$. Since $n_1 > q$, we need to choose four independent defining words: two whole-plot defining words that contain whole-plot treatment factors only and two subplot defining words that contain some subplot treatment factors. Suppose we choose the basic factors to be A, B, C, and U, and let $D = -AB$, $E = -AC$, $V = -AU$, and $W = BCU$. Then we have the following design with each pair having the same level combination of A, B, C, D, E assigned to one whole-plot.

A	B	C	$D = -AB$	$E = -AC$	U	$V = -AU$	$W = BCU$
-1	-1	-1	-1	-1	-1	-1	-1
-1	-1	-1	-1	-1	1	1	1
-1	-1	1	-1	1	-1	-1	1
-1	-1	1	-1	1	1	1	-1
-1	1	-1	1	-1	-1	-1	1
-1	1	-1	1	-1	1	1	-1
-1	1	1	1	1	-1	-1	-1
-1	1	1	1	1	1	1	1
1	-1	-1	1	1	-1	1	-1
1	-1	-1	1	1	1	-1	1
1	-1	1	1	-1	-1	1	1
1	-1	1	1	-1	1	-1	-1
1	1	-1	-1	1	-1	1	1
1	1	-1	-1	1	1	-1	-1
1	1	1	-1	-1	-1	1	-1
1	1	1	-1	-1	1	-1	1

With the treatment combinations written in shorthand notation, the eight whole-plots are

(1)		cew		bdw		bcde
uvw		ceuv		bduv		bcdeuvw

adev		acdvw		abevw		abcv
adeuw		acdu		abeu		abcuw

From the independent defining words ABD, ACE, AUV, and $BCUW$, we obtain the defining relation

$$I = -ABD = -ACE = -AUV = BCUW = BCDE$$
$$= BDUV = -ACDUW = CEUV = -ABEUW = -ABCVW$$
$$= -ABCDEUV = DEUW = CDVW = BEVW = -ADEVW.$$

None of the defining words contains only one subplot treatment factor. Therefore (S3) is satisfied. If we assume that all the three-factor and higher-order interactions are negligible, then the following summarizes the aliasing and confounding of the treatment main effects and two-factor interactions, with the signs ignored:

Strata	Aliasing
Whole-plot	$A = BD = CE = UV$, $B = AD$, $C = AE$, $D = AB$, $E = AC$, $BC = DE = UW$, $BE = CD = VW$
Subplot	$U = AV$, $V = AU$, W, AW, $BU = CW = DV$, $BV = DU = EW$, $BW = CU = EV$, $CV = DW = EU$

This design can also be obtained by using the design key matrix

$$\mathbf{K} = \begin{array}{cccc} \mathcal{S}_1 & \mathcal{P}_1 & \mathcal{P}_2 & \mathcal{P}_3 \\ \left[\begin{array}{cccc} 1 & 0 & 0 & 0 \\ 0 & 1 & 0 & 0 \\ 0 & 0 & 1 & 0 \\ 0 & 0 & 0 & 1 \\ 0 & 1 & 1 & 0 \\ 0 & 1 & 0 & 1 \\ 1 & 1 & 0 & 0 \\ 1 & 0 & 1 & 1 \end{array}\right] & \begin{array}{c} U \\ A \\ B \\ C \\ D \\ E \\ V \\ W \end{array} \end{array}$$

In this example, since $n_1 > q$, we can use (7.14) to write down a design key for the basic factors A, B, C, and U. We need to add one row for each added factor, subject to the constraint that the first entries of the rows corresponding to V and W must be 1, and they must be 0 for D and E. Or, if the defining words have already been chosen as in this example, then the row corresponding to each added factor can be determined as the sum of those corresponding to the basic factors whose interactions are used to define the added factors, but we need to verify that the first entry of each of the resulting rows satisfies the constraint mentioned above.

Example 14.4. Suppose there are eight whole-plots, each containing two subplots, and an experiment with two two-level whole-plot treatment factors (A, B) and five two-level subplot treatment factors (S, T, U, V, W) is to be conducted. This calls for two replications of the complete factorial of A and B on the whole-plots. Then we have $m = 4$, $q = 3$, $n_1 = 2$, $n_2 = 5$, $p_1 = 0$, and $p_2 = 3$. We need to choose four independent words: three subplot defining words and one splitting word. Let A, B, S, and T be basic factors, define U, V, and W by $U = -AS$, $V = -BS$, $W = ABT$, and use ST as the splitting word. Then we have the following design, with the last column showing the values of the splitting word.

A	B	S	T	$U = -AS$	$V = -BS$	$W = ABT$	ST
-1	-1	-1	-1	-1	-1	-1	1
-1	-1	1	1	1	1	1	1
-1	-1	1	-1	1	1	-1	-1
-1	-1	-1	1	-1	-1	1	-1
-1	1	-1	-1	-1	1	1	1
-1	1	1	1	1	-1	-1	1
-1	1	1	-1	1	-1	1	-1
-1	1	-1	1	-1	1	-1	-1
1	-1	-1	-1	1	-1	1	1
1	-1	1	1	-1	1	-1	1
1	-1	1	-1	-1	1	1	-1
1	-1	-1	1	1	-1	-1	-1
1	1	-1	-1	1	1	-1	1
1	1	1	1	-1	-1	1	1
1	1	1	-1	-1	-1	-1	-1
1	1	-1	1	1	1	1	-1

Each set of four treatment combinations with A and B at constant levels is divided into two sets according to the values of ST, producing the following design layout:

(1)	suv	bvw	$bsuw$
$stuvw$	tw	$bstu$	btv

auw	$asvw$	$abuv$	abs
$astv$	atu	$abstw$	$abtuvw$

The factorial effects that are estimated in the whole-plot stratum consist of A, B, ST

and all their generalized interactions and aliases. This design can also be obtained by using a design key matrix:

$$
\mathbf{K} =
\begin{array}{cccc}
\mathcal{S}_1 & \mathcal{P}_1 & \mathcal{P}_2 & \mathcal{P}_3 \\
\left[\begin{array}{cccc}
1 & 0 & 0 & 0 \\
0 & 1 & 0 & 0 \\
0 & 0 & 1 & 0 \\
1 & 0 & 0 & 1 \\
1 & 1 & 0 & 0 \\
1 & 0 & 1 & 0 \\
1 & 1 & 1 & 1
\end{array}\right]
\begin{array}{c}
S \\ A \\ B \\ T \\ U \\ V \\ W
\end{array}
\end{array}
$$

We use (7.13) to write down the first four rows of $\mathbf{K}$ corresponding to the basic factors S, T, A, and B. Then we add one row for each of the three added factors. The first entry of each of these three rows must be 1. The splitting word ST can be identified from the fourth row.

14.4 Blocked fractional factorial split-plot designs

In the setting of Section 14.3, suppose the whole-plots are to be grouped into s^g blocks of size s^{q-g}. As in Section 13.2, for $n_1 > q$ we require the design on the whole-plots to be an $s^{n_1-(n_1-q)}$ fraction of the whole-plot treatment factors. For $q \geq n_1 > q-g$ (respectively, $n_1 \leq q-g$), we require an $s^{n_1-(n_1-q+g)}$ fraction (respectively, s^{q-g-n_1} replicates of the complete factorial) of the whole-plot treatment factors in each block. We first use the procedure presented in Section 14.3 to construct an unblocked design with s^q whole-plots, each containing s^{m-q} subplots; then we choose g independent words, each of which is a splitting word or involves the whole-plot treatment factors only, to partition the whole-plots into s^g blocks. If $n_1 \geq q$, then these g words involve whole-plot treatment factors only. On the other hand, if $n_1 < q$, then as many independent splitting words as possible should double as blocking words. Specifically, for $g \leq q-n_1$, all the blocking words are splitting words, and for $g > q-n_1$, besides the $q-n_1$ independent splitting words, $g-(q-n_1)$ additional independent blocking words that involve the whole-plot treatment factors only are also needed. The g independent blocking effects and their generalized interactions, as well as all the aliases, are confounded with blocks, and we need to check that none of them are main effects if confounding main effects with blocks is to be avoided.

For design key construction, we use (13.9), with n replaced by $n-p$, as the template for a complete factorial of basic factors.

Example 14.5. In Example 14.3, suppose the eight whole-plots are to be divided into two blocks. Since $n_1 > q$, one blocking word involving the whole-plot treatment factors only is needed. Suppose the three-factor interaction ABC is used. Then we have the following design.

First block	(1)	bcde	acdvw	abevw
	uvw	bcdeuvw	acdu	abeu

Second block	cew	bdw	adev	abcv
	ceuv	bduv	adeuw	abcuw

The first block consists of all the treatment combinations that have even numbers of letters in common with *ABC*. This three-factor interaction and its aliases (*CD*, *BE*, *VW*, and some other higher-order interactions), which are estimated in the whole-plot stratum under the unblocked design, are now estimated in the block stratum. The same design can also be constructed by using the following design key matrix, where $\mathcal{B}$ is a two-level unit factor that defines the two blocks.

$$\mathbf{K} = \begin{matrix} & \mathcal{S}_1 & \mathcal{P}_1 & \mathcal{P}_2 & \mathcal{B} & \\ & \begin{bmatrix} 1 & 0 & 0 & 0 \\ 0 & 1 & 0 & 0 \\ 0 & 0 & 1 & 0 \\ 0 & 1 & 1 & 1 \\ 0 & 1 & 1 & 0 \\ 0 & 0 & 1 & 1 \\ 1 & 1 & 0 & 0 \\ 1 & 1 & 0 & 1 \end{bmatrix} & \begin{matrix} U \\ A \\ B \\ C \\ D \\ E \\ V \\ W \end{matrix} \end{matrix}$$

We use the template in (13.9) to write down the first four rows of **K** corresponding to the basic factors *U*, *A*, *B*, and *C*. The second and third entries of the fourth row are the only entries in the template that need to be filled. In this example, we make both entries equal to 1. This amounts to using *ABC* as the blocking word. Rows corresponding to the added factors are supplemented. The generators identified from the columns of the design key matrix can be used to obtain the design constructed earlier.

Example 14.6. (Examples 13.1 and 1.2 revisited) In Example 13.1, we have already constructed a complete 2^5 factorial for three two-level whole-plot treatment factors *A*, *B*, *C* and two two-level subplot treatment factors *S* and *T*, with the block structure (4 blocks)/(4 whole-plots)/(2 subplots). A design key was given in Example 13.9. The original Example 1.2 has an additional subplot treatment factor *U*. The construction is completed by defining *U* in terms of the basic factors subject to the constraint that the main effect of *U* is confounded with neither whole-plot nor block contrasts. Suppose we define *U* by $U = ABT$. Expanding the design key in Example 13.9 by an additional row that is the sum of the three rows corresponding to *A*, *B*, and *T*, we obtain the following design key for a 2^{6-1} design as required.

$$
\begin{array}{ccccc}
\mathcal{S} & \mathcal{P}_1 & \mathcal{P}_2 & \mathcal{B}_1 & \mathcal{B}_2 \\
\begin{bmatrix}
1 & 0 & 0 & 0 & 0 \\
0 & 1 & 0 & 0 & 0 \\
0 & 0 & 1 & 0 & 0 \\
0 & 1 & 1 & 1 & 0 \\
1 & 0 & 1 & 0 & 1 \\
1 & 1 & 0 & 0 & 1
\end{bmatrix}
&
\begin{matrix}
S \\ A \\ B \\ C \\ T \\ U
\end{matrix}
\end{array}
$$

This design has defining word $ABTU$, splitting word BST, which is also a blocking word, and an additional independent blocking word ABC.

14.5   Fractional factorial strip-plot designs

Suppose the block structure is $s^{n_1-p_1} \times s^{n_2-p_2}$, where s is a prime number or power of a prime number, $p_1 \geq 0$, $p_2 \geq 0$, and $p_1 + p_2 > 0$. Furthermore, there are n_1 row treatment factors and n_2 column treatment factors, all with s levels. Then we have an $s^{(n_1+n_2)-(p_1+p_2)}$ fractional factorial for the treatment factors. We impose the requirement that no treatment main effect is confounded with blocks and no level combination of row (respectively, column) treatment factors is replicated in the row (respectively, column) design. Then the row and column designs are, respectively, $s^{n_1-p_1}$ and $s^{n_2-p_2}$ designs.

We first choose p_1 independent defining words $\mathbf{a}_1, \ldots, \mathbf{a}_{p_1}$ from $EG(n_1, s)$ to construct an $s^{n_1-p_1}$ fractional factorial design of the row treatment factors and choose p_2 independent defining words $\mathbf{a}_{p_1+1}, \ldots, \mathbf{a}_{p_1+p_2}$ from $EG(n_2, s)$ to construct an $s^{n_2-p_2}$ fractional factorial design of the column treatment factors. Then we form the product of the row and column designs, assigning each of the $s^{n_1-p_1}$ level combinations of the row design to all the units in one row, and assigning each of the $s^{n_2-p_2}$ level combinations of the column design to all the units in one column. Then the factorial effects of row treatment factors are estimated in the row stratum, those of the column treatment factors are estimated in the column stratum, and the remaining factorial effects are estimated in the bottom stratum.

In the notations of Section 14.1, $m = 2$, $l_1 = n_1 - p_1$, $l_2 = n_2 - p_2$, $n = n_1 + n_2$, and $p = p_1 + p_2$. Denote each treatment combination by $\mathbf{x} = (x_1, \ldots, x_{n_1}, x_{n_1+1}, \ldots, x_{n_1+n_2})^T$, where $x_1, \ldots, x_{n_1}$ are levels of the row treatment factors, and $x_{n_1+1}, \ldots, x_{n_1+n_2}$ are levels of the column treatment factors. Then the defining words in Δ_0 are $\binom{\mathbf{a}_1}{\mathbf{0}}, \ldots,$ $\binom{\mathbf{a}_{p_1}}{\mathbf{0}}, \binom{\mathbf{0}}{\mathbf{a}_{p_1+1}}, \ldots, \binom{\mathbf{0}}{\mathbf{a}_{p_1+p_2}}$, Δ_1 consists of the main effects of $n_1 - p_1$ basic factors for the row design, and Δ_2 consists of the main effects of $n_2 - p_2$ basic factors for the column design. Effectively there is no need to choose Δ_1 and Δ_2.

Now consider the construction of blocked fractional factorial strip-plot designs. Suppose the block structure is $s^g/(s^{n_1-p_1-g} \times s^{n_2-p_2-g})$. Then the total run size is $s^{(n_1+n_2)-(p_1+p_2+g)}$. We assume that $p_1 \geq 0$, $p_2 \geq 0$, and $p_1 + p_2 + g > 0$. In this case, $m = 3$, $l_1 = g$, $l_2 = n_1 - p_1 - g$, $l_3 = n_2 - p_2 - g$, $n = n_1 + n_2$, $p = p_1 + p_2 + g$, and we need four sets of independent words Δ_0, Δ_1, Δ_2, and Δ_3 to complete the construction, where Δ_0

consists of $p_1 + p_2 + g$ independent words for defining the treatment fraction, and Δ_1 consists of g independent words for partitioning the treatment combinations into s^g blocks. There is no need to choose Δ_2 and Δ_3: within each block we simply put all the treatment combinations with the same level combination of the row treatment factors in the same row, and those with the same level combination of the column treatment factors in the same column. The row and column designs are, respectively, blocked $s^{n_1-p_1}$ and $s^{n_2-p_2}$ designs, both with s^g blocks.

The g independent words in Δ_0 are blocking words for both the row design and column design. This implies that the factorial effect defined by each of these g words is simultaneously a row treatment factorial effect and a column treatment factorial effect. This is possible only if these two effects are aliased. Suppose $\mathbf{a}_1, \ldots, \mathbf{a}_{p_1}$, $\mathbf{b}_1, \ldots, \mathbf{b}_g$ are, respectively, $n_1 \times 1$ independent defining words and blocking words for the row design, and $\mathbf{a}_{p_1+1}, \ldots, \mathbf{a}_{p_1+p_2}, \mathbf{c}_1, \ldots, \mathbf{c}_g$ are, respectively, $n_2 \times 1$ independent defining words and blocking words for the column design. Without loss of generality, we may assume that $\binom{\mathbf{b}_i}{\mathbf{0}}^T \mathbf{x} + \binom{\mathbf{0}}{\mathbf{c}_i}^T \mathbf{x} = 0$, that is, $\binom{\mathbf{b}_i}{\mathbf{c}_i}^T \mathbf{x} = 0$ for all $i = 1, \ldots, g$, and all the $s^{(n_1+n_2)-(p_1+p_2+g)}$ treatment combinations $\mathbf{x}$. Then each of $\binom{\mathbf{b}_1}{\mathbf{c}_1}, \ldots, \binom{\mathbf{b}_g}{\mathbf{c}_g}$ is a defining word for the $s^{(n_1+n_2)-(p_1+p_2+g)}$ fraction. These g words together with $\binom{\mathbf{a}_1}{\mathbf{0}}, \ldots, \binom{\mathbf{a}_{p_1}}{\mathbf{0}}$ and $\binom{\mathbf{0}}{\mathbf{a}_{p_1+1}}, \ldots, \binom{\mathbf{0}}{\mathbf{a}_{p_1+p_2}}$ constitute the $p_1 + p_2 + g$ independent defining words in Δ_0, and Δ_1 is either $\{\binom{\mathbf{b}_1}{\mathbf{0}}, \ldots, \binom{\mathbf{b}_g}{\mathbf{0}}\}$ or $\{\binom{\mathbf{0}}{\mathbf{c}_1}, \ldots, \binom{\mathbf{0}}{\mathbf{c}_g}\}$.

In other words, we first choose blocked fractional factorial designs for the row and column treatment factors separately. The product of the row and column designs, with $\binom{\mathbf{a}_1}{\mathbf{0}}, \ldots, \binom{\mathbf{a}_{p_1}}{\mathbf{0}}, \binom{\mathbf{0}}{\mathbf{a}_{p_1+1}}, \ldots, \binom{\mathbf{0}}{\mathbf{a}_{p_1+p_2}}$ as independent defining words and $\binom{\mathbf{b}_1}{\mathbf{0}}, \ldots, \binom{\mathbf{b}_g}{\mathbf{0}}, \binom{\mathbf{0}}{\mathbf{c}_1}, \ldots, \binom{\mathbf{0}}{\mathbf{c}_g}$ as blocking words, can be viewed as an $s^{(n_1+n_2)-(p_1+p_2)}$ fractional factorial design with the block structure $s^{2g}/(s^{n_1-p_1-g} \times s^{n_2-p_2-g})$. This design is obtained by coupling each block of the row design with every block of the column design to form a total of s^{2g} blocks of $s^{n_1-p_1-g}$ rows and $s^{n_2-p_2-g}$ columns. The construction is completed by choosing s^g of the s^{2g} blocks in the product design, which is achieved by aliasing the row treatment factorial effect defined by each $\mathbf{b}_i$ with the column treatment factorial effect defined by a certain $\mathbf{c}_j$. This is the construction proposed by Miller (1997). Then the row and column treatment factorial effects that are confounded with blocks under the row and column designs, respectively, and their aliases, are confounded with blocks under the constructed design. All the factorial effects of the row treatment factors (and their aliases), except those that are confounded with blocks, are estimated in the row stratum. Likewise all the factorial effects of the column treatment factors (and their aliases), except those that are confounded with blocks, are estimated in the column stratum. All the other treatment factorial effects are estimated in the bottom stratum.

Example 14.7. (Example 1.3 revisited) The experiment discussed in Miller (1997) has six two-level row treatment factors denoted by A, B, C, D, E, F, and four column treatment factors denoted by S, T, U, and V. The block structure is (2 blocks)/[(4 rows) × (4 columns)]. The row design is a 2^{6-3} design with two blocks of size four, and the column design is a 2^{4-1} design also with two blocks of size four. Suppose we pick the defining relations $I = -ABD = -BCF = ABCE = ACDF = -CDE = -AEF$ $= BDEF$ and $I = STUV$ for the row and column designs, respectively, use AC to block the row design, and use SU to block the column design. Each block of the row design can be coupled with every block of the column design to form a block of four rows and four columns. Altogether there are four such blocks in the resulting product design, from which two are to be chosen. Suppose we choose the two blocks that satisfy $AC = SU$. Then we have the 2^{10-5} design with independent defining words $-ABD$, $-BCF$, $ABCE$, $STUV$, and $ACSU$, and with the blocks defined by AC (or equivalently, by SU). The blocked row and column designs are shown below. We can write down the row design by using A, B, C as basic factors, with the four level combinations satisfying $AC = 1$ in the first block and those with $AC = -1$ in the second block, and define D, E, and F by $D = -AB$, $E = ABC$, $F = -BC$. For the column design, we use S, T, U as basic factors, with the four level combinations satisfying $SU = 1$ in the first block and those with $SU = -1$ in the second block, and then define V by $V = STU$.

Washer	A	B	C	D $-AB$	E ABC	F $-BC$	Dryer	S	T	U	V STU
Block 1											
1	-1	-1	-1	-1	-1	-1	1	-1	-1	-1	-1
2	$+1$	-1	$+1$	$+1$	-1	$+1$	2	$+1$	-1	$+1$	-1
3	-1	$+1$	-1	$+1$	$+1$	$+1$	3	-1	$+1$	-1	$+1$
4	$+1$	$+1$	$+1$	-1	$+1$	-1	4	$+1$	$+1$	$+1$	$+1$
Block 2											
1	-1	-1	$+1$	-1	$+1$	$+1$	1	-1	-1	$+1$	$+1$
2	$+1$	-1	-1	$+1$	$+1$	-1	2	$+1$	-1	-1	$+1$
3	-1	$+1$	$+1$	$+1$	-1	-1	3	-1	$+1$	$+1$	-1
4	$+1$	$+1$	-1	-1	-1	$+1$	4	$+1$	$+1$	-1	-1

In each block, each combination of the washer factors is coupled with each combination of the dryer factors to produce 16 runs. With the treatment combinations written in shorthand notation, we have the following design before randomization.

Block 1

(1)	su	tv	$stuv$
$acdf$	$acdfsu$	$acdftv$	$acdfstuv$
$bdef$	$bdefsu$	$bdeftv$	$bdefstuv$
$abce$	$abcesu$	$abcetv$	$abcestuv$

Block 2

$cefuv$	$cefsv$	$ceftu$	$cefst$
$adeuv$	$adesv$	$adetu$	$adest$
$bcduv$	$bcdsv$	$bcdtu$	$bcdst$
$abfuv$	$abfsv$	$abftu$	$abfst$

This design is the same as the one constructed in Miller (1997) subject to factor relabeling.

14.6   Design key construction of blocked strip-plot designs

We first write down a design key as in (13.12) for the complete factorial of $n_1 + n_2 - p_1 - p_2 - g$ basic factors. We call it a basic design key. Suppose u rows of the basic design key matrix correspond to row treatment factors and the other $v = n_1 + n_2 - p_1 - p_2 - g - u$ rows correspond to column treatment factors. If $u < n_1 - p_1$, then we need to expand the u row treatment factors in the basic design key to a set of $n_1 - p_1$ basic factors for the row design. On the other hand, if $v < n_2 - p_2$, then we need to add $n_2 - p_2 - v$ column treatment factors to form a set of basic factors for the column design. Altogether we need to add $(n_1 - p_1 - u) + (n_2 - p_2 - v) = g$ factors to serve as basic factors for the row or column design. These factors must be defined by g independent words of the $u + v$ factors in the basic design key. The following result is useful for this purpose.

Lemma 14.1. *(Cheng and Tsai, 2013) Suppose neither the row nor the column design is a replicated fraction. Then each additional basic factor for the row or column design must be defined by a word of the form XY, where both X and Y involve at least one factor, and X (respectively, Y) involves the row (respectively, column) treatment factors in the basic design key only. If XY defines a row (respectively, column) treatment factor, then Y (respectively, X) is a blocking word.*

Proof. If an additional basic factor for the row design is defined by a word that involves row treatment factors only, then it would not be possible to generate $s^{n_1 - p_1}$ distinct level combinations of the row treatment factors. This contradicts the assumption that the row design is not a replicated fraction. Write such a word in the form XY, where X involves row treatment factors only and Y contains at least one column treatment factor. Then since XY defines a row treatment factor, the unit alias of Y must not involve any of $C_1, \ldots, C_{n_2 - p_2 - g}$. Furthermore, since Y involves column treatment factors only, its unit alias also does not involve any of $\mathcal{R}_1, \ldots, \mathcal{R}_{n_1 - p_1 - g}$. This implies that Y is a blocking word. The same argument applies to additional basic factors for the column design.   □

An implication of Lemma 14.1 is that each additional basic factor for the row (respectively, column) design is defined by aliasing a row (respectively, column) treatment interaction with a blocking word that involves column (respectively, row) treatment factors only. This yields g independent defining words, corresponding to g

pairs of aliased blocking words for the row and column designs. One can see that this is equivalent to selecting s^g of the s^{2g} blocks in the product of a blocked row design and a blocked column design discussed in the previous section.

Example 14.8. (Example 14.7 continued) A design key for the five basic factors A, B, C, S, T, with one blocking word AC, was given in (13.7). We have enough number (3) of basic factors for the row design, but need one more basic factor, say U, for the column design. It follows from Lemma 14.1 that we must define U as ACX, where X is a factorial effect of S and T. Suppose we define U as ACS. Then $ACSU$ is a defining word. It follows that SU is aliased with AC and is also a blocking word. We can complete the construction by choosing three independent interactions of A, B, and C to define D, E, and F, and one interaction of S, T, U to define V. For example, if we define D, E, F, and V by $D = -AB$, $E = ABC$, $F = -BC$, and $V = STU$ as in Example 14.7, then we have the following design key matrix.

$$
\begin{array}{ccccc}
\mathcal{R}_1 & \mathcal{R}_2 & \mathcal{C}_1 & \mathcal{C}_2 & \mathcal{B} \\
\end{array}
$$

$$
\begin{bmatrix}
1 & 0 & 0 & 0 & 0 \\
0 & 1 & 0 & 0 & 0 \\
0 & 0 & 1 & 0 & 0 \\
0 & 0 & 0 & 1 & 0 \\
1 & 0 & 0 & 0 & 1 \\
1 & 1 & 0 & 0 & 0 \\
0 & 1 & 0 & 0 & 1 \\
1 & 1 & 0 & 0 & 1 \\
0 & 0 & 1 & 0 & 1 \\
0 & 0 & 0 & 1 & 1 \\
\end{bmatrix}
\begin{array}{l}
A \\
B \\
S \\
T \\
C \\
D \\
E \\
F \\
U \\
V \\
\end{array}
$$

14.7 Post-fractionated strip-plot designs

A strip-plot design in which the row design is an $s^{n_1-p_1}$ fractional factorial design and the column design is an $s^{n_2-p_2}$ fractional factorial design, as constructed at the beginning of Section 14.5, requires $s^{(n_1+n_2)-(p_1+p_2)}$ experimental runs. Vivacqua and Bisgaard (2009) considered the situation where in order to save time or experimental cost, only a $1/s^g$ fraction of the $s^{(n_1+n_2)-(p_1+p_2)}$ treatment combinations is to be observed. To choose such a fraction, we need g independent defining words, in addition to p_1 independent defining words $\mathbf{a}_1, \ldots, \mathbf{a}_{p_1}$ for the row design and p_2 independent defining words $\mathbf{a}_{p_1+1}, \ldots, \mathbf{a}_{p_1+p_2}$ for the column design. We only consider the case where each of these g independent words involves at least one row treatment factor and at least one column treatment factor. This is because if k of the g words involve row treatment factors only, then the row design would be reduced to an $s^{n_1-p_1-k}$ fraction; likewise, if l of them involve column treatment factors only, then the column design would be reduced to an $s^{n_2-p_2-l}$ fraction. Then the g additional independent words must be of the form $\binom{\mathbf{b}_i}{\mathbf{c}_i}$, $1 \leq i \leq g$, where $\mathbf{b}_i \neq \mathbf{0}$, $\mathbf{c}_i \neq \mathbf{0}$, $\mathbf{b}_i$ is $n_1 \times 1$ and $\mathbf{c}_i$ is $n_2 \times 1$. Such designs are called post-fractionated strip-block designs in Vivacqua and

Bisgaard (2009).

Example 14.9. Suppose there are six two-level first-stage (row) treatment factors A, B, C, D, E, F, and four two-level second-stage (column) treatment factors S, T, U, and V. An 8×8 strip-plot design with the row design defined by $I = -ABD = -BCF = ABCE = ACDF = -CDE = -AEF = BDEF$ and the column design defined by $I = STUV$ requires 64 experimental runs. To construct a design with 32 runs, suppose we choose the defining word $ACSU$. Then we end up with the following design.

(1)	su	tv	$stuv$				
$acdf$	$acdfsu$	$acdftv$	$acdfstuv$				
$bdef$	$bdefsu$	$bdeftv$	$bdefstuv$				
$abce$	$abcesu$	$abcetv$	$abcestuv$				
				$cefuv$	$cefsv$	$ceftu$	$cefst$
				$adeuv$	$adesv$	$adetu$	$adest$
				$bcduv$	$bcdsv$	$bcdtu$	$bcdst$
				$abfuv$	$abfsv$	$abftu$	$abfst$

We see that the 32 runs can be arranged in two 4×4 pseudo blocks. The pseudo blocks result from post-fractionation and, unlike in Example 14.7, are not part of the original block structure. Note that these pseudo blocks are the same as the two blocks of the design constructed in Example 14.7.

The g words $\mathbf{b}_1, \ldots, \mathbf{b}_g$ can be used to block the row design defined by $\mathbf{a}_1, \ldots, \mathbf{a}_{p_1}$ into s^g blocks of size $s^{n_1 - p_1 - g}$, and $\mathbf{c}_1, \ldots, \mathbf{c}_g$ can be used to block the column design into s^g blocks of size $s^{n_2 - p_2 - g}$. As in Section 14.5, by coupling each pseudo block of the row design with every pseudo block of the column design, we obtain s^{2g} pseudo blocks of $s^{n_1 - p_1 - g}$ rows and $s^{n_2 - p_2 - g}$ columns. Using $\binom{\mathbf{b}_1}{\mathbf{c}_1}, \ldots, \binom{\mathbf{b}_g}{\mathbf{c}_g}$ as defining words amounts to choosing s^g of the s^{2g} pseudo blocks by aliasing the row treatment factorial effects defined by $\mathbf{b}_1, \ldots, \mathbf{b}_g$ with the column treatment factorial effects defined by $\mathbf{c}_1, \ldots, \mathbf{c}_g$, respectively. This results in a design that is combinatorially equivalent to a blocked strip-plot design with the block structure $s^g / (s^{n_1 - p_1 - g} \times s^{n_2 - p_2 - g})$. It turns out that the two designs also have the same aliasing/confounding pattern of treatment factorial effects. Let $\mathcal{R}$ and $\mathcal{C}$ be the $s^{n_1 - p_1}$-level and $s^{n_2 - p_2}$-level factors on the $s^{n_1 + n_2 - p_1 - p_2 - g}$ units defined by the row and column partitions, respectively. Then $\mathcal{R}$ and $\mathcal{C}$ are uniform and orthogonal. We have $\mathcal{R} \vee \mathcal{C} = \mathcal{B}$, where $\mathcal{B}$ is the s^g-level factor defined by the pseudo blocks. Since $\mathcal{B}$ is uniform, by the discussions in Sections 12.13 and 13.12, under model (12.21) with one set of random effects for each of $\mathcal{R}$, $\mathcal{C}$, and $\mathcal{E}$, there is an extra stratum corresponding to the pseudo factor $\mathcal{B}$. Furthermore, $\xi_{\mathcal{B}}$ is greater than both $\xi_{\mathcal{R}}$ and $\xi_{\mathcal{C}}$. All the factorial effects defined by nonzero linear combinations of $\mathbf{b}_1, \ldots, \mathbf{b}_g$ (or nonzero linear combinations of $\mathbf{c}_1, \ldots, \mathbf{c}_g$) and their aliases are estimated in $W_{\mathcal{B}} = V_{\mathcal{B}} \ominus V_{\mathcal{U}}$. All the row treatment factorial effects (and their aliases), except those that are estimated in $W_{\mathcal{B}}$, are estimated in $W_{\mathcal{R}} = V_{\mathcal{R}} \ominus V_{\mathcal{B}}$.

Likewise all the column treatment factorial effects (and their aliases), except those that are estimated in $W_{\mathcal{B}}$, are estimated in $W_{\mathcal{C}} = V_{\mathcal{C}} \ominus V_{\mathcal{B}}$. All the other factorial effects are estimated in the bottom stratum. The $p_1 + p_2 + g$ independent defining words $\binom{\mathbf{a}_1}{\mathbf{0}}, \ldots, \binom{\mathbf{a}_{p_1}}{\mathbf{0}}, \binom{\mathbf{0}}{\mathbf{a}_{p_1+1}}, \ldots, \binom{\mathbf{0}}{\mathbf{a}_{p_1+p_2}}, \binom{\mathbf{b}_1}{\mathbf{c}_1}, \ldots, \binom{\mathbf{b}_g}{\mathbf{c}_g}$ can be used to determine aliasing of the treatment factorial effects.

14.8   Criteria for selecting blocked fractional factorial designs based on modified wordlength patterns

In the rest of this chapter, we consider criteria for selecting multi-stratum fractional factorial designs. For simplicity, we restrict to two-level designs, though many results can be extended to the case where the number of levels is a prime or power of a prime. We start with the selection of blocked fractional factorial designs.

The construction of a blocked fractional factorial design requires the selection of appropriate treatment and block defining effects. If we have knowledge about the effects that should not be confounded with blocks, then the algorithm of Franklin and Bailey (1977) and Franklin (1985) can be applied to search for appropriate defining effects. As in Chapter 10, we consider here the selection of blocked fractional factorial designs under the hierarchical assumption that the lower-order effects are more important than higher-order effects and effects of the same order are equally important. We further require that no main effect is aliased with another main effect or confounded with blocks.

We have seen in Section 14.2 that a regular 2^{n-p} fractional factorial design in 2^q blocks of size 2^{n-p-q} can be viewed as a $2^{(n+q)-(p+q)}$ design with p independent treatment defining words and q independent block defining words. Since the treatment and block defining words play very different roles, it is not appropriate to apply the minimum aberration criterion directly to the wordlength pattern of this $2^{(n+q)-(p+q)}$ design. For example, based on the hierarchical assumption, a shorter treatment defining word has more severe consequences than a longer one, but the two block defining words $BC\mathcal{B}_1\mathcal{B}_2$ and $CD\mathcal{B}_1$, though of unequal lengths, have the same consequence of confounding a two-factor interaction with blocks. Thus the definition of the length of a block defining word needs to be modified. Two block defining words should be regarded as having the same length if they contain the same number of treatment letters. Bisgaard (1994) defined the length of a block defining word as the number of treatment letters it contains plus 1. According to this definition, $BC\mathcal{B}_1\mathcal{B}_2$ has length 3 and $ACD\mathcal{B}_1$ has length 4, correctly reflecting the order of desirability for these two defining words. However, $ABCE$ would have the same length as $ACD\mathcal{B}_1$, even though $ABCE$ causes the aliasing of some two-factor interactions with each other, while $ACD\mathcal{B}_1$ leads to the confounding of a three-factor interaction with blocks. These two defining words have different implications and should not be treated equally.

Sun, Wu and Chen (1997) proposed to consider treatment and block defining words separately. For each design d, let $A_{i,0}(d)$ (respectively, $A_{i,1}(d)$) be the number of treatment defining words (respectively, block defining words) containing i treatment letters. Under the requirement that no main effect is aliased with another main effect or confounded with blocks, $A_{1,0}(d) = A_{2,0}(d) = A_{1,1}(d) = 0$. Then there are two different wordlength patterns: the *treatment wordlength pattern* $W_t(d) = (A_{3,0}(d), \ldots, A_{n,0}(d))$ and *block wordlength pattern* $W_b(d) = (A_{2,1}(d), \ldots, A_{n,1}(d))$. The minimum aberration criterion can be applied to W_t and W_b separately, but a design that has minimum aberration with respect to W_t does not have minimum aberration with respect to W_b, and vice versa (Zhang and Park, 2000). Sun, Wu, and Chen (1997) proposed a concept of admissible designs: a design is admissible if there is no other design that is at least as good under both W_t and W_b and is better under at least one of the two wordlength patterns. However, there may be too many admissible designs.

In order to combine the two wordlength patterns so that the designs can be rank-ordered, Sitter, Chen and Feder (1997) modified the length of a block defining word to be the number of treatment letters it contains plus 1.5. This leads to the following wordlength pattern:

$$W_{SCF} = (A_{3,0}, A_{2,1}, A_{4,0}, A_{3,1}, \ldots, A_{n,0}, A_{n-1,1}, A_{n,1}).$$

To save space, the dependence on d is suppressed. The minimum aberration criterion based on this wordlength pattern has an undesirable consequence. For example, it considers block defining words containing four treatment letters to be less desirable than treatment defining words of length six. If $A_{6,0}$ is nonzero, then some three-factor interactions are aliased with other three-factor interactions. However, a nonzero $A_{4,1}$ implies the confounding of a certain less important four-factor interaction with blocks. Therefore the ordering of the components in W_{SCF} is not consistent with the hierarchical assumption.

Instead, Chen and Cheng (1999) and Cheng and Wu (2002) proposed the following wordlength pattern:

$$W_2 = (A_{3,0}, A_{2,1}, A_{4,0}, A_{5,0}, A_{3,1}, A_{6,0}, A_{7,0}, A_{4,1}, A_{8,0}, \ldots).$$

Cheng and Wu (2002) further proposed an additional wordlength pattern:

$$W_1 = (A_{3,0}, A_{4,0}, A_{2,1}, A_{5,0}, A_{6,0}, A_{3,1}, A_{7,0}, \ldots).$$

In both W_1 and W_2, for each j, $A_{j,1}$ is placed after $A_{2j-1,0}$ and before $A_{2j+1,0}$. This is because (i) each treatment defining word of length $2j-1$ causes $\binom{2j-1}{j}$ j-factor interactions to be aliased with $(j-1)$-factor interactions, (ii) a block defining word containing j treatment letters causes only one single j-factor interaction to be confounded with blocks, and (iii) no treatment defining word of length $2j+1$ can cause the aliasing of j-factor interactions with other j-factor interactions or lower-order effects. Under the hierarchical assumption, (i) is less desirable than (ii), which is in turn less desirable than (iii).

It is not clear-cut whether $A_{j,1}$ should appear before or after $A_{2j,0}$. The difference between W_1 and W_2 is that $A_{j,1}$ precedes $A_{2j,0}$ in W_2, and it is the opposite in W_1. That $A_{2j,0}$ appears before $A_{j,1}$ in W_1 is based on the rationale that each treatment defining word of length $2j$ leads to $\frac{1}{2}\binom{2j}{j}$ pairs of aliased j-factor interactions, while a block defining word containing j treatment letters causes only one single j-factor interaction to be confounded with blocks. On the other hand, suppose the block effects are unknown constants, i.e., they are fixed, instead of random effects; then the factorial effects that are confounded with blocks cannot be estimated. In this case, confounding j-factor interactions with blocks is more serious than aliasing among such interactions. Even if some j-factor interactions are aliased with other j-factor interactions, their information is not completely lost. For example, de-aliasing can be performed in a follow-up experiment. Then it is reasonable to have $A_{j,1}$ appear before $A_{2j,0}$, leading to wordlength pattern W_2. Chen and Cheng (1999) further proposed a modification of W_2 based on the consideration of estimation capacity, to be discussed in the next section.

14.9 Fixed block effects: surrogate for maximum estimation capacity

It was shown in Section 10.4 that for regular designs with unstructured units, minimum aberration is a good surrogate for the criterion of maximum estimation capacity. Minimum aberration designs tend to produce large, if not the maximum, number of estimable models containing all the main effects and a given number of two-factor interactions. Chen and Cheng (1999) extended this result to blocked factorial designs in the case of fixed block effects and derived a surrogate criterion that can be viewed as an extension of the minimum aberration criterion to blocked designs.

Under a model with fixed block effects, the treatment factorial effects that are confounded with blocks are not estimable. Suppose the three-factor and higher-order interactions are negligible. As in Section 10.4, one can define the estimation capacity of a blocked fractional factorial design d as the number of models containing all the main effects and k two-factor interactions that can be estimated under d. We denote this number by $E_k^{\mathcal{E}}(d)$ and reserve $E_k(d)$ for the estimation capacity under the model without block effects. Under a blocked regular design constructed by the method presented in Section 7.2, the treatment factorial effects are either not estimable or are estimated with full efficiency. Thus $E_k^{\mathcal{E}}(d)$ is a reasonable measure of the performance of a design when it is not known a priori which k two-factor interactions are active.

Cheng and Mukerjee (2001) studied designs that maximize $E_k^{\mathcal{E}}(d)$. Let

$$h = 2^{n-p} - n - 2^q \qquad (14.3)$$

be the number of alias sets of factorial effects that are neither aliased with main effects nor confounded with blocks, and $m_1(d), \ldots, m_h(d)$ be the numbers of two-factor interactions in these alias sets. Then $E_k^{\mathcal{E}}(d) = 0$ for $k > h$, and analogous to (10.4), we have

$$E_k^{\mathcal{E}}(d) = \sum_{1 \le i_1 < \cdots < i_k \le h} \prod_{j=1}^{k} m_{i_j}(d), \text{ if } k \le h.$$

Thus to have large $E_k^{\mathcal{E}}(d)$, $\sum_{i=1}^{h} m_i(d)$ should be as large as possible, and $m_1(d), \ldots, m_h(d)$ should be as equal as possible. Since $\sum_{i=1}^{h} m_i(d)$ is the number of two-factor interactions that are neither aliased with main effects nor confounded with blocks, $\sum_{i=1}^{h} m_i(d) = \binom{n}{2} - 3A_{3,0}(d) - A_{2,1}(d)$. Therefore maximizing $\sum_{i=1}^{h} m_i(d)$ is equivalent to minimizing $3A_{3,0}(d) + A_{2,1}(d)$. Furthermore, (10.2) continues to hold.

Thus a design that has small $3A_{3,0} + A_{2,1}$ and, subject to that, has small $A_{4,0}$, is expected to have large estimation capacity. This suggests a wordlength pattern with $3A_{3,0} + A_{2,1}$ and $A_{4,0}$ as the two leading terms. We note that $3A_{3,0} + A_{2,1}$ is a linear combination of the two leading terms of W_2.

We have the following result similar to Corollary 10.3.

Proposition 14.2. *If a blocked regular fractional factorial design d^* minimizes $3A_{3,0}(d) + A_{2,1}(d)$, and $m_1(d^*), \dots, m_h(d^*)$ differ from one another by at most one, then it maximizes $E_k^{\mathcal{E}}(d)$ for all k.*

Example 14.10. (Examples 14.1 and 14.2 revisited) Consider the case $n = 6$, $p = 2$, and $q = 2$. Then $h = 6$. It can be shown that the design constructed in Example 14.1, denoted by d^*, has the maximum number (12) of two-factor interactions that are neither aliased with main effects nor confounded with blocks. From the alias sets displayed in Example 14.2, one can see that there are two two-factor interactions in each of the six alias sets of effects that are neither aliased with main effects nor confounded with blocks. It follows from Proposition 14.2 that d^* maximizes $E_k^{\mathcal{E}}(d)$ for all k.

With $3A_{3,0} + A_{2,1}$ and $A_{4,0}$ as the leading terms, Chen and Cheng (1999) proposed the wordlength pattern

$$W_{CC} = \left(\binom{3}{1} A_{3,0} + A_{2,1}, A_{4,0}, \binom{5}{2} A_{5,0} + A_{3,1}, A_{6,0}, \binom{7}{3} A_{7,0} + A_{4,1}, A_{8,0}, \dots \right).$$

This is obtained from W_2 by replacing $(A_{2j-1,0}, A_{j,1})$ with the linear combination $\binom{2j-1}{j-1} A_{2j-1,0} + A_{j,1}$, which is the number of j-factor interactions that are aliased with $(j-1)$-factor interactions or confounded with blocks when the interactions involving at most $j-1$ factors are neither aliased among themselves nor confounded with blocks.

Example 14.11. Consider 32-run designs in 8 blocks of size 4 for 13 two-level factors. Let d_1 be the design with independent defining words **126**, **137**, **148**, **2349**, **1234t$_{10}$**, **235t$_{11}$**, **245t$_{12}$**, **345t$_{13}$**, **23B_1**, **24B_2**, **15B_3**, where $t_{10}, \dots, t_{13}$ are factors $10, \dots, 13$. This design has the maximum number (44) of two-factor interactions that are neither aliased with main effects nor confounded with blocks. Furthermore, $h = 11$, and $m_i(d_1) = 4$ for all $i = 1, \dots, 11$. Thus d_1 maximizes $E_k^{\mathcal{E}}(d)$ for all k. We note that d_1 is a resolution III design; thus it does not have minimum aberration with respect to the usual wordlength pattern, although it has minimum aberration with respect to W_{CC}. In this case, all the resolution IV designs d have $A_{2,1}(d) > 3A_{3,0}(d_1) + A_{2,1}(d_1)$.

Construction of blocked regular fractional factorial designs with maximum estimation capacity was studied in Cheng and Mukerjee (2001). In particular, they considered the construction of designs under which the number of two-factor interactions that are neither aliased with main effects nor confounded with blocks is maximized, and these interactions are distributed among the alias sets uniformly or nearly uniformly.

Xu (2006) and Xu and Lau (2006) carried out a study of minimum aberration blocked two-level fractional factorial designs with 8, 16, and 32 runs and those with 64 runs and up to 32 factors with respect to W_1, W_2, and W_{CC}. In most cases these three wordlength patterns produce the same minimum aberration designs.

Ai, Kang, and Joseph (2009) used the Bayesian approach discussed in Section 10.11 to study optimal blocked fractional factorial designs.

Chen and Cheng (1999) derived a complementary design theory for blocked fractional factorial designs. Their approach can also be applied to experiments with two distinct types of treatment factors and thus two different wordlength patterns. Zhu (2003) developed a general approach to the study of fractional factorial designs involving multiple groups of treatment factors and applied his results to experiments for robust parameter designs in quality improvement, where control and noise factors play different roles.

14.10 Information capacity and its surrogate

We have seen in the previous section that for models with fixed block effects, minimum aberration with respect to W_{CC} is a good surrogate for maximum estimation capacity. We further extend this connection to randomization models or models with random block effects. Under such models, a treatment interaction that is confounded with blocks is estimable, albeit the variance of its estimator depends on the interblock variance ξ_B, whereas for treatment interactions that are orthogonal to blocks, the variances of their estimators depend on the intrablock variance $\xi_{\mathcal{E}}$. As a result, the factorial effects in different estimable treatment models may be estimated with different efficiencies. In this case, counting the number of estimable potential treatment models to evaluate the performance of a design is not enough. One should also take the efficiencies into account, and the variance ratio $\xi_{\mathcal{E}}/\xi_B$ should play a role. Cheng and Tsai (2009) used the criterion of information capacity introduced by Sun (1993) for design selection. This criterion evaluates the performance of a design under model uncertainty by considering its average efficiency over a class of potential treatment models.

We assume that

$$\xi_{\mathcal{E}} \leq \xi_B; \tag{14.4}$$

see (12.50). Denote the ratio $\xi_{\mathcal{E}}/\xi_B$ by γ; then $0 \leq \gamma \leq 1$. Assume that three-factor and higher-order interactions are negligible and the two-factor interactions are equally important. In view of (14.4), we require that the main effects be estimated in the intrablock stratum. Then the efficiencies of their estimators are fixed. Thus it is sufficient to consider the efficiencies of the estimators of two-factor interactions. Without loss of generality, we assume that the factorial effects are normalized so that the estimator of each estimable factorial effect is of the form $\mathbf{a}^T \mathbf{y}$, where $\mathbf{y}$ is the vector of observations and $\|\mathbf{a}\| = 1$. It follows that if a factorial effect is estimated in the intrablock (respectively, interblock) stratum, then its estimator has variance $\xi_{\mathcal{E}}$ (respectively, $\xi_{\mathcal{B}}$). Suppose there are k two-factor interactions in the treatment model. For each design d, let $D(d)$ be the determinant of the covariance matrix of the estimators of these two-factor interactions under d, provided that they are all estimable; otherwise, $D(d) = \infty$. Then

$$D(d) \geq \xi_{\mathcal{E}}^k,$$

where the equality holds if and only if all the k two-factor interactions are estimated in the intrablock stratum. Define the D-efficiency $e(d)$ of d as

$$e(d) = [\xi_{\mathcal{E}}^k / D(d)]^{1/k} = \xi_{\mathcal{E}} / [D(d)]^{1/k}. \tag{14.5}$$

Then $0 \leq e(d) \leq 1$, where $e(d) = 0$ if not all the k two-factor interactions are estimable, and $e(d) = 1$ if all of them are estimable and are estimated in the intrablock stratum. If all these interactions are estimated in the interblock stratum, then $e(d) = \gamma$.

Given that we do not know which k two-factor interactions are active, we define the information capacity $I_k(d)$ as the average of $e(d)$ over the treatment models containing all the main effects and k two-factor interactions. The objective is to find a design for which $I_k(d)$ is as large as possible.

In the case of fixed block effects, a treatment model is estimable if and only if all the k two-factor interactions in the model are orthogonal to blocks; otherwise the model is not estimable. Thus $e(d)$ is either 1 or 0, and it follows that

$$I_k(d) = E_k^{\mathcal{E}}(d) \bigg/ \binom{n(n-1)/2}{k},$$

the proportion of estimable models. Therefore, for models with fixed block effects, maximizing $I_k(d)$ is equivalent to maximizing $E_k^{\mathcal{E}}(d)$. This shows that $I_k(d)$ can be viewed as an extension of the estimation capacity to models with random block effects.

Like $E_k^{\mathcal{E}}(d)$, $I_k(d)$ can be expressed as a function of the numbers of two-factor interactions in the alias sets. Under a 2^{n-p} design with 2^q blocks of size 2^{n-p-q}, there are $f = 2^{n-p} - n - 1$ alias sets that do not contain main effects. Among these, $2^q - 1$ contain effects that are estimated in the interblock stratum, and the other $2^{n-p} - n - 2^q$ contain effects that are estimated in the intrablock stratum. Let $h_{\mathcal{B}} = 2^q - 1$ and $h_{\mathcal{E}} = 2^{n-p} - n - 2^q$. Note that $h_{\mathcal{E}}$ is the same as the h in (14.3). Let $m_1^{\mathcal{B}}(d), \ldots, m_{h_{\mathcal{B}}}^{\mathcal{B}}(d)$ and $m_1^{\mathcal{E}}(d), \ldots, m_{h_{\mathcal{E}}}^{\mathcal{E}}(d)$ be the numbers of two-factor interactions in these alias sets.

Lemma 14.3. *(Cheng and Tsai, 2009) If $k > f$, then $I_k(d) = 0$. For $k \leq f$,*

$$I_k(d) = E_k \left(\gamma^{1/k} m_1^{\mathcal{B}}(d), \ldots, \gamma^{1/k} m_{h_{\mathcal{B}}}^{\mathcal{B}}(d), m_1^{\mathcal{E}}(d), \ldots, m_{h_{\mathcal{E}}}^{\mathcal{E}}(d) \right) \bigg/ \binom{n(n-1)/2}{k}$$

where, as in (10.4),

$$E_k(x_1, \ldots, x_f) = \sum_{1 \le i_1 < \cdots < i_k \le f} \prod_{j=1}^{k} x_{i_j}.$$

Proof. For each treatment model containing all the main effects and k two-factor interactions, if l of these two-factor interactions are estimated in the interblock stratum and the other $k - l$ are estimated in the intrablock stratum, then the determinant of the covariance matrix of the estimators of the k two-factor interactions is equal to $\xi_{\mathcal{E}}^{k-l} \xi_{\mathcal{B}}^{l}$ when the interaction contrasts are properly normalized. For such models, by (14.5), $e(d) = \xi_{\mathcal{E}}/[\xi_{\mathcal{E}}^{k-l} \xi_{\mathcal{B}}^{l}]^{1/k} = \gamma^{l/k}$. Since the total number of treatment models with l two-factor interactions estimated in the interblock stratum and the other $k - l$ estimated in the intrablock stratum is given by

$$\left[\sum_{1 \le i_1 < \cdots < i_l \le h_{\mathcal{B}}} \prod_{u=1}^{l} m_{i_u}^{\mathcal{B}}(d) \right] \left[\sum_{1 \le j_1 < \cdots < j_{k-l} \le h_{\mathcal{E}}} \prod_{v=1}^{k-l} m_{j_v}^{\mathcal{E}}(d) \right],$$

we have that $\binom{n(n-1)/2}{k} I_k(d)$ is equal to

$$\sum_{l=\max(0, k-f+h_{\mathcal{B}})}^{\min(k, h_{\mathcal{B}})} \gamma^{l/k} \cdot \left[\sum_{1 \le i_1 < \cdots < i_l \le h_{\mathcal{B}}} \prod_{u=1}^{l} m_{i_u}^{\mathcal{B}}(d) \right] \left[\sum_{1 \le j_1 < \cdots < j_{k-l} \le h_{\mathcal{E}}} \prod_{v=1}^{k-l} m_{j_v}^{\mathcal{E}}(d) \right]$$

$$= \sum_{l=\max(0, k-f+h_{\mathcal{B}})}^{\min(k, h_{\mathcal{B}})} \left[\sum_{1 \le i_1 < \cdots < i_l \le h_{\mathcal{B}}} \prod_{u=1}^{l} (\gamma^{1/k} m_{i_u}^{\mathcal{B}}(d)) \right] \left[\sum_{1 \le j_1 < \cdots < j_{k-l} \le h_{\mathcal{E}}} \prod_{v=1}^{k-l} m_{j_v}^{\mathcal{E}}(d) \right]$$

$$= E_k(\gamma^{1/k} m_1^{\mathcal{B}}(d), \ldots, \gamma^{1/k} m_{h_{\mathcal{B}}}^{\mathcal{B}}(d), m_1^{\mathcal{E}}(d), \ldots, m_{h_{\mathcal{E}}}^{\mathcal{E}}(d)).$$

□

By Lemma 14.3 and the fact that E_k is a Schur concave function that is non-decreasing in each component of its argument, $I_k(d)$ is large if $\gamma^{1/k} \sum_{i=1}^{h_{\mathcal{B}}} m_i^{\mathcal{B}}(d) + \sum_{j=1}^{h_{\mathcal{E}}} m_j^{\mathcal{E}}(d)$ is large, and $\gamma^{1/k} m_1^{\mathcal{B}}(d), \ldots, \gamma^{1/k} m_{h_{\mathcal{B}}}^{\mathcal{B}}(d), m_1^{\mathcal{E}}(d), \ldots, m_{h_{\mathcal{E}}}^{\mathcal{E}}(d)$ are as equal as possible. This suggests that a good surrogate for maximizing $I_k(d)$ is

maximizing

$$\gamma^{1/k} \sum_{i=1}^{h_{\mathcal{B}}} m_i^{\mathcal{B}}(d) + \sum_{j=1}^{h_{\mathcal{E}}} m_j^{\mathcal{E}}(d), \tag{14.6}$$

and minimizing

$$\gamma^{2/k} \sum_{i=1}^{h_{\mathcal{B}}} \left[m_i^{\mathcal{B}}(d) \right]^2 + \sum_{j=1}^{h_{\mathcal{E}}} \left[m_j^{\mathcal{E}}(d) \right]^2 \tag{14.7}$$

among those that maximize (14.6).

This surrogate criterion depends on the variance ratio γ. The two extreme cases $\gamma = 0$ and 1 correspond to, respectively, $\xi_{\mathcal{B}} = \infty$ (fixed block effects) and $\xi_{\mathcal{E}} = \xi_{\mathcal{B}}$ (no block effects, that is, experiments with unstructured units). We have

$$\gamma^{1/k} \sum_{i=1}^{h_{\mathcal{B}}} m_i^{\mathcal{B}}(d) + \sum_{j=1}^{h_{\mathcal{E}}} m_j^{\mathcal{E}}(d)$$

$$= \gamma^{1/k} \left[\sum_{i=1}^{h_{\mathcal{B}}} m_i^{\mathcal{B}}(d) + \sum_{j=1}^{h_{\mathcal{E}}} m_j^{\mathcal{E}}(d) \right] + (1 - \gamma^{1/k}) \sum_{j=1}^{h_{\mathcal{E}}} m_j^{\mathcal{E}}(d)$$

$$= \gamma^{1/k} \left[\binom{n}{2} - 3A_{3,0}(d) \right] + (1 - \gamma^{1/k}) \left[\binom{n}{2} - 3A_{3,0}(d) - A_{2,1}(d) \right]$$

$$= \binom{n}{2} - 3A_{3,0}(d) - (1 - \gamma^{1/k})A_{2,1}(d),$$

which is equal to $\binom{n}{2} - 3A_{3,0}(d)$ when $\gamma = 1$, and $\binom{n}{2} - 3A_{3,0}(d) - A_{2,1}(d)$ when $\gamma = 0$. For $\gamma = 1$, (14.6) and (14.7) reduce to the minimization of $A_3(d)$ and, subject to that, the minimization of $\sum_{i=1}^{f} [m_i(d)]^2$. We have already seen in Section 10.4 that this is a good surrogate for maximizing $E_k(d)$. For $\gamma = 0$, (14.6) and (14.7) reduce to the minimization of $3A_{3,0}(d) + A_{2,1}(d)$ and, subject to that, the minimization of $\sum_{j=1}^{h_{\mathcal{E}}} [m_j^{\mathcal{E}}(d)]^2$, which is a good surrogate for maximizing $E_k^{\mathcal{E}}(d)$. If we further replace the minimization of $\sum_{i=1}^{f} [m_i(d)]^2$ and $\sum_{j=1}^{h_{\mathcal{E}}} [m_j^{\mathcal{E}}(d)]^2$ with that of $A_{4,0}(d)$, then for $\gamma = 1$, we have the successive minimization of $A_3(d)$ and $A_4(d)$ as in the usual minimum aberration criterion, and for $\gamma = 0$, we have the successive minimization of $3A_{3,0}(d) + A_{2,1}(d)$ and $A_{4,0}(d)$, as in the minimum aberration criterion with respect to W_{CC}.

The following result provides a useful tool for drawing the strong conclusion that a design is optimal with respect to all γ from its optimality in the two extreme cases $\gamma = 0$ and 1.

Theorem 14.4. *(Cheng and Tsai, 2009) Suppose a design d^**

(i) maximizes $\sum_{i=1}^{h_{\mathcal{B}}} m_i^{\mathcal{B}}(d) + \sum_{j=1}^{h_{\mathcal{E}}} m_j^{\mathcal{E}}(d)$ and minimizes $\sum_{i=1}^{h_{\mathcal{B}}} [m_i^{\mathcal{B}}(d)]^2 + \sum_{j=1}^{h_{\mathcal{E}}} [m_j^{\mathcal{E}}(d)]^2$ among those that maximize $\sum_{i=1}^{h_{\mathcal{B}}} m_i^{\mathcal{B}}(d) + \sum_{j=1}^{h_{\mathcal{E}}} m_j^{\mathcal{E}}(d)$;

(ii) maximizes $\sum_{j=1}^{h_{\mathcal{E}}} m_j^{\mathcal{E}}(d)$ and minimizes $\sum_{j=1}^{h_{\mathcal{E}}} [m_j^{\mathcal{E}}(d)]^2$ among those that maximize $\sum_{j=1}^{h_{\mathcal{E}}} m_j^{\mathcal{E}}(d)$.

Then for all $0 \le \gamma \le 1$ and all k, design d^* maximizes (14.6) and minimizes (14.7) among those that maximize (14.6).

Proof. By (i) and (ii), d^* maximizes both $\sum_{i=1}^{h_{\mathcal{B}}} m_i^{\mathcal{B}}(d) + \sum_{j=1}^{h_{\mathcal{E}}} m_j^{\mathcal{E}}(d)$ and $\sum_{j=1}^{h_{\mathcal{E}}} m_j^{\mathcal{E}}(d)$; therefore it maximizes $\gamma^{1/k} [\sum_{i=1}^{h_{\mathcal{B}}} m_i^{\mathcal{B}}(d) + \sum_{j=1}^{h_{\mathcal{E}}} m_j^{\mathcal{E}}(d)] + (1 - \gamma^{1/k}) \sum_{j=1}^{h_{\mathcal{E}}} m_j^{\mathcal{E}}(d) =$

$\gamma^{1/k} \sum_{i=1}^{h_\mathcal{B}} m_i^\mathcal{B}(d) + \sum_{j=1}^{h_\mathcal{E}} m_j^\mathcal{E}(d)$. Suppose d_1 is another design that also maximizes $\gamma^{1/k} \sum_{i=1}^{h_\mathcal{B}} m_i^\mathcal{B}(d) + \sum_{j=1}^{h_\mathcal{E}} m_j^\mathcal{E}(d)$. Then, since d^* maximizes both $\sum_{i=1}^{h_\mathcal{B}} m_i^\mathcal{B}(d) + \sum_{j=1}^{h_\mathcal{E}} m_j^\mathcal{E}(d)$ and $\sum_{j=1}^{h_\mathcal{E}} m_j^\mathcal{E}(d)$, d_1 must also maximize both of these two quantities. By assumptions (i) and (ii), $\sum_{j=1}^{h_\mathcal{E}} [m_j^\mathcal{E}(d^*)]^2 \leq \sum_{j=1}^{h_\mathcal{E}} [m_j^\mathcal{E}(d_1)]^2$ and $\sum_{i=1}^{h_\mathcal{B}} [m_i^\mathcal{B}(d^*)]^2 + \sum_{j=1}^{h_\mathcal{E}} [m_j^\mathcal{E}(d^*)]^2 \leq \sum_{i=1}^{h_\mathcal{B}} [m_i^\mathcal{B}(d_1)]^2 + \sum_{j=1}^{h_\mathcal{E}} [m_j^\mathcal{E}(d_1)]^2$. Then since $\gamma^{2/k} \sum_{i=1}^{h_\mathcal{B}} [m_i^\mathcal{B}(d)]^2 + \sum_{j=1}^{h_\mathcal{E}} [m_j^\mathcal{E}(d)]^2 = \gamma^{2/k} \{ \sum_{i=1}^{h_\mathcal{B}} [m_i^\mathcal{B}(d)]^2 + \sum_{j=1}^{h_\mathcal{E}} [m_j^\mathcal{E}(d)]^2 \} + (1 - \gamma^{2/k}) \cdot \sum_{j=1}^{h_\mathcal{E}} [m_j^\mathcal{E}(d)]^2$, we have $\gamma^{2/k} \sum_{i=1}^{h_\mathcal{B}} [m_i^\mathcal{B}(d^*)]^2 + \sum_{j=1}^{h_\mathcal{E}} [m_j^\mathcal{E}(d^*)]^2 \leq \gamma^{2/k} \sum_{i=1}^{h_\mathcal{B}} [m_i^\mathcal{B}(d_1)]^2 + \sum_{j=1}^{h_\mathcal{E}} [m_j^\mathcal{E}(d_1)]^2$. □

Roughly speaking, a blocked regular fractional factorial design performs well for all ratios of intrablock and interblock variances as long as it is a good design for unblocked experiments (say it has minimum aberration with respect to the usual wordlength pattern) and for blocked experiments with fixed block effects (say it has minimum aberration with respect to W_{CC}).

We say that a design d_1 dominates another design d_2 if and only if d_1 is at least as good as d_2 with respect to the surrogate criterion for all $0 \leq \gamma \leq 1$ and is better for at least one γ. A design is called admissible if it is not dominated by any other design; otherwise it is inadmissible. The inadmissible designs can be eliminated from consideration, at least for the surrogate criterion. Theorem 14.4 also provides a useful tool for eliminating inadmissible designs without having to know the stratum variances. Often there is only a small number of inadmissible designs, which can be compared directly based on what is known about the stratum variances or other information.

Two designs are said to be equivalent if they have the same m_i values in each stratum. This definition of design equivalence also applies to the situation where there are more than two strata, to be studied later in this chapter. By Lemma 14.3 (and its extension to multi-stratum experiments), equivalent designs have the same performance under the information capacity criterion with respect to all possible stratum variances.

Example 14.12. (Example 14.10 continued) For $n = 6$, $p = 2$, and $q = 2$, design d^* constructed in Example 14.1 is a minimum aberration design for unblocked experiments. It maximizes $\sum_{i=1}^{h_\mathcal{B}} m_i^\mathcal{B}(d) + \sum_{j=1}^{h_\mathcal{E}} m_j^\mathcal{E}(d)$ and minimizes $\sum_{i=1}^{h_\mathcal{B}} [m_i^\mathcal{B}(d)]^2 + \sum_{j=1}^{h_\mathcal{E}} [m_j^\mathcal{E}(d)]^2$ among those that maximize $\sum_{i=1}^{h_\mathcal{B}} m_i^\mathcal{B}(d) + \sum_{j=1}^{h_\mathcal{E}} m_j^\mathcal{E}(d)$. In Example 14.10, we saw that it also maximizes $\sum_{j=1}^{h_\mathcal{E}} m_j^\mathcal{E}(d)$ and minimizes $\sum_{j=1}^{h_\mathcal{E}} [m_j^\mathcal{E}(d)]^2$ among those that maximize $\sum_{j=1}^{h_\mathcal{E}} m_j^\mathcal{E}(d)$. Thus by Theorem 14.4, it is optimal with respect to the surrogate criterion for all $0 \leq \gamma \leq 1$ and all k. Any other design either has the same performance as d^* or is dominated by d^*.

Example 14.13. (Example 14.11 continued) For $n = 13$, $p = 8$, and $q = 3$, design d_1 in Example 14.11 minimizes $3A_{3,0}(d) + A_{2,1}(d)$ or, equivalently, maximizes $\sum_{j=1}^{h_\mathcal{E}} m_j^\mathcal{E}(d)$ and minimizes $\sum_{j=1}^{h_\mathcal{E}} [m_j^\mathcal{E}(d)]^2$ among those that maximize $\sum_{j=1}^{h_\mathcal{E}} m_j^\mathcal{E}(d)$. As pointed out in Example 14.11, d_1 does not have minimum aberration with respect to the usual wordlength pattern and does not maximize $\sum_{i=1}^{h_\mathcal{B}} m_i^\mathcal{B}(d) + \sum_{j=1}^{h_\mathcal{E}} m_j^\mathcal{E}(d)$. On the other hand, all the designs that maximize $\sum_{i=1}^{h_\mathcal{B}} m_i^\mathcal{B}(d) + \sum_{j=1}^{h_\mathcal{E}} m_j^\mathcal{E}(d)$ are of resolution IV and cannot maximize $\sum_{j=1}^{h_\mathcal{E}} m_j^\mathcal{E}(d)$. This is an example where there is no design that is the best with respect to the surrogate criterion for all $0 \leq \gamma \leq 1$. Let d_2 be defined by the independent words **1236, 1247, 1348, 2349, 125t$_{10}$, 135t$_{11}$, 235t$_{12}$, 145t$_{13}$, 13$\mathcal{B}_1$, 14$\mathcal{B}_2$**, and **15$\mathcal{B}_3$**; then a search of all the designs shows that d_1 and d_2 are admissible, and all the other designs are dominated by or have the same performance as at least one of d_1 and d_2. It follows that for each $0 \leq \gamma \leq 1$, one of these two designs is optimal with respect to the surrogate criterion. In particular, d_1 is optimal for $\gamma = 0$ and the γ's that are sufficiently small ($\xi_\mathcal{B}$ is sufficiently greater than $\xi_\mathcal{E}$), and d_2 is optimal for the other γ's, including the case $\gamma = 1$. Indeed, d_2 has minimum aberration with respect to the usual wordlength pattern.

It was mentioned at the end of Section 14.9 that in most cases of blocked two-level fractional factorial designs with 8, 16, 32 runs and those with 64 runs and up to 32 factors, minimum aberration designs with respect to W_1, W_2, and W_{CC} are the same. For 16- and 32-run designs, except for $(n,p,q) = (5,1,1), (5,1,2), (6,1,2), (7,2,2), (8,3,3), (9,4,2), (9,4,3), (10,5,1), (10,5,2), (10,5,3), (11,6,3), (12,7,3), (13,8,3), (21,16,1)$, and $(21,16,2)$, W_1 and W_{CC} have the same minimum aberration designs, which are also optimal with respect to the surrogate criterion for all $0 \leq \gamma \leq 1$. The 15 cases listed above are the only cases where there is more than one admissible design. Except for $(n,p,q) = (9,4,2)$, in all the other 14 cases the W_{CC} (respectively, W_1) minimum aberration designs are optimal with respect to the surrogate criterion for smaller (respectively, larger) γ values, and they are the only admissible designs up to equivalence. For $(n,p,q) = (9,4,2)$, there are three admissible designs up to equivalence. The design defined by the independent words **12456, 2359, 2578, 2689, 14$\mathcal{B}_1$**, and **15$\mathcal{B}_2$** is dominated by neither the W_{CC} nor the W_1 minimum aberration design. Readers are referred to Cheng and Tsai (2009) for details.

Remark 14.1. We emphasize that no single criterion is suitable for all the situations. Like the minimum aberration criterion, the results developed here and in the rest of this chapter are appropriate under the hierarchical assumption and, as noted by Fries and Hunter (1980), for "the situation in which prior knowledge is diffuse concerning the possible greater importance of certain effects."

14.11   Selection of fractional factorial split-plot designs

The results in the previous section can be applied to split-plot designs with the inter- and intrablock strata replaced by whole-plot and subplot strata, respectively, except that the main effects of whole-plot treatment factors must be confounded with whole-plot contrasts, whereas for block designs, we have required that no treatment main effect be confounded with blocks.

We follow the notations in Section 14.3. Suppose there are n_1 whole-plot treatment factors, $n_2 = n - n_1$ subplot treatment factors, and 2^{n-p} whole-plots each consisting of 2^{n-p-q} subplots. Let $h_\mathcal{P}$ (respectively, $h_\mathcal{S}$) be the number of alias sets of effects that are estimated in the whole-plot stratum (respectively, subplot stratum) but are not aliased with main effects. Then $h_\mathcal{P} = 2^q - 1 - n_1$ and $h_\mathcal{S} = 2^{n-p} - 2^q - n_2$. Note that in Section 14.10, $h_\mathcal{B} = 2^q - 1$ and $h_\mathcal{E} = 2^{n-p} - n - 2^q$. Let $m_1^P(d), \ldots, m_{h_\mathcal{P}}^P(d)$ and $m_1^S(d), \ldots, m_{h_\mathcal{S}}^S(d)$ be the numbers of two-factor interactions in these alias sets. Then the results in Section 14.10 continue to hold with $\mathcal{B}$ and $\mathcal{E}$ replaced by $\mathcal{P}$ and $\mathcal{S}$, respectively.

The usual minimum aberration criterion for unblocked designs was used by Huang, Chen, and Voelkel (1998), Bingham and Sitter (1999), and Bingham, Schoen, and Sitter (2004) to compare split-plot designs. This criterion does not take the block structure into account and often leads to more than one nonisomorphic minimum aberration design. Bingham and Sitter (2001) and Mukerjee and Fang (2002) adopted additional secondary criteria to break the ties. For example, for nonisomorphic split-plot designs that are tied under the usual minimum aberration criterion, Bingham and Sitter (2001) proposed the minimization of the number of subplot treatment two-factor interactions that are estimated in the whole-plot stratum. This approach yields a unique optimal design in each of the 16-run cases in Table 3 and all but 12 of the 32-run cases in Table 4 of their paper; see Cheng and Tsai (2009) for details. The secondary criterion proposed by Bingham and Sitter (2001) is similar to (though not exactly the same as) the first half of condition (ii) in Theorem 14.4. The second half of the condition may be needed to identify a unique optimal design.

Example 14.14. Consider the case of 16 whole-plots each containing two subplots, five whole-plot treatment factors A, B, C, D, E, and two subplot treatment factors P and Q. We have $n_1 = 5$, $n_2 = 2$, and $p_1 = p_2 = 1$. Table 4 of Bingham and Sitter (2001) listed two designs that are the best and are equally good according to the dual criteria of minimum aberration and minimizing the number of subplot treatment two-factor interactions estimated in the whole-plot stratum among the minimum aberration designs:

$$d_1 : I = ABCDE = ABPQ = CDEPQ; \quad d_2 : I = ABCE = ABDPQ = CDEPQ.$$

We have $h_\mathcal{P} = 10$ and $h_\mathcal{S} = 14$. The following shows the numbers of treatment two-factor interactions in the 24 alias sets for d_1 and d_2:

	d_1	d_2
$W_\mathcal{P}$	1 1 1 1 1 1 1 1 1 2	2 2 2 1 1 1 1 1 0 0
$W_\mathcal{S}$	2 2 1 1 1 1 1 1 0 0 0 0 0 0	1 1 1 1 1 1 1 1 1 0 0 0 0

The values of $\sum_{i=1}^{h_\mathcal{P}} m_i^P(d) + \sum_{j=1}^{h_\mathcal{S}} m_j^S(d)$, $\sum_{i=1}^{h_\mathcal{P}} [m_i^P(d)]^2 + \sum_{j=1}^{h_\mathcal{S}} [m_j^S(d)]^2$, $\sum_{j=1}^{h_\mathcal{S}} m_j^S(d)$, and $\sum_{j=1}^{h_\mathcal{S}} [m_j^S(d)]^2$ for d_1 are 21, 27, 10, and 14, and those for d_2 are 21, 27, 10 and 10, respectively. By Theorem 14.4, d_2 dominates d_1. Thus d_1 is inadmissible. The two designs are tied with respect to the surrogate criterion when $\gamma = 1$, but d_2 is better than d_1 for all $0 \leq \gamma < 1$. It turns out that d_2 is optimal with respect to the surrogate criterion for all $0 \le \gamma \le 1$. Although d_1 and d_2 have the same number of treatment two-factor interactions that are not aliased with main effects and are estimated in the subplot stratum, the distribution of these two-factor interactions over the alias sets is more uniform under d_2. Both designs have three pairs of aliased two-factor interactions, but under d_2, all such aliasing occurs in the less precise whole-plot stratum. This leads to the higher overall efficiency of d_2.

Except for the three cases $(n_1, n_2, p_1, p_2) = (4, 2, 1, 1)$, $(7, 3, 3, 2)$, and $(8, 2, 4, 1)$, the best design or one of the best designs in Tables 3 and 4 of Bingham and Sitter (2001) is optimal with respect to the surrogate criterion for all $0 \le \gamma \le 1$. In each of the three exceptional cases, there is an additional admissible design up to equivalence. For the designs tabulated in Bingham, Schoen, and Sitter (2004) for the situation where the number of whole-plots is greater than the number of whole-plot treatment factor level combinations, in many cases the approach presented here can be used to find better designs under the criteria considered here.

Example 14.15. Suppose there are eight whole-plots each containing four subplots, one whole-plot treatment factor A, and 13 subplot treatment factors $N, O, P, Q, R, S, T, U, V, W, X, Y, Z$. The design listed in Table 2 of Bingham, Schoen, and Sitter (2004), denoted by d_3, is not admissible with respect to the surrogate criterion and is dominated by the design d_4 defined by $R = AN$, $S = ANP$, $T = AOP$, $U = NOP$, $V = ANQ$, $W = AOQ$, $X = NOQ$, $Y = NPQ$, $Z = OPQ$, and with NO and PQ as independent splitting words. In this case $h_\mathcal{P} = 6$ and $h_\mathcal{S} = 11$. The following shows the numbers of treatment two-factor interactions in the 17 alias sets of factorial effects that are not aliased with main effects:

	d_3	d_4
$W_\mathcal{P}$	5 5 5 5 5 6	1 1 1 6 6 6
$W_\mathcal{S}$	1 1 1 5 5 5 5 5 5 6 6	5 5 5 5 5 5 5 5 5 5 5

The two designs have the same m_i values: three 1's, eleven 5's and three 6's. It follows that they are tied under (i) of Theorem 14.4 and have the same $E_k(d)$ (equivalently, the same $I_k(d)$ when $\xi_\mathcal{P} = \xi_\mathcal{S}$) for all k. In other words, they perform equally well in the case of unstructured units. However, d_4 is better than d_3 with respect to (ii) of Theorem 14.4. Thus d_4 is a better design for $\xi_\mathcal{P} > \xi_\mathcal{S}$. One can see that d_4 has ten more two-factor interactions estimated in the subplot stratum than d_3, and under d_4 the distribution of the two-factor interactions over the alias sets in the subplot stratum is the most uniform possible. In fact, d_4 is optimal with respect to the surrogate criterion for all $0 \le \gamma \le 1$.

14.12   A general result on multi-stratum fractional factorial designs

We consider experiments in which the treatment design is a regular fractional factorial design and the experimental units have a block structure $\mathcal{B}$ that is an orthogonal block structure or satisfies the conditions in Theorem 12.8. Furthermore, the conditions of orthogonal designs as given in Section 13.1 are satisfied by $\mathfrak{B}$ and the complete factorial of a set of basic treatment factors. As in Section 14.10, assume that three-factor and higher-order interactions are negligible and the two-factor interactions are equally important.

If the main effect of a treatment factor is required to be estimated in a certain stratum, then the efficiency of its estimator is fixed. When it is not subject to such a constraint, we impose the requirement that it be estimated in the bottom stratum to gain the best precision. Then the efficiency of its estimator is also fixed. Therefore it is sufficient to consider two-factor interactions, and the information capacity $I_k(d)$ can be defined in the same way as in Section 14.10. An argument similar to the proof of Lemma 14.3 shows that $I_k(d)$ is a function of the counts of two-factor interactions in the alias sets of effects that are not aliased with main effects.

For each $\mathcal{F} \in \mathfrak{B}, \mathcal{F} \neq \mathcal{U}$, let $h_\mathcal{F}$ be the number of alias sets of factorial effects that are not aliased with main effects and are estimated in stratum $W_\mathcal{F}$, and let $m_1^\mathcal{F}, \ldots, m_{h_\mathcal{F}}^\mathcal{F}$ be the numbers of two-factor interactions contained in these alias sets. Furthermore, let $\mathbf{m}(d)$ be the $(2^{n-p} - 1 - n) \times 1$ vector whose components consist of $\{[\xi_\varepsilon/\xi_\mathcal{F}]^{1/k} m_i^\mathcal{F} : 1 \leq i \leq h_\mathcal{F}, \mathcal{F} \in \mathfrak{B}, \mathcal{F} \neq \mathcal{U}\}$. Then we have

$$I_k(d) = E_k[\mathbf{m}(d)] \Big/ \binom{n(n-1)/2}{k},$$

where E_k is as in Lemma 14.3. By the same argument as in Section 14.10, a good surrogate for maximizing $I_k(d)$ is to

$$\text{maximize} \quad \sum_{\mathcal{F} \in \mathfrak{B}, \mathcal{F} \neq \mathcal{U}} \xi_\mathcal{F}^{-1/k} \sum_{i=1}^{h_\mathcal{F}} m_i^\mathcal{F}, \tag{14.8}$$

and

$$\text{minimize} \quad \sum_{\mathcal{F} \in \mathfrak{B}, \mathcal{F} \neq \mathcal{U}} \xi_\mathcal{F}^{-2/k} \sum_{i=1}^{h_\mathcal{F}} [m_i^\mathcal{F}]^2 \quad \text{among those maximizing} \quad \sum_{\mathcal{F} \in \mathfrak{B}, \mathcal{F} \neq \mathcal{U}} \xi_\mathcal{F}^{-1/k} \sum_{i=1}^{h_\mathcal{F}} m_i^\mathcal{F}. \tag{14.9}$$

The following extension of Theorem 14.4 to general multi-stratum experiments provides a useful sufficient condition for a design to be optimal with respect to (14.8) and (14.9) for all possible stratum variances that satisfy (12.50).

Theorem 14.5. *Suppose d is a design that maximizes $\sum_{\mathcal{F} \in \mathfrak{G}} \sum_{i=1}^{h_\mathcal{F}} m_i^\mathcal{F}$ and minimizes $\sum_{\mathcal{F} \in \mathfrak{G}} \sum_{i=1}^{h_\mathcal{F}} [m_i^\mathcal{F}]^2$ among those that maximize $\sum_{\mathcal{F} \in \mathfrak{G}} \sum_{i=1}^{h_\mathcal{F}} m_i^\mathcal{F}$, for all subsets $\mathfrak{G}$ of $\mathfrak{B} \backslash \{\mathcal{U}\}$ that satisfy the condition*

$$\mathcal{F} \in \mathfrak{G} \text{ and } \mathcal{F}' \prec \mathcal{F} \Rightarrow \mathcal{F}' \in \mathfrak{G}. \tag{14.10}$$

Then d is optimal with respect to (14.8) and (14.9) for all k and all $\{\xi_\mathcal{F}\}_{\mathcal{F} \in \mathfrak{B}, \mathcal{F} \neq \mathcal{U}}$ that satisfy (12.50).

See Cheng and Tsai (2011) for a proof of Theorem 14.5.

Like Theorem 14.4, Theorem 14.5 can be used to verify the strong optimality with respect to all possible values of the stratum variances that satisfy (12.50) and to eliminate inadmissible designs, without having to know the stratum variances. However, Theorem 14.5 is practically useful only when there are not too many strata. In the next two sections we apply this tool to the selection of blocked fractional factorial split-plot and strip-plot designs.

14.13 Selection of blocked fractional factorial split-plot designs

Consider the same setup as in Section 14.4, with the block structure $2^g/2^{q-g}/2^{n-p-q}$, n_1 two-level whole-plot treatment factors, and n_2 two-level subplot treatment factors. McLeod and Brewster (2004) used the minimum aberration criterion for blocked factorials proposed by Sitter, Chen, and Feder (1997) to compare blocked split-plot designs, with a penalty of 0.5 added to the length of each block defining word. In the current setting, $\mathcal{B} = \{\mathcal{U}, \mathcal{B}, \mathcal{P}, \mathcal{S}\}$, and the subsets $\mathfrak{G}$ of $\mathcal{B} \setminus \{\mathcal{U}\}$ that satisfy condition (14.10) are $\{\mathcal{B}, \mathcal{P}, \mathcal{S}\}$, $\{\mathcal{P}, \mathcal{S}\}$, and $\{\mathcal{S}\}$. By Theorem 14.5, a sufficient condition for a design d to be optimal with respect to the surrogate criterion (14.8)–(14.9) for all $\xi_{\mathcal{B}} \geq \xi_{\mathcal{P}} \geq \xi_{\mathcal{S}}$ is that it

(i) maximizes $\sum_{i=1}^{2^g-1} m_i^{\mathcal{B}} + \sum_{i=1}^{2^q-2^g-n_1} m_i^{\mathcal{P}} + \sum_{i=1}^{2^{n-p}-2^q-n_2} m_i^{\mathcal{S}}$ and, subject to that, minimizes $\sum_{i=1}^{2^g-1} \left[m_i^{\mathcal{B}}\right]^2 + \sum_{i=1}^{2^q-2^g-n_1} \left[m_i^{\mathcal{P}}\right]^2 + \sum_{i=1}^{2^{n-p}-2^q-n_2} \left[m_i^{\mathcal{S}}\right]^2$;

(ii) maximizes $\sum_{i=1}^{2^q-2^g-n_1} m_i^{\mathcal{P}} + \sum_{i=1}^{2^{n-p}-2^q-n_2} m_i^{\mathcal{S}}$ and, subject to that, minimizes $\sum_{i=1}^{2^q-2^g-n_1} \left[m_i^{\mathcal{P}}\right]^2 + \sum_{i=1}^{2^{n-p}-2^q-n_2} \left[m_i^{\mathcal{S}}\right]^2$;

(iii) maximizes $\sum_{i=1}^{2^{n-p}-2^q-n_2} m_i^{\mathcal{S}}$ and, subject to that, minimizes $\sum_{i=1}^{2^{n-p}-2^q-n_2} \left[m_i^{\mathcal{S}}\right]^2$.

A design that satisfies condition (i) has the maximum number of two-factor interactions that are not aliased with main effects and tends to have these interactions nearly uniformly distributed over the alias sets. This assures a good design for unstructured units ($\xi_{\mathcal{B}} = \xi_{\mathcal{P}} = \xi_{\mathcal{S}}$). If condition (ii) is also satisfied, then the design maximizes the number of two-factor interactions that are neither aliased with main effects nor confounded with blocks, and also tends to have these interactions nearly uniformly distributed over the alias sets. This assures that the design is a good block design for models with fixed block effects and unstructured units within each block ($\xi_{\mathcal{B}} = \infty$ and $\xi_{\mathcal{P}} = \xi_{\mathcal{S}}$). Condition (iii) further assures a good design at the subplot level ($\xi_{\mathcal{B}} = \xi_{\mathcal{P}} = \infty$). If a design is good for all these three extreme cases, then it is a good blocked split-plot design for all $\xi_{\mathcal{B}} \geq \xi_{\mathcal{P}} \geq \xi_{\mathcal{S}}$.

Example 14.16. (Example 1.2 revisited) This calls for a 32-run blocked split-plot design with the block structure (4 weeks)/(4 days)/(2 runs), three two-level whole-plot treatment factors A, B, and C, and three two-level subplot treatment factors S, T, and U. Table 1 of McLeod and Brewster (2004) listed two designs for this case, hereafter denoted by d_1 and d_2, respectively, where d_1 is the design constructed in Example 14.6, and d_2 is defined by $C = AB$, with both $ABST$ and BSU doubling as splitting and blocking words. The following shows the numbers of two-factor interactions in the 25 alias sets of factorial effects that are not aliased with main effects.

$$d_1 \qquad\qquad\qquad\qquad d_2$$

	d_1	d_2
$W_{\mathcal{B}}$	0 0 0	0 0 0
$W_{\mathcal{P}}$	0 0 0 0 1 1 1 1 2	0 0 0 0 0 0 1 1 1
$W_{\mathcal{S}}$	0 0 0 0 0 0 1 1 1 1 1 1 2 2	0 0 0 0 1 1 1 1 1 1 1 1 1 1

Neither design dominates the other. Design d_1 is better than d_2 with respect to (i) and (ii). For (iii), both designs have $\sum_i m_i^S = 9$, but d_2 has a smaller $\sum_i [m_i^S]^2$ (9 vs. 13). Since d_2 has a whole-plot defining word ABC, it does not satisfy the requirement imposed in the first paragraph of Section 14.4. Indeed, it is dominated by a design d_3 that is defined by $U = ABCS$, with AST as a splitting word, and AST, AC as blocking words. The numbers of two-factor interactions in the alias sets for d_3 are as follows.

$W_{\mathcal{B}}$	0 0 1
$W_{\mathcal{P}}$	0 0 0 0 1 1 1 1 1
$W_{\mathcal{S}}$	0 0 0 0 1 1 1 1 1 1 1 1 1 1

Thus d_2 can be eliminated. A complete search shows that d_1 and d_3 are the only admissible designs up to equivalence. One may need more information to choose between d_1 and d_3. We note that d_1 is better than d_3 with respect to (ii) since one two-factor interaction is confounded with blocks under d_3. However, d_3 is better with respect to (i) and (iii), and has higher resolution as well as more clear two-factor interactions than d_1. We might prefer d_1 if $\xi_{\mathcal{B}}$ is sufficiently greater than $\xi_{\mathcal{P}}$ and $\xi_{\mathcal{P}}$ is not much greater than $\xi_{\mathcal{S}}$. Otherwise we expect d_3 to be a better design.

14.14 Selection of blocked fractional factorial strip-plot designs

Suppose the block structure is $2^g/(2^{n_1-p_1-g} \times 2^{n_2-p_2-g})$, and there are n_1 two-level row treatment factors and n_2 two-level column treatment factors. In this case, $\mathfrak{B} = \{\mathcal{U}, \mathcal{B}, \mathcal{R}', \mathcal{C}', \mathcal{E}\}$, and the subsets $\mathfrak{G}$ of $\mathfrak{B}\backslash\{\mathcal{U}\}$ that satisfy (14.10) are $\{\mathcal{B}, \mathcal{R}', \mathcal{C}', \mathcal{E}\}$, $\{\mathcal{R}', \mathcal{C}', \mathcal{E}\}$, $\{\mathcal{R}', \mathcal{E}\}$, $\{\mathcal{C}', \mathcal{E}\}$, and $\{\mathcal{E}\}$. To apply Theorem 14.5, we need to check the condition in the theorem for all these five choices of $\mathfrak{G}$.

For $\mathfrak{G} = \{\mathcal{B}, \mathcal{R}', \mathcal{C}', \mathcal{E}\}$, the condition implies that the design is good for unstructured units ($\xi_{\mathcal{B}} = \xi_{\mathcal{R}'} = \xi_{\mathcal{C}'} = \xi_{\mathcal{E}}$). For $\mathfrak{G} = \{\mathcal{R}', \mathcal{C}', \mathcal{E}\}$, the condition implies a good block design for models with fixed block effects and unstructured units within each block ($\xi_{\mathcal{B}} = \infty$ and $\xi_{\mathcal{R}'} = \xi_{\mathcal{C}'} = \xi_{\mathcal{E}}$). Similarly, for $\mathfrak{G} = \{\mathcal{R}', \mathcal{E}\}$ (respectively, $\{\mathcal{C}', \mathcal{E}\}$), it calls for a good blocked row (respectively, column) design for the case $\xi_{\mathcal{B}} = \xi_{\mathcal{R}'} = \infty$, $\xi_{\mathcal{C}'} = \xi_{\mathcal{E}}$ (respectively, $\xi_{\mathcal{B}} = \xi_{\mathcal{C}'} = \infty$, $\xi_{\mathcal{R}'} = \xi_{\mathcal{E}}$). The condition for $\mathfrak{G} = \{\mathcal{E}\}$ assures a good design for models with fixed block, row, and column effects.

If a design is good for all these extreme cases, then it is a good blocked strip-plot design for all stratum variances with $\xi_\mathcal{B} \geq \xi_{\mathcal{R}'}, \xi_{\mathcal{C}'} \geq \xi_\mathcal{E}$.

As discussed in Section 14.7, blocked strip-plot designs can be converted into post-fractionated strip-plot designs, with $\mathcal{F}_1, \mathcal{F}_2$ (partitions at the two stages) and the added pseudo factor corresponding to $\mathcal{R}'$, $\mathcal{C}'$, and $\mathcal{B}$, respectively. Thus the results in this section can also be applied to post-fractionated strip-plot designs as long as we require that the main effects of all the treatment factors not be confounded with pseudo blocks.

Example 14.17. (Example 1.3 revisited) This calls for the selection and construction of a blocked strip-plot design with six two-level first-stage treatment factors A, B, C, D, E, F, four two-level second-stage treatment factors S, T, U, V, and 32 experimental units with the block structure $2/(4 \times 4)$. A complete search shows that the design constructed by Miller (1997) (see Example 14.7), denoted by d_1 hereafter, is one of two admissible designs up to equivalence. The other admissible design, denoted by d_2, has independent defining words $ABD, ACE, ABCF$ (for the row design), SUV (for the column design), and blocking word $ST(= BC)$. The following are the m_i values for d_1 and d_2.

	d_1	d_2
$W_\mathcal{B}$	5	4
$W_{\mathcal{R}'}$		
$W_{\mathcal{C}'}$	2 2	1 1
$W_\mathcal{E}$	2 2 2 2 2 2 2 2 2 2 0 0 0 0 0 0	2 2 2 2 2 2 1 1 1 1 1 1 1 1 1 1

Neither design dominates the other, and one may need more information to choose between them. Design d_1 is better than d_2 for $\mathfrak{G} = \{\mathcal{B}, \mathcal{R}', \mathcal{C}', \mathcal{E}\}$, $\{\mathcal{R}', \mathcal{C}', \mathcal{E}\}$, and $\{\mathcal{C}', \mathcal{E}\}$, mainly because it has higher resolution and has more (33) two-factor interactions that are not aliased with main effects than d_2 (30). On the other hand, d_2 is better for $\mathfrak{G} = \{\mathcal{R}', \mathcal{E}\}$ and $\{\mathcal{E}\}$. This is because d_1 and d_2 have the same number (24) of two-factor interactions estimated in the bottom stratum, but d_2 has less severe aliasing of these interactions. Thus d_1 is a better design if $\xi_\mathcal{E}$ is not much smaller than $\xi_{\mathcal{C}'}$, and we may prefer d_2 if $\xi_\mathcal{E}$ is substantially smaller than $\xi_{\mathcal{C}'}$. We also note that there are 14 clear two-factor interactions under d_2, but none under d_1.

14.15   Geometric formulation*

Let P be the set of the $(s^{n-p} - 1)/(s - 1)$ distinct points of PG$(n - p - 1, s)$. For any subset S of P, let $\mathbf{V}_S$ be the matrix with the points in S as its columns. We have seen in Section 9.11 that constructing an (unblocked) s^{n-p} regular fractional factorial design of resolution III+ is equivalent to choosing a set T of n points from P such that the $(n - p) \times n$ matrix $\mathbf{V}_T$ has rank $n - p$. Let $\mathbf{a}_1, \ldots, \mathbf{a}_p \in$ EG(n, s) be linearly independent treatment defining words of the resulting design. Then we have

$$\mathbf{V}_T \mathbf{a}_i = \mathbf{0}, \; i = 1, \ldots, p.$$

To construct a design with s^q blocks of equal size, we need q blocking words $\mathbf{b}_1, \ldots, \mathbf{b}_q \in EG(n, s)$ such that $\mathbf{a}_1, \ldots, \mathbf{a}_p, \mathbf{b}_1, \ldots, \mathbf{b}_q$ are linearly independent. Let

$$\mathbf{V}_B = \mathbf{V}_T[\mathbf{b}_1 \cdots \mathbf{b}_q]. \tag{14.11}$$

Lemma 14.6. *The columns of* $\mathbf{V}_B$ *are linearly independent.*

Proof. Suppose $\mathbf{V}_B\mathbf{u} = \mathbf{V}_T[\mathbf{b}_1 \cdots \mathbf{b}_q]\mathbf{u} = \mathbf{0}$ for some $\mathbf{u} \neq \mathbf{0}$. Then $[\mathbf{b}_1 \cdots \mathbf{b}_q]\mathbf{u}$ belongs to the contrast subgroup of the design. Since the contrast subgroup is generated by $\mathbf{a}_1, \ldots, \mathbf{a}_p$, and $\mathbf{a}_1, \ldots, \mathbf{a}_p, \mathbf{b}_1, \ldots, \mathbf{b}_q$ are linearly independent, we must have

$$[\mathbf{b}_1 \cdots \mathbf{b}_q]\mathbf{u} = \mathbf{0}.$$

It follows from the linear independence of $\mathbf{b}_1, \ldots, \mathbf{b}_q$ that $\mathbf{u} = \mathbf{0}$. $\square$

Since $\mathbf{V}_B$ is $(n - p) \times q$, Lemma 14.6 implies that the columns of $\mathbf{V}_B$ correspond to a set of q linearly independent points of $PG(n - p - 1, s)$. By reversing the argument in the proof of Lemma 14.6, it is easy to see that if the columns of $\mathbf{V}_B$ defined in (14.11) are linearly independent, then $\mathbf{a}_1, \ldots, \mathbf{a}_p, \mathbf{b}_1, \ldots, \mathbf{b}_q$ are also linearly independent. Therefore choosing q independent blocking words is equivalent to choosing q linearly independent points from $PG(n - p - 1, s)$ or, equivalently, choosing a $(q - 1)$-flat.

Let F_B be the $(q - 1)$-flat generated by the q columns of $\mathbf{V}_B$. Then by the discussions in Sections 7.2 and 14.2, a nonzero vector $\mathbf{b}$ in $EG(n, s)$ does not appear in the defining contrast subgroup but the contrasts defined by the pencil $P(\mathbf{b})$ are confounded with blocks if and only if $\mathbf{V}_T\mathbf{b}$ represents a point in F_B. Therefore in the one-to-one correspondence between P and the alias sets established in Section 9.11, each of the $(s^q - 1)/(s - 1)$ points in F_B corresponds to an alias set of treatment factorial effects that are confounded with blocks, while the n points in T correspond to the n alias sets each of which contains a main effect pencil.

In summary, constructing a blocked regular resolution III+ s^{n-p} fractional factorial design in s^q blocks of size s^{n-p-q} is equivalent to choosing a pair of subsets (T, F_B) of P such that

(i) F_B is a $(q - 1)$-flat;
(ii) T contains n points, and rank$(\mathbf{V}_T) = n - p$.

If we further require that no main effect be confounded with blocks, then

(iii) T and F_B are disjoint.

This geometric formulation was discussed in Mukerjee and Wu (1999) and Cheng and Mukerjee (2001). By counting the number of points in F_B and T, it is clear that such a design exists if and only if $(s^q - 1)/(s - 1) + n \leq (s^{n-p} - 1)/(s - 1)$. Let $\tilde{T} = P \setminus (F_B \cup T)$; then the $(s^{n-p} - s^q)/(s - 1) - n$ points in $\tilde{T}$ correspond to the alias sets that neither contain a main effect pencil nor are confounded with blocks.

Example 14.18. (Example 14.1 revisited) We use the design constructed in Example 14.1 to illustrate the geometric connection. For $n = 6$ and $p = q = 2$, we need two subsets F_B and T of PG(3,2), where F_B is a 1-flat and T contains six points. Let

$$T = \{(1,0,0,0)^T, (0,1,0,0)^T, (0,0,1,0)^T, (0,0,0,1)^T, (1,0,1,1)^T, (1,1,1,0)^T\}.$$

To satisfy the rank condition (ii), T must contain four linearly independent points. This is clearly satisfied. Identify the six points in T with treatment factors A, B, C, D, E, F in the order as given. Then A, B, C, D form a set of basic factors. If E and F are to be defined by $E = ACD$ and $F = ABC$, then the last two points of T can be determined as the sum of the first, third, and fourth points and that of the first three points, respectively. To satisfy (i) and (iii), we need F_B to be a 1-flat that is disjoint with T. One choice is

$$F_B = \{(1,0,1,0)^T, (1,1,0,1)^T, (0,1,1,1)^T\}.$$

Identify the first two points of F_B with blocking factors B_1 and B_2, respectively; note that the third point of F_B is uniquely determined by the first two as their sum. Write down the 4×8 matrix with the eight points corresponding to $A, B, C, D, E, F, B_1, B_2$ as its columns, and form a 16×8 matrix whose rows are all the linear combinations of the rows of the 4×8 matrix constructed above. Then by changing 0 and 1 to 1 and -1, respectively, we obtain the 16×8 matrix displayed in Example 14.1. Since the sum of the points corresponding to A and C, $(1,0,0,0)^T + (0,0,1,0)^T = (1,0,1,0)^T$, belongs to F_B, we see that the interaction of factors A and C is confounded with blocks. For the same reason, the three-factor interactions BCD and ABD are also confounded with blocks.

In the case of two-level designs, the interaction of two treatment factors A and B is aliased with a main effect if the third point of the line determined by the two points corresponding to A and B belongs to T. This fact has been used in Section 10.7 to show that the condition in Corollary 10.3 (or condition (i) of Theorem 14.4) holds if the number of lines with one point in $\overline{T} = P \setminus T$ and two points in T is maximized, and that the numbers of such lines passing through the points in $\overline{T}$ differ from one another by at most one. If this is true, then the resulting fractional factorial design maximizes $E_k(d)$ $(I_k(d)$ for $\gamma = 1)$ for all k.

The result reviewed in the previous paragraph can be extended to blocked fractional factorial designs as follows. The interaction of two treatment factors is confounded with blocks if the third point of the line determined by the two points corresponding to these two factors belongs to F_B. Thus each two-factor interaction that is neither aliased with main effects nor confounded with blocks corresponds to a line with two points in T and one point in $\tilde{T} = P \setminus (T \cup F_B)$. Therefore $\sum_{j=1}^{h_\mathcal{E}} m_j^\mathcal{E}(d)$ is equal to the total number of such lines. It follows that the condition in Proposition 14.2 (or condition (ii) of Theorem 14.4) holds if the number of lines with two points in T and one point in $\tilde{T}$ is maximized, and the numbers of such lines passing through the $h_\mathcal{E}$ points in $\tilde{T}$ differ from one another by at most one. If this is true, then the resulting blocked fractional factorial design maximizes $E_k^\mathcal{E}(d)$ $(I_k(d)$ for $\gamma = 0)$ for all

k. By Theorem 14.4, if the geometric conditions in this and the previous paragraph hold simultaneously, then we have a design that is optimal with respect to the surrogate criterion (14.6)–(14.7) for all $0 \leq \gamma \leq 1$. Some results on designs that satisfy the condition in the previous paragraph can be found in Section 10.7. We now present a simple result on blocked fractional factorial designs that satisfy the geometric condition in this paragraph.

For each subset H of P, let $n(H)$ be the number of lines in H, and for any two subsets H_1 and H_2 of P, let $m(H_1, H_2)$ be the number of lines with two points in H_1 and one point in H_2. The complementary design theory in Section 10.7 effectively shows the equivalence of (i) minimizing $n(T)$, (ii) maximizing $n(\overline{T})$, (iii) maximizing $m(T, \overline{T})$, and (iv) minimizing $m(\overline{T}, T)$. In the present setting, we want to maximize $m(T, \tilde{T})$. It can be shown that

$$\text{maximizing } m(T, \tilde{T}) \text{ is equivalent to maximizing } 3n(\tilde{T}) + 2m(\tilde{T}, F_B). \quad (14.12)$$

We leave this as an exercise.

We also write $h_{\mathcal{E}}$, the number of points in $\tilde{T}$, as $\bar{n}$. Since each pair of points determines a line, we must have

$$m(\tilde{T}, F_B) \leq \bar{n}(\bar{n} - 1)/2 - 3n(\tilde{T}). \quad (14.13)$$

Proposition 14.7. *Suppose $n = 2^{n-p} - 2^{q+1}$, $q < n - p - 1$. If a blocked fractional factorial design d^* is constructed by choosing an n-subset T_{d^*} and a $(q-1)$-flat F_B of P such that $\overline{T}_{d^*}$ is a q-flat, $F_B \subset \overline{T}_{d^*}$, and $\operatorname{rank}(\mathbf{V}_{T_{d^*}}) = n - p$, then d^* maximizes $\sum_{i=1}^{h_{\mathcal{E}}} m_i^{\mathcal{E}}(d)$, and all the $m_i^{\mathcal{E}}(d^*), 1 \leq i \leq h_{\mathcal{E}}$, are equal.*

Proof. We first show that T_{d^*} maximizes $3n(\tilde{T}) + 2m(\tilde{T}, F_B)$. By (14.13), we have

$$3n(\tilde{T}) + 2m(\tilde{T}, F_B) \leq \bar{n}(\bar{n} - 1) - 3n(\tilde{T}) \leq \bar{n}(\bar{n} - 1). \quad (14.14)$$

We show that this upper bound is achieved by $\tilde{T}_{d^*}$. Since F_B is a $(q-1)$-flat contained in $\overline{T}_{d^*}$, which is a q-flat, we must have $\overline{T}_{d^*} = F_B \cup \{\mathbf{y}\} \cup \{\mathbf{x} + \mathbf{y} : \mathbf{x} \in F_B\}$, where $\mathbf{y} \notin F_B$. It follows that $\tilde{T}_{d^*} = \{\mathbf{y}\} \cup \{\mathbf{x} + \mathbf{y} : \mathbf{x} \in F_B\}$. Clearly for any two distinct points $\mathbf{a}$ and $\mathbf{b}$ in $\tilde{T}_{d^*}$, we have $\mathbf{a} + \mathbf{b} \in F_B$. Thus $n(\tilde{T}_{d^*}) = 0$ and $m(\tilde{T}_{d^*}, F_B) = \bar{n}(\bar{n} - 1)/2$; therefore the upper bound in (14.14) is achieved. This shows that d^* maximizes $\sum_{i=1}^{h_{\mathcal{E}}} m_i^{\mathcal{E}}(d)$. To show that all the $m_i^{\mathcal{E}}(d^*), 1 \leq i \leq h_{\mathcal{E}}$, are equal, we need to show that for each of the $h_{\mathcal{E}}$ points $\mathbf{x}$ in $\tilde{T}_{d^*}$, the number of lines passing through $\mathbf{x}$ and two points in T_{d^*} is a constant. We note that since $\overline{T}_{d^*}$ is a flat and $\tilde{T}_{d^*} \subset \overline{T}_{d^*}$, for each line passing through a point $\mathbf{x} \in \tilde{T}_{d^*}$, either both of the other two points are in $\overline{T}_{d^*}$, or both are in T_{d^*}; also, the number of lines passing through $\mathbf{x}$ and two points in $\overline{T}_{d^*}$ is a constant. It follows that the number of lines passing through $\mathbf{x}$ and two points in T_{d^*} is also a constant since the total number of lines passing through $\mathbf{x}$ is a constant. $\square$

By Theorem 10.6 (or by the same argument as in the proof of Proposition 14.7), the unblocked 2^{n-p} fractional factorial design d^* in Proposition 14.7 maximies $\sum_{i=1}^{f} m_i(d)$ and has all the $m_i(d)$'s equal. That is, it also satisfies condition (i) in Theorem 14.4. We have proved the following result.

Corollary 14.8. *The design d^* in Proposition 14.7 is optimal with respect to the surrogate criterion (14.6)–(14.7) for all $0 \leq \gamma \leq 1$.*

Among other results, Cheng and Mukerjee (2001) showed that the design in Proposition 14.7 has maximum estimation capacity $E_k^{\mathcal{E}}(d)$ for all k, and Chen and Cheng (1999) showed that it has minimum aberration with respect to W_{CC}. We refer readers to these two papers for more results in this direction.

Similar to the construction of block designs, for constructing a split-plot design with n_1 whole-plot treatment factors, n_2 subplot treatment factors, and s^q whole-plots each containing s^{n-p-q} subplots, we need to choose from $\mathrm{PG}(n-p-1,s)$ a set T of n points for the treatment factors and a $(q-1)$-flat $F_{\mathcal{P}}$ for determining the design on the whole-plots. However, instead of being disjoint as in the case of block designs, T and $F_{\mathcal{P}}$ must have exactly n_1 points in common, where the n_1 points in $T \cap F_{\mathcal{P}}$ correspond to the whole-plot treatment factors. Each point in $F_{\mathcal{P}}$ corresponds to an alias set of treatment factorial effects that are estimated in the whole-plot stratum, and each of the other points in the geometry corresponds to an alias set of treatment factorial effects that are estimated in the subplot stratum. We need $\mathrm{rank}(\mathbf{V}_T) = n - p$, and the requirements described in the first paragraph of Section 14.3 are equivalent to that $\mathrm{rank}(\mathbf{V}_{T \cap F_{\mathcal{P}}}) = q$ if $n_1 \geq q$, and is equal to n_1 if $n_1 < q$.

If a split-plot design as in the previous paragraph is to be blocked, with the s^q whole-plots grouped into s^g blocks of equal size, then in addition to T and $F_{\mathcal{P}}$, we need to choose a $(g-1)$-flat $F_{\mathcal{B}} \subset F_{\mathcal{P}}$ to perform blocking. Each point in $F_{\mathcal{B}}$ (respectively, $F_{\mathcal{P}} \setminus F_{\mathcal{B}}$) corresponds to an alias set of treatment factorial effects that are estimated in the block (respectively, whole-plot) stratum, whereas each of the other points in the geometry corresponds to an alias set of treatment factorial effects that are estimated in the subplot stratum. Therefore, if the treatment main effects are not to be confounded with blocks, then $T \cap F_{\mathcal{P}}$ must be disjoint with $F_{\mathcal{B}}$.

To construct a regular s^{n-p} row-column design with s^{k_1} rows and s^{k_2} columns, where $n - p = k_1 + k_2$, we need a (k_1-1)-flat $F_{\mathcal{R}}$, a (k_2-1)-flat $F_{\mathcal{C}}$, and a set T of n points from $\mathrm{PG}(k_1 + k_2 - 1, s)$ such that $F_{\mathcal{R}} \cap F_{\mathcal{C}} = \emptyset$. Each point in $F_{\mathcal{R}}$ (respectively, $F_{\mathcal{C}}$) corresponds to an alias set of treatment factorial effects that are estimated in the row (respectively, column) stratum, whereas each of the other points in the geometry corresponds to an alias set of treatment factorial effects that are estimated in the bottom stratum. We need $\mathrm{rank}(\mathbf{V}_T) = n - p$. If the treatment main effects are required to be confounded with neither rows nor columns, then we must have $T \cap F_{\mathcal{R}} = T \cap F_{\mathcal{C}} = \emptyset$. On the other hand, if the main effects of n_1 of the treatment factors are to be confounded with rows, and those of the other $n_2 = n - n_1$ treatment factors are to be confounded with columns (so we have a strip-plot design), then $T \cap F_{\mathcal{R}}$ and $T \cap F_{\mathcal{C}}$ must contain exactly n_1 and n_2 points, respectively.

For constructing a design with the block structure $s^g / (s^{n_1 - p_1 - g} \times s^{n_2 - p_2 - g})$, n_1 row treatment factors, and n_2 column treatment factors, we need to choose from $\mathrm{PG}(n - p - g - 1, s)$, where $n = n_1 + n_2$ and $p = p_1 + p_2$, a set T of n points with $\mathrm{rank}(\mathbf{V}_T) = n - p - g$, a $(g-1)$-flat $F_{\mathcal{B}}$, an $(n_1 - p_1 - 1)$-flat $F_{\mathcal{R}}$, and an $(n_2 - p_2 - 1)$-flat $F_{\mathcal{C}}$ such that $F_{\mathcal{B}} = F_{\mathcal{R}} \cap F_{\mathcal{C}}$, $T \cap F_{\mathcal{R}}$ contains n_1 points, and $T \cap F_{\mathcal{C}}$ contains n_2

points. Furthermore, if no main effect is to be confounded with blocks, then we must have $T \cap F_{\mathcal{B}} = \emptyset$.

For experiments with multiple processing stages discussed in Section 13.11, suppose the treatment design is a regular s^{n-p} design, and the treatment combinations are partitioned into s^{r_i} disjoint classes of equal size at the ith-stage, $i = 1, \ldots, h$. The geometric connection is that the selection of an s^{n-p} design is again equivalent to choosing from $PG(n - p - 1, s)$ a set T of n points with $\text{rank}(\mathbf{V}_T) = n - p$, and the partition at stage i is equivalent to choosing an $(r_i - 1)$-flat, say F_i. If stage i is nested in stage j, then we have $F_j \subset F_i$. If n_i factors are to have constant levels assigned to all the units in the same class at the ith stage, then $T \cap (F_i \setminus \bigcup_{j:\text{stage } i \text{ is nested in stage } j} F_j)$ must contain exactly n_i points. In the nonoverlapping case, $F_1, \ldots, F_h$ are mutually disjoint.

The condition in Theorem 14.5 can also be rephrased geometrically for unblocked and blocked split-plot and strip-plot designs, and can be used to derive results on families of optimal designs in these settings.

Exercises

14.1   Show that the design constructed in Example 14.1 maximizes the number of two-factor interactions that are neither aliased with main effects nor confounded with blocks.

14.2   Repeat the previous exercise for the design in Example 14.11.

14.3* Extend the result in Example 14.12 to a family of blocked 2^{n-p} designs in 2^q blocks with $n = 2^{n-p-1} - 2$. Show that the design constructed in Example 14.1 is a special case with $n = 6$, $p = 2$, and $q = 2$.

14.4   Suppose a blocked fractional factorial design d^* successively minimizes $A_3(d)$ and $A_4(d)$, and also successively minimizes $3A_{3,0}(d) + A_{2,1}(d)$ and $A_{4,0}(d)$. Show that it successively minimizes $3A_{3,0}(d) + (1 - \gamma^{1/k})A_{2,1}(d)$ and $A_{4,0}(d)$ for all $0 \le \gamma \le 1$.

14.5* Prove (14.12).

Chapter 15

Nonregular Designs

The run sizes of regular fractional factorial designs must be prime numbers or powers of prime numbers. Nonregular designs are more flexible in run sizes and are more abundant. For example, if the Hadamard conjecture is true, then Hadamard designs can be constructed for every run size that is a multiple of 4. The economy in run sizes makes these designs suitable for factor-screening experiments where the primary objective is to identify important factors for further exploration. In this chapter we briefly discuss some properties of nonregular designs and how some of the results presented in previous chapters can be extended to nonregular designs. The aliasing of factorial effects under a nonregular design is more complex than that under a regular design. This, however, leads to interesting projection properties with important implications. A design with good projections onto small subsets of factors can provide useful information in factor-screening experiments. For two-level designs, Deng and Tang (1999) proposed a generalized minimum aberration criterion as an extension of minimum aberration to nonregular designs. A simpler version, called minimum G_2-aberration, was proposed in Tang and Deng (1999). When applied to regular designs, both versions are the same as minimum aberration. We concentrate on minimum G_2-aberration and its generalizations, including Xu and Wu's (2001) extension to mixed-level designs. When regular designs are available, there often exist better nonregular designs than minimum aberration regular designs under the generalized minimum aberration criterion. We also present some results on two-level supersaturated designs. Such designs, with more factors than the available degrees of freedom, are useful for factor screening. Other topics covered include parallel flats designs and saturated designs for hierarchical models.

15.1 Indicator functions and J-characteristics

Given an N-run design d for n two-level factors, let the two levels be represented by 1 and -1 and let $\mathbf{X}_d$ be the $N \times n$ matrix whose (l,i)th entry $x_{li}(d)$ is the level of the ith factor at the lth observation, $1 \le l \le N$. Let the ith column of $\mathbf{X}_d$ be $\mathbf{x}_i(d)$, and for each $1 \le k \le n$ and any subset $S = \{i_1,\dots,i_k\}$ of $\{1,\dots,n\}$, let $\mathbf{x}_S(d) = \mathbf{x}_{i_1}(d) \odot \cdots \odot \mathbf{x}_{i_k}(d)$. Then $\mathbf{x}_i(d)$ and $\mathbf{x}_S(d)$ are columns of the model matrix that correspond to, respectively, the main effect of the ith factor and the interaction of factors $i_1,\dots,i_k$.

329

Tang (2001) defined the *J-characteristics* as

$$J_S(d) = \sum_{l=1}^{N} x_{li_1}(d) \cdots x_{li_k}(d),$$

which is the sum of all the entries in the column of the model matrix corresponding to the interaction of factors $i_1, \ldots, i_k$. The *J*-characteristics were originally defined in Deng and Tang (1999) and Tang and Deng (1999) as absolute values of the $J_S(d)$'s defined above. These quantities contain important information about the design. As we will see later, they are crucial for extending the concepts of wordlength, resolution, and aberration to nonregular designs. We first state a simple result.

Proposition 15.1. *If d is a two-level orthogonal array of strength t, then $J_S(d) = 0$ for all S with $|S| \leq t$.*

The converse of Proposition 15.1 is also true; see Theorem 15.15 in Section 15.7.

Tang (2001) also showed that each two-level design is uniquely determined by its *J*-characteristics. His proof was based on a property of Hadamard matrices. Ye (2003) obtained the result by using the approach of indicator functions first proposed by Fontana, Pistone, and Rogantin (2000) to study fractional factorial designs. This result can be extended to general multi-level and mixed-level designs. Given an N-run design d for an $s_1 \times \cdots \times s_n$ experiment, let $n_d(x_1, \ldots, x_n)$ be the number of observations on treatment combination $(x_1, \ldots, x_n)^T$ and let $\mathbf{n}_d$ be the $(s_1 \cdots s_n) \times 1$ vector whose entries are all the $n_d(x_1, \ldots, x_n)$'s. Then d is completely determined by $\mathbf{n}_d$, and we have $\mathbf{n}_d^T \mathbf{1} = N$. In the special case where $n_d(x_1, \ldots, x_n) = 0$ or 1, a fractional factorial design can be considered as a subset of the set of all the $s_1 \cdots s_n$ treatment combinations, and $\mathbf{n}_d$ can be viewed as the *indicator function* of the selected treatment combinations.

For $\mathbf{z} \in S_1 \times \cdots \times S_n$, where $S_i = \{0, 1, \ldots, s_i - 1\}$, let $\mathbf{x}^{\mathbf{z}}$ be the column of the full model matrix $\mathbf{X}_T \mathbf{P}$ corresponding to $\beta^{\mathbf{z}}$; see (8.3).

Theorem 15.2. *A fractional factorial design is uniquely determined by the $s_1 \cdots s_n - 1$ quantities $\{\mathbf{1}_N^T \mathbf{x}^{\mathbf{z}}, \mathbf{z} \in S_1 \times \cdots \times S_n, \mathbf{z} \neq \mathbf{0}\}$.*

Proof. We have seen that the column vectors $\{\mathbf{p}^{\mathbf{z}} : \mathbf{z} \in S_1 \times \cdots \times S_n\}$ in (6.22) form an orthogonal basis of $\mathbb{R}^{s_1 \cdots s_n}$. Therefore for any design d of run size N, the vector $\mathbf{n}_d$ can be expressed as

$$\mathbf{n}_d = \sum_{\mathbf{z} \in S_1 \times \cdots \times S_n} \frac{1}{\|\mathbf{p}^{\mathbf{z}}\|^2} (\mathbf{n}_d^T \mathbf{p}^{\mathbf{z}}) \mathbf{p}^{\mathbf{z}}. \tag{15.1}$$

We have $(\mathbf{n}_d^T \mathbf{p}^{\mathbf{0}}) \mathbf{p}^{\mathbf{0}} = (\mathbf{n}_d^T \mathbf{1}_{s_1 \cdots s_n}) \mathbf{1}_{s_1 \cdots s_n} = N \mathbf{1}_{s_1 \cdots s_n}$, and

$$\mathbf{n}_d^T \mathbf{p}^{\mathbf{z}} = \mathbf{1}_N^T \mathbf{x}^{\mathbf{z}} \text{ for } \mathbf{z} \neq \mathbf{0}. \tag{15.2}$$

Equation (15.2) holds because the components of $\mathbf{n}_d$ corresponding to the treatment combinations not included in d are 0, and for those that are selected, the corresponding entries of $\mathbf{x}^\mathbf{z}$ and $\mathbf{p}^\mathbf{z}$ are the same. The result then follows from (15.1) and (15.2). $\qquad\qquad\qquad\qquad\qquad\qquad\qquad\qquad\qquad\qquad\qquad\qquad\qquad\qquad$ $\square$

Recall that if $\mathbf{z}$ has exactly one nonzero entry, then $\mathbf{x}^\mathbf{z}$ is a main-effect column, and if $\mathbf{z}$ has k nonzero entries, $k > 1$, then $\mathbf{x}^\mathbf{z}$ corresponds to a k-factor interaction and is the Hadamard product of k main-effect columns. Since $\mathbf{1}_N^T \mathbf{x}^\mathbf{z}$ is the sum of the entries in the column of the model matrix corresponding to $\beta^\mathbf{z}$, for two-level designs, the quantities $\mathbf{1}_N^T \mathbf{x}^\mathbf{z}$ are the same as the J-characteristics. Therefore, Theorem 15.2 is an extension of Tang's (2001) result that two-level fractional factorial designs are uniquely determined by their J-characteristics.

Under a two-level regular fractional factorial design, if an interaction appears in the defining relation, then the corresponding column of the full model matrix consists of all 1's or all -1's; otherwise half of the entries are equal to 1 and the other half are equal to -1. The sum of the entries of the column is N or $-N$ in the former case, and is 0 in the latter case.

Proposition 15.3. *Under a two-level regular fractional factorial design d, for any set S of distinct factors, $J_S(d) = N$ or $-N$ if the factors in S constitute a defining word. Otherwise, $J_S(d) = 0$.*

Many orthogonal arrays are obtained by taking the Kronecker products of smaller arrays. The following result from Tang (2006) is useful for studying the J-characteristics of such arrays.

Proposition 15.4. *Let d_i, $i = 1, 2$, be a design with n_i two-level factors and N_i runs. Let $d_1 \otimes d_2$ be the design with $\mathbf{X}_{d_1 \otimes d_2} = \mathbf{X}_{d_1} \otimes \mathbf{X}_{d_2}$. Suppose the Kronecker product of the ith column of $\mathbf{X}_{d_1}$ and jth column of $\mathbf{X}_{d_2}$ is labeled by (i, j). Then for any $S = \{(i_1, j_1), \ldots, (i_k, j_k)\}$, we have $J_S(d_1 \otimes d_2) = J_{\{i_1, \ldots, i_k\}}(d_1) J_{\{j_1, \ldots, j_k\}}(d_2)$.*

For an application of Proposition 15.4, see Exercise 15.2.

15.2   Partial aliasing

For two-level factors, the concept of defining words can easily be extended to non-regular designs. We say that a defining word of length k exists if there are k factors, say factors $i_1, \ldots, i_k$, such that $\mathbf{x}_{i_1} \odot \cdots \odot \mathbf{x}_{i_k} = \mathbf{1}$ or $-\mathbf{1}$, where $\mathbf{1}$ is the $N \times 1$ vector of 1's. This is equivalent to that $\frac{1}{N}|J_S| = 1$, where $S = \{i_1, \ldots, i_k\}$.

For any two subsets S and T of $\{1, \ldots, n\}$, we have

$$\mathbf{x}_S(d) \odot \mathbf{x}_T(d) = \mathbf{x}_{S \triangle T}(d), \qquad\qquad\qquad (15.3)$$

where $S \triangle T = (S \cup T) \setminus (S \cap T)$. Therefore

$$\frac{1}{N}[\mathbf{x}_S(d)]^T \mathbf{x}_T(d) = \frac{1}{N} J_{S \triangle T}(d).$$

We can think of $\frac{1}{N}[\mathbf{x}_S(d)]^T\mathbf{x}_T(d)$ as the "correlation" of two columns of the full model matrix. It follows from Proposition 15.3 that under a two-level regular design, for any two subsets S and T of $\{1,\ldots,n\}$, $\left|\frac{1}{N}[\mathbf{x}_S(d)]^T\mathbf{x}_T(d)\right|$ is either 1 or 0. We say that the two corresponding factorial effects are fully aliased in the former case and orthogonal in the latter. Therefore, under two-level regular designs, any two factorial effects are either fully aliased or orthogonal; also see the discussion in Section 9.3. Ye (2004) showed that the converse is also true: any two-level factorial design that has $\left|\frac{1}{N}J_S(d)\right| = 0$ or 1 for all S must be a regular 2^{n-p} design or replicates of a regular 2^{n-p} design. This gives a characterization of regular designs.

For nonregular designs, $\left|\frac{1}{N}J_{S\triangle T}(d)\right|$ and hence $\left|\frac{1}{N}[\mathbf{x}_S(d)]^T\mathbf{x}_T(d)\right|$ can be strictly between 0 and 1, leading to the so-called *partial aliasing*. For example, it can be verified that under the 12-run Plackett–Burman design, $\left|\frac{1}{N}J_S(d)\right| = \frac{1}{3}$ for all S of size three.

When $n_d(x_1,\ldots,x_n) = 0$ or 1, by restricting (15.1) to the treatment combinations in d, we have

$$\frac{1}{N}\mathbf{1}_N = \sum_{S:S\subseteq\{1,\ldots,n\}} \left[\frac{1}{N}J_S(d)\right][2^{-n}\mathbf{x}_S(d)], \qquad (15.4)$$

where $x_\emptyset(d)$ is the vector of ones. For any subset T of $\{1,\ldots,n\}$, by taking the Hadamard product of $\mathbf{x}_T(d)$ with both sides of (15.4) and using (15.3), we obtain

$$\frac{1}{N}\mathbf{x}_T(d) = \sum_{S:S\subseteq\{1,\ldots,n\}} \left[\frac{1}{N}J_S(d)\right][2^{-n}\mathbf{x}_{S\triangle T}(d)]. \qquad (15.5)$$

Equation (15.5) shows how a given factorial effect is aliased with the other effects under design d.

For two-level regular designs, (15.4) is equivalent to the defining relation, and (15.5) is the usual method for determining the aliasing structure from the defining relation of the design; see Section 9.2.

Under a nonregular design, for given T, in addition to having S's with $0 < \left|\frac{1}{N}[\mathbf{x}_S(d)]^T\mathbf{x}_T(d)\right| < 1$, there are *many* such S's. For example, under the 12-run Plackett–Burman design, even under the assumption that all the three-factor and higher order interactions are negligible, since $\left|\frac{1}{N}J_S(d)\right| = \frac{1}{3}$ for all S of size three, by (15.5), each main effect is partially aliased with all the two-factor interactions that do not involve that factor. A consequence of such *complex aliasing* is that in order to estimate all the main effects, all the interactions must be assumed negligible. We will continue the discussion of this issue in the next section.

15.3   Projectivity

Definition 8.1 imposes a projection property on orthogonal arrays: the projection of an orthogonal array of strength t onto *any* t factors consists of one or more replicates of the complete factorial of the t factors. A statistical implication of this property is that if there are no more than t active factors, then under the design, one is able to estimate all the factorial effects of the active factors regardless of which factors are active. Note that a factor is said to be active if it is involved in at least one nonnegligible factorial effect.

At the initial stage of experimentation, the number of potentially important factors is usually large. The limited number of experimental runs cannot provide an adequate coverage of the design space, nor can complicated models be entertained. However, often only a small number of factors are expected to be active. Under such *factor sparsity* (Box and Meyer, 1986), it is useful to study properties of the design when it is projected onto small subsets of factors. Knowledge about projection properties is useful for helping incorporate prior knowledge into designs for factor-screening experiments. It is also sensible to use a design with good low-dimensional projections. Such designs have the advantage that when they are restricted to the small subset of factors identified to be important, the design points have a good coverage of the reduced design space, and more complicated models in the smaller subset of factors can be entertained. This is useful for follow-up analyses and experiments.

Box and Tyssedal (1996) defined a design to be of *projectivity p* if in all its *p*-factor projections, all the treatment combinations appear at least once. Projections onto *all* subsets of *p* factors are considered since it is unknown a priori which factors are active. This concept, useful for designing factor-screening experiments, was also discussed on p. 363 of Constantine (1987).

According to this definition, an orthogonal array of strength t has projectivity t. How about projections onto more than t factors? The answer to this question is straightforward for regular designs. The projection property of a regular fractional factorial design can be studied via its defining relation; see Chen (1998) and Exercise 9.2. Theorem 9.5 shows that a regular design of resolution R has projectivity at least $R-1$. On the other hand, when the design is projected onto the factors that appear in a defining word, not all the level combinations of these factors are covered. Therefore the maximum projectivity of a regular design is exactly its resolution minus one. A nonregular orthogonal array of strength t, however, may have projectivity greater than t.

Lin and Draper (1991, 1992) examined all the projections of 12-, 16-, 20-, 24-, 28-, 32- and 36-run Plackett–Burman designs onto three factors. For each of these designs, the projection onto any three factors consists of one or more copies of the complete 2^3 factorial and/or one or more copies of a half-replicate of 2^3. This projection property is a consequence of Lemma 8.12 in Section 8.7. Let $\mathbf{X}$ be an $OA(N, 2^n, t)$ with $n \geq t+1$. Then by Lemma 8.12, in the projection of $\mathbf{X}$ onto any $t+1$ factors, there exist nonnegative integers α and β such that each combination $\mathbf{x} = (x_1, x_2, \ldots, x_{t+1})^T$ with $x_1 x_2 \cdots x_{t+1} = 1$ appears α times and each of those with $x_1 x_2 \cdots x_{t+1} = -1$ appears β times, where $x_1, \ldots, x_{t+1}$ are the levels of the $t+1$ factors onto which $\mathbf{X}$ is projected. Clearly $\mathbf{X}$ has projectivity $t+1$ if and only if both α and β are positive for all its $(t+1)$-factor projections. If $\alpha = 0$ or $\beta = 0$, then the projection consists of $2^{-t}N$ copies of a half-replicate $2^{(t+1)-1}$. In this case the design has a defining word of length $t+1$ and cannot have projectivity $t+1$. We can rephrase this as follows.

Lemma 15.5. *An $OA(N, 2^n, t)$ with $n \geq t+1$ has projectivity $t+1$ if and only if it has no defining word of length $t+1$.*

Theorem 15.6. *(Cheng, 1995) Let* $\mathbf{X}$ *be an* $OA(N, 2^n, t)$ *with* $n \geq t+2$. *If* N *is not a multiple of* 2^{t+1}, *then* $\mathbf{X}$ *has no defining word of length* $t+1$ *and thus has projectivity* $t+1$.

Proof. It suffices to show that if there is a defining word of length $t+1$, then N must be a multiple of 2^{t+1}. Without loss of generality, by permuting the columns and/or changing the signs of all the entries in the same column of the orthogonal array if necessary, we assume that every row vector $(x_1, x_2, \ldots, x_n)$ of $\mathbf{X}$ satisfies $x_1 \cdots x_{t+1} = 1$. Then

$$x_1 = x_2 \cdots x_{t+1}. \tag{15.6}$$

By Lemma 8.12, there are nonnegative integers α and β such that α rows of $\mathbf{X}$ satisfy $x_{t+2} = x_2 \cdots x_{t+1}$, and β rows of $\mathbf{X}$ satisfy $x_{t+2} = -x_2 \cdots x_{t+1}$. Since $\mathbf{X}$ has strength t, all the combinations of 1's and -1's appear equally often as row vectors of the projection of $\mathbf{X}$ onto columns $2, \ldots, t+1$. If N is not a multiple of 2^{t+1}, then $\alpha \neq \beta$. This and (15.6) imply that in the projection of $\mathbf{X}$ onto the first and $(t+2)$th columns, the four pairs $(1, 1)$, $(1, -1)$, $(-1, 1)$, and $(-1, -1)$ do not appear equally often as its rows. This is a contradiction. □

Thus, e.g., if N is not a multiple of 8, then an $OA(N, 2^n, 2)$ with $n \geq 4$ has no defining word of length three and hence has projectivity three.

15.4 Hidden projection properties of orthogonal arrays

Lin and Draper (1993) also studied projections of 12-, 16-, 20-, and 24-run Plackett–Burman designs onto four and five factors. The 12-run Plackett–Burman design does not have projectivity four; to have projectivity four, at least 16 runs are needed. However, Lin and Draper (1993) observed that the projection of the 12-run Plackett–Burman design onto any four factors has the desirable property that all the main effects and two-factor interactions of the four factors are estimable if the higher-order interactions are negligible. Wang and Wu (1995) also observed this property for 12- and 20-run Plackett–Burman designs and coined the term *hidden projection*.

Due to complex aliasing of factorial effects, Plackett–Burman designs had been used for factor-screening experiments under the assumption that the interactions are negligible. There are obvious pitfalls in ignoring interactions. The hidden projection property indicates that if there is only a small number of active factors, then one may be able to estimate some interactions.

Hamada and Wu (1992) proposed an analysis strategy to entertain interactions in addition to main effects and successfully applied it to some Plackett–Burman and other nonregular designs. The success of their strategy can be attributed to good projection properties of nonregular designs. Box and Bisgaard (1993) also commented, "the interesting projective properties of Plackett–Burman designs, which the experimenters have sometimes been reluctant to use for industrial experimentation due to their complicated alias structures, provide a compelling rationale for their use." It is the complex aliasing of nonregular designs that contributes to their good projection properties.

We refer readers to Wu and Hamada (2009) for detailed discussions of data-analysis strategies for entertaining and estimating two-factor interactions from Plackett–Burman type designs. Cheng and Wu (2001) discussed a strategy of using three-level nonregular designs to perform factor screening and response surface exploration after the active factors have been identified. Chipman (2006) discussed Bayesian analysis of screening experiments. Morris (2006) provided an overview of strategies for group factor screening. These two papers are part of a collection of articles in Dean and Lewis (2006) that covers many aspects of design and analysis of screening experiments.

In the rest of this section, we concentrate on the study of hidden projection properties. It is clear that if a design has the property that all the main effects and two-factor interactions are estimable in all its four-factor projections, then it cannot have defining words of length three or four. We show that the converse is also true for two-level orthogonal arrays of strength two.

Lemma 15.7. *Let* $\mathbf{X}$ *be an* $OA(N, 2^4, 2)$. *Then a necessary and sufficient condition for all the main effects and two-factor interactions to be estimable under the assumption that the higher-order interactions are negligible is that there is no defining word of length three or four.*

Proof. The necessity of the condition is trivial. To prove the sufficiency, suppose $\mathbf{X}$ has no defining word of length three or four. By Theorem 8.11, the array

$$\tilde{\mathbf{X}} = \begin{bmatrix} \mathbf{X} \\ -\mathbf{X} \end{bmatrix}$$

has strength three and also has no defining word of length three or four. By applying Lemma 15.5 to $\mathbf{X}$ and $\tilde{\mathbf{X}}$, we conclude that

(i) each $\mathbf{x} = (x_1, x_2, x_3)$ with $x_i = 1$ or -1 appears at least once as a row vector in any three columns of $\mathbf{X}$;

(ii) for each $\mathbf{x} = (x_1, x_2, x_3, x_4)$ with $x_i = 1$ or -1, at least one of $\mathbf{x}$ and $-\mathbf{x}$ appears as a row vector of $\mathbf{X}$.

If all the 1×4 vectors of 1's and -1's appear at least once as row vectors of $\mathbf{X}$, then every factorial effect can be estimated and there is nothing to prove. Therefore we only need to consider the case where the row vectors of $\mathbf{X}$ do not cover all these 16 vectors. Without loss of generality (by changing signs of all the entries in the same column if necessary), we may assume that at least one vector with $x_1 x_2 x_3 x_4 = 1$ is absent. Since there are eight such vectors in four pairs of mirror images, by (ii), no more than four vectors with $x_1 x_2 x_3 x_4 = 1$ can be absent, and no two of them are mirror images of each other.

Suppose exactly one $\mathbf{x}$ with $x_1 x_2 x_3 x_4 = 1$ does not appear as a row vector of $\mathbf{X}$. Without loss of generality, we may assume that it is $(-1, -1, -1, -1)$. Then by (i), all four vectors $(1, -1, -1, -1)$, $(-1, 1, -1, -1)$, $(-1, -1, 1, -1)$, and $(-1, -1, -1, 1)$

must be row vectors of $\mathbf{X}$. These four vectors and the seven vectors other than $(-1,-1,-1,-1)$ that satisfy $x_1x_2x_3x_4 = 1$ provide at least 11 distinct rows of $\mathbf{X}$. Let $\mathbf{F}$ be the 11×11 model matrix for the 11 runs identified above, where the 11 columns correspond to the mean, four main effects, and six two-factor interactions. It can be verified that the information matrix $\mathbf{F}^T\mathbf{F}$ is nonsingular.

In each of the cases where exactly two, three, or four $\mathbf{x}$'s with $x_1x_2x_3x_4 = 1$ do not appear as row vectors of $\mathbf{X}$, we can use (i) and (ii) to identify a set of distinct rows that must appear. In each case we can verify that these rows of $\mathbf{X}$ define a subdesign with a nonsingular information matrix for the main effects and two-factor interactions. The details are omitted here.   □

The main result of this section follows immediately from Lemma 15.7.

Theorem 15.8. *(Cheng, 1995) Let* $\mathbf{X}$ *be an* $OA(N, 2^n, 2)$ *with* $n \geq 4$. *Then the following are equivalent.*

  (i) *In all the four-factor projections of* $\mathbf{X}$, *all the main effects and two-factor interactions are estimable under the assumption that the higher-order interactions are negligible.*

 (ii) $\mathbf{X}$ *has no defining word of length three or four.*

Corollary 15.9. *Suppose* N *is not a multiple of 8. Then in the projection of an* $OA(N, 2^n, 2)$ *with* $n \geq 4$ *onto any four factors, all the main effects and two-factor interactions are estimable if the higher-order interactions are negligible.*

Proof. By Theorem 15.6, $\mathbf{X}$ has no defining word of length three and $\tilde{\mathbf{X}}$ has no defining word of length four. It is clear that the latter implies that $\mathbf{X}$ also does not have a defining word of length four.   □

For a regular fractional factorial design to have the hidden projection property as described in Theorem 15.8, the strength must be at least four.

Theorem 15.8 settles the four-factor hidden projection properties of two-level orthogonal arrays of strength two. It can be extended to five-factor projections of orthogonal arrays of strength three.

Lemma 15.10. *Let* $\mathbf{X}$ *be an* $OA(N, 2^5, 3)$. *Then a necessary and sufficient condition for all the main effects and two-factor interactions to be estimable under the assumption that the higher-order interactions are negligible is that there is no defining word of length four.*

Proof. Again it is enough to show the sufficiency. Assume that $\mathbf{X}$ has no defining word of length four. Let

$$\mathbf{F} = [\mathbf{1} \ \mathbf{X} \ \mathbf{Y}]$$

be the model matrix, where $\mathbf{1}$ is the $N \times 1$ vector of ones, $\mathbf{X}$ is an $N \times 5$ matrix whose columns correspond to main effects of the five factors, and $\mathbf{Y}$ is an $N \times 10$ matrix whose columns are all the pairwise Hadamard products of the columns of $\mathbf{X}$. Since $\mathbf{X}$ has strength three, the columns of $\mathbf{X}$ are orthogonal to those of $\mathbf{Y}$. It follows that

$$\mathbf{F}^T \mathbf{F} = \begin{bmatrix} N\mathbf{I}_6 & \mathbf{0} \\ \mathbf{0} & \mathbf{Y}^T\mathbf{Y} \end{bmatrix}.$$

It is sufficient to show that $\mathbf{Y}^T\mathbf{Y}$ is nonsingular.

Suppose $\mathbf{X} = [\mathbf{x}_1 \ \mathbf{x}_2 \ \mathbf{x}_3 \ \mathbf{x}_4 \ \mathbf{x}_5]$. Let $\mathbf{Z} = [\mathbf{x}_1 \odot \mathbf{x}_2 \ \mathbf{x}_1 \odot \mathbf{x}_3 \ \mathbf{x}_1 \odot \mathbf{x}_4 \ \mathbf{x}_1 \odot \mathbf{x}_5]$. Then it can easily be verified that $\mathbf{Z}$ is an $OA(N, 2^4, 2)$ and, when considered as an N-run design for four two-level factors, it has no defining word of length three or four. Let $\mathbf{G}$ be the $N \times 10$ matrix whose columns consist of the four columns of $\mathbf{Z}$ and their six pairwise Hadamard products. Then by Lemma 15.7, $\mathbf{G}^T\mathbf{G}$ is nonsingular. We finish the proof by noting that $\mathbf{G} = \mathbf{Y}$. $\square$

The following is an immediate consequence of Lemma 15.10.

Theorem 15.11. *(Cheng, 1998a) Let* $\mathbf{X}$ *be an* $OA(N, 2^n, 3)$ *with* $n \geq 5$. *Then the following are equivalent.*

(i) *In the projection of* $\mathbf{X}$ *onto any five factors, all the main effects and two-factor interactions are estimable under the assumption that the higher-order interactions are negligible.*

(ii) $\mathbf{X}$ *has no defining word of length four.*

By Theorem 15.6, if $\mathbf{X}$ is an $OA(N, 2^n, 3)$ with $n \geq 5$ such that N is not a multiple of 16, then it does not have a defining word of length 4. Thus we have the following corollary of Theorem 15.11.

Corollary 15.12. *Let* $\mathbf{X}$ *be an* $OA(N, 2^n, 3)$ *with* $n \geq 5$. *If* N *is not a multiple of 16, then in the projection of* $\mathbf{X}$ *onto any five factors, all the main effects and two-factor interactions are estimable under the assumption that the higher-order interactions are negligible.*

In particular, the hidden projection property described in Corollary 15.12 holds for the designs obtained by applying the foldover construction (8.18) to orthogonal arrays of strength two whose run sizes are not multiples of 8.

The condition that N is not a multiple of 2^{t+1} is needed in a general result such as Theorem 15.6. However, it does not preclude the existence of $OA(N, 2^n, 2)$'s that have projectivity three when N is a multiple of 8. Likewise, there are $OA(N, 2^n, 3)$'s with the hidden projection property described in Corollary 15.12, where N is a multiple of 16. For example, the Paley design of size 24, an $OA(24, 2^{23}, 2)$, has the hidden projection property in Theorem 15.8, and the $OA(48, 2^{24}, 3)$ constructed from this design by the foldover method has the property in Corollary 15.12. We state a result in this regard from Bulutoglu and Cheng (2003).

Theorem 15.13. *All Paley designs* $\mathbf{P}_N$ *with* $N \geq 12$ *have no defining words of length three or four, and hence have projectivity three and the projection properties described in Theorem 15.8.*

We note that under the 32-run Plackett–Burman design, all the main effects and two-factor interactions can be estimated in all its six-factor projections.

Loeppky, Sitter, and Tang (2007) used a criterion called *projection estimation capacity* to rank two-level orthogonal arrays. For each design d and each k with $k \leq n$, let $p_k(d) = \rho_k(d)/\binom{n}{k}$, where $\rho_k(d)$ is the number of estimable models that contain the main effects and two-factor interactions of k factors. This is equivalent to the estimation capacity criterion defined in Section 10.4 with the set $\mathfrak{M}$ of potential models consisting of those that contain all the main effects and two-factor interactions in k-factor projections. Loeppky, Sitter, and Tang (2007) proposed to sequentially maximize $p_1(d), \ldots, p_n(d)$ if the lower-dimensional projections are more important. Some theoretical results were obtained to help develop a computationally efficient algorithm, and useful designs with 16, 20, 24, and 28 runs were tabulated. Since not all estimable models are estimated with the same efficiency under nonregular designs, Loeppky, Sitter, and Tang (2007) also considered a criterion called *projection information capacity*, which is equivalent to the information capacity criterion discussed in Section 14.10, again with the potential models consisting of those that contain all the main effects and two-factor interactions in k-factor projections.

Projection properties of designs with more than two levels were studied, e.g., by Wang and Wu (1995), Tsai, Gilmour, and Mead (2000), Cheng and Wu (2001), Xu, Cheng, and Wu (2004), Dey (2005), and Evangelaras, Koukouvinos, Dean, and Dingus (2005). Xu, Cheng, and Wu (2004) proposed criteria to evaluate the performance of designs for factor screening, projection, and interaction detection. They also developed techniques for searching and constructing good designs, and applied these techniques to search for the best three-level designs with 18 and 27 runs. They were able to find an OA$(27, 3^{13}, 2)$ such that the following quadratic model can be estimated in all of its five-factor projections:

$$\mu + \sum_{i=1}^{5} \beta_i x_i + \sum_{i=1}^{5} \beta_{ii} x_i^2 + \sum_{1 \leq i < j \leq 5} \beta_{ij} x_i x_j,$$

where $\mu, \beta_i, \beta_{ii}, \beta_{ij}$ are unknown constants.

15.5 Generalized minimum aberration for two-level designs

The concept of resolution and the minimum aberration criterion cannot be applied to nonregular designs directly since nonregular designs do not have defining relations. Given a two-level design d, let r^* be the smallest integer r such that $\max_{S:|S|=r} |J_S(d)| > 0$. Deng and Tang (1999) defined the *generalized resolution R* of d as

$$R = r^* + 1 - \max_{S:|S|=r^*} \frac{|J_S(d)|}{N}. \tag{15.7}$$

Since $0 < \max_{S:|S|=r^*}[|J_S(d)|/N] \leq 1$, we have $r^* \leq R < r^* + 1$. If d is a regular design of resolution R, then it follows from Proposition 15.3 that the generalized resolution of d is exactly equal to R. For a nonregular orthogonal array d of strength t but not of strength $t + 1$, we have $r^* = t + 1$. By Lemma 8.12, the projection of d onto any set S of $t + 1$ factors consists of, say γ copies of a complete 2^{t+1} factorial and δ copies of a $2^{(t+1)-1}$ fraction, where $\gamma \geq 0$ and $\delta \geq 0$. It is easy to see that $\delta = |J_S(d)|/2^t$. Ideally we would want to have $\delta = 0$ for all $(t+1)$-factor projections. This is not possible since d is not of strength $t + 1$. Then the worst projection corresponds to the largest $|J_S(d)|$. This leads to the definition in (15.7). For example, since the 12-run Plackett–Burman design is of strength two and $|J_S(d)| = 4$ for all three-factor projections, its generalized resolution is equal to $3\frac{2}{3}$.

For any two-level design d, let

$$B_k(d) = N^{-2} \sum_{S:|S|=k} [J_S(d)]^2. \tag{15.8}$$

By Proposition 15.3, under a two-level regular design, if the interaction of the factors in S appears in the defining relation, then $\left|\frac{1}{N}J_S(d)\right|$ is equal to 1; otherwise, it is equal to zero. Therefore, for two-level regular designs, $B_k(d)$ is equal to the number of defining words of length k. We call $(B_1(d), B_2(d), \ldots, B_n(d))$ the *generalized wordlength pattern*. This extends the notion of wordlength pattern from regular designs to the nonregular case. The minimum G_2-aberration criterion proposed by Tang and Deng (1999) is to sequentially minimize $B_1(d), B_2(d), \ldots, B_n(d)$, which reduces to the usual minimum aberration criterion when d is a regular design.

It is easy to see that Equation (10.7) still holds when A_{j+1} and A_{j-1} are replaced by B_{j+1} and B_{j-1}, respectively. Therefore, as in Section 10.5, minimum G_2-aberration can be interpreted as a criterion for minimizing the bias of the estimates of lower-order effects in the presence of higher-order effects.

In view of the discussions following (15.3), for two-level nonregular designs, the $B_k(d)$'s can also be viewed as overall measures of partial aliasing among various factorial effects. By sequentially minimizing $B_1(d), B_2(d), \ldots, B_n(d)$, the minimum G_2-aberration criterion seeks to minimize aliasing among the more important lower-order effects.

Remark 15.1. While minimum aberration regular designs always have maximum resolution, nonregular minimum G_2-aberration designs may not have maximum generalized resolution. We also note that the notion of generalized resolution is different from the extension of resolution to nonregular designs by Webb (1968). Webb's definition mimics the estimability properties of regular designs. A design is said to be of resolution $R = 2k + 1$ if all the factorial effects involving up to k factors are estimable under the assumption that all those involving more than k factors are negligible. It is of resolution $2k$ if all the factorial effects involving up to $k - 1$ factors are estimable under the assumption that all those involving more than k factors are negligible.

15.6   Generalized minimum aberration for multiple and mixed levels

Xu and Wu (2001) extended the minimum G_2-aberration criterion to designs with more than two levels and mixed levels. Consider a design d of run size N for n factors with $s_1, \ldots, s_n$ levels, respectively. For each subset $S = \{i_1, \ldots, i_k\}$ of $\{1, \ldots, n\}$, the set of vectors $\{\mathbf{p}^{\mathbf{z}} : \mathbf{z} \in S_1 \times \cdots \times S_n, z_{i_1} \neq 0, \ldots, z_{i_k} \neq 0$, and all the other entries of $\mathbf{z}$ are equal to zero$\}$ as defined in (6.22) is an orthogonal basis of $W_{A_{i_1} \wedge \cdots \wedge A_{i_k}}$, the space of all the contrasts representing the interaction of factors $i_1, \ldots, i_k$. Let $W_k = \oplus_{1 \leq i_1 < \cdots < i_k \leq n} W_{A_{i_1} \wedge \cdots \wedge A_{i_k}}$, the space of all the contrasts representing k-factor interactions. For simplicity, suppose each $\mathbf{p}^{\mathbf{z}}$ is normalized to have unit length. Then $\{\mathbf{p}^{\mathbf{z}} : \mathbf{z} \in S_1 \times \cdots \times S_n, w(\mathbf{z}) = k\}$ is an orthonormal basis of W_k, where $w(\mathbf{z})$ is the Hamming weight of $\mathbf{z}$.

Let

$$B_k(d) = \frac{s_1 \cdots s_n}{N^2} \sum_{\mathbf{z} \in S_1 \times \cdots \times S_n: \, w(\mathbf{z}) = k} (\mathbf{1}^T \mathbf{x}^{\mathbf{z}})^2, \tag{15.9}$$

where $\mathbf{x}^{\mathbf{z}}$ is the column of the model matrix corresponding to the parameter $\beta^{\mathbf{z}}$ as defined in Section 8.1. Each $\mathbf{1}^T \mathbf{x}^{\mathbf{z}}$ is the sum of all the entries in a column of the model matrix corresponding to a factorial effect involving k factors, and the sum on the right side of (15.9) is over all such columns. It is easy to see that for two-level designs (15.9) reduces to (15.8). We continue to call $(B_1(d), B_2(d), \ldots, B_n(d))$ defined here the generalized wordlength pattern. To extend the minimum G_2 aberration criterion to the general case, we sequentially minimize the $B_k(d)$'s defined in (15.9). This is called a *generalized minimum aberration criterion*.

Proposition 15.14. $B_1(d)$, $B_2(d)$, $\ldots$, $B_n(d)$ *do not depend on orthonormal bases of the spaces W_1, W_2, $\ldots$, W_n.*

Proof. Since $\{\mathbf{p}^{\mathbf{z}} : \mathbf{z} \in S_1 \times \cdots \times S_n, w(\mathbf{z}) = k\}$ is an orthonormal basis of W_k, by (15.2), we have

$$\left\| \mathbf{P}_{W_k} \mathbf{n}_d \right\|^2 = \sum_{\mathbf{z} \in S_1 \times \cdots \times S_n: \, w(\mathbf{z}) = k} (\mathbf{n}_d^T \mathbf{p}^{\mathbf{z}})^2 = \sum_{\mathbf{z} \in S_1 \times \cdots \times S_n: \, w(\mathbf{z}) = k} (\mathbf{1}^T \mathbf{x}^{\mathbf{z}})^2.$$

By (15.9),

$$B_k(d) = \frac{s_1 \cdots s_n}{N^2} \left\| \mathbf{P}_{W_k} \mathbf{n}_d \right\|^2.$$

Apart from a multiplicative constant, $B_k(d)$ is the squared length of the orthogonal projection of $\mathbf{n}_d$ onto W_k. Therefore it does not depend on the choices of orthonormal bases. $\square$

We will see in the next section that for regular s-level designs the generalized minimum aberration also reduces to the usual minimum aberration criterion.

A modification for the case where the factors are quantitative is due to Cheng and Ye (2004). The notion of generalized minimum aberration defined above treats all the contrasts that represent k-factor interactions equally. When the factors are quantitative and the factorial effects are defined by using orthogonal polynomials, (15.9) may not be appropriate if the effects corresponding to polynomials of lower degrees are considered to be more important than those of higher degrees. For instance, in the case of three-level designs, one might assign higher priority to linear $\times$ linear effects than quadratic $\times$ quadratic effects. Suppose $\mathbf{z}$ has k nonzero entries $z_{i_1}, \ldots, z_{i_k}$. Then $\mathbf{x}^{\mathbf{z}} = \mathbf{x}^{z_{i_1} \mathbf{e}_{i_1}} \odot \cdots \odot \mathbf{x}^{z_{i_k} \mathbf{e}_{i_k}}$, where the ith entry of $\mathbf{e}_i$ is equal to 1, and all its other entries are zero; see (8.5). If the column $\mathbf{x}^{z_{i_j} \mathbf{e}_{i_j}}$ of the model matrix consists of the values of a polynomial of degree z_{i_j} on the s_{i_j} levels of the i_jth factor, then the k-factor interaction column $\mathbf{x}^{\mathbf{z}}$ represents a polynomial of k variables with total degree $\sum_{j=1}^{k} z_{i_j} = \mathbf{1}_n^T \mathbf{z}$. For each positive integer $1 \leq l \leq \sum_{i=1}^{n}(s_i - 1)$, let

$$B_l' = \sum_{\mathbf{z} : \mathbf{1}_n^T \mathbf{z} = l} (\mathbf{1}_N^T \mathbf{x}^{\mathbf{z}})^2.$$

Cheng and Ye (2004) proposed to sequentially minimize the B_l's. Under this criterion, all the terms that correspond to polynomials of the same degree are treated equally.

We have seen in Section 10.4 that minimum aberration is a good surrogate for maximum estimation capacity. For nonregular designs, since different factorial effects can be estimated with different precisions, counting the number of estimable models is not enough for assessing the performance of a design. This is similar to the situation with designs for multi-stratum experiments; see Section 14.10. Cheng, Deng, and Tang (2002) showed that for two-level nonregular designs, minimum G_2-aberration is a good surrogate for the information capacity. Mandal and Mukerjee (2005) and Ai, Li, and Zhang (2005) extended this to the case of more than two levels.

Tsai, Gilmour, and Mead (2000) also used the average efficiency over a set of candidate models to select three-level designs under model uncertainty. Unlike the information capacity studied in Sun (1993) and Section 14.10, which is the average efficiency with respect to a determinant-based criterion over the models containing all the main effects and a certain number of two-factor interactions, the Q-criterion introduced by Tsai, Gilmour, and Mead (2000) is the average of a surrogate for a trace-based criterion over lower-dimensional projections. A more general version, called Q_B-criterion, that incorporates prior information about model uncertainty was proposed by Tsai, Gilmour, and Mead (2007). Tsai and Gilmour (2010) applied this criterion to different settings and compared it with other criteria. It was found that the Q_B-criterion is closely related to generalized minimum aberration.

The projection justification of minimum aberration and the relation between uniformity and minimum aberration reviewed in Section 10.5 can be extended to the generalized minimum aberration criterion (Ma and Fang, 2001; Ai and Zhang, 2004). We refer readers to Xu, Phoa, and Wong (2009), a review article on nonregular designs, for more references.

15.7   Connection with coding theory

In this and the next two sections, readers are referred to Section 9.9 for some definitions and notations from coding theory.

Suppose $s_1 = \cdots = s_n = s$. Each design d of size N can be considered as a code.

If d is a regular design, then it is a linear code, and the dual code $d^{\perp}$ is the defining contrast subgroup of d. The MacWilliams identities (9.19) and (9.20) connect the weight distribution of a linear code to that of its dual code. Following (9.20), for a nonlinear code (nonregular design) d, define the *MacWilliams transform* of the distance distribution $(W_0(d), \ldots, W_n(d))$ of d as

$$W_i'(d) = N^{-1} \sum_{j=0}^{n} W_j(d) P_i(j; n, s) \text{ for } 0 \leq i \leq n. \tag{15.10}$$

By the orthogonality of Krawtchouk polynomials,

$$W_i(d) = N s^{-n} \sum_{j=0}^{n} W_j'(d) P_i(j; n, s) \text{ for } 0 \leq i \leq n. \tag{15.11}$$

The Delsarte theory (Delsarte, 1973) shows that

$$B_i(d) = W_i'(d); \tag{15.12}$$

see Xu and Wu (2001). So the generalized wordlength pattern of d is the MacWilliams transform of its distance distribution. Therefore the relationship between the distance distribution of a nonregular design and its generalized wordlength pattern is the same as that between the weight distribution of a regular design and its wordlength pattern. In particular, by (9.18), (9.20), and (15.10), if d is a regular design, then $W_i'(d) = W_i(d^{\perp}) = (s-1)A_i(d)$. It follows from (15.12) that $B_i(d) = (s-1)A_i(d)$. Therefore, for regular designs, generalized minimum aberration also reduces to the usual minimum aberration criterion.

Independently, for s^n experiments, Ma and Fang (2001) defined generalized wordlength pattern directly as the MacWilliams transform of the distance distribution of a design.

The Delsarte theory has the following important implication.

Theorem 15.15. *An N-run design for an s^n experiment is an orthogonal array of strength t if and only if $B_i(d) = 0$ for all $i \leq t$.*

Sloane and Stufken (1996) extended this result to mixed-level orthogonal arrays; see Xu and Wu (2001).

Theorem 15.16. *An N-run design for an $s_1 \times \cdots \times s_n$ experiment is an orthogonal array of strength t if and only if $B_i(d) = 0$ for all $i \leq t$.*

Now we see that Theorem 9.5 is a special case of Theorem 15.15 and that the converse of Proposition 15.1 holds.

Xu (2005b) studied properties of nonregular designs obtained from the Nordstrom–Robinson code (Nordstrom and Robinson, 1967). When considered as a two-level factorial design, the Nordstrom–Robinson code is a nonregular $OA(256, 2^{16}, 5)$ of projectivity 7. A minimum aberration regular design of the same size, however, only has strength and projectivity 4. Taking the 128 rows with entry 0 at a fixed column and deleting that column, one obtains a design with 128 runs and 15 two-level factors. Repeated applications of this procedure produce an $OA(128, 2^{15}, 4)$ of projectivity 6, an $OA(64, 2^{14}, 3)$ of projectivity 5, and an $OA(32, 2^{13}, 2)$ of projectivity 4. All the corresponding minimum aberration regular designs have strength/projectivity 3. Except for the $OA(32, 2^{13}, 2)$, many projections of these nonregular designs also have superior properties in terms of strength, projectivity, and aberration. Xu (2005b) used linear programming to show that in the following 13 cases nonregular designs derived from the Nordstrom–Robinson code are the best among all designs with respect to the generalized minimum aberration criterion: $N = 256$, $n = 14, 15, 16$; $N = 128$, $n = 13, 14, 15$; $N = 64$, $n = 8, 9, 12, 13, 14$; $N = 32$, $n = 7, 8$. They are tied with the best regular designs for $N = 64$, $n = 8, 9$, 12 and $N = 32$, $n = 7, 8$, and are better than the best regular designs in the other eight cases. Xu (2005b) also showed that for $N = 2^{n-2}$, a regular minimum aberration design is the best among all possible designs with respect to the generalized minimum aberration criterion.

In another application of using nonlinear codes to construct nonregular designs, Xu and Wong (2007) studied nonregular two-level designs constructed from quaternary codes. Quaternary codes are linear codes over $\mathbb{Z}_4 = \{0, 1, 2, 3\}$, the ring of integers modulo 4. Take a $k \times n$ matrix with entries from $\mathbb{Z}_4$; then the 4^k linear combinations of the row vectors form a quaternary code. This is similar to the construction of regular designs except that modulo 4 arithmetic is used. Replacing the four entries 0, 1, 2, and 3 with $(0,0)$, $(0,1)$, $(1,1)$, and $(1,0)$, respectively, we obtain a $4^k \times 2n$ matrix of 1's and 0's, which defines a nonregular design of size $N = 4^k$ for $2n$ two-level factors. Such designs are relatively straightforward to construct, have simple design representation, and often have better projectivity/strength/aberration than their regular counterparts. Tables of designs with 16, 32, 64, 128, and 256 runs constructed by a systematic method were provided. Phoa and Xu (2009) carried out a comprehensive study of quarter fractions of two-level designs constructed from quarternary codes. Some of the designs were shown to have generalized minimum aberration and maximum projectivity over all designs. Zhang, Phoa, Mukerjee, and Xu (2011) used a trigonometric approach to study designs that are more highly fractionated, in particular the one-eighth and one-sixteenth fractions. Phoa (2012) developed a theory to characterize the wordlengths and aliasing indexes for general $(1/4)^p$-fraction quarternary code designs, and applied the theory to $(1/64)$-fractions.

15.8   Complementary designs

Throughout this section we assume $s_1 = \cdots = s_n = s$. We have seen in Section 10.6 how the technique of complementary designs can be used to construct minimum aberration designs, based on identities derived by Chen and Hedayat (1996), Tang and Wu (1996), and Suen, Chen, and Wu (1997) that relate the wordlength pattern of a regular design to that of its complement in a saturated regular design. In the case where s is not a prime number or power of a prime number, let $\mathbf{H}$ be a saturated orthogonal array of strength two. Then the number of factors in $\mathbf{H}$ is $q = (N-1)/(s-$

1). Suppose d is a design consisting of n columns of $\mathbf{H}$. Then the design $\bar{d}$ that consists of the remaining $\bar{n} = q - n$ columns of $\mathbf{H}$ is called the complementary design of d in $\mathbf{H}$. For $s = 2$, Tang and Deng (1999) derived identities relating the generalized wordlength pattern of a design to that of its complementary design.

The key to the derivation of the complementary design theory for regular designs in Suen, Chen, and Wu (1997) is the MacWilliams identities that connect the weight distribution of a regular design to its wordlength pattern. Another key fact is that in a saturated orthogonal array of strength two, the pairwise Hamming distances of the treatment combinations are a constant (Theorem 8.6). Since the same relationship holds between the distance distribution of a nonregular design and its generalized wordlength pattern (see Section 15.7), one can see that the argument used by Suen, Chen, and Wu (1997) to derive the relationship between the wordlength pattern of a design and that of its complement in a saturated regular design (see Section 10.6 for a sketch) can also be used to establish the same relationship between the generalized wordlength pattern of a nonregular design and that of its complement in a saturated orthogonal array of strength two. This was observed by Xu and Wu (2001).

Theorem 15.17. *Suppose the columns of d and $\bar{d}$ form a partition of the columns of a saturated s-level orthogonal array of strength two. Then*

$$B_k(d) = (C_k + C_{k0}) + \sum_{j=3}^{k-2} C_{kj} B_j(\bar{d}) + (-1)^k [1 + (s-2)(k-1)] B_{k-1}(\bar{d}) + (-1)^k B_k(\bar{d}),$$

for $0 \le k \le n$, where

$$C_k = N^{-1} [P_k(0; n, s) - P_k(Ns^{-1}; n, s)],$$

and

$$C_{kj} = s^{-\bar{n}} \sum_{i=0}^{\bar{n}} P_k(Ns^{-1} - i; n, s) P_i(j; \bar{n}, s) \text{ for } 0 \le j \le \bar{n}.$$

Since $B_k(d) = (-1)^k B_k(\bar{d})$ + lower-order terms, the same rule as that given in Section 10.6 applies: a generalized minimum aberration projection d of a saturated orthogonal array of strength two can be obtained by sequentially minimizing $(-1)^k B_k(\bar{d})$.

Remark 15.2. Although every regular design can be expanded to a saturated regular design, which is unique up to isomorphism, a nonregular orthogonal array of strength two may not be a projection of a saturated orthogonal array. Furthermore, saturated orthogonal arrays may not be unique. Thus there is no guarantee that a generalized minimum aberration projection of a saturated orthogonal array obtained by using the complementary design theory presented in this section has generalized minimum aberration over all possible designs.

15.9 Minimum moment aberration

The generalized minimum aberration criterion is based on the correlation (or similarity) of the columns of the model matrix, which correspond to various factorial effects. Suppose a two-level design is represented by an $N \times n$ matrix $\mathbf{X}$ with ± 1 entries. Then minimizing B_2 is equivalent to minimizing the sum of squares of all the entries of $\mathbf{X}^T\mathbf{X}$, which is equal to $\text{tr}[(\mathbf{X}^T\mathbf{X})^2]$. Since $\text{tr}(\mathbf{AB}) = \text{tr}(\mathbf{BA})$, we have $\text{tr}[(\mathbf{X}^T\mathbf{X})^2] = \text{tr}(\mathbf{X}^T\mathbf{X}\mathbf{X}^T\mathbf{X}) = \text{tr}(\mathbf{X}\mathbf{X}^T\mathbf{X}\mathbf{X}^T) = \text{tr}[(\mathbf{X}\mathbf{X}^T)^2]$. Now the roles of columns and rows are switched. This idea was used by Nguyen (1996) and Cheng (1997) in their studies of optimal supersaturated designs (see Section 15.15). Let $\mathbf{r}_1, \ldots, \mathbf{r}_N$ be the rows of $\mathbf{X}$. Then

$$
\begin{aligned}
\text{tr}\left[(\mathbf{X}\mathbf{X}^T)^2\right] &= \sum_{i=1}^{N}\sum_{j=1}^{N}\left(\mathbf{r}_i\mathbf{r}_j^T\right)^2 \\
&= n^2 N + 2\sum_{1 \le i < j \le N}\left(\mathbf{r}_i\mathbf{r}_j^T\right)^2 \\
&= n^2 N + 2\sum_{1 \le i < j \le N}[2\delta(\mathbf{r}_i, \mathbf{r}_j) - n]^2,
\end{aligned}
$$

where $\delta(\mathbf{r}_i, \mathbf{r}_j) = n - \text{dist}(\mathbf{r}_i, \mathbf{r}_j)$ is the number of components where $\mathbf{r}_i$ and $\mathbf{r}_j$ agree. On the other hand, minimizing B_1 is clearly equivalent to minimizing $\sum_{1 \le i < j \le N}\delta(\mathbf{r}_i, \mathbf{r}_j)$. Therefore minimizing B_1, B_2 successively is equivalent to minimizing $\sum_{1 \le i < j \le N}\delta(\mathbf{r}_i, \mathbf{r}_j)$ and $\sum_{1 \le i < j \le N}[\delta(\mathbf{r}_i, \mathbf{r}_j)]^2$ successively. This is part of a more general result that Xu (2003) obtained by using coding theoretic tools.

Xu (2003) formulated a general row-based minimum moment aberration criterion that can be applied to regular, nonregular, symmetric, and asymmetrical designs. Again express each design d as an $N \times n$ matrix $\mathbf{X}_d$. Let the N rows be $\mathbf{r}_1(d), \ldots, \mathbf{r}_N(d)$, and for $t = 1, 2, \ldots$, define

$$
K_t(d) = [N(N-1)/2]^{-1}\sum_{1 \le i < j \le N}[\delta(\mathbf{r}_i(d), \mathbf{r}_j(d))]^t.
$$

A design is said to have *minimum moment aberration* if it sequentially minimizes $K_1(d)$, $K_2(d)$, \ldots, etc.

Theorem 15.18. *A design with $s_1 = \cdots = s_n$ has generalized minimum aberration if and only if it has minimum moment aberration.*

The proof of Theorem 15.18 is deferred to the next section. We have seen the regular design version of this result as Theorem 10.10. For two-level designs, Butler (2003a,b) proposed an equivalent form of minimum moment aberration based on pairwise inner products of the runs, and showed its equivalence to generalized minimum aberration. We have also seen the regular design version of this in Section 10.10

Besides the theoretical interest, the equivalence of (generalized) minimum aberration and minimum moment aberration provides powerful tools for the theoretical

characterization and algorithmic construction of minimum aberration and generalized minimum aberration designs. The determination of minimum moment aberration designs involves much less computational complexity. Another advantage of the
minimum moment aberration criterion is that it allows for the possibility of imposing
different weights on different factors. Suppose $\mathbf{w} = (w_1, \ldots, w_n)$ are weights associated with the n factors, $w_k > 0$ for all k. Then one can define

$$K_t(d; \mathbf{w}) = [N(N-1)/2]^{-1} \sum_{1 \leq i < j \leq N} [\delta_{\mathbf{w}}(\mathbf{r}_i(d), \mathbf{r}_j(d))]^t,$$

with

$$\delta_{\mathbf{w}}(\mathbf{r}_i(d), \mathbf{r}_j(d)) = \sum_{k=1}^{n} w_k \delta(x_{ik}(d), x_{jk}(d)),$$

where $\delta(x_{ik}(d), x_{jk}(d)) = 1$ if $x_{ik}(d) = x_{jk}(d)$, and is 0 otherwise. One can choose a
design by sequentially minimizing $K_t(d; \mathbf{w})$ for $t = 1, 2, \ldots$; see Xu (2003).

Let $\boldsymbol{\delta}(d)$ be the $(N(N-1)/2) \times 1$ vector whose entries consist of $\delta(\mathbf{r}_i(d), \mathbf{r}_j(d))$,
$1 \leq i < j \leq N$. Then $K_t(d) = [N(N-1)/2]^{-1} \sum_{1 \leq i < j \leq N} [\delta(\mathbf{r}_i(d), \mathbf{r}_j(d))]^t$ is a Schur
convex function of $\boldsymbol{\delta}(d)$. Suppose $s_1 = \cdots = s_n = s$, and we impose the constraint that
for all the treatment factors, the numbers of times the s levels appear differ from one
another by at most 1. Then it is easy to see that the sum of all the components of $\boldsymbol{\delta}(d)$
is a constant. It follows that if $\boldsymbol{\delta}(d_1)$ is majorized by $\boldsymbol{\delta}(d_2)$, then $K_t(d_1) \leq K_t(d_2)$ for
all t; see Definition 10.2. By this and Theorem 15.18, we have the following result.

Theorem 15.19. *(Xu, 2003) Suppose $s_1 = \cdots = s_n = s$. Among the designs with all the
treatment factor levels appearing $\lfloor N/s \rfloor$ or $\lfloor N/s \rfloor + 1$ times, if there is a design d^*
such that all the $\delta(\mathbf{r}_i(d^*), \mathbf{r}_j(d^*))$, $1 \leq i < j \leq N$, differ from one another by at most 1,
then d^* minimizes $K_t(d)$ for all t; in particular, it has minimum moment aberration,
and hence also has generalized minimum aberration.*

Theorem 15.19 can be used to establish lower bounds on $K_t(d)$. For example, if
N is a multiple of s, then

$$K_t(d) \geq \left[\frac{n(N-s)}{s(N-1)}\right]^t. \tag{15.13}$$

This lower bound is obtained by computing $K_t(d)$ for a design such that each treatment factor level appears N/s times and $\boldsymbol{\delta}(d)$ has constant entries. When it is not
possible for $\boldsymbol{\delta}(d)$ to have constant integer entries, a better bound can be obtained
by assuming that the entries of $\boldsymbol{\delta}(d)$ differ from one another by at most 1. Liu and
Hickernell (2002) used this technique to derive lower bounds on two criteria for supersaturated designs that are Schur convex functions of $\boldsymbol{\delta}(d)$; see Remark 15.6 in
Section 15.15. Zhang, Fang, Li, and Sudjianto (2005) studied other Schur convex
functions of $\boldsymbol{\delta}(d)$.

By Theorem 8.6, if d^* is a saturated $OA(N, s^{(N-1)/(s-1)}, 2)$, then all the com-
ponents of $\boldsymbol{\delta}(d^*)$ are equal. A design with the components of $\boldsymbol{\delta}(d)$ differing from
one another by at most 1 can be obtained by deleting one factor from a saturated
$OA(N, s^{(N-1)/(s-1)}, 2)$.

15.10   Proof of Theorem 15.18*

To prove Theorem 15.18, define $M_t(d), t = 1, 2, \ldots$, as in (10.12). We note that (10.10) and (10.11) hold for both regular and nonregular designs. Therefore the argument in Section 10.10 shows that minimum moment aberration is equivalent to the sequential minimization of $(-1)^t M_t(d), t = 1, 2, \ldots$, etc. We need to show that sequential minimization of $B_t(d) = W_t'(d)$ is also equivalent to that of $(-1)^t M_t(d)$. This is a consequence of Lemma 15.20 below. This result, relating the moments of the distance distribution to the generalized wordlength pattern, was proved by Xu (2003) using the Pless power moment identities in coding theory (Pless, 1963).

Lemma 15.20. *For any s^n design d of size N,*

$$M_t(d) = \sum_{i=0}^{n} (-1)^i W_i'(d) \left[\sum_{j=0}^{t} j! S(t, j) s^{-j} (s-1)^{j-i} \binom{n-i}{j-i} \right] \quad \text{for } t \geq 0,$$

where, for nonnegative integers t and j, $S(t, j)$ is a Stirling number of the second kind.

We now prove Lemma 15.20. By (10.12) and (15.11), we have

$$M_t(d) = N^{-1} \sum_{j=0}^{n} j^t W_j(d) = N^{-1} \sum_{j=0}^{n} j^t N s^{-n} \sum_{i=0}^{n} W_i'(d) P_j(i; n, s).$$

Thus

$$M_t(d) = \sum_{i=0}^{n} W_i'(d) \left[\sum_{j=0}^{n} s^{-n} j^t P_j(i; n, s) \right]. \tag{15.14}$$

For each integer $0 \leq i \leq n$, let $f(z) = (1 - z)^i [1 + (s - 1)z]^{n-i}$. It is a property of the Krawtchouk polynomials that $P_j(i; n, s)$ is the coefficient of z^j in $f(z)$; that is,

$$f(z) = \sum_{j=0}^{n} P_j(i; n, s) z^j. \tag{15.15}$$

Let D_z be the differentiation operator with respect to z. By (15.15), $(zD_z)^t f(z)|_{z=1} = \sum_{j=0}^{n} j^t P_j(i; n, s)$. Combining this with the fact that $(zD_z)^t = \sum_{j=0}^{t} S(t, j) z^j (D_z)^j$, we have

$$\sum_{j=0}^{n} j^t P_j(i; n, s) = \sum_{j=0}^{t} S(t, j) z^j (D_z)^j f(z)|_{z=1} = \sum_{j=0}^{t} S(t, j)(D_z)^j f(z)|_{z=1}. \tag{15.16}$$

On the other hand, since

$$f(z) = (1 - z)^i [s + (s - 1)(z - 1)]^{n-i} = (-1)^i \sum_{l=0}^{n-i} \binom{n-i}{l} s^{n-i-l} (s-1)^l (z-1)^{i+l},$$

$$\tag{15.17}$$

we have $(D_z)^j f(z)|_{z=1} = (-1)^i j! \binom{n-i}{j-i} s^{n-j}(s-1)^{j-i}$; this is because when the operator D_z is applied to the sum in (15.17) j times, all the terms except the one with $i+l=j$ vanish. Then by (15.16),

$$\sum_{j=0}^{n} j^t P_j(i;n,s) = (-1)^i \sum_{j=0}^{t} j! S(t,j) \binom{n-i}{j-i} s^{n-j}(s-1)^{j-i}.$$

Substituting this into (15.14), we finish the proof of the lemma.

15.11   Even designs and foldover designs

Regular two-level even designs are defined as those with no defining words of odd lengths. The notion of wordlength patterns was extended from regular designs to nonregular ones in Sections 15.5 and 15.6. In general, we say that a design d is even if $B_k(d) = 0$ for all odd k. Throughout this section by foldover we mean full foldover in which the levels of all the factors are reversed.

It was shown in Section 9.6 that regular designs constructed by the foldover method are even. It is easy to see that this also holds for nonregular designs. Let d be an N-run design for n factors with levels 1 and -1, and let

$$\mathbf{X}_{d^*} = \begin{bmatrix} \mathbf{X}_d \\ -\mathbf{X}_d \end{bmatrix}. \tag{15.18}$$

Then for any $S = \{i_1, \ldots, i_k\} \subseteq \{1, \ldots, n\}$, where k is odd,

$$J_S(d) = \sum_{l=1}^{N} x_{li_1}(d) \cdots x_{li_k}(d) + \sum_{l=1}^{N} (-1)^k x_{li_1}(d) \cdots x_{li_k}(d) = 0.$$

It follows that all two-level designs constructed by the foldover method are even. The converse is also true.

Theorem 15.21. *If a two-level design d^* is even, then it can be constructed by the foldover method; that is, $\mathbf{X}_{d^*}$ must be of the form in (15.18).*

Proof. Let $n_{d^*}(x_1, \ldots, x_n)$ be the number of times the treatment combination $(x_1, \ldots, x_n)$ appears in design d^*. By (15.1),

$$n_{d^*}(x_1, \ldots, x_n) = \frac{N}{2^n} + \sum_{1 \le k \le n} \sum_{1 \le i_1 < \cdots < i_k \le n} \frac{J_{\{i_1, \ldots, i_k\}}(d^*)}{2^n} x_{i_1} \cdots x_{i_k}.$$

By (15.8), if $B_k(d^*) = 0$, then $J_S(d^*) = 0$ for all S such that $|S| = k$. Therefore if d^* is an even design, then $J_S(d^*) = 0$ for all S such that $|S|$ is odd. In this case,

$$n_{d^*}(x_1, \ldots, x_n) = \frac{N}{2^n} + \sum_{1 \le k \le n,\, k \text{ is even}} \sum_{1 \le i_1 < \cdots < i_k \le n} \frac{J_{\{i_1, \ldots, i_k\}}(d^*)}{2^n} x_{i_1} \cdots x_{i_k}.$$

It follows that $n_{d^*}(x_1, \ldots, x_n) = n_{d^*}(-x_1, \ldots, -x_n)$ for any $(x_1, \ldots, x_n)$. $\qquad \square$

Theorem 15.21 appeared in Balakrishnan and Yang (2006). It was also obtained by the author contemporaneously and was reported in Cheng, Mee, and Yee (2008).

Since regular 2^{n-p} even designs are projections of maximal even designs, such designs have all the two-factor interactions appear in $N/2 - 1$ alias sets and therefore can entertain at most $N/2 - 1$ two-factor interactions. This is also true in the nonregular case. An N-run even design $\mathbf{Y}$ with n two-level factors is of the form $\begin{bmatrix} \mathbf{X} \\ -\mathbf{X} \end{bmatrix}$, where $\mathbf{X}$ is $N/2 \times n$. Since the columns in the model matrix that correspond to two-factor interactions are pairwise Hadamard products of main effect columns, the two-factor interaction columns form an $N \times \binom{n}{2}$ matrix of the form $\begin{bmatrix} \mathbf{Z} \\ \mathbf{Z} \end{bmatrix}$, which has rank $\leq N/2 - 1$ if $\mathbf{Y}$ is an orthogonal array of strength two.

By Theorem 11.14, all resolution IV two-level regular designs with $5N/16 < n$ must be projections of maximal even designs and hence must be even. Butler (2007) extended this result to nonregular designs.

Theorem 15.22. *For $N/3 \leq n \leq N/2$, even designs are the only $OA(N, 2^n, t)$'s with $t \geq 3$.*

By combining Theorems 15.21 and 15.22, we see that for for $N/3 \leq n \leq N/2$, all the two-level orthogonal arrays of strength three, in particular the two-level generalized minimum aberration designs, can be constructed by the method of foldover. This substantially reduces the search of generalized minimum aberration designs.

Butler (2004) showed that designs constructed by the foldover method are the only two-level 24-run orthogonal arrays of strength at least three, and that for all $n \leq 12$, 24-run generalized minimum aberration designs with n factors are n-factor projections of the design obtained by applying the foldover construction (8.18) to the 12-run Plackett–Burman design. Note that the optimality here is over all possible designs of the same size; see Remark 15.2.

Butler (2004) also studied generalized minimum aberration even designs with 32, 48, and 64 runs.

15.12   Parallel flats designs

Recall that a regular s^{n-p} fractional factorial design consists of all the treatment combinations $\mathbf{x}$ that satisfy

$$\mathbf{A}^T \mathbf{x} = \mathbf{b}, \qquad (15.19)$$

where $\mathbf{A}$ is an $n \times p$ matrix of rank p and $\mathbf{b}$ is a $p \times 1$ vector, both with entries from GF(s). The solutions to this equation constitute an $(n - p)$-flat in EG(n, s). There are s^p such disjoint (parallel) $(n - p)$-flats, corresponding to s^p different choices of $\mathbf{b}$. A parallel flats design is the union of f of these flats. Note that here the word "union" is interpreted broadly that some of the f flats may be the same. Such a design has size fs^{n-p} and is determined by $\mathbf{A}$ and f vectors $\mathbf{b}_1, \ldots, \mathbf{b}_f$ for the right side of the equation in (15.19). Let $\mathbf{B} = [\mathbf{b}_1 \cdots \mathbf{b}_f]$. Then we also say that the parallel flats design is determined by $\mathbf{A}$ and $\mathbf{B}$. The column space $R(\mathbf{A})$ of $\mathbf{A}$ is the defining contrast subgroup of each of the s^{n-p} regular fractions in the pencil. We continue to call $R(\mathbf{A})$ the defining contrast subgroup of the parallel flats design, and the nonzero vectors in $R(\mathbf{A})$ defining words. Regular fractional factorial designs are those with $f = 1$. Parallel flats designs provide more flexibility in run sizes since they are not necessarily powers of prime numbers.

A theory of parallel flats designs was developed in Srivastava, Anderson, and Mardekian (1984), Srivastava (1987), and Srivastava and Li (1996). Earlier work on parallel flats designs appeared in, e.g., Connor and Young (1959), Addelman (1961), Daniel (1962), and John (1962). John (1962) proposed three-quarter replicates of two-level regular fractional factorial designs. Designs constructed from semifolding of regular designs discussed in Section 11.6 can also be considered as parallel flats designs.

Example 15.1. (a three-quarter replicate) Take the 2^{8-2} design defined by $G = -ABCD$ and $H = -ACEF$. Divide this design into four quarters by using ABE and CDF as additional independent defining words. Delete the quarter that is defined by $I = -ABE = -CDF$; then we have a three-quarter replicate of the 2^{8-2} design that is a 48-run parallel flats design with $n = 8$, $p = 4$, and $f = 3$. Under this design, all the main effects and two-factor interactions are estimable if all the higher-order interactions are negligible. For eight two-level factors, a regular design of resolution V requires at least 64 runs.

We refer readers to John (1971) and Mee (2009) for more detailed discussions of three-quarter replicates of two-level designs.

Let $\mathbf{V} = [\mathbf{x}_1 \cdots \mathbf{x}_{n-p}]$ be an $n \times (n - p)$ matrix such that its columns $\mathbf{x}_1, \ldots,$ and $\mathbf{x}_{n-p}$ are linearly independent solutions to the equation $\mathbf{A}^T \mathbf{x} = \mathbf{0}$. We have

$$\mathbf{A}^T \mathbf{V} = \mathbf{0} \text{ and } \text{rank}(\mathbf{V}) = n - p.$$

Furthermore, for each $1 \leq i \leq f$, let $\mathbf{z}_i$ be a particular solution to the equation

$$\mathbf{A}^T \mathbf{z}_i = \mathbf{b}_i.$$

Then the parallel flats design determined by $\mathbf{A}$ and $\mathbf{B}$ consists of all the treatment combinations $\mathbf{x}$ that are of the form

$$\sum_{j=1}^{n-p} \lambda_j \mathbf{x}_j + \mathbf{z}_i,$$

where $\lambda_j \in \text{GF}(s)$.

For any two nonzero vectors $\mathbf{g}_1, \mathbf{g}_2 \in \text{EG}(n, s)$ such that $\mathbf{g}_1 - \lambda \mathbf{g}_2 \notin R(\mathbf{A})$ for all nonzero $\lambda \in \text{GF}(s)$, it is clear that the columns of the model matrix corresponding to the factorial-effect contrasts defined by $\mathbf{g}_1$ are orthogonal to those corresponding to the factorial-effect contrasts defined by $\mathbf{g}_2$. Such orthogonality can be extended to some factorial effects that are aliased under the individual s^{n-p} regular fractions. If $\mathbf{g}_1 - \lambda_0 \mathbf{g}_2 \in R(\mathbf{A})$ for some $\lambda_0 \in GF(s)$, where $\mathbf{g}_1, \mathbf{g}_2 \notin R(\mathbf{A})$, then for each $1 \leq i \leq f$, $\mathbf{g}_1$ and $\mathbf{g}_2$ give the same partition of the s^{n-p} solutions to $\mathbf{A}^T \mathbf{x} = \mathbf{b}_i$ into s disjoint sets according to the values of $\mathbf{g}_1^T \mathbf{x}$ and $\mathbf{g}_2^T \mathbf{x}$, respectively. For each treatment combination $\mathbf{x} = \sum_{j=1}^{n-p} \lambda_j \mathbf{x}_j + \mathbf{z}_i$ in the parallel flats design, we have

$$\left(\mathbf{g}_1 - \lambda_0 \mathbf{g}_2\right)^T \mathbf{x} = \left(\mathbf{g}_1 - \lambda_0 \mathbf{g}_2\right)^T \left(\sum_{j=1}^{n-p} \lambda_j \mathbf{x}_j + \mathbf{z}_i\right) = \left(\mathbf{g}_1 - \lambda_0 \mathbf{g}_2\right)^T \mathbf{z}_i.$$

Thus, if $(\mathbf{g}_1 - \lambda_0 \mathbf{g}_2)^T \mathbf{z}_i$, $1 \leq i \leq f$, together cover the s elements of $GF(s)$ equally often, then the values of $\mathbf{g}_1^T \mathbf{x}$ and $\mathbf{g}_2^T \mathbf{x}$ jointly partition the treatment combinations in the parallel flats design into s^2 disjoint sets of equal size. It follows that the column of the model matrix corresponding to any factorial-effect contrast defined by $\mathbf{g}_1$ is orthogonal to that corresponding to any factorial-effect contrast defined by $\mathbf{g}_2$.

Likewise, under a regular fraction, for any $\mathbf{g} \in R(\mathbf{A})$, the factorial effect defined by $\mathbf{g}$ is aliased with the mean and cannot be estimated. Such effects may also become orthogonal to the mean under a parallel flats design. For each treatment combination $\mathbf{x} = \sum_{j=1}^{n-p} \lambda_j \mathbf{x}_j + \mathbf{z}_i$ in the parallel flats design, we have

$$\mathbf{g}^T \mathbf{x} = \mathbf{g}^T \left(\sum_{j=1}^{n-p} \lambda_j \mathbf{x}_j + \mathbf{z}_i\right) = \mathbf{g}^T \mathbf{z}_i.$$

Thus, if $\mathbf{g}^T \mathbf{z}_i$, $1 \leq i \leq f$, together cover the s elements of $GF(s)$ equally often, then the treatment combinations $\mathbf{x}$ in the parallel flats design are partitioned into s disjoint sets of equal size according to the values of $\mathbf{g}^T \mathbf{x}$. It follows that the column of the model matrix corresponding to any factorial-effect contrast defined by $\mathbf{g}$ is orthogonal to the vector of 1's.

In general, we have the following necessary and sufficient condition for the columns of the model matrix of a parallel flats design to be orthogonal.

Theorem 15.23. *Given a parallel flats design,*

    (i) all the columns of the full model matrix corresponding to the factorial-effect contrasts defined by a nonzero $\mathbf{g} \in EG(n, s)$ are orthogonal to the vector of 1's if and only if $\mathbf{g} \notin R(\mathbf{A})$, or $\mathbf{g} \in R(\mathbf{A})$ and $\mathbf{g}^T \mathbf{z}_i$, $1 \leq i \leq f$, together cover the s elements of $GF(s)$ equally often;

    (ii) for any two vectors $\mathbf{g}_1$ and $\mathbf{g}_2$ satisfying the condition in (i), the columns of the full model matrix corresponding to all the factorial-effect contrasts defined by $\mathbf{g}_1$ are orthogonal to those corresponding to the factorial-effect contrasts defined by $\mathbf{g}_2$ if and only if for each nonzero $\lambda \in GF(s)$ such that $\mathbf{g}_1 - \lambda \mathbf{g}_2 \in R(\mathbf{A})$, $(\mathbf{g}_1 - \lambda \mathbf{g}_2)^T \mathbf{z}_i$, $1 \leq i \leq f$, together cover the s elements of $GF(s)$ equally often.

Proof. For part (i), suppose n_i is the number of times the ith element of $GF(s)$ appears among $\mathbf{g}^T\mathbf{z}_1,\ldots,\mathbf{g}^T\mathbf{z}_f$, $1 \le i \le s$. If $\mathbf{g} \notin R(\mathbf{A})$, then clearly all the columns of the full model matrix corresponding to the factorial-effect contrasts defined by $\mathbf{g}$ are orthogonal to the vector of 1's. Otherwise, the inner product of the column of the model matrix corresponding to a factorial-effect contrast defined by $\mathbf{g}$ and the vector of 1's is of the form $\sum_{i=1}^{s} c_i n_i$ for some c_i's such that $\sum_{i=1}^{s} c_i = 0$. Then the conclusion of part (i) is a consequence of the following simple fact: $\sum_{i=1}^{s} c_i n_i = 0$ for all c_i's such that $\sum_{i=1}^{s} c_i = 0$ if and only if $n_1 = \cdots = n_s$.

For part (ii), consider the $N \times 2$ matrix $\mathbf{X}$ with the two columns consisting of the values of $\mathbf{g}_1^T\mathbf{x}$ and $\mathbf{g}_2^T\mathbf{x}$, respectively, where the $\mathbf{x}$'s are the treatment combinations in the parallel flats design. Then the columns of the model matrix of the parallel flats design corresponding to the factorial-effect contrasts defined by $\mathbf{g}_1$ are orthogonal to those corresponding to the factorial-effect contrasts defined by $\mathbf{g}_2$ if and only if $\mathbf{X}$ has zero B_2, where B_2 is a term of the generalized wordlength pattern defined in Section 15.6. By Proposition 15.14, B_2 does not depend on orthogonal bases of the space of contrasts representing the interaction of the two factors defined by the columns of $\mathbf{X}$. An alternative orthogonal basis consists of $(s-1)^2$ orthogonal contrasts, with $s-1$ of which defined by each $\mathbf{g}_1 - \lambda\mathbf{g}_2$, where $\lambda \in GF(s)$ and $\lambda \ne 0$. Therefore $B_2 = 0$ if and only if for each $\lambda \ne 0$, the columns of the full model matrix corresponding to all the factorial-effect contrasts defined by $(\mathbf{g}_1 - \lambda\mathbf{g}_2)^T\mathbf{x}$ are orthogonal to the vector of 1's. By the same argument as in the proof of part (i), this holds if and only if $(\mathbf{g}_1 - \lambda\mathbf{g}_2)^T\mathbf{z}_i$, $1 \le i \le f$, together cover the s elements of $GF(s)$ equally often for each nonzero λ such that $\mathbf{g}_1 - \lambda\mathbf{g}_2 \in R(\mathbf{A})$.

$\square$

Remark 15.3. The necessary and sufficient condition in part (ii) of Theorem 15.23 is satisfied if $\mathbf{g}_1 - \lambda\mathbf{g}_2 \notin R(\mathbf{A})$ for all nonzero $\lambda \in GF(s)$.

Theorem 15.23 is the main result in Srivastava and Li (1996), stated there in a somewhat different form. The form as stated in Theorem 15.23 was given by Liao (1999) for the case of three-level designs. We provide a different proof here. Liao and Iyer (1998) extended the result to the mixed-level case.

Theorem 15.15 can be used to provide an alternative proof of the following result, which first appeared in Srivastava and Chopra (1973) and was proved in Srivastava and Throop (1990).

Theorem 15.24. *A necessary and sufficient condition for a parallel flats design to be an orthogonal array of strength t is that for all defining words $\mathbf{g}$ of length no more than t, $\mathbf{g}^T\mathbf{z}_i$, $1 \le i \le f$, together cover the s elements of $GF(s)$ equally often.*

Based on the result in Theorem 15.23, Srivastava and Li (1996) tabulated some two-level parallel flats designs under which the main effects and certain two-factor interactions are orthogonally estimated. Liao (1999) provided tables of such designs for the three-level case. Liao, Iyer, and Vecchia (1996) proposed an algorithm to search parallel flats designs under which some required effects are orthogonally estimated, as in Franklin and Bailey (1977). Liao and Chai (2009) considered partially replicated parallel flats designs under which some specified possibly active effects are estimated optimally under the D-criterion. In such a design, one flat is replicated once to provide estimates of pure error variance for identifying truly active effects.

15.13 Saturated designs for hierarchical models: an application of algebraic geometry

Throughout this section, we follow the definition of resolution for nonregular designs given in Webb (1968); see Remarrk 15.1. In Example 15.1 we saw a nonregular resolution V design with one-quarter fewer runs than a regular design of resolution V. For experiments with n two-level factors, a resolution V design requires at least $1 + n + \binom{n}{2}$ runs and is called saturated if this lower bound is achieved. In general, we say that a design is saturated if the run size is equal to the number of parameters in the model. Rechtschaffner (1967) proposed a simple way of constructing two-level saturated designs of resolution V as well as three-level saturated designs for the following second-order model:

$$E(y_{\mathbf{x}}) = \beta_0 + \sum_{i=1}^{n} \beta_i x_i + \sum_{i=1}^{n} \beta_{ii} x_i^2 + \sum_{1 \leq i < j \leq n} \beta_{ij} x_i x_j, \qquad (15.20)$$

where $y_{\mathbf{x}}$ is the response at $\mathbf{x} = (x_1, \ldots, x_n)^T$.

Pistone and Wynn (1996) showed how the theory of Gröbner bases in algebraic geometry can be used to find estimable saturated linear polynomial models for a given design. Suppose a design d consists of the solutions to a set of polynomial equations. Let I be the set of all the polynomial functions f such that $f(\mathbf{x}) = 0$ for all $\mathbf{x} \in d$. For any two functions f and g, define $f + g$ as the function such that $(f + g)(\mathbf{x}) = f(\mathbf{x}) + g(\mathbf{x})$. Then for any g, $g(\mathbf{x}) = (f + g)(\mathbf{x})$ for all $x \in d$ and all $f \in I$. In other words, a model defined by $E(y_{\mathbf{x}}) = g(\mathbf{x})$ is totally aliased with the one defined by $g + f$ for all $f \in I$. We can think of I as an extension of the defining contrast subgroup of a regular fractional factorial design; then for any $g \notin I$, $\{g + f \colon f \in I\}$ is a set of aliased models. For regular designs, an estimable saturated model can be obtained by choosing one effect from each alias set of factorial effects. In general, the theory of Gröbner bases can be used to determine identifiable monomial terms of the models, and thus all estimable saturated models. We refer readers to Pistone and Wynn (1996) for the details and Pistone, Riccomagno, and Wynn (2001) for a general introduction to statistical applications of algebraic geometry.

Let $S_i = \{0, 1, \ldots, s_i - 1\}$, $1 \leq i \leq n$, where s_i is a positive integer, and let $H \subseteq S_1 \times \cdots \times S_n$, so H is a set of vectors $\mathbf{z} = (z_1, \ldots, z_n)^T$ with $z_i = 0, 1, \ldots, s_i - 1$. We say that a polynomial model of the form

$$y_{\mathbf{x}} = \sum_{\mathbf{z} \in H} \beta_{\mathbf{z}} x_1^{z_1} \cdots x_n^{z_n} \qquad (15.21)$$

is *hierarchical* if for any $(z_1, \ldots, z_n)^T \in H$, all $(z_1', \ldots, z_n')^T$ with $z_i' \leq z_i$, $1 \leq i \leq n$, are also in H. Clearly the model in (15.20) is hierarchical.

Let Z_+ be the set of all nonnegative integers. We say that a design $d \subseteq Z_+^n$ is an *echelon design* if for any $(x_1, \ldots, x_n)^T \in d$, all $(x'_1, \ldots, x'_n)^T$ with $x'_i \leq x_i, 1 \leq i \leq n$, are also in d. The following is a corollary of a result on echelon designs in Caboara, Pistone, Riccomagno, and Wynn (1997).

Theorem 15.25. *Any hierarchical model as in (15.21) is estimable under the echelon design consisting of all* $\mathbf{x} \in H$. *Such a design is saturated.*

The saturated design and the hierarchical model in Theorem 15.25 have the same pattern. This result was also proved by Dursun Bulutoglu and the author in an unpublished work without using Gröbner bases. The construction proposed by Rechtschaffner (1967) was for the *largest* two- and three-level second-order designs, and no proof was provided in the paper.

The nonnegative integer factor levels in Theorem 15.25 can be re-scaled. For example, 0 and 1 can be replaced by -1 and 1, respectively, for two-level factors, and 0, 1, and 2 can be replaced by -1, 0, and 1, respectively, for three-level factors. It is easy to see that under a hierarchical model, if for each factor the levels are converted according to a linear function (say $x \rightarrow 2x - 1$ is used to convert 0 and 1 to -1 and 1, respectively), then the parameters $\boldsymbol{\beta}$ and $\boldsymbol{\beta}^*$ before and after the conversion are such that

$$\boldsymbol{\beta} = \mathbf{A}\boldsymbol{\beta}^* \tag{15.22}$$

for some nonsingular matrix $\mathbf{A}$. If $\boldsymbol{\beta}$ is estimable, then $\boldsymbol{\beta}^*$ is also estimable. Therefore the conclusion of Theorem 15.25 also holds under alternative level codings.

Example 15.2. It follows from Theorem 15.25 that a design consisting of the following three types of treatment combinations is a saturated two-level design of resolution V: the single combination with all the factors at level -1 (corresponding to the mean in the model), the n combinations with one factor at level 1 and the other factors at level -1 (corresponding to the main effects in the model), and the $\binom{n}{2}$ combinations with two factors at level 1 and the other factors at level -1 (corresponding to the two-factor interactions in the model).

The designs given in Theorem 15.25 tend to be not very efficient. In Example 15.2, each factor appears $1 + n(n-1)/2$ times at level -1, but only n times at level 1. Rechtschaffner (1967) recommended switching the two levels in the n runs corresponding to main effects; that is, the n runs with exactly one factor at level 1 are replaced by those with exactly one factor at level -1. One can also switch the two levels in the combination corresponding to the mean and/or all the treatment combinations corresponding to two-factor interactions. Qu (2007) compared the eight classes of designs resulting from such level switching and showed that the designs recommended by Rechtschaffner (1967) are the best under the D- and A-criteria. Tobias (1996) used a computer search to identify a class of two-level saturated designs of resolution V that outperforms Rechtschaffner's designs under the D- and A-criteria for $n > 6$.

15.14   Search designs

The need to consider projections also arises in the study of *search designs*. Srivastava (1975) introduced such designs for model identification and discrimination. Consider the model

$$y = X_1\beta_1 + X_2\beta_2 + \varepsilon, \qquad (15.23)$$

with $E(\varepsilon) = 0$ and $cov(\varepsilon) = \sigma^2 I_N$. For example, β_1 may consist of the grand mean and main effects of all treatment factors, and β_2 may consist of the two-factor interactions. Suppose we need to estimate the parameters in β_1 and at most k parameters in β_2 are nonnegligible, but we do not know which the nonnegligible parameters are. One can define the estimation capacity of a design as in Section 10.4, with respect to the potential models that contain all the parameters in β_1 and k parameters in β_2. Under a design with the largest estimation capacity, we can estimate the largest number of potential models, but the capability of estimating potential models is not the same as that of discriminating between different models. Srivastava (1975) proved the following result.

Theorem 15.26. *Under (15.23) and the assumption that no more than k parameters in β_2 are nonnegligible, a necessary condition that the nonnegligible parameters in β_2 can be identified and estimated along with the required effects in β_1 is that*

$$\begin{bmatrix} X_1 & X_2^0 \end{bmatrix} \text{ is of full column rank for all } N \times 2k \text{ submatrices } X_2^0 \text{ of } X_2. \qquad (15.24)$$

This condition is also sufficient if $\sigma = 0$.

Designs that satisfy condition (15.24) are said to have resolving power $\{\beta_1; \beta_2, k\}$.

It can be seen that if the rank condition (15.24) does not hold, then either X_1 does not have full column rank, in which case not all the parameters in β_1 are estimable, or else there are at least two indistinguishable models with at most k parameters from β_2. On the other hand, in the noiseless case, condition (15.24) implies that the true model with up to k parameters from β_2 is the only such model with a perfect fit and can be identified as the model with zero residual sum of squares.

Srivastava (1975) proposed to identify the true model by minimizing the residual sum of squares. In the noisy case where $\sigma \neq 0$, it does not necessarily lead to correct identification, and the probability of correctly identifying nonnegligible effects needs to be considered; see Shirakura, Takahashi and Srivastava (1996). Miller and Sitter (2004) showed that under condition (15.24) the expected residual sum of squares under the true model is strictly smaller than that under any other model with up to k parameters from β_2.

Example 15.3. Consider 12-run designs for four two-level factors. Let β_1 consist of the grand mean and main effects, and β_2 consist of all the interactions. Srivastava and Arora (1987) showed that in this case, for $k = 2$, a two-level design satisfying (15.24) requires a minimum of 11 runs, and this lower bound is achieved by the design consisting of all the treatment combinations with 1, 2, or 4 factors at the high level. Each four-factor projection of the 12-run Plackett–Burman design has a pair of identical runs. Removing one such run yields the 11-run design described above. The design with the repeated run has one more run than the minimum, but it has the nice structure of an orthogonal array. Cheng (1998b) discussed the method of constructing search designs by drawing columns from nonregular orthogonal arrays. Such designs do not require ingenious construction, have comparable sizes to those in the literature, and have the advantage of being orthogonal arrays themselves.

Miller and Sitter (2004) called a set of columns of $\mathbf{X}_2$ a *dependent set* if these columns and all the columns in $\mathbf{X}_1$ together are linearly dependent. Such sets provide information about models that cannot be discriminated. A dependent set is called *minimal* if, when any one of its columns is removed, the remaining columns and those in $\mathbf{X}_1$ together are linearly independent. Each dependent set is either minimal or contains a minimal dependent set (MDS). The negative impact of an MDS on the resolvability decreases as its size increases. Lin, Miller, and Sitter (2008) defined MDS-resolution as the size of the smallest MDS. It is desirable to have large MDS-resolution. For a design with MDS-resolution r, (15.24) is satisfied for all k such that $k < r/2$. Let A_i be the number of MDS of size i. Lin, Miller, and Sitter (2008) called $(A_1, \ldots, A_n)$ the MDS wordlength pattern and proposed the minimum MDS-aberration criterion that sequentially minimizes $A_1, A_2, \ldots$, etc.

Chapter 16 of Hinkelmann and Kempthorne (2005) contains a more detailed treatment of search designs.

15.15   Supersaturated designs

An experiment involving n two-level treatment factors requires at least $n+1$ runs to estimate all the main effects. Two-level supersaturated designs are those with $N < n+1$. Under the assumption of factor sparsity that only a small number of the factors are active, a supersaturated design can provide considerable cost saving in factor screening.

Let $\mathbf{X}$ be an $N \times n$ matrix of 1's and -1's, where $N < n+1$, again with each column of $\mathbf{X}$ corresponding to a factor and each row representing a factor-level combination. It is necessary that no two columns of $\mathbf{X}$ are completely aliased, that is, there are no two columns $\mathbf{x}$ and $\mathbf{y}$ with $\mathbf{x} = \mathbf{y}$ or $\mathbf{x} = -\mathbf{y}$. Throughout this section it is assumed that the numbers of occurrences of 1 and -1 in each column of $\mathbf{X}$ differ by at most one (see Remark 15.4 later in this section). In particular, when N is even, each column of $\mathbf{X}$ is a *balanced* column that contains the same number of 1's and -1's. For odd N, without loss of generality (subject to level relabeling), assume that each column is *nearly balanced*, with level 1 appearing $(N-1)/2$ times and -1 appearing $(N+1)/2$ times.

Ideally one would want the columns of $\mathbf{X}$ to be mutually orthogonal, which is not possible in the current situation. This is clear when N is odd. For even N, the constraint that 1 and -1 appear the same number of times in each column of $\mathbf{X}$ forces the rank of $\mathbf{X}$ to be no more than $N-1$. Then $\text{rank}(\mathbf{X}^T \mathbf{X}) \leq N-1 < n$; therefore not all the off-diagonal entries of $\mathbf{X}^T\mathbf{X}$ can be zero. In choosing a supersaturated design, it is desirable to have the columns of $\mathbf{X}$ as orthogonal as possible. The most commonly used measure of nonorthogonality among the columns of a supersaturated design is

$$E(s^2) = \frac{\sum_{i<j} s_{ij}^2}{\binom{n}{2}},$$

with $s_{ij} = \mathbf{x}_i^T\mathbf{x}_j$, where $\mathbf{x}_i$ and $\mathbf{x}_j$ are the ith and jth columns of $\mathbf{X}$. A design minimizing $E(s^2)$ is said to be $E(s^2)$-optimal. This criterion was first proposed by Booth and Cox (1962) for the construction of supersaturated designs. Another criterion that has been considered is to minimize $\max_{i<j}|s_{ij}|$. As a combination of this criterion with the $E(s^2)$-criterion, one can minimize $E(s^2)$ subject to an upper bound on $\max_{i<j}|s_{ij}|$. Cheng and Tang (2001) studied upper bounds on the number of factors such that $\frac{1}{N}\max_{i<j}|s_{ij}| \leq e$ can be achieved, where $1 < e < 1$ is the degree of nonorthogonality the experimenter is willing to sacrifice.

We only present results on the construction of $E(s^2)$-optimal designs here. Readers are referred to Gilmour (2006) for discussions of analysis and other issues including the use of supersaturated designs.

Remark 15.4. Cheng, Deng, and Tang (2002) suggested that the generalized minimum aberration criterion can be applied to the selection of supersaturated designs as a refinement of the $E(s^2)$-criterion. By (15.8),

$$B_1 = N^{-2} \sum_{j=1}^{n} \left[\sum_{i=1}^{N} x_{ij}\right]^2.$$

It is clear that B_1 is minimized when the numbers of occurrences of 1 and -1 in each column of $\mathbf{X}$ differ by at most one. This is the constraint imposed earlier. Then minimizing B_2 is the same as minimizing $E(s^2)$. One can go on to sequentially minimize B_k for $k > 2$.

A number of authors have generalized the $E(s^2)$-criterion to multi-level supersaturated designs. We refer readers to Xu and Wu (2005) for additional references as well as general construction and optimality results on generalized minimum aberration supersaturated designs. Many results for two-level $E(s^2)$-supersaturated designs to be discussed in the rest of this section can be obtained as special cases of the results in Xu (2003) and Xu and Wu (2005). We restrict to the two-level case here for simplicity of presentation and for the practical reason that two-level designs are the most useful in the context of supersaturated designs.

Theorem 15.27. *For any supersaturated design with n two-level factors and N runs, where the numbers of occurrences of 1 and -1 in each column differ by at most one,*

$$E(s^2) \geq \frac{n^2 + (N-1)x^2 - nN}{n(n-1)}N, \qquad (15.25)$$

where

$$x = \begin{cases} -n/(N-1), & \text{for even } N; \\ -n/N, & \text{for odd } N. \end{cases} \qquad (15.26)$$

The equality in (15.25) is achieved if and only if

$$\mathbf{X}\mathbf{X}^T = (n-x)\mathbf{I}_N + x\mathbf{J}_N. \qquad (15.27)$$

Proof. Since all the diagonal entries of $\mathbf{X}^T\mathbf{X}$ are equal to N and the sum of squares of all the entries of $\mathbf{X}^T\mathbf{X}$ can be computed as $\text{tr}(\mathbf{X}^T\mathbf{X}\mathbf{X}^T\mathbf{X})$, we have

$$E(s^2) = \left[\text{tr}(\mathbf{X}^T\mathbf{X}\mathbf{X}^T\mathbf{X}) - nN^2 \right] / [n(n-1)].$$

By the identity $\text{tr}(\mathbf{AB}) = \text{tr}(\mathbf{BA})$,

$$E(s^2) = \left[\text{tr}(\mathbf{X}\mathbf{X}^T\mathbf{X}\mathbf{X}^T) - nN^2 \right] / [n(n-1)]. \qquad (15.28)$$

The sum of the entries of $\mathbf{X}\mathbf{X}^T$ is equal to

$$\mathbf{1}^T\mathbf{X}\mathbf{X}^T\mathbf{1} = \begin{cases} 0, & \text{for even } N; \\ n, & \text{for odd } N. \end{cases}$$

On the other hand, all the diagonal entries of $\mathbf{X}\mathbf{X}^T$ are equal to n. It follows that for even (respectively, odd) N, the sum of the off-diagonal entries of $\mathbf{X}\mathbf{X}^T$ is equal to $-nN$ (respectively, $-n(N-1)$). Thus $\text{tr}(\mathbf{X}\mathbf{X}^T\mathbf{X}\mathbf{X}^T)$, the sum of squares of the entries of $\mathbf{X}\mathbf{X}^T$, attains the minimum if all the off-diagonal entries of $\mathbf{X}\mathbf{X}^T$ are equal; that is, they are equal to $-n/(N-1)$ (respectively, $-n/N$) for even (respectively, odd) N. In this case $\mathbf{X}\mathbf{X}^T$ is of the form in (15.27) with x given in (15.26), and, by (15.28), the lower bound on $E(s^2)$ in (15.25) can readily be verified. □

For even N, (15.25) reduces to

$$E(s^2) \geq \frac{n-N+1}{(n-1)(N-1)}N^2, \qquad (15.29)$$

and for odd N,

$$E(s^2) \geq \frac{n(N^2+N-1)-N^3}{N(n-1)}. \qquad (15.30)$$

The bound in (15.29) was derived independently by Nguyen (1996) and Tang and Wu (1997). The bound in (15.30) and the more general version in Theorem 15.27 were given by Nguyen and Cheng (2008).

We present a result on the existence and construction of $E(s^2)$-optimal designs that attain the lower bound in (15.25).

Theorem 15.28. *Suppose $N < n + 1$. Then a necessary condition for the existence of a design attaining the lower bound in (15.25) is that there is a positive integer m such that $n = m(N - 1)$ when N is even and $n = mN$ when N is odd; furthermore, for $N \geq 3$, m must be even if $N \equiv 1$ or 2 (mod 4). A design $\mathbf{X}$ achieves the lower bound in (15.25) if and only if any two rows of the matrix $\tilde{\mathbf{X}}$ obtained by adding m columns of 1's to $\mathbf{X}$ are orthogonal.*

Proof. For even (respectively, odd) N, the lower bound in (15.25) is attained if and only if all the off-diagonal entries of $\mathbf{X}\mathbf{X}^T$ are equal to $-n/(N - 1)$ (respectively, $-n/N$), and so a necessary condition for the existence of such designs is that n is a multiple of $N - 1$ (respectively, a multiple of N). In this case, let $m = n/(N - 1)$ or n/N for even or odd N, respectively, and let $\tilde{\mathbf{X}}$ be obtained by adding m columns of 1's to $\mathbf{X}$. Then all the off-diagonal entries of $\mathbf{X}\mathbf{X}^T$ are equal to $-n/(N - 1)$ (respectively, $-n/N$) if and only if any two rows of $\tilde{\mathbf{X}}$ are orthogonal. Without loss of generality, by changing the signs of all the entries in the same column if necessary, we may normalize $\tilde{\mathbf{X}}$ so that all the entries in its first row are equal to 1. By Corollary 8.8, as long as $N \geq 3$, the rows of such an $\tilde{\mathbf{X}}$ are pairwise orthogonal if and only if the transpose of the array obtained from $\tilde{\mathbf{X}}$ by deleting the row of 1's is an orthogonal array of strength two. Then the number of rows of the resulting orthogonal array, which is mN for even N and $m(N + 1)$ for odd N, must be a multiple of 4. Thus, if N is even and is not a multiple of 4, or N is odd and $N \equiv 1$ (mod 4), then m must be even. $\qquad\square$

Remark 15.5. For the $E(s^2)$-optimal designs that achieve the bound in (15.25), all the off-diagonal entries of $\mathbf{X}\mathbf{X}^T$ are equal. In other words, the pairwise Hamming distances of the rows of $\mathbf{X}$ are a constant. By Theorem 15.19, such designs have generalized minimum aberration. Thus Theorem 15.27 also follows from Theorem 15.19. Cheng (1997) showed that for even N, if we add a balanced column to (or remove such a column from) a design that achieves the bound in (15.25), then the resulting design is $E(s^2)$-optimal. Since the pairwise Hamming distances of the rows of such a design differ from one another by at most 1, by Theorem 15.19, it also has generalized minimum aberration. For odd N, the same conclusion holds for the design obtained by adding a nearly balanced column to (or removing such a column from) a design that achieves the bound in (15.25).

Remark 15.6. By Theorem 15.19, when the bound in (15.25) cannot be achieved, a better lower bound on $E(s^2)$ can be obtained by computing the $E(s^2)$ value for a hypothetical design in which the pairwise Hamming distances of the runs differ from one another by at most 1. This is based on the equivalence of generalized minimum aberration and minimum moment aberration. Liu and Hickernell (2002) derived such a bound by directly observing that $E(s^2)$ is itself a convex function of the pairwise Hamming distances of the runs. They also obtained similar results for a discrepancy criterion for uniformity.

Theorem 15.28 provides a simple method of constructing $E(s^2)$-optimal supersaturated designs that attain the bound in (15.25). Some applications are given below.

Lin (1993) proposed a method of using a $k \times k$ Hadamard matrix to construct a supersaturated design with $k - 2$ factors and $k/2$ runs. Normalizing a Hadamard matrix so that all the entries of its first column are equal to 1, and suitably permuting the rows, we can express a $k \times k$ Hadamard matrix $\mathbf{H}$ as

$$\mathbf{H} = \begin{bmatrix} 1 & 1 & \mathbf{H}^h \\ 1 & -1 & * \end{bmatrix}, \tag{15.31}$$

where $\mathbf{1}$ is a $(k/2) \times 1$ vector of 1's. Since the submatrix consisting of the last $k - 1$ columns of $\mathbf{H}$ is an orthogonal array of strength two, each column of the $(k/2) \times (k - 2)$ matrix $\mathbf{H}^h$ contains the same number of 1's and -1's. Therefore $\mathbf{H}^h$ is a supersaturated design with $N = k/2$ and $n = k - 2$ provided that it does not contain completely aliased columns. For example, from a 12×12 Hadamard matrix, one can construct a supersaturated design with 10 factors in six runs.

The $E(s^2)$-optimality of such half-Hadamard matrices follows from Theorem 15.28, as shown below.

Theorem 15.29. *If the half-Hadamard matrix $\mathbf{H}^h$ in (15.31) contains no completely aliased columns, then it is $E(s^2)$-optimal.*

Proof. Since $\mathbf{H}$ is a Hadamard matrix, its rows are pairwise orthogonal. In particular, any two rows of $[\mathbf{1} \ \mathbf{1} \ \mathbf{H}^h]$ are orthogonal. The $E(s^2)$-optimality of $\mathbf{H}^h$ follows from Theorem 15.28 with $m = 2$.

$\square$

Theorem 15.29 requires $\mathbf{H}^h$ to have no completely aliased columns. If the orthogonal array

$$\mathbf{H}' = \begin{bmatrix} 1 & \mathbf{H}^h \\ -1 & * \end{bmatrix}$$

has no defining word of length three, then by Lemma 15.5, $\mathbf{H}'$ has projectivity at least three. It follows that $\mathbf{H}^h$ has no completely aliased columns. We have mentioned in Section 15.4 that Paley designs of sizes greater than or equal to 12 have no defining words of length three. Therefore by applying Lin's method to Paley matrices of order greater than or equal to 12, we always obtain $E(s^2)$-optimal supersaturated designs with no completely aliased factors.

Tang and Wu (1997) also proposed a method of using Hadamard matrices to construct $E(s^2)$-optimal supersaturated designs.

Theorem 15.30. *Suppose $n = m(N - 1)$ and for $i = 1, \ldots, m$, there exists an $N \times N$ Hadamard matrix $\mathbf{H}_i = [\mathbf{1} \ \mathbf{H}_i^*]$ such that no two of the $m(N - 1)$ columns in $\mathbf{H}_1^*, \ldots, \mathbf{H}_m^*$ are completely aliased. Then $[\mathbf{H}_1^* \ \cdots \ \mathbf{H}_m^*]$ achieves the lower bound in (15.25).*

Proof. Since each $\mathbf{H}_i$ is a Hadamard matrix, any two rows of $\mathbf{H}_i$ are orthogonal. It follows that any two rows of

$$[\underbrace{\mathbf{1} \cdots \mathbf{1}}_{m} \quad \mathbf{H}_1^* \ \cdots \ \mathbf{H}_m^*]$$

are orthogonal. The $E(s^2)$-optimality of $[\mathbf{H}_1^* \ \cdots \ \mathbf{H}_m^*]$ follows from Theorem 15.28.
□

Wu (1993) proposed another method of using Hadamard matrices to construct supersaturated designs. Let $\mathbf{X}$ be obtained by deleting a column of 1's from an $N \times N$ Hadamard matrix. Supplementing $\mathbf{X}$ with f pairwise Hadamard products of its columns, where $f \leq \binom{N-1}{2}$, we obtain a supersaturated design with $N-1+f$ factors. If $\mathbf{X}$ has no defining word of length three or four (e.g., a Paley design of size $N \geq 12$), then the resulting array has no completely aliased columns. The supersaturated designs constructed by this method may not be $E(s^2)$-optimal. Bulutoglu and Cheng (2004) reported several cases where the $E(s^2)$-optimality can be established by using Theorem 15.28.

Supersaturated designs achieving the lower bound in (15.25) can also be constructed by using balanced incomplete block designs (Nguyen, 1996; Cheng, 1997). Given a supersaturated design $\mathbf{X}$ with even N, multiply the entries in the same column by -1 if necessary to make all the entries in the first row equal to 1. Deleting this row of 1's, and replacing all the -1's with 0's, we obtain an $(N-1) \times n$ matrix $\mathbf{Y}$ with 0 and 1 entries. We can think of $\mathbf{Y}$ as the incidence matrix of an incomplete block design $d(\mathbf{X})$ with $N-1$ treatments and n blocks: each row is associated with a treatment, each column corresponds to a block, and the ith treatment appears in the jth block if and only if the (i, j)th entry of $\mathbf{Y}$ is equal to 1. Since each column of $\mathbf{Y}$ has $N/2-1$ ones, $d(\mathbf{X})$ has block size $N/2-1$. Conversely, given an incomplete block design with $N-1$ treatments and n blocks of size $N/2-1$, we can reverse the above process to construct a supersaturated design with n factors and size N. This is the same construction used in Section 8.5 to establish the equivalence of an $N \times N$ Hadamard matrix and a balanced incomplete block design with $N-1$ treatments and $N-1$ blocks of size $(N-1)/2$. It can be shown that $\mathbf{X}$ attains the lower bound in (15.25) if and only if $d(\mathbf{X})$ is a balanced incomplete block design; this is left as an exercise. The supersaturated design does not have completely aliased columns if and only if the corresponding block design has no identical blocks. Bulutoglu and Cheng (2004) used an algebraic method to construct families of balanced incomplete block designs with $N-1$ treatments and $m(N-1)$ distinct blocks of size $N/2-1$, useful for constructing $E(s^2)$-optimal supersaturated designs.

The construction using balanced incomplete block designs is more flexible than that based on Hadamard matrices. The construction of supersaturated designs by combining m Hadamard matrices in Theorem 15.30 is equivalent to combining m balanced incomplete block designs with $N-1$ treatments and $N-1$ blocks of size $(N-1)/2$ to form a balanced incomplete block design with $N-1$ treatments and $m(N-1)$ blocks of size $(N-1)/2$. This construction is applicable only when N is a multiple of 4, and it is not clear how one can avoid duplicated blocks. Furthermore, a balanced incomplete block design with $N-1$ treatments and $m(N-1)$ blocks of size $(N-1)/2$ may not be the union of m balanced incomplete block designs with $N-1$ treatments and $N-1$ blocks of size $(N-1)/2$.

The $E(s^2)$-optimal supersaturated designs constructed in Bulutoglu and Cheng (2004) are based on difference families and hence can be generated cyclically. Nguyen (1996), Liu and Zhang (2000), Eskridge, Gilmour, Mead, Butler, and Travnicek (2004), and Liu and Dean (2004) also used cyclic methods to construct supersaturated designs. Liu and Dean (2004) provided tables of generators of $E(s^2)$-optimal or nearly optimal supersaturated designs that are at least 97.8% efficient.

For odd N, given a supersaturated design $\mathbf{X}$ with n two-level factors in which each factor appears at level 1 (respectively, -1) $(N-1)/2$ (respectively, $(N+1)/2$) times, one can convert it to the incidence matrix of a block design $d'(\mathbf{X})$ with N treatments and n blocks of size $(N-1)/2$ by replacing all the -1's with 0's. Note that this differs from the construction for the case of even N in that we do not delete a row of 1's here. It can be shown that $\mathbf{X}$ achieves the lower bound in (15.25) if and only if $d'(\mathbf{X})$ is a balanced incomplete block design. This is also left as an exercise.

Some improved lower bounds for $E(s^2)$ when the bound given in (15.25) is not achievable can be found in Butler, Mead, Eskridge, and Gilmour (2001), Bulutoglu and Cheng (2004), and Nguyen and Cheng (2008).

Chen and Lin (1998) studied the probability that estimates of the active effects are the largest among the estimates of all the effects in a supersaturated design.

The $E(s^2)$-criterion does not address the identification of active factors. The approach of search designs discussed in Section 15.14 can also be applied to factor screening. Suppose there are n two-level potential factors, at most k of which are active, but we do not know which these factors are. Cheng (1998b) observed that in the noiseless case ($\sigma = 0$), a design can be used to identify the active factors and determine all their main effects and interactions if and only if for all pairs (A_1, A_2) of k-subsets of potential factors,

$$[\mathbf{1}_N \ \mathbf{X}(A_1, A_2)] \text{ has full column rank,} \tag{15.32}$$

where $\mathbf{X}(A_1, A_2)$ is the matrix whose columns are all the columns in the model matrix corresponding to the main effects of the factors in $A_1 \cup A_2$, interactions of the factors in A_1, and interactions of the factors in A_2. If it can be assumed that some higher-order interactions are negligible, then the columns associated with such interactions are not included in $\mathbf{X}(A_1, A_2)$. For example, Ghosh (1979) and Ghosh and Avila (1985) studied factor screening under the assumption that all the interactions are negligible. In this case, for any design $\mathbf{X}$, (15.32) reduces to

$$\text{rank}([\mathbf{1} \ \mathbf{X}_0]) = 2k + 1 \tag{15.33}$$

for every $N \times 2k$ submatrix $\mathbf{X}_0$ of $\mathbf{X}$. Clearly condition (15.33) holds if $[\mathbf{1} \ \mathbf{X}]$ has full column rank. Thus it is interesting only for the designs with $n > N - 1$, that is, supersaturated designs.

For even N, since in each column of a design $\mathbf{X}$, half of the entries are equal to 1 and the other half are equal to -1, (15.33) holds if and only if $\text{rank}(\mathbf{X}_0) = 2k$ for every $N \times 2k$ submatrix $\mathbf{X}_0$ of $\mathbf{X}$. Deng, Lin and Wang (1999) defined the *resolution rank* of $\mathbf{X}$ as $\max\{c :$ any c columns of $\mathbf{X}$ are linearly independent$\}$. Then (15.33) states that the resolution rank of $\mathbf{X}$ is at least $2k$.

Miller and Tang (2012) applied the minimum MDS-aberration criterion (see Lin, Miller, and Sitter (2008) and Section 15.14) to supersaturated designs with even N. In this case, an MDS is a set of linearly dependent columns of $\mathbf{X}$ such that any of its subsets is linearly independent, and the MDS resolution is equal to the resolution rank plus 1. Miller and Tang (2012) investigated the number and structure of minimal dependent sets and constructed some supersaturated designs with large minimal dependent sets. For each even $N \geq 6$, they constructed an N-run supersaturated design for N factors that has MDS-resolution N. These designs, however, have large $E(s^2)$-values. On the other hand, the $E(s^2)$-optimal designs obtained by adding a column to a saturated design of size N have MDS-resolution equal to 5, 10, and 11 for $N = 8$, 12, and 16, respectively. This calls for the construction of designs that perform well under both the MDS-resolution and $E(s^2)$-criteria.

Exercises

15.1   Prove Proposition 15.4.

15.2   Use Proposition 15.4 and Theorem 15.15 to prove Theorem 8.11, Exercise 8.2, and Exercise 8.3.

15.3   Complete the proof of Theorem 15.19 by showing that if, for all the treatment factors, the numbers of times the s levels appear differ from one another by at most 1, then the sum of all the pairwise Hamming distances of the runs is a constant.

15.4   Prove (15.13).

15.5   Use Theorem 15.15 to prove Theorem 15.24.

15.6   Prove (15.22).

15.7   As described in Remark 15.6, use Theorem 15.19 to derive a better lower bound on $E(s^2)$ than (15.25).

15.8   The supersaturated designs constructed by Wu (1993) consist of all the main-effect columns and some two-factor interaction columns of a saturated design based on a Hadamard matrix. Assume that the designs constructed have no completely aliased columns. Use Theorem 15.28 to show the $E(s^2)$-optimalty of the supersaturated designs consisting of the following columns: (i) all the main-effect and two-factor interaction columns, (ii) all the main-effect columns and the two-factor interaction columns involving a given factor, and (iii) all the two-factor interaction columns only.

[Bulutoglu and Cheng (2004)]

15.9   Let $d(\mathbf{X})$ (respectively $d'(\mathbf{X})$) be the block design constructed from a supersaturated design $\mathbf{X}$ of even (respectively, odd) run size by the method presented on p. 361 and p. 362. Show that $\mathbf{X}$ achieves the lower bound in (15.25) if and only if $d(\mathbf{X})$ (respectively $d'(\mathbf{X})$) is a balanced incomplete block design.

Appendix

A.1 Groups

A binary operation $*$ on a set S is a function $* : S \times S \to S$, where $S \times S$ is the set of all the ordered pairs (a, b) of the elements of S. The operation is called *commutative* if $a * b = b * a$ for all $a, b \in S$. It is called *associative* if $(a * b) * c = a * (b * c)$ for all $a, b, c \in S$. An element $e \in S$ is called an *identity element* if $e * a = a * e = a$ for all $a \in S$. An element b is an *inverse* of a if $a * b = b * a = e$.

A set G with an operation $*$ is called a *group* if $*$ is associative, an identity element exists, and each element of G has an inverse. The group is called commutative (or *Abelian*) if $*$ is commutative.

A subset H of G is called a subgroup if H is also a group under $*$. For each $g \in G$, let $g * H = \{g * h : h \in H\}$. Then for all $g_1, g_2 \in G$, either $g_1 * H = g_2 * H$ or they are disjoint. We have $g_1 * H = g_2 * H$ if and only if $g_2^{-1} * g_1 \in H$. Each $g * H$ different from H is called a *coset* of H. If G and H are finite and contain g and h elements, respectively, then g is a multiple of h. Each subgroup H and its cosets together partition G into g/h disjoint sets of size h.

For each g in an Abelian group G, we denote $g * g$ by g^2, and for any positive integer $k > 1$, define g^k inductively as $g * g^{k-1}$. The smallest subgroup containing g consists of all the powers of g. Such a group is called a *cyclic* group, and g is called a generator. We denote the cyclic subgroup generated by g as $< g >$. For finite G, the *order* of g is the smallest positive integer k such that $g^k = e$. Then $< g >= \{g^0 = e, g^1 = g, \ldots, g^{k-1}\}$. A power g^r of g is also a generator of $< g >$ if and only if r is coprime to k, that is, r and k have no common prime divisor. Therefore the number of distinct generators of $< g >$ is equal to the Euler function $\phi(k)$, which is the number of integers $1 \leq l \leq k - 1$ that are coprime to k.

When the group operation is denoted by $+$, the elements of $< g >$, where g has order k, are written as $0, g, \ldots, (k-1)g$.

A.2 Finite fields

A set F with two binary operations, which are usually denoted by "+" and "·", is called a ring if the following conditions are satisfied:

(a) F is an Abelian group under "+".

(b)   The operation "·" is associative.

(c)   The distributive laws $a \cdot (b+c) = a \cdot b + a \cdot c$ and $(a+b) \cdot c = a \cdot c + b \cdot c$ hold for all $a, b, c \in F$.

(d) F contains an identity element for the operation "$\cdot$".

The two operations "$+$" and "$\cdot$" are called addition and multiplication, and their identity elements are denoted by 0 and 1, respectively. If, in addition, the elements of F other than 0 form an Abelian group under multiplication, then F is called a *field*. The additive inverse of an element a is denoted by $-a$, and if a is nonzero, its multiplicative inverse is denoted by a^{-1}.

Both the set of all the real numbers and the set of all the rational numbers are fields under the usual addition and multiplication.

For any positive integer $s > 1$, let $\mathbb{Z}_s$ be the set $\{0, 1, \ldots, s-1\}$ of nonnegative integers less than s. Two integers a and b are said to be congruent to each other modulo s, denoted $a \equiv b \pmod{s}$, if $a - b$ is divisible by s. For any two elements a and b of $\mathbb{Z}_s$, the usual sum $a+b$ is not in $\mathbb{Z}_s$ if $a+b \geq s$, but if we divide $a+b$ by s, call the remainder the sum of a and b, and still denote it as $a+b$, then "$+$" is an operation on $\mathbb{Z}_s$. Similarly, $a \cdot b$ is defined as the remainder when the usual $a \cdot b$ is divided by s. These operations are called addition and multiplication modulo s, respectively. It can be seen that, equipped with these two operations, $\mathbb{Z}_s$ is a ring, and it is a field if and only if s is a prime number.

Two fields F_1 and F_2 are said to be isomorphic if there is a one-to-one function $\phi : F_1 \to F_2$ such that $\phi(x+y) = \phi(x) + \phi(y)$ and $\phi(x \cdot y) = \phi(x) \cdot \phi(y)$ for all $x, y \in F_1$. A *finite field*, or *Galois field*, is a field that contains finitely many elements. It can been shown that the number of elements in a finite field must be of the form p^k, where p is a prime number and $k \geq 1$, and all the fields with p^k elements are isomorphic. Such a field is denoted by $\mathrm{GF}(p^k)$. In particular, for each prime number p, $\mathrm{GF}(p)$ is isomorphic to the field $\mathbb{Z}_p$. The construction of a finite field with p^k elements, where p is a prime number and $k > 1$, can be found in, e.g., the appendix of Hedayat, Sloane and Stufken (1999). The addition and multiplication tables of $\mathrm{GF}(4)$ are shown below.

$+$	0	1	x	$1+x$
0	0	1	x	$1+x$
1	1	0	$1+x$	x
x	x	$1+x$	0	1
$1+x$	$1+x$	x	1	0

$\cdot$	0	1	x	$1+x$
0	0	0	0	0
1	0	1	x	$1+x$
x	0	x	$1+x$	1
$1+x$	0	$1+x$	1	x

The multiplicative group of $GF(p^k)$ is cyclic: there is an element $x \in GF(p^k)$, called a primitive element of the field, such that the $p^k - 1$ nonzero elements of $GF(p^k)$ can be expressed as $x^0, x^1, \ldots, x^{p^k-2}$.

A.3   Vector spaces

Let F be a field with operations "+" and "$\cdot$". Suppose V is a another set, and there are two operations, which we also denote as "+" and "$\cdot$" without danger of confusion, such that $+ : V \times V \to V$ and $\cdot : F \times V \to V$. We say that V is a vector space over F if V is an Abelian group under "+", and for all $a, b \in F$ and $\mathbf{x}, \mathbf{y} \in V$,

(a) $(a+b) \cdot \mathbf{x} = a \cdot \mathbf{x} + b \cdot \mathbf{x}$,

(b) $a \cdot (\mathbf{x}+\mathbf{y}) = a \cdot \mathbf{x} + b \cdot \mathbf{y}$,

(c) $(a \cdot b) \cdot \mathbf{x} = a \cdot (b \cdot \mathbf{x})$,

(d) $1 \cdot \mathbf{x} = \mathbf{x}$.

Each element in V is called a vector, and we denote the additive identity of V by $\mathbf{0}$.

Let $F^n = \{(x_1, \ldots, x_n)^T : x_i \in F\}$ be the set of all the ordered n-tuples of the elements of F, written as columns, where T means "transpose." For $a \in F$ and $\mathbf{x} = (x_1, \ldots, x_n)^T$, $\mathbf{y} = (y_1, \ldots, y_n)^T \in F^n$, define

$$\mathbf{x}+\mathbf{y} = (x_1 + y_1, \ldots, x_n + y_n)^T,$$

$$a \cdot \mathbf{x} = (a \cdot x_1, \ldots, a \cdot x_n)^T.$$

Then F^n is a vector space over F. In particular, when $F = \mathbb{R}$, the set of the real numbers, we have $F^n = \mathbb{R}^n$, the usual n-dimensional Euclidean space.

A subset U of V is called a subspace if U is itself a vector space under the same operations. A set of vectors $\{\mathbf{x}_1, \ldots, \mathbf{x}_k\}$ is said to be linearly independent if and only if the equation $a_1 \cdot \mathbf{x}_1 + \cdots + a_k \cdot \mathbf{x}_k = \mathbf{0}$ implies that $a_1 = \cdots = a_k = 0$; therefore, if $b_1 \cdot \mathbf{x}_1 + \cdots + b_k \cdot \mathbf{x}_k = c_1 \cdot \mathbf{x}_1 + \cdots + c_k \cdot \mathbf{x}_k$, then $b_i = c_i$ for all i. A set of linearly independent vectors is called a *basis* of V if it satisfies the additional property that each vector in V can be expressed as a linear combination $a_1 \cdot \mathbf{x}_1 + \cdots + a_k \cdot \mathbf{x}_k$, where $a_1, \ldots, a_k \in F$. The linear independence of $\mathbf{x}_1, \ldots, \mathbf{x}_k$ implies that the coefficients $a_1, \ldots, a_k$ are uniquely determined. If V has a finite basis, then all of its bases contain the same number of vectors. This common number of vectors in a basis, denoted by $\dim(V)$, is called the dimension of V. The vector space F^n defined above has dimension n.

If $F = GF(s)$, then a k-dimensional vector space V over F contains s^k vectors. This is because all the vectors in V can be uniquely expressed as $a_1 \cdot \mathbf{x}_1 + \cdots + a_k \cdot \mathbf{x}_k$, where $\{\mathbf{x}_1, \ldots, \mathbf{x}_k\}$ is a basis, and there are s choices for each $a_i \in F$, $i = 1, \ldots, k$.

The inner product of two vectors $\mathbf{x} = (x_1, \ldots, x_n)^T$ and $\mathbf{y} = (y_1, \ldots, y_n)^T$ in F^n is defined as $\langle \mathbf{x}, \mathbf{y} \rangle = \sum_{i=1}^{n} x_i y_i$, which can also be expressed as $\mathbf{x}^T \mathbf{y}$. If $\langle \mathbf{x}, \mathbf{y} \rangle = 0$, then we say that $\mathbf{x}$ and $\mathbf{y}$ are orthogonal to each other. For $\mathbf{x} \in \mathbb{R}^n$, we denote the length (norm) of $\mathbf{x}$ by $\|\mathbf{x}\|$. Then for $\mathbf{x} = (x_1, \ldots, x_n)^T$, $\|\mathbf{x}\|^2 = \sum_{i=1}^n x_i^2 = \mathbf{x}^T \mathbf{x}$.

For simplicity, we also write $a \cdot b$ and $a \cdot \mathbf{x}$ as ab and $a\mathbf{x}$, respectively.

A.4 Finite Euclidean geometry

Let F be a finite field with s elements. Then the vector space F^n over F is also denoted as $\mathrm{EG}(n,s)$ and is called an n-dimensional *Euclidean geometry* based on F. When viewed in this way, each vector $\mathbf{x} = (x_1, \ldots, x_n)^T \in F^n$ is also called a point. There are a total of s^n points in $\mathrm{EG}(n,s)$.

Suppose $\mathbf{a}_1, \ldots, \mathbf{a}_k$ are k linearly independent vectors in F^n. Then $C(\mathbf{a}_1, \ldots, \mathbf{a}_k) = \{\mathbf{x} \in \mathrm{EG}(n,s) : \mathbf{a}_i^T \mathbf{x} = 0 \text{ for all } 1 \leq i \leq k\}$ is an $(n-k)$-dimensional subspace. For each $\mathbf{y} \in \mathrm{EG}(n,s)$, $C(\mathbf{a}_1, \ldots, \mathbf{a}_k) + \mathbf{y} = \{\mathbf{x} + \mathbf{y} : \mathbf{x} \in C(\mathbf{a}_1, \ldots, \mathbf{a}_k)\}$ is called an $(n-k)$-*flat* defined by $\mathbf{a}_1, \ldots, \mathbf{a}_k$. Since $C(\mathbf{a}_1, \ldots, \mathbf{a}_k)$ is an $(n-k)$-dimensional subspace, each $(n-k)$-flat contains s^{n-k} points. In particular, each 1-flat (a line) contains s points.

When $C(\mathbf{a}_1, \ldots, \mathbf{a}_k)$ is considered as a subgroup of F^n, each flat $C(\mathbf{a}_1, \ldots, \mathbf{a}_k) + \mathbf{y}$ is a coset. Therefore these flats are mutually disjoint, and we say that they are parallel. The set of all the distinct flats defined by $\mathbf{a}_1, \ldots, \mathbf{a}_k$ is denoted by $P(\mathbf{a}_1, \ldots, \mathbf{a}_k)$ and is called the *pencil* of $(n-k)$-flats defined by $\mathbf{a}_1, \ldots, \mathbf{a}_k$. Each of these $(n-k)$-flats is the solution set of the equations

$$\mathbf{a}_i^T \mathbf{x} = b_i, 1 \leq i \leq k, \tag{A.1}$$

where $b_i \in \mathrm{GF}(s)$. There are s^k parallel $(n-k)$-flats in the pencil defined by each set of k linearly independent vectors, corresponding to the s^k choices of the b_i's on the right side of (A.1). These s^k parallel $(n-k)$-flats form a partition of the points of $\mathrm{EG}(n,s)$.

A.5 Finite projective geometry

Let F be a finite field with s elements. Then there are $s^{n+1} - 1$ nonzero vectors in F^{n+1}. Two nonzero vectors $\mathbf{x} = (x_1, \ldots, x_{n+1})^T$ and $\mathbf{y} = (y_1, \ldots, y_{n+1})^T$ in F^{n+1} are said to be equivalent if there exists a nonzero $\lambda \in F$ such that $\mathbf{x} = \lambda \mathbf{y}$. An n-dimensional finite *projective geometry* based on $\mathrm{GF}(s)$, denoted $\mathrm{PG}(n,s)$, is obtained by considering each set of equivalent nonzero vectors in F^{n+1} as one point. Each point in $\mathrm{PG}(n,s)$ can be represented by any of the $s-1$ equivalent nonzero vectors. Thus there are $(s^{n+1} - 1)/(s-1)$ points in $\mathrm{PG}(n,s)$.

Let $\mathbf{a}_1, \ldots, \mathbf{a}_k$ be k linearly independent vectors in F^{n+1}. Then all the points $\mathbf{x}$ in $\mathrm{PG}(n,s)$ satisfying the equations $\mathbf{a}_i^T \mathbf{x} = 0$ for all $1 \leq i \leq k$ are said to constitute an $(n-k)$-flat. The number of points in an $(n-k)$-flat is $(s^{n-k+1} - 1)/(s-1)$. This is because there are $s^{n-k+1} - 1$ nonzero $\mathbf{x}$ that satisfy $\mathbf{a}_i^T \mathbf{x} = 0$ for all $1 \leq i \leq k$, and all the $s-1$ nonzero multiples of each $\mathbf{x}$ represent the same point in $\mathrm{PG}(n,s)$. In particular, a 1-flat (a line) contains $s+1$ points, one more point than in a Euclidean geometry.

The relation between finite Euclidean and projective geometries can be described as follows. Each point $\mathbf{x} = (x_1, \ldots, x_{n+1})^T$ in $\mathrm{PG}(n,s)$ with $x_{n+1} \neq 0$ is called a finite point. Since all nonzero multiples of $\mathbf{x}$ represent the same point, we can normalize $\mathbf{x}$ so that $x_{n+1} = 1$. Therefore there are s^n finite points in $PG(n,s)$ corresponding to the s^n choices of $x_1,\ldots,x_n$. These s^n points can be put into a one-to-one correspondence with the points in $EG(n,s)$. The points $(x_1,\ldots,x_{n+1})^T$ in $PG(n,s)$ with $x_{n+1} = 0$ are called points at infinity. These points, $(s^n - 1)/(s - 1)$ in total, are solutions to the single equation $x_{n+1} = 0$. Therefore the points at infinity form an $(n-1)$-flat, which is called the $(n-1)$-flat at infinity. If all the points in the $(n-1)$-flat at infinity are removed, then we obtain an n-dimensional Euclidean geometry. Conversely, an n-dimensional projective geometry can be obtained by adding $(s^n - 1)/(s - 1)$ points at infinity to $EG(n,s)$. Specifically, a common point at infinity is added to all the parallel lines in the same pencil of $EG(n,s)$, making each line in $PG(n,s)$ have one more point than in $EG(n,s)$. It also makes each pair of lines in a projective geometry intersect. Last, it can be seen that $EG(n,s)$ has $(s^n - 1)/(s - 1)$ pencils of 1-flats, contributing a total of $(s^n - 1)/(s - 1)$ points at infinity to $PG(n,s)$.

A.6   Orthogonal projections and orthogonal complements

For any subspace A of $\mathbb{R}^N$ and any $\mathbf{y} \in \mathbb{R}^N$, the orthogonal projection $P_A(\mathbf{y})$ of $\mathbf{y}$ onto A is the unique $\mathbf{x}$ in A such that $\|\mathbf{x} - \mathbf{y}\|$ is minimized. Then $P_A(\mathbf{y})$ is orthogonal to $\mathbf{y} - P_A(\mathbf{y})$. Furthermore, there exists a matrix $\mathbf{P}_A$ such that $P_A(\mathbf{y}) = \mathbf{P}_A\mathbf{y}$. The $N \times N$ matrix $\mathbf{P}_A$ is called the orthogonal projection matrix onto A. We have

$$\mathbf{P}_A^T = \mathbf{P}_A; \tag{A.2}$$

$$\mathbf{P}_A\mathbf{P}_A = \mathbf{P}_A; \tag{A.3}$$

$$\text{tr}(\mathbf{P}_A) = \text{rank}(\mathbf{P}_A) = \dim(A). \tag{A.4}$$

Given two subspaces A and B of $\mathbb{R}^N$ such that $B \subseteq A$, define

$$A \ominus B = \{\mathbf{x} \in A : \mathbf{x} \text{ is orthogonal to all the vectors in } B\};$$

then $A \ominus B$ is a subspace of A and is called the orthogonal complement of B relative to A. If $\dim(A) = n$ and $\dim(B) = m$, then $\dim(A \ominus B) = n - m$, and $\mathbf{P}_{A \ominus B} = \mathbf{P}_A - \mathbf{P}_B$. The orthogonal complement $\mathbb{R}^N \ominus A$ of A relative to $\mathbb{R}^N$ is denoted by $A^\perp$. Then $\dim(A^\perp) = N - \dim(A)$, and $\mathbf{P}_{A^\perp} = \mathbf{I} - \mathbf{P}_A$, where $\mathbf{I}$ is the identity matrix.

A.7   Expectation of a quadratic form

Suppose $\mathbf{y}$ is a random vector $(y_1,\ldots,y_n)^T$. Let $\boldsymbol{\mu} = E(\mathbf{y}) = (E(y_1),\ldots,E(y_n))^T$ and let $\mathbf{V} = \text{cov}(\mathbf{y})$ be the covariance matrix of $\mathbf{y}$. Then for any $n \times n$ symmetric matrix $\mathbf{A}$,

$$E\left(\mathbf{y}^T\mathbf{A}\mathbf{y}\right) = \boldsymbol{\mu}^T\mathbf{A}\boldsymbol{\mu} + \text{tr}(\mathbf{A}\mathbf{V}). \tag{A.5}$$

A.8   Balanced incomplete block designs

A block design is an assignment of treatments to experimental units arranged in blocks. Suppose there are t treatments and b blocks of size k. In general, we allow the same treatment to appear more than once in a block. If $k < t$ and each treatment appears in each block at most once, then it is called a binary incomplete block design. Such a design is said to be equireplicate if all the treatments appear the same number of times. Balanced incomplete block designs are equireplicate binary incomplete block designs in which each pair of treatments appears together in the same number of blocks.

An important method of constructing balanced incomplete block designs is to use difference sets. We provide a sample result here. Let V be an Abelian group under an operation "$+$". Suppose V contains t elements, and there are s subsets $M_1, \ldots, M_s$ of V, each containing k elements, such that among all the $sk(k-1)$ differences $a - b$, where a and b belong to the same set M_i, every nonzero element of V appears the same number of times. Then the st sets $\{M_i + v : 1 \le i \le s, \, v \in V\}$ constitute the blocks of a balanced incomplete block design with t treatments and st blocks of size k. For example, suppose t is a prime number or power of a prime number such that $t \equiv 3$ (mod 4). Let V be a GF(t) and $M = \{x \in V : \ x = y^2$ for some $y \in V\}$. The elements of M are called quadratic residues. Then it can be shown that each nonzero element of GF(t) appears the same number of times among the differences $\{a - b : a, \, b \in M, \, a \ne b\}$. Since M contains $(t-1)/2$ elements, this implies the existence of a balanced incomplete block design with t treatments and t blocks of size $(t-1)/2$ for every prime number or prime power t such that $t \equiv 3$ (mod 4). For example, for $t = 11$, GF(11) is the same as $\mathbb{Z}_{11}$. In this case, the quadratic residues are $1^2 = 1$, $2^2 = 4$, $3^2 = 9$, $4^2 = 16 \equiv 5$ (mod 11), and $5^2 = 25 \equiv 3$ (mod 11). Adding all the elements of $\mathbb{Z}_{11}$ to all the elements in the initial block $\{1, 3, 4, 5, 9\}$, we obtain a balanced incomplete block design with 11 treatments and 11 blocks of size 5: $\{1, 3, 4, 5, 9\}$, $\{2, 4, 5, 6, 10\}$, $\{3, 5, 6, 7, 0\}$, $\{4, 6, 7, 8, 1\}$, $\{5, 7, 8, 9, 2\}$, $\{6, 8, 9, 10, 3\}$, $\{7, 9, 10, 0, 4\}$, $\{8, 10, 0, 1, 5\}$, $\{9, 0, 1, 2, 6\}$, $\{10, 1, 2, 3, 7\}$, $\{0, 2, 3, 4, 8\}$.

References

Addelman, S. (1961). Irregular fractions of the 2^n factorial experiments. *Technometrics*, **3**, 479–496.

Addelman, S. (1962a). Orthogonal main-effect plans for asymmetrical factorial experiments. *Technometrics*, **4**, 21–46.

Addelman, S. (1962b). Symmetrical and asymmetrical fractional factorial plans. *Technometrics*, **4**, 47–58.

Ai, M., Kang, L., and Joseph, V. R. (2009). Bayesian optimal blocking of factorial designs. *J. Statist. Plann. Inference*, **139**, 3319–3328.

Ai, M. Y. and Zhang, R. C. (2004). Projection justification of generalized minimum aberration for asymmetrical fractional factorial designs. *Metrika*, **60**, 279–285.

Ai, M. Y., Li, P. F., and Zhang, R. R. (2005). Optimal criteria and equivalence for nonregular fractional factorial designs. *Metrika*, **62**, 73–83.

André, J. (1954). Uber nicht-Desarguessche Ebenen mit transitiver Translationsgruppe. *Math. Z.*, **60**, 156–186.

Bailey, R. A. (1977). Patterns of confounding in factorial designs. *Biometrika*, **64**, 597–603.

Bailey, R. A. (1981). A unified approach to design of experiments. *J. R. Statist. Soc. Ser. A*, **144**, 214–223.

Bailey, R. A. (1985). Factorial design and Abelian groups. *Linear Algebra Appl.*, **70**, 349–368.

Bailey, R. A. (1990). Cyclic designs and factorial designs. In R. R. Bahadur, editor, *Probability, Statistics and Design of Experiments (Proceedings of the R. C. Bose Symposium)*, pages 51–74. Wiley Eastern, New Delhi.

Bailey, R. A. (1991). Strata for randomized experiments (with discussion). *J. R. Statist. Soc. Ser. B*, **53**, 27–78.

Bailey, R. A. (1992). Efficient semi-Latin squares. *Statist. Sinica*, **2**, 413–437.

Bailey, R. A. (1996). Orthogonal partitions in designed experiments. *Des. Codes Cryptogr.*, **8**, 45–77.

Bailey, R. A. (2004). *Association Schemes: Designed Experiments, Algebra and Combinatorics*. Cambridge University Press, Cambridge, U.K.

Bailey, R. A. (2008). *Design of Comparative Experiments*. Cambridge University Press, Cambridge, U.K.

Bailey, R. A. and Brien, C. J. (2010). Data analysis for multitiered experiments using randomization-based models. I. A chain of randomizations, Preprint.

Bailey, R. A., Gilchrist, F. H. L., and Patterson, H. D. (1977). Identification of effects and confounding patterns in factorial designs. *Biometrika*, **64**, 347–354.

Bailey, R. A., Praeger, C. E., Rowley, C. A., and Speed, T. P. (1983). Generalized wreath products of permutation groups. *Proc. Lond. Math. Soc.*, **47**, 69–82.

Bailey, R. A., Cheng, C. S., and Kipnis, P. (1992). Construction of trend-resistant factorial designs. *Statist. Sinica*, **2**, 393–411.

Balakrishnan, N. and Yang, P. (2006). Connections between the resolutions of general two-level factorial designs. *Ann. Inst. Statist. Math.*, **58**, 609–618.

Balakrishnan, N. and Yang, P. (2009). De-aliasing effects using semifoldover techniques. *J. Statist. Plann. Inference*, **139**, 3102–3111.

Banerjee, T. and Mukerjee, R. (2008). Optimal factorial designs for cDNA microarray experiments. *Ann. Appl. Statist.*, **2**, 366–385.

Barnett, J., Czitrom, V., John, P. W. M., and Leon, R. V. (1997). Using fewer wafers to resolve confounding in screening experiments. In V. Czitrom and P. Q. Spagon, editors, *Statistical Case Studies for Industrial Process Improvement*, pages 235–250. SIAM, Philadelphia.

Bingham, D. R. and Sitter, R. R. (1999). Minimum-aberration two-level fractional factorial split-plot designs. *Technometrics*, **41**, 62–70.

Bingham, D. R. and Sitter, R. R. (2001). Design issues in fractional factorial split-plot experiments. *J. Qual. Technol.*, **33**, 2–15.

Bingham, D. R., Schoen, E. D., and Sitter, R. R. (2004). Designing fractional factorial split-plot experiments with few whole-plot factors. *Appl. Statist.*, **53**, 325–339.

Bingham, D. R., Sitter, R. R., Kelly, E., Moore, L., and Olivas, J. D. (2008). Factorial designs with multiple levels of randomization. *Statist. Sinica*, **18**, 493–513.

Bisgaard, S. (1994). A note on the definition of resolution for blocked 2^{k-p} designs. *Technometrics*, **36**, 308–311.

Block, R. M. (2003). *Theory and Construction Methods for Large Regular Resolution IV Designs*. Ph.D. Dissertation, Univ. Tennessee, Knoxville.

Block, R. M. and Mee, R. W. (2003). Second order saturated resolution IV designs. *J. Stat. Theory Appl.*, **2**, 96–112.

Block, R. M. and Mee, R. W. (2005). Resolution IV designs with 128 runs. *J. Qual. Technol.*, **37**, 282–293.

Booth, E. H. and Cox, D. R. (1962). Some systematic supersaturated designs. *Technometrics*, **4**, 489–495.

Bose, R. C. (1947). Mathematical theory of the symmetrical factorial design. *Sankhyā*, **8**, 107–166.

Bose, R. C. (1961). On some connections between the design of experiments and information theory. *Bull. Inst. Internat. Statist.*, **38**, 257–271.

Bose, R. C. and Bush, K. A. (1952). Orthogonal arrays of strength two and three. *Ann. Math. Statist.*, **23**, 508–524.

Bose, R. C. and Kishen, K. (1940). On the problem of confounding in the general symmetrical factorial design. *Sankhyā*, **5**, 21–36.

Bose, R. C. and Ray-Chaudhuri, D. K. (1960). On a class of error correcting binary group codes. *Inform. Control*, **3**, 68–79.

Box, G. E. P. and Bisgaard, S. (1993). What can you find out from 12 experimental runs? *Qual. Engr.*, **5**, 663–668.

Box, G. E. P. and Hunter, J. S. (1961). The 2^{k-p} fractional factorial designs Part I. *Technometrics*, **3**, 311–351.

Box, G. E. P. and Jones, S. (1992). Split-plot designs for robust product experimentation. *J. Appl. Statist.*, **19**, 3–26.

Box, G. E. P. and Meyer, R. D. (1986). An analysis for unreplicated fractional factorials. *Technometrics*, **28**, 11–18.

Box, G. E. P. and Tyssedal, J. (1996). Projective properties of certain orthogonal arrays. *Biometrika*, **83**, 950–955.

Box, G. E. P. and Wilson, K. B. (1951). On the experimental attainment of optimum conditions. *J. R. Statist. Soc. Ser. B*, **13**, 1–45.

Box, G. E. P., Hunter, J. S., and Hunter, W. G. (2005). *Statistics for Experimenters: Design, Innovation, and Discovery , Second Edition*. Wiley, New York.

Brickell, E. F. (1984). A few results in message authentication. *Congressus Numerantium*, **43**, 141–154.

Brien, C. J. and Bailey, R. A. (2006). Multiple randomizations (with discussion). *J. R. Statist. Soc. Ser. B*, **68**, 571–609.

Brien, C. J. and Bailey, R. A. (2009). Decomposition tables for experiments I. A chain of randomizations. *Ann. Statist.*, **37**, 4184–4213.

Brien, C. J. and Bailey, R. A. (2010). Decomposition tables for experiments II. Two-one randomizations. *Ann. Statist.*, **38**, 3164–3190.

Bruen, A. and Wehlau, D. L. (1999). Long binary linear codes and large caps in projective space. *Des. Codes Cryptogr.*, **17**, 37–60.

Bruen, A., Haddad, L., and Wehlau, D. L. (1998). Binary codes and caps. *J. Combin. Designs*, **6**, 275–284.

Bulutoglu, D. A. and Cheng, C. S. (2003). Hidden projection properties of some nonregular fractional factorial designs and their applications. *Ann. Statist.*, **31**, 1012–1026.

Bulutoglu, D. A. and Cheng, C. S. (2004). Construction of $E(s^2)$-optimal supersaturated designs. *Ann. Statist.*, **32**, 1662–1678.

Bulutoglu, D. A. and Margot, F. (2008). Classification of orthogonal arrays by integer programming. *J. Statist. Plann. Inference*, **138**, 654–666.

Butler, N. A. (2003a). Some theory for constructing minimum aberration fractional factorial designs. *Biometrika*, **90**, 233–238.

Butler, N. A. (2003b). Minimum aberration construction results for nonregular two-level fractional factorial designs. *Biometrika*, **90**, 891–898.

Butler, N. A. (2004). Minimum G_2-aberration properties of two-level foldover designs. *Statist. Probab. Lett.*, **67**, 121–132.

Butler, N. A. (2005). Classification of efficient two-level fractional factorial designs of resolution IV or more. *J. Statist. Plann. Inference*, **131**, 145–159.

Butler, N. A. (2007). Results for two-level fractional factorial designs of resolution IV or more. *J. Statist. Plann. Inference*, **137**, 317–323.

Butler, N. A., Mead, R., Eskridge, K. M., and Gilmour, S. G. (2001). A general method of constructing $E(s^2)$-optimal supersaturated designs. *J. R. Statist. Soc. Ser. B*, **63**, 621–632.

Caboara, M., Pistone, G., Riccomagno, E., and Wynn, H. P. (1997). The fan of an experimental design, unpublished manuscript.

Chacko, A. and Dey, A. (1981). Some orthogonal main effect plans for asymmetrical factorials. *Sankhyā, Ser. B*, **43**, 384–391.

Chakravarti, I. M. (1956). Fractional replication in asymmetrical factorial designs and partially balanced arrays. *Sankhyā*, **17**, 143–164.

Chakravarti, I. M. (1976). Optimal linear mapping of a Burnside group and its applications. *Atti dei Convegni Lincei*, **17**, 171–181.

Chen, H. (1998). Some projective properties of fractional factorial designs. *Statist. Probab. Lett.*, **40**, 185–188.

Chen, H. and Cheng, C. S. (1999). Theory of optimal blocking of 2^{n-m} designs. *Ann. Statist.*, **27**, 1948–1973.

Chen, H. and Hedayat, A. S. (1996). 2^{n-l} designs with weak minimum aberration. *Ann. Statist.*, **24**, 2536–2548.

Chen, H. and Hedayat, A. S. (1998). 2^{n-m} designs with resolution III or IV containing clear two-factor interactions. *J. Statist. Plann. Inference*, **75**, 147–158.

Chen, H. H. and Cheng, C. S. (2004). Aberration, estimation capacity and estimation index. *Statist. Sinica*, **14**, 203–215.

Chen, H. H. and Cheng, C. S. (2006). Doubling and projection: a method of constructing two-level designs of resolution IV. *Ann. Statist.*, **34**, 546–558.

Chen, H. H. and Cheng, C. S. (2009). Some results on 2^{n-m} designs of resolution IV with (weak) minimum aberration. *Ann. Statist.*, **37**, 3600–3615.

Chen, J. (1992). Some results on 2^{n-k} fractional factorial designs and search for minimum aberration designs. *Ann. Statist.*, **20**, 2124–2141.

Chen, J. and Lin, D. K. J. (1998). On the identifiability of a supersaturated design. *J. Statist. Plann. Inference*, **72**, 99–107.

Chen, J. and Wu, C. F. J. (1991). Some results on s^{n-k} fractional factorial designs with minimum aberration or optimal moments. *Ann. Statist.*, **19**, 1028–1041.

Chen, J., Sun, D. X., and Wu, C. F. J. (1993). A catalogue of two-level and three-level fractional factorial designs with small runs. *Internat. Statist. Rev.*, **61**, 131–145.

Cheng, C. S. (1995). Some projection properties of orthogonal arrays. *Ann. Statist.*, **23**, 1223–1233.

Cheng, C. S. (1997). E(s^2)-optimal supersaturated designs. *Statist. Sinica*, **7**, 929–939.

Cheng, C. S. (1998a). Hidden projection properties of orthogonal arrays with strength three. *Biometrika*, **85**, 491–495.

Cheng, C. S. (1998b). Projectivity and resolving power. *J. Comb. Inf. Syst. Sci. (special issue in honor of J. N. Srivastava)*, **23**, 47–58.

Cheng, C. S. and Jacroux, M. (1988). The construction of trend-free run orders of two-level factorial designs. *J. Amer. Statist. Assoc.*, **83**, 1152–1158.

Cheng, C. S. and Mukerjee, R. (1998). Regular fractional factorial designs with minimum aberration and maximum estimation capacity. *Ann. Statist.*, **26**, 2289–2300.

Cheng, C. S. and Mukerjee, R. (2001). Blocked regular fractional factorial designs with maximum estimation capacity. *Ann. Statist.*, **29**, 530–548.

Cheng, C. S. and Tang, B. (2001). Upper bounds on the number of columns in supersaturated designs. *Biometrika*, **88**, 1169–1174.

Cheng, C. S. and Tang, B. (2005). A general theory of minimum aberration and its applications. *Ann. Statist.*, **33**, 944–958.

Cheng, C. S. and Tsai, P. W. (2009). Optimal two-level regular fractional factorial block and split-plot designs. *Biometrika*, **96**, 83–93.

Cheng, C. S. and Tsai, P. W. (2011). Multistratum fractional factorial designs. *Statist. Sinica*, **21**, 1001–1021.

Cheng, C. S. and Tsai, P. W. (2013). Templates for design key construction. *Statist. Sinica*, **23**, 1419–1436.

Cheng, C. S., Steinberg, D. M., and Sun, D. X. (1999). Minimum aberration and model robustness for two-level fractional factorial designs. *J. R. Statist. Soc. Ser. B*, **61**, 85–93.

Cheng, C. S., Deng, L. Y., and Tang, B. (2002). Generalized minimum aberration and design efficiency for nonregular fractional factorial designs. *Statist. Sinica*, **12**, 991–1000.

Cheng, C. S., Mee, R. W., and Yee, O. (2008). Second order saturated orthogonal arrays of strength three. *Statist. Sinica*, **18**, 105–119.

Cheng, S. W. and Wu, C. F. J. (2001). Factor screening and response surface exploration (with discussions). *Statist. Sinica*, **11**, 553–604.

Cheng, S. W. and Wu, C. F. J. (2002). Choice of optimal blocking schemes in two-level and three-level designs. *Technometrics*, **44**, 269–277.

Cheng, S. W. and Ye, K. Q. (2004). Geometric isomorphism and minimum aberration for factorial designs with quantitative factors. *Ann. Statist.*, **32**, 2168–2185.

Cheng, Y. and Zhang, R. (2010). On construction of general minimum lower order confounding 2^{n-m} designs with $N/4 + 1 \leq n \leq 9N/32$. *J. Statist. Plann. Inference*, **140**, 2384–2394.

Chipman, H. (2006). Prior distributions for Bayesian analysis of screening experiments. In A. M. Dean and S. Lewis, editors, *Screening: Methods for Experimentation in Industry, Drug Discovery, and Genetics*, pages 235–267. Springer, New York.

Connor, W. S. and Young, S. (1959). *Fractional Factorial Designs for Experiments with Factors at Two and Three levels*. Applied Mathematics Series, National Bureau of Standards, U.S. Government Printing Office, Washington, DC.

Constantine, G. M. (1987). *Combinatorial Theory and Statistical Design*. Wiley, New York.

Cotter, S. C. (1974). A general method of confounding for symmetrical factorial experiments. *J. R. Statist. Soc. Ser. B*, **36**, 267–276.

Cox, D. R. (1958). *Planning of Experiments*. Wiley, New York.

Daniel, C. (1959). Use of half-normal plots in interpreting factorial two-level experiments. *Technometrics*, **1**, 311–341.

Daniel, C. (1962). Sequences of fractional replicates in the 2^{p-q} series. *J. Amer. Statist. Assoc.*, **57**, 403–429.

Daniel, C. (1976). *Applications of Statistics to Industrial Experimentation*. Wiley, New York.

Das, M. N. (1964). A somewhat alternative approach for construction of symmetrical factorial designs and obtaining the maximum number of factors. *Calcutta Statist. Ass. Bull.*, **13**, 1–17.

Davydov, A. A. and Tombak, L. M. (1990). Quasiperfect linear binary codes with distance 4 and complete caps in projective geometry. *Problems of Inform. Transmission*, **25**, 265–275.

Dean, A. M. (1990). Designing factorial experiments: a survey of the use of generalized cyclic designs. In S. Ghosh, editor, *Statistical Design and Analysis of Industrial Experiments*, pages 479–516. Marcel Dekker, New York.

Dean, A. M. and John, J. A. (1975). Single replicate factorial experiments in generalized cyclic designs: II. Asymmetrical arrangements. *J. R. Statist. Soc. Ser. B*, **37**, 72–76.

Dean, A. M. and Lewis, S. (2006). *Screening: Methods for Experimentation in Industry, Drug Discovery, and Genetics*. Springer, New York.

Delsarte, P. (1973). An algebraic approach to the association schemes of coding theory. *Philips Res. Reports Suppl.*, **10**.

Deng, L., Lin, D., and Wang, J. (1999). A resolution rank criterion for supersaturated designs. *Statist. Sinica*, **9**, 605–610.

Deng, L. Y. and Tang, B. (1999). Generalized resolution and minimum aberration criteria for Plackett-Burman and other nonregular factorial designs. *Statist. Sinica*, **9**, 1071–1082.

Dey, A. (2005). Projection properties of some orthogonal arrays. *Statist. Probab. Lett.*, **75**, 298–306.

Dey, A. and Midha, C. K. (1996). Construction of some asymmetrical orthogonal arrays. *Statist. Probab. Lett.*, **28**, 211–217.

Dey, A. and Mukerjee, R. (1999). *Fractional Factorial Plans*. Wiley, New York.

Draper, N. R. and Mitchell, T. J. (1967). The construction of saturated 2_R^{k-p} designs. *Ann. Math. Statist.*, **38**, 1110–1126.

Edwards, D. J. (2011). Optimal semifoldover plans for two-level orthogonal designs. *Technometrics*, **53**, 274–284.

Ehrenfeld, S. (1956). Complete class theorems in experimental design. In *Proc. Third Berkeley Symp. on Math. Statist. and Prob.*, volume 1, pages 57–67. Univ. of Calif. Press.

Eskridge, K. M., Gilmour, S. G., Mead, R., Butler, N. A., and Travnicek, D. A. (2004). Large supersaturated designs. *J. Stat. Comput. Simul.*, **74**, 525–542.

Evangelaras, H., Koukouvinos, C., Dean, A. M., and Dingus, C. A. (2005). Projection properties of certain three level orthogonal arrays. *Metrika*, **62**, 241–257.

Fang, K. T. (1980). The uniform design: Application of number-theoretic methods in experimental design. *Acta Mathematicae Applagatae Sinica*, **3**, 363–372.

Fang, K. T. and Mukerjee, R. (2000). A connection between uniformity and aberration in regular fractions of two-level factorials. *Biometrika*, **87**, 193–198.

Fang, K. T., Lin, D. K. J., Winker, P., and Zhang, Y. (2000). Uniform design: theory and application. *Technometrics*, **42**, 237–248.

Fang, K. T., Li, R., and Sudjianto, A. (2006). *Design and Modeling for Computer Experiments*. Chapman & Hall/CRC, New York.

Federer, W. T. and King, F. (2006). *Variations on Split Plot and Split Block Experiment Designs*. Wiley, New York.

Finney, D. J. (1945). The fractional replication of factorial experiments. *Ann. Eugen.*, **12**, 291–301.

Fisher, R. A. (1926). The arrangement of field experiments. *J. Ministry Agri. G. Br.*, **33**, 503–513.

Fisher, R. A. (1942). The theory of confounding in factorial experiments in relation to the theory of groups. *Ann. Eugen.*, **11**, 341–353.

Fisher, R. A. (1945). A system of confounding for factors with more than two alternatives giving completely orthogonal cubes and higher powers. *Ann. Eugen.*, **12**, 283–290.

Fontana, R., Pistone, G., and Rogantin, M. P. (2000). Classication of two-level factorial fractions. *J. Statist. Plann. Inference*, **87**, 149–172.

Franklin, M. F. (1984). Constructing tables of minimum aberration p^{n-m} designs. *Technometrics*, **26**, 225–232.

Franklin, M. F. (1985). Selecting defining contrasts and confounded effects in p^{n-m} factorial experiments. *Technometrics*, **27**, 165–172.

Franklin, M. F. and Bailey, R. A. (1977). Selection of defining contrasts and confounded effects in two-level experiments. *Appl. Statist.*, **26**, 321–326.

Fries, A. and Hunter, W. G. (1980). Minimum aberration 2^{k-p} designs. *Technometrics*, **22**, 601–608.

Fujii, Y., Namikawa, T., and Yamamoto, S. (1989). Classification of two-symbol orthogonal arrays of strength t, $t+3$ constraints and index 4. II. *SUT J. Math.*, **25**, 161–177.

Ghosh, S. (1979). On single and multistage factor screening procedures. *J. Comb. Inf. Syst. Sci.*, **4**, 275–284.

Ghosh, S. and Avila, D. (1985). Some new factor screening designs using the search linear model. *J. Statist. Plann. Inference*, **11**, 259–266.

Gilmour, S. G. (2006). Factor screening via supersaturated designs. In A. M. Dean and S. Lewis, editors, *Screening: Methods for Experimentation in Industry, Drug Discovery, and Genetics*, pages 169–190. Springer, New York.

Govaerts, P. (2005). Small maximal partial t-spreads. *Bull. Belg. Math. Soc. Simon Stevin*, **12**, 607–615.

Greenfield, A. A. (1976). Selection of defining contrasts in two-level experiments. *Appl. Statist.*, **25**, 64–67.

Großmann, H. (2014). Automating the analysis of variance of orthogonal designs. *Comput. Stat. Data Anal.*, **to appear**.

Grundy, P. M. and Healy, M. J. R. (1950). Restricted randomization and quasi-Latin squares. *J. R. Statist. Soc. Ser. B*, **12**, 286–291.

Hall, M. (1986). *Combinatorial Theory, 2nd edition*. Wiley, New York.

Hamada, M. and Wu, C. F. J. (1992). Analysis of designed experiments with complex aliasing. *J. Qual. Technol.*, **24**, 130–137.

He, Y. and Tang, B. (2013). Strong orthogonal arrays and associated Latin hypercubes for computer experiments. *Biometrika*, **100**, 254–260.

Hedayat, A. S., Pu, K., and Stufken, J. (1992). On the construction of asymmetrical orthogonal arrays. *Ann. Statist.*, **20**, 2142–2152.

Hedayat, A. S., Stufken, J., and Su, G. (1996). On difference schemes and orthogonal arrays of strength t. *J. Statist. Plann. Inference*, **56**, 307–324.

Hedayat, A. S., Seiden, E., and Stufken, J. (1997). On the maximal number of factors and the enumeration of 3-symbol orthogonal arrays of strength 3 and index 2. *J. Statist. Plann. Inference*, **58**, 43–63.

Hedayat, A. S., Sloane, N. J. A., and Stufken, J. (1999). *Orthogonal Arrays: Theory and Applications*. Springer-Verlag, New York.

Hickernell, F. J. (1998). A generalized discrepancy and quadrature error bound. *Math. Comp.*, **67**, 299–322.

Hinkelmann, K. and Kempthorne, O. (2005). *Design and Analysis of Experiments, Vol. 2*. Wiley, New York.

Hocquenghem, A. (1959). Codes correcteurs d'erreurs. *Chiffres*, **2**, 147–156.

Hooper, P. M. (1989). Experimental randomization and the validity of normal-theory inference. *J. Amer. Statist. Assoc.*, **84**, 576–586.

Huang, P., Chen, D., and Voelkel, J. O. (1998). Minimum aberration two-level split-plot designs. *Technometrics*, **40**, 314–326.

Jacroux, M. (2004). A modified minimum aberration criterion for selecting regular 2^{m-k} fractional factorial designs. *J. Statist. Plann. Inference*, **126**, 325–336.

John, J. A. and Dean, A. M. (1975). Single replicate factorial experiments in generalized cyclic designs: I. Symmetrical arrangements. *J. R. Statist. Soc. Ser. B*, **37**, 63–71.

John, P. W. M. (1962). Three quarter replicates of 2^n designs. *Biometrics*, **18**, 172–184.

John, P. W. M. (1971). *Statistical Design and Analysis of Experiments*. Macmillan, New York.

Johnson, M. E., Moore, L. M., and Ylvisaker, D. (1990). Minimax and maximin distance designs. *J. Statist. Plann. Inference*, **26**, 131–148.

Joseph, V. R. (2006). A Bayesian approach to the design and analysis of fractionated experiments. *Technometrics*, **48**, 219–229.

Joseph, V. R. and Delaney, J. D. (2007). Functionally induced priors for the analysis of experiments. *Technometrics*, **49**, 1–11.

Joseph, V. R., Ai, M., and Wu, C. F. J. (2009). Bayesian-inspired minimum aberration two- and four-level designs. *Biometrika*, **96**, 95–106.

Kang, L. and Joseph, V. R. (2009). Bayesian optimal single arrays for robust parameter design. *Technometrics*, **51**, 250–261.

Katsaounis, T. I. and Dean, A. M. (2008). A survey and evaluation of methods for determination of combinatorial equivalence of factorial designs. *J. Statist. Plann. Inference*, **138**, 245–258.

Ke, W. and Tang, B. (2003). Selecting 2^{m-p} designs using a minimum aberration criterion when some two-factor interactions are important. *Technometrics*, **45**, 352–360.

Kempthorne, O. (1947). A simple approach to confounding and fractional replication in factorial experiments. *Biometrika*, **34**, 255–272.

Kerr, K. F. (2006). Efficient 2^k factorial designs for blocks of size 2 with microarray applications. *J. Qual. Technol.*, **38**, 309–318.

Kerr, M. K. (2001). Bayesian optimal fractional factorials. *Statist. Sinica*, **11**, 605–630.

Khatirinejad, M. and Lisoněk, P. (2006). Classification and constructions of complete caps in binary spaces. *Des. Codes Cryptogr.*, **39**, 17–31.

Kiefer, J. (1958). On the nonrandomized optimality and randomized nonoptimality of symmetrical designs. *Ann. Math. Statist.*, **29**, 675–699.

Kiefer, J. (1975). Construction and optimality of generalized Youden designs. In J. N. Srivastava, editor, *A Survey of Statistical Design and Linear Models*, pages 333–353. North Holland, Amsterdam.

Kobilinsky, A. (1985). Confounding in relation to duality of finite Abelian groups. *Linear Algebra Appl.*, **70**, 321–347.

Kurkjian, B. and Zelen, M. (1962). A calculus for factorial arrangements. *Ann. Math. Statist.*, **33**, 600–619.

Laycock, P. J. and Rowley, P. J. (1995). A method for generating and labelling all regular fractions or blocks for q^{n-m}-designs. *J. R. Statist. Soc, Ser. B*, **57**, 191–204.

Lenth, R. V. (1989). Quick and easy analysis of unpreplicated factorials. *Technometrics*, **31**, 469–473.

Li, H. and Mee, R. W. (2002). Better foldover fractions for resolution III 2^{k-p} designs. *Technometrics*, **44**, 278–283.

Li, W. and Lin, D. K. J. (2003). Optimal foldover plans for two-level fractional factorial designs. *Technometrics*, **45**, 142–149.

Liao, C. T. (1999). Orthogonal three-level parallel flats designs for user-specified resolution. *Comm. Statist. Theory Methods*, **28**, 1945–1960.

Liao, C. T. and Chai, F. S. (2009). Design and analysis of two-level factorial experiments with partial replication. *Technometrics*, **51**, 66–74.

Liao, C. T. and Iyer, H. K. (1998). Orthogonal parallel-flats designs for $s_1^{n_1} \times s_2^{n_2}$ mixed factorial experiments. *J. Statist. Plann. Inference*, **69**, 175–191.

Liao, C. T., Iyer, H. K., and Vecchia, D. F. (1996). Construction of orthogonal two-level designs of user-specified resolution where $N \neq 2^k$. *Technometrics*, **38**, 342–353.

Liau, P. H. (2006). The existence of the strong combined-optimal design. *Statist. Papers*, **48**, 143–150.

Lin, C. D., Miller, A., and Sitter, R. R. (2008). Folded over non-orthogonal designs. *J. Statist. Plann. Inference*, **138**, 3107–3124.

Lin, D. K. J. (1993). A new class of supersaturated designs. *Technometrics*, **35**, 28–31.

Lin, D. K. J. and Draper, N. R. (1991). Projection properties of Plackett and Burman designs. Technical report, Department of Statistics, University of Wisconsin, Madison.

Lin, D. K. J. and Draper, N. R. (1992). Projection properties of Plackett and Burman designs. *Technometrics*, **34**, 423–428.

Lin, D. K. J. and Draper, N. R. (1993). Generating alias relationships for two-level Plackett and Burman designs. *Comput. Statist. Data Anal.*, **15**, 147–157.

Liu, M. and Zhang, R. (2000). Construction of $E(s^2)$-optimal supersaturated designs using cyclic BIBDs. *J. Statist. Plann. Inference*, **91**, 139–150.

Liu, M. Q. and Hickernell, F. J. (2002). $E(s^2)$-optimality and minimum discrepancy in 2-level supersaturated designs. *Statist. Sinica*, **12**, 931–939.

Liu, Y. and Dean, A. M. (2004). k-circulant supersaturated designs. *Technometrics*, **46**, 32–43.

Loeppky, J. L., Sitter, R. R., and Tang, B. (2007). Nonregular designs with desirable projection properties. *Technometrics*, **49**, 454–467.

Loughin, T. M. (2005). Seven error terms!! are you kidding??. Invited talk at the 2005 Joint Statistical Meetings, Minneapolis.

Ma, C. X. and Fang, K. T. (2001). A note on generalized aberration in factorial designs. *Metrika*, **53**, 85–93.

MacWilliams, F. J. (1963). A theorem on the distribution of weights in a systematic code. *Bell Systems Tech. Journal*, **42**, 79–94.

MacWilliams, F. J. and Sloane, N. J. A. (1977). *The Theory of Error-Correcting Codes*. North-Holland, Amsterdam.

Mandal, A. and Mukerjee, R. (2005). Design efficiency under model uncertainty for nonregular fractions of general factorials. *Statist. Sinica*, **15**, 697–707.

Margolin, B. H. (1969). Resolution IV fractional factorial designs. *J. R. Statist. Soc. Ser. B*, **31**, 514–523.

Marshall, A. W. and Olkin, I. (1979). *Inequalities: Theory of Majorization and Its Applications*. Academic Press, New York.

Masuyama, M. (1957). On difference sets for constructing orthogonal arrays of index two and of strength two. *Rep. Statist. Appl. Res. Un. Japan Sci. Engrs.*, **5**, 27–34.

McKay, M., Beckman, R. J., and Conover, W. J. (1979). A comparison of three methods of selecting values of input variables in the analysis of output from a computer code. *Technometrics*, **21**, 239–245.

McLeod, R. G. and Brewster, J. F. (2004). The design of blocked fractional factorial split-plot experiments. *Technometrics*, **46**, 135–146.

Mee, R. W. (2009). *A Comprehensive Guide to Factorial Two-Level Experimentation*. Springer, New York.

Mee, R. W. and Bates, R. L. (1998). Split-lot designs: experiments for multistage batch processes. *Technometrics*, **40**, 127–140.

Mee, R. W. and Peralta, M. (2000). Semifolding 2^{k-p} designs. *Technometrics*, **42**, 122–134.

Meyer, R. D., Steinberg, D. M., and Box, G. E. P. (1996). Follow-up designs to resolve confounding in multifactor experiments. *Technometrics*, **38**, 303–313.

Miller, A. (1997). Strip-plot configurations of fractional factorials. *Technometrics*, **39**, 153–161.

Miller, A. and Sitter, R. R. (2004). Choosing columns from the 12-run Plackett-Burman design. *Statist. Probab. Lett.*, **67**, 193–201.

Miller, A. and Tang, B. (2012). Minimal dependent sets for evaluating supersaturated designs. *Statist. Sinica*, **22**, 1273–1285.

Mitchell, T. J., Morris, M. D., and Ylvisaker, D. (1995). Two-level fractional factorials and Bayesian prediction. *Statist. Sinica*, **15**, 559–573.

Montgomery, D. C. and Runger, G. C. (1996). Foldovers of 2^{k-p} resolution IV experimental designs. *J. Qual. Technol.*, **28**, 446–450.

Morris, M. D. (2006). An overview of group factor screening. In A. M. Dean and S. Lewis, editors, *Screening: Methods for Experimentation in Industry, Drug Discovery, and Genetics*, pages 191–206. Springer, New York.

Mukerjee, R. and Fang, K. T. (2002). Fractional factorial split-plot designs with minimum aberration and maximum estimation capacity. *Statist. Sinica*, **12**, 885–903.

Mukerjee, R. and Wu, C. F. J. (1995). On the existence of saturated and nearly saturated asymmetrical orthogonal arrays. *Ann. Statist.*, **23**, 2102–2115.

Mukerjee, R. and Wu, C. F. J. (1999). Blocking in regular fractional factorials: a projective geometric approach. *Ann. Statist.*, **27**, 1256–1271.

Mukerjee, R. and Wu, C. F. J. (2006). *A Modern Theory of Factorial Designs*. Springer, New York.

Nelder, J. A. (1965a). The analysis of randomized experiments with orthogonal block structure. I. Block structure and the null analysis of variance. *Proc. R. Soc. Lond. Ser. A*, **283**, 147–162.

Nelder, J. A. (1965b). The analysis of randomized experiments with orthogonal block structure. II. Treatment structure and the general analysis of variance. *Proc. R. Soc. Lond. Ser. A*, **283**, 163–178.

Nguyen, N. K. (1996). An algorithmic approach to constructing supersaturated designs. *Technometrics*, **38**, 69–73.

Nguyen, N. K. and Cheng, C. S. (2008). New $E(s^2)$-optimal supersaturated designs constructed from incomplete block designs. *Technometrics*, **50**, 26–31.

Niederreiter, H. (1987). Point sets and sequences with small discrepancy. *Monatsh. Math.*, **104**, 273–337.

Nordstrom, A. W. and Robinson, J. P. (1967). An optimum nonlinear code. *Info. Contr.*, **11**, 613–616.

Owen, A. B. (1992). Orthogonal arrays for computer experiments, integration and visualization. *Statist. Sinica*, **2**, 439–452.

Owen, A. B. (1995). Randomly permuted (t,m,s)-nets and (t,s)-sequences. In H. Niederreiter and P. J. Shiue, editors, *Monte Carlo and Quasi-Monte Carlo Methods in Scientific Computing*, pages 299–317. Springer, New York.

Paley, R. E. A. C. (1933). On orthogonal matrices. *J. Math. Phys.*, **12**, 311–320.

Patterson, H. D. (1965). The factorial combination of treatments in rotation experiments. *J. Agric. Sci.*, **65**, 171–182.

Patterson, H. D. (1976). Generation of factorial designs. *J. R. Statist. Soc, Ser. B*, **38**, 175–179.

Patterson, H. D. and Bailey, R. A. (1978). Design keys for factorial experiments. *Appl. Statist.*, **27**, 335–343.

Phoa, F. K. H. (2012). A code arithmetic approach for quaternary code designs and its application to (1/64)th-fractions. *Ann. Statist.*, **40**, 3161–3175.

Phoa, F. K. H. and Xu, H. (2009). Quarter-fraction factorial designs constructed via quaternary codes. *Ann. Statist.*, **37**, 2561–2581.

Pistone, G. and Wynn, H. P. (1996). Generalised confounding with Gröbner bases. *Biometrika*, **83**, 653–666.

Pistone, G., Riccomagno, E., and Wynn, H. P. (2001). *Algebraic Statistics: Computational Commutative Algebra in Statistics*. Chapman & Hall/CRC.

Plackett, R. L. and Burman, J. P. (1946). The design of optimum multifactorial experiments. *Biometrika*, **33**, 305–325.

Pless, V. (1963). Power moment identities on weight distributions in error correcting codes. *Inform. Control*, **6**, 147–152.

Preece, D. A., Bailey, R. A., and Patterson, H. D. (1978). A randomization problem in forming designs with superimposed treatments. *Austral. J. Statist.*, **20**, 111–125.

Pu, K. (1989). *Contributions to Fractional Factorial Designs*. Ph.D. Dissertation, University of Illinois at Chicago.

Qian, P. Z. G. and Wu, C. F. J. (2009). Sliced space-filling designs. *Biometrika*, **96**, 945–956.

Qian, P. Z. G., Ai, M., and Wu, C. F. J. (2009). Construction of nested space-filling designs. *Ann. Statist.*, **37**, 3616–3643.

Qian, P. Z. G., Tang, B., and Wu, C. F. J. (2009). Nested space-filling designs for computer experiments with two levels of accuracy. *Statist. Sinica*, **19**, 287–300.

Qu, X. (2007). Statistical properties of Rechtschaffner designs. *J. Statist. Plann. Inference*, **137**, 2156–2164.

Ranjan, P., Bingham, D. R., and Dean, A. M. (2009). Existence and construction of randomization defining contrast subspaces for regular factorial designs. *Ann. Statist.*, **37**, 3580–3599.

Ranjan, P., Bingham, D., and Mukerjee, R. (2010). Stars and regular fractional factorial designs with randomization restrictions. *Statist. Sinica*, **20**, 1637–1653.

Rao, C. R. (1946). Hypercubes of strength d leading to confounded designs in factorial experiments. *Bull. Calcutta Math. Soc.*, **38**, 67–78.

Rao, C. R. (1947). Factorial experiments derivable from combinatorial arrangements of arrays. *J. R. Statist. Soc. Suppl.*, **9**, 128–139.

Rao, C. R. (1973). Some combinatorial problems of arrays and applications to design of experiments. In J. N. Srivastava, editor, *A Survey of Combinatorial Theory*, pages 349–359. North-Holland, Amsterdam.

Rechtschaffner, R. L. (1967). Saturated fractions of 2^n and 3^n factorial design. *Technometrics*, **9**, 569–575.

Rosenbaum, P. R. (1994). Dispersion effects from fractional factorials in Taguchi's method of quality design. *J. R. Statist. Soc., Ser. B*, **56**, 641–652.

Ryan, K. J. and Bulutoglu, D. A. (2010). Minimum aberration fractional factorial designs with large N. *Technometrics*, **52**, 250–255.

Sacks, J., Welch, W. J., Mitchell, T. J., and Wynn, H. P. (1989). Design and analysis of computer experiments (with comments). *Statist. Sci.*, **4**, 409–435.

Santner, T. J., Williams, B. J., and Notz, W. I. (2003). *The Design and Analysis of Computer Experiments*. Springer, New York.

Schoen, E. D. (1999). Designing fractional two-level experiments with nested error structures. *J. Appl. Statist.*, **26**, 495–508.

Schoen, E. D., Eendebak, P. T., and Nguyen, M. V. M. (2010). Complete enumeration of pure-level and mixed-level orthogonal arrays. *J. Combi. Des.*, **18**, 123–140.

Seiden, E. (1954). On the problem of construction of orthogonal arrays. *Ann. Math. Statist.*, **25**, 151–156.

Seiden, E. and Zemach, R. (1966). On orthogonal arrays. *Ann. Math. Statist.*, **37**, 1355–1370.

Shirakura, T., Takahashi, T., and Srivastava, J. N. (1996). Searching probabilities for nonzero effects in search designs for the noisy case. *Ann. Statist.*, **24**, 2560–2568.

Sitter, R. R. (1993). Balanced repeated replications based on orthogonal multiarrays. *Biometrika*, **80**, 211–221.

Sitter, R. R., Chen, J., and Feder, M. (1997). Fractional resolution and minimum aberration in blocked 2^{n-k} designs. *Technometrics*, **39**, 382–390.

Sloane, N. J. A. and Stufken, J. (1996). A linear programming bound for orthogonal arrays with mixed levels. *J. Statist. Plann. Inference*, **56**, 295–305.

Speed, T. P. and Bailey, R. A. (1982). On a class of association schemes derived from lattices of equivalence relations. In P. Schultz, C. E. Praeger, and R. P. Sullivan, editors, *Algebraic Structures and Applications*, pages 55–74. Marcel Dekker, New York.

Srivastava, J., Anderson, D., and Mardekian, J. (1984). Theory of factorial designs of the parallel flats type. 1: The coefficient matrix. *J. Statist. Plann. Inference*, **9**, 229–252.

Srivastava, J. N. (1975). Designs for searching non-negligible effects. In J. N. Srivastava, editor, *A Survey of Statistical Design and Linear Models*, pages 507–519. North-Holland, Amsterdam.

Srivastava, J. N. (1987). Advances in the general theory of factorial designs based on parallel pencils in Euclidean n-space. *Util. Math.*, **32**, 75–94.

Srivastava, J. N. and Arora, S. (1987). A minimal search design of resolution III.2 for the 2^4 factorial experiment. *Indian J. Math.*, **29**, 309–320.

Srivastava, J. N. and Chopra, D. V. (1971). On the characteristic roots of the information matrix of 2^m balanced factorial designs of resolution V, with applications. *Ann. Math. Statist.*, **42**, 722–734.

Srivastava, J. N. and Chopra, D. V. (1973). Balanced arrays and orthogonal arrays. In J. N. Srivastava, editor, *A Survey of Combinatorial Theory*, pages 411–428. North-Holland, Amsterdam.

Srivastava, J. N. and Li, J. (1996). Orthogonal designs of parallel flats type. *J. Statist. Plann. Inference*, **53**, 261–283.

Srivastava, J. N. and Throop, D. (1990). Orthogonal arrays obtainable as solutions to linear equations over finite fields. *Linear Algebra Appl.*, **127**, 283–300.

Stein, M. L. (1987). Large sample properties of simulations using Latin hypercube sampling. *Technometrics*, **29**, 143–151.

Stufken, J. and Tang, B. (2007). Complete enumeration of two-level orthogonal arrays of strength d with $d+2$ constraints. *Ann. Statist*, **35**, 793–814.

Suen, C. y., Chen, H., and Wu, C. F. J. (1997). Some identities on q^{n-m} designs with application to minimum aberration designs. *Ann. Statist.*, **25**, 1176–1188.

Sun, D. X. (1993). *Estimation Capacity and Related Topics in Experimental Designs*. Ph.D. Dissertation, University of Waterloo.

Sun, D. X., Wu, C. F. J., and Chen, Y. (1997). Optimal blocking schemes for 2^n and 2^{n-p} designs. *Technometrics*, **39**, 298–307.

Taguchi, G. (1987). *System of Experimental Design: Engineering Methods to Optimize Quality and Minimize Costs*. UNIPUB/Kraus International Publications, White Plains, New York and Dearborn, Michigan.

Tang, B. (1993). Orthogonal array-based Latin hypercubes. *J. Amer. Statist. Assoc.*, **88**, 1392–1397.

Tang, B. (2001). Theory of J-characteristics for fractional factorial designs and projection justification of minimum G_2-aberration. *Biometrika*, **88**, 401–407.

Tang, B. (2006). Orthogonal arrays robust to nonnegligible two-factor interactions. *Biometrika*, **93**, 137–146.

Tang, B. (2007). Construction results on minimum aberration blocking schemes for 2^m designs. *J. Statist. Plann. Inference*, **137**, 2355–2361.

Tang, B. and Deng, L. Y. (1999). Minimum G_2-aberration for nonregular fractional factorial designs. *Ann. Statist.*, **27**, 1914–1926.

Tang, B. and Wu, C. F. J. (1996). Characterization of minimum aberration 2^{n-k} designs in terms of their complementary designs. *Ann. Statist.*, **24**, 2549–2559.

Tang, B. and Wu, C. F. J. (1997). A method for constructing supersaturated designs and its $E(s^2)$-optimality. *Canad. J. Statist.*, **25**, 191–201.

Tang, B., Ma, F., Ingram, D., and Wang, H. (2002). Bounds on the maximum number of clear two factor interactions for 2^{m-p} designs of resolution III and IV. *Can. J. Statist.*, **30**, 127–136.

Throckmorton, T. N. (1961). *Structures of Classification Data*. Ph.D. Dissertation, Iowa State University.

Tjur, T. (1984). Analysis of variance models in orthogonal designs. *Int. Statist. Rev.*, **52**, 33–81.

Tobias, R. D. (1996). Saturated second-order two-level designs: an empirical approach. Technical report, SAS Institute.

Tsai, P. W. and Gilmour, S. G. (2010). A general criterion for factorial designs under model uncertainty. *Technometrics*, **52**, 231–242.

Tsai, P. W., Gilmour, S. G., and Mead, R. (2000). Projective three-level main effects designs robust to model uncertainty. *Biometrika*, **87**, 467–475.

Tsai, P. W., Gilmour, S. G., and Mead, R. (2007). Three-level main-effects designs exploiting prior information about model uncertainty. *J. Statist. Plann. Inference*, **137**, 619–627.

Vivacqua, C. A. and Bisgaard, S. (2009). Post-fractionated strip-block designs. *Technometrics*, **51**, 47–55.

Wang, J. C. and Wu, C. F. J. (1991). An approach to the construction of asymmetrical orthogonal arrays. *J. Amer. Statist. Assoc.*, **86**, 450–456.

Wang, J. C. and Wu, C. F. J. (1995). A hidden projection property of Plackett-Burman and related designs. *Statist. Sinica*, **15**, 235–250.

Webb, S. R. (1968). Non-orthogonal designs of even resolution. *Technometrics*, **10**, 291–299.

Wu, C. F. J. (1989). Construction of $2^m 4^n$ designs via a grouping scheme. *Ann. Statist.*, **17**, 1880–1885.

Wu, C. F. J. (1993). Construction of supersaturated designs through partially aliased interactions. *Biometrika*, **80**, 661–669.

Wu, C. F. J. and Chen, Y. (1992). A graph-aided method for planning two-level experiments when certain interactions are important. *Technometrics*, **34**, 162–175.

Wu, C. F. J. and Hamada, M. (2009). *Experiments: Planning, Analysis, and Optimization, 2nd edition*. Wiley, New York.

Wu, C. F. J. and Zhang, R. (1993). Minimum aberration designs with two-level and four-level factors. *Biometrika*, **80**, 203–209.

Wu, C. F. J., Zhang, R., and Wang, R. (1992). Construction of asymmetrical orthogonal arrays of the type $OA(s^k, s^m(s^{r_1})^{n_1} \cdots (s^{r_t})^{n_t})$. *Statist. Sinica*, **2**, 203–219.

Wu, H. and Wu, C. F. J. (2002). Clear two-factor interactions and minimum aberration. *Ann. Statist.*, **30**, 1496–1511.

Xu, H. (2003). Minimum moment aberration for nonregular designs and supersaturated designs. *Statist. Sinica*, **13**, 691–708.

Xu, H. (2005a). A catalogue of three-level regular fractional factorial designs. *Metrika*, **62**, 259–281.

Xu, H. (2005b). Some nonregular designs from the Nordstrom-Robinson code and their statistical properties. *Biometrika*, **92**, 385–397.

Xu, H. (2006). Blocked regular fractional factorial designs with minimum aberration. *Ann. Statist.*, **34**, 2534–2553.

Xu, H. (2009). Algorithmic construction of efficient fractional factorial designs with large run sizes. *Technometrics*, **51**, 262–277.

Xu, H. and Cheng, C. S. (2008). A complementary design theory for doubling. *Ann. Statist.*, **36**, 445–457.

Xu, H. and Lau, S. (2006). Minimum aberration blocking schemes for two- and three-level fractional factorial designs. *J. Statist. Plann. Inference*, **136**, 4088–4118.

Xu, H. and Wong, A. (2007). Two-level nonregular designs from quaternary linear codes. *Statist. Sinica*, **17**, 1191–1213.

Xu, H. and Wu, C. F. J. (2001). Generalized minimum aberration for asymmetrical fractional factorial designs. *Ann. Statist.*, **29**, 1066–1077.

Xu, H. and Wu, C. F. J. (2005). Construction of optimal multi-level supersaturated designs. *Ann. Statist.*, **33**, 2811–2836.

Xu, H., Cheng, S. W., and Wu, C. F. J. (2004). Optimal projective three-level designs for factor screening and interaction detection. *Technometrics*, **46**, 280–292.

Xu, H., Phoa, F. K. H., and Wong, W. K. (2009). Recent developments in nonregular fractional factorial designs. *Stat. Surv.*, **3**, 18–46.

Yang, Y. H. and Speed, T. P. (2002). Design issues for cDNA microarray experiments. *Nat. Rev. Genet.*, **3**, 579–588.

Yates, F. (1935). Complex experiments. *J. R. Statist. Soc. Suppl.*, **2**, 181–247.

Yates, F. (1937). *The Design and Analysis of Factorial Experiments.* Technical Communication, 35, Imperial Bureau of Soil Science, Harpenden.

Yates, F. (1940). The recovery of interblock information in balanced incomplete block designs. *Ann. Eugen.*, **10**, 317–325.

Ye, K. Q. (2003). Indicator function and its application in two-level factorial designs. *Ann. Statist.*, **31**, 984–994.

Ye, K. Q. (2004). A note on regular fractional factorial designs. *Statist. Sinica*, **14**, 1069–1074.

Ye, Q. and Hamada, M. S. (2000). Critical values of the Lenth method for unreplicated factorial designs. *J. Qual. Technol.*, **32**, 57–66.

Zhang, A., Fang, K. T., Li, R., and Sudjianto, A. (2005). Majorization framework for balanced lattice designs. *Ann. Statist.*, **33**, 2837–2853.

Zhang, R. and Park, D. K. (2000). Optimal blocking of two-level fractional factorial designs. *J. Statist. Plann. Inference*, **91**, 107–121.

Zhang, R., Li, P., Zhao, S., and Ai, M. (2008). A general minimum lower-order confounding criterion for two-level regular designs. *Statist. Sinica*, **18**, 1689–1705.

Zhang, R., Phoa, F. K. H., Mukerjee, R., and Xu, H. (2011). A trigonometric approach to quaternary code designs with application to one-eighth and one-sixteenth fractions. *Ann. Statist.*, **39**, 931–955.

Zhu, Y. (2003). Structure function for aliasing patterns in 2^{l-n} design with multiple groups of factors. *Ann. Statist.*, **31**, 995–1011.

Zyskind, G. (1962). On structure, relation, sigma, and expectation of mean squares. *Sankhyā, A*, **24**, 115–148.

Index

Milton Keynes UK
Ingram Content Group UK Ltd.
UKHW021826071024
449327UK00021B/1445